*Determination of
Organic Substances in Water*
Volume 2

Determination of Organic Substances in Water
Volume 2

T. R. CROMPTON

A Wiley–Interscience Publication

JOHN WILEY & SONS
Chichester · New York · Brisbane · Toronto · Singapore

Copyright © 1985 by John Wiley & Sons Ltd.

All rights reserved.

No part of this book may be reproduced by any means, nor
transmitted, nor translated into a machine language
without the written permission of the publisher.

Library of Congress Cataloging in Publication Data:
(Revised for volume 2)

Crompton, T. R. (Thomas Roy)
   Determination of organic substances in water.

   'A Wiley-Interscience publication.'
   Includes index.
   1. Water chemistry.   2. Organic compounds—Analysis.
   3. Organic water pollutants—Analysis.   I. Title.
   GB855.C76   1985   628.1'61   84-7443
   ISBN 0 471 90468 6 (v.1)
   ISBN 0 471 90469 4 (v.2)

British Library Cataloguing in Publication Data:

Crompton, T. R.
   Determination of organic subtances in water.
   Vol. 2
   1. Organic water pollutants—Analysis
   2. Water—Analysis
   I. Title
   628.1'68   TD427.07
   ISBN 0 471 90469 4

Printed and bound in Great Britain by
Anchor Brendon Ltd, Tiptree, Essex

Contents

Preface ix

1. Organometallic Compounds 1
   Organomercury compounds 1
   Organoarsenic compounds 49
   Organotin compounds 68
   Organolead compounds 90
   Organogermanium compounds 98
   Organoantimony compounds 98
   Silicones 98
   References 98

2. Oxygen Compounds 105
   Fatty Acids 105
   Alcohols 129
   Glycols 130
   Aldehydes 130
   Ketones 131
   Esters 131
   Carbohydrates 144
   Lactams 148
   Dioxans 148
   Quinones 149
   Phenols 149
   References 183

3. Nitrogen, Phosphorus, and Sulphur Compounds 190
   Nitrogen compounds 190
   Aliphatic amines 190
   Aromatic amines 191
   Heterocyclic nitrogen compounds 191
   Amino acids 192
   Nitrosamines 199
   Acrylamide 206
   Nitriles 212

Urea and substituted ureas 215
Nitro compounds 215
Ethylene diamine tetraacetic acid 216
Nitriloacetic acid 221
Melamine and cyanuric acid 227
Nucleic acids 227
Determination of organic nitrogen......................... 227
Phosphorus Compounds 240
Adenosine triphosphate 240
Glyphosphate residues 246
Inositol phosphate esters 246
Triaryl phosphate esters 246
Determination of organic phosphorus...................... 247
Sulphur Compounds.................................... 259
Organic sulphides and disulphides 259
Benzthiazole and 2-mercaptobenzothiazole.................. 270
Hydroxymethanesulphinite salts 272
Dimethyl sulphoxide 272
Tetramethylthiuram disulphide (Thiram) 276
Calcium lignosulphonate 276
Organosulphur compounds in marine sediments.............. 276
References... 276

4. Halogen Compounds 283
   Aliphatic halogen compounds............................. 283
   Saturated polychloro compounds.......................... 292
   Chlorolefins including vinyl chloride...................... 298
   Miscellaneous aromatic chloro compounds 313
   Haloforms 328
   References... 417

5. Miscellaneous Compounds and Ozonization Products 424
   Squoxin piscicide (1,1′-methylenedi-2-napththol) 424
   Coprostanol 427
   Fluorescent Whitening Agents 434
   Mestranol (17α-ethynyl-3-methoxyestra-1,3,5(10)-trien-17β-ol) and
      Ethynyloestradoil..................................... 438
   Carboxymethoxy succinate 444
   Isopropylmethylphosphonofluoridate (GB) and S-2-(di-isopropyl-
      amino) ethyl O-ethylmethylphosphonothioate 444
   2,3-Dichloro-1,4-naphthaquinone 445
   Benethonium Salts 445
   ααα-Trifluoro-4-nitro-*m*-cresol 445
   Pyrethrins 445
   Toxaphene (camphechlor) 446

Carbofuran 446
Geosmin (*trans*-1,10-dimethyl-*trans*-decalol) 446
Ozonation of Water 448
References 459

6. Natural Pigments in Water 463
Humic and Fulvic acids 463
Chlorophyll and other plant pigments in waters and algae 483
Cobalamin (vitamin B_{12}) 504
References 505

INDEX 509

Preface

The presence of concentrations of organic substances in water is a matter of increasing concern to the water industry, environmentalists and the general public alike from the point of view of possible health hazards presented to both human and animal life as represented by domesticated and wild animals and bird and fish life. This awareness hinges on three facts: the increasing interest by the scientist and the public alike in matters environmental, an increased usage of organic materials in commerce coupled with the much wider variety of organic substances used nowadays, and finally, the availability of analytical methods sensitive to determine very low concentrations of these substances, the presence of which we formerly were unaware.

It has been estimated that river waters can contain up to two thousand different organic substances over a wide concentration range and many of these survive processing in the water works and occur in potable water with possible health implications. The Food and Drug Administration in America, amongst others, is systematically working its way through screening tests on the substances so far identified in water, but this is a process that will take many years to complete.

As well as organics occurring in water as a direct result of industrial activity there are those which occur more indirectly from other causes, such as haloforms produced in the chlorination stage of the water treatment process, organometallic compounds produced by conversion of inorganic salts by biological activity in rivers and nitrosamine formation by conversion of inorganic nitrites. There are also, of course, naturally occurring organic substances in water.

The purpose of this work is to draw together and systemize the body of information available throughout the world on the occurrence and determination of organics in all types of water and effluents. A particular feature is the presentation of detailed procedures in the case of many of the more important procedures, so that reference to a very scattered literature can in many cases be avoided. Complete coverage is given of all the major instrumental techniques now available.

The contents are presented in as logical a fashion as possible, starting in Volume 1 Chapter 1 with a discussion of hydrocarbons and polyaromatic hydrocarbons. Chapter 2 deals with the various types of surface active agents whilst Chapters 3 and 4 respectively discuss the numerous types of organochlorine and organophosphorus insecticides and herbicides, polychlorinated biphenyls, now in use in agriculture and which are finding their way into the water courses. Volume 2 deals with organometallic compounds, types of compounds classified under elements including carbon, oxygen, halogen, nitrogen, sulphur, and phosphorus and concludes with chapters on ozonization products and natural pigments.

Particular groups of substances which are causing concern by their presence in the environment are discussed in detail, e.g. polychromatic hydrocarbons, chlorine and phosphoric insecticides, herbicides, polychlorinated biphenyls, haloforms, and organometallic compounds.

As well as discussing the analysis of river surface and underground waters and potable water the various sections include discussion, where relevant, of ocean and beach waters, sewage and trade effluents and muds and sediments. In certain instances, the analysis of fish, crustaceae and plant life for organic pollutant is also discussed. Such measurements are very useful as these will reflect the general level of pollution that has occurred over a period of time, as opposed to spot measurements obtained by analysis of water samples.

Examination for organic substances combines all the exciting features of analytical chemistry. First, the analysis must be successful and in many cases, such as spillages, must be completed quickly. Often the nature of the substances to be analysed for is unknown, might occur at exceedingly low concentrations and might, indeed, be a complex mixture. To be successful in such an area requires analytical skills of a high order and the availability of sophisticated instrumentation.

The work has been written with the interest of the following groups of people in mind: management and scientists in all aspects of the water industry, river management, fishery industries, sewage effluent treatment and disposal, land drainage and water supply; also management and scientists in all branches of industry which produce aqueous effluents. It will also be of interest to agricultural chemists, agriculturalists concerned with the ways in which organic chemicals used in crop or soil treatment permeate through the ecosystem, the biologists and scientists involved in fish, plant, insect and plant life, and also to the medical profession, toxicologists and public health workers and public analysts. Other groups or workers to whom the work will be of interest include oceanographers, environmentalists and, not least, members of the public who are concerned with the protection of our environment.

Finally, it is hoped that the work will act as a spur to students of all subjects mentioned and assist them in the challenge that awaits them in ensuring that the pollution of the environment is controled so as to ensure that by the turn of the century we are left with a worthwhile environment to protect.

T. R. CROMPTON
BEECHCROFT

Chapter 1
Organometallic Compounds

ORGANOMERCURY COMPOUNDS

Mercury in natural water

The two main techniques for the determination of mercury in natural water samples are atomic absorption spectrophotometry and gas chromatography. Mercury in water samples can exist in inorganic or organic forms or both. The preliminary section is concerned with methods for the preliminary degradation of organomercury compounds in the sample to inorganic mercury preparatory to analysis. This is necessary because the normal methods of reducing inorganic mercury compounds to mercury with reagents such as stannous chloride do not work with organomercury compounds and hence organomercury compounds are not included in such determinations. Owing to the conversion of Hg^{2+} to CH_3Hg^+ in natural water and owing to the presence of mercury in a large number of organic pollutants, it is often observed that a high percentage of the mercury is present in the form of organic compounds. Some organic mercurials like CH_3HgCl and $(CH_3)_2Hg$ may be reduced by a combination of cadmous chloride and stannous chloride, but this method requires large quantities of reductants and the use of strong acid and strong alkali.[1]

Organic mercury compounds can be decomposed by heating with strong oxidizing agents such as potassium chromate or nitric acid–perchloric acid, followed by reduction of the formed divalent mercury to mercury vapour.[2,3] Both methods are rather time-consuming and not very suitable for automation. Potassium persulphate has also been used to aid the oxidation of organomercury to inorganic mercury compounds and this forms the basis of an automated method.[4]

Goulden and Afghan[5] have used ultraviolet irradiation as a means of decomposition, following the original proposal of Armstrong et al.[6] After the photochemical oxidation, the formed inorganic mercury is reduced to metallic mercury in the usual way by stannous chloride. This method reduces the consumption by oxidizing agents and thus diminishes considerably the risk of contamination; it also leads to shorter analysis times. Determination with and without irradiation enable the separate determination of total and inorganic mercury respectively. Bennett et al.[7] later showed that acid permanganate alone did not recover three methyl mercuric compounds, while the addition of a potassium persulphate oxidation step increased recoveries to 100%. El-Awady et al.[8] confirmed the low recoveries of methylmercury by acid permanganate. They showed that only about 3% of methylmercury could be recovered by this method, while the use of potassium persulphate produced complete recovery.

Atomic absorption spectroscopy

Kalb[9] used concentrated nitric acid to decompose organomercury compounds in water samples prior to estimation by flameless atomic absorption spectroscopy. Stannous chloride was used to liberate elementary mercury, which is then vaporized by passing a stream of air (1360 ml min^{-1}) through the solution. The air stream passes over silver foil, where mercury is retained by amalgamation and other volatile substances pass out of the system. The foil is heated at 350 °C in an induction coil, and the air stream carries the mercury vapour through a cell with quartz windows. The atomic absorption at 253.65 nm is measured, and the mercury concentration (up to 0.02 ppm) is determined by reference to a calibration graph.

Umezaki and Iwamoto[10] differentiated between organic and inorganic mercury in river samples. They used the reduction–aeration technique described by Kimura and Miller.[11] By using stannous chloride solution in hydrochloric acid only inorganic mercury is reduced, whereas stannous chloride in sodium hydroxide medium and in the presence of cupric copper reduces both organic and inorganic mercury. The mercury vapour is measured conventionally at 254 nm. Ions that form insoluble salts or stable complexes with Hg(II) interfere.

Doherty and Dorsett[12] analysed environmental water samples by separating the total organic and inorganic mercury by electrodeposition for 60–90 min on a copper coil in 0.1 M nitric acid medium and then determined it directly by flameless atomic absorption spectrophotometry.[13,14] The precision and accuracy are within ± 10% for the range 0.1–10 parts per 10^9. The sensitivity is 0.1 part per 10^9 (50 ml sample).

Graf et al.[15] used sulphuric acid acidified potassium permanganate to decompose organically bound mercury, prior to reduction with stannous chloride and determination of the evolved mercury by atomic absorption spectroscopy.

Other workers who have made earlier contributions to the determination of organically bound mercury compounds and inorganic mercury compounds by flameless atomic absorption spectroscopy include Baltisberger and Knudson,[16] Bisagni and Lawrence,[17] Frimmel and Winckler,[18] Chan and Saitoh,[20] Umezaki and Iwamoto,[21] Stainton,[22] Carr et al.,[23] Fitzgerald et al.,[24] and Watling and Watling[19] who carried out a literature survey on this subject (34 references).

Breakdown of organomercury compounds preparatory to analysis

Kiemeneij and Kloosterboer[25] have described an improvement on the Goulden and Afghan[5] photochemical decomposition of organomercury compounds in the ppb range in natural water prior to determination by cold vapour atomic absorption spectrophotometry. Decomposition of the organomercurials is carried out by means of ultraviolet radiation of a suitable wavelength from small, low-pressure lamps containing either Zn, Cd, Hg, or a mixture of these metals in their cathodes. The formed inorganic mercury is determined in the usual way by cold vapour atomic absorption after reduction of divalent mercury to mercury vapour. Determinations with and without irradiation make possible separate determinations of total and inorganic mercury, respectively, in about 20 minutes.

Method

Photolysis of organomercurials was carried out with small low-pressure spectral lamps (12–15 W) emitting either the Hg, Cd, or Zn spectrum which have their strongest lines at 254, 229, and 214 nm, respectively (Philips spectral lamps No. 93109, 93107, and 93106).

Kiemeneij and Kloosterboer[25] used toroidal silica irradiation cells which can be placed around the lamp (Figure 1A). The use of the cell shown in Figure 1B obviates the transfer of the sample solution to the aeration bottle of the AAS detection system, since a glass frit is mounted near the bottom of the cell, giving the possibility of aeration through the side tube. The outer walls of the cells were coated with aluminium which in turn was protected by a varnish layer.

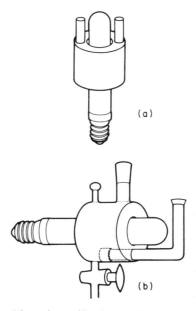

Figure 1 View of two silica irradiation cells with the light source. Dimensions: height 50 mm, internal diameter 33 mm, and external diameter 70 mm. Reprinted with permission from Kiemeneij and Kloosterboer.[25] Copyright (1976) American Chemical Society

The analytical determination was carried out by the Hatch and Ott[26] method, i.e. the mercury vapour, formed upon the reduction of Hg^{2+} is transferred from the sample solution to an optical cell of the atomic absorption spectrophotometer by bubbling air through the sample solution.

To prevent condensation of moisture in the optical cell, the latter was heated to 60 °C with an infrared lamp; further drying of the vapour proved to be unnecessary. Glass tubing was used to prevent loss of mercury vapour by diffusion through the tubing. The optical cell had a path length of 19 cm.

Atomic absorption measurements were performed on a Pye-Unicam SP 1900 AA spectrophotometer using a deuterium background corrector. Separate measurements showed that the background corrections were negligible for all the samples.

Procedure

Sampling

Samples of natural waters were acidified to 0.25 M hydrochloric acid immediately after being taken in order to prevent adsorption of divalent mercury on the wall of the vessel; the samples were not filtered.

Photolysis

One hundred ml of the samples were transferred to an irradiation cell (Figure 1A) and irradiated for 10 min with either the Cd or the Zn–Cd–Hg lamp. After irradiation, the solution was transferred to the aeration bottle of the mercury

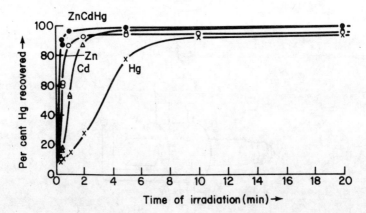

Figure 2 Percentage recovery of mercury from CH_3HgCl added to deionized water as a function of the time of irradiation for various light sources. Concentration: 4.1 µg Hg l^{-1}. Reference $Hg(NO_3)_2$. The small recovery observed for $t = 0$ is probably caused by the thermal or daylight decomposition; it was not observed with diphenylmercury. Reprinted with permission from Kiemeneij and Kloosterboer.[25] Copyright (1976) American Chemical Society

detection system and 0.5 ml of 10% stannous chloride in 1 M hydrochloric acid was added. Together with the sample solutions a number of test solutions were irradiated and analysed. When the cell of Figure 1B was used, the stannous chloride was added directly to the contents of the irradiation cell.

The recovery of mercury from aqueous solutions of methylmercury chloride as a function of the time of irradiation for a number of light sources is shown in Figure 2. Clearly, the efficiency of destruction increases in the order Hg < Cd < Zn although the relative intensities of the lamps were in the opposite order, namely, Hg:Cd:Zn = 6:2:1. In going from cadmium to zinc, the initial slope of the curves increases more than is accounted for by the increase in extinction coefficient as gauged from Figure 1.

The same order Hg < Cd < Zn was observed in the case of diphenylmercury. This compound even has a slightly lower extinction coefficient at 214 nm (Zn) than at 229 nm (Cd). This shows that for both methylmercury chloride and diphenylmercury, the quantum efficiency for photodecomposition increases with decreasing wavelength, in accord with what might be expected. Both absorbance and quantum yield, therefore, favour the use of short wavelength light sources. In natural waters, on the other hand, the background absorption of dissolved organic compounds which increases towards the ultraviolet (Figure 3) favours the use of the more intense long wavelength sources. The cadmium lamp offers a good compromise (Figure 4). If the heating of the sample is not considered as disadvantageous, the more powerful combined Zn-Cd-Hg lamp offers the best results.

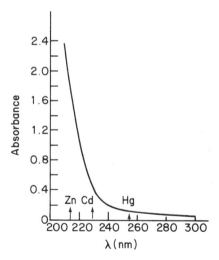

Figure 3 Absorption spectrum of a natural water sample from the river Waal. The sample contains 0.25 M HCl for preservation. Optical path length: 1 cm. Reprinted with permission from Kiemeneij and Kloosterboer.[25] Copyright (1976) American Chemical Society

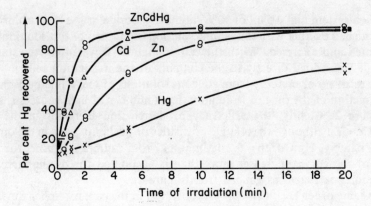

Figure 4 Percentage recovery of mercury from CH_3HgCl added to natural water from the river Waal as a function of the time of irradiation for various light sources. Concentration: 4.1 µg Hg l^{-1}. Reference: $Hg(NO_3)_2$. Reprinted with permission from Kiemeneij and Kloosterboer.[25] Copyright (1976) American Chemical Society

In Figure 5 calibration curves are shown for the standard addition of organomercurials to deionized and natural water, respectively. They contained approximately 0.8 µg Hg l^{-1}. The calibration curve for the natural water sample has a slightly smaller slope than that for dionized water or for inorganic mercury in natural water (not shown) pointed to a slight incompleteness of the

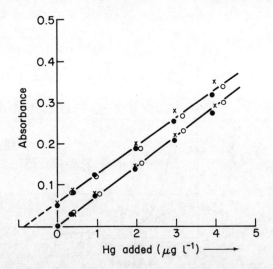

Figure 5 Calibration curves for the standard addition of CH_3HgCl (●) and $(C_6H_5)_2Hg$ (○) to deionized water (lower curve) and natural water (upper curve). Light source: combined Zn–Cd–Hg lamp. Irradiation time: 10 min. For comparison some results obtained with $Hg(NO_3)_2$ are also given (x). Reprinted with permission from Kiemeneij and Kloosterboer.[25] Copyright (1976) American Chemical Society

decomposition of the added organomercurials. However, even if the added compounds are completely decomposed, this may not be the case for the compounds originally present in the sample, especially if the latter has not been filtered.

Comparison of the photochemical with a wet-chemical method (Table 1) showed that the results of prolonged irradiation compare well with the results obtained after complete wet-chemical destruction.

Table 1 Comparison of photochemical and wet-chemical decomposition of mercury compounds in an acidified natural water sample (River Waal)

Sample treatment		Hg found (μg l^{-1})*	
Unirradiated		0.31	
Irradiated	10 min	1.01	1.00
(Zn–Cd–Hg lamp)	30 min	1.15	1.11
	30 min	1.05	1.12
Stored with KMnO$_4$	2% KMnO$_4$	0.98 (0.06)	1.02 (0.03)
	4% (KMnO$_4$)	1.04 (0.08)	0.97
	4% KMnO$_4$	1.07	1.01
Stored with KMnO$_4$,	2% KMnO$_4$	1.00 (0.24)	1.06 (0.23)
partly evaporated	4% KMnO$_4$	1.07 (0.27)	1.06 (0.31)
and rediluted			
Stored with KMnO$_4$,	2% KMnO$_4$	1.12 (0.38)	0.99 (0.39)
partly evaporated,	4% KMnO$_4$	1.08 (0.42)	1.09 (0.43)
digested in bomb			
and rediluted			

*Blank values were obtained from deionized water. When they exceeded the detection limit (0.03 μg l^{-1}) they were subtracted from the results. Subtracted blanks are given in parentheses. Reprinted with permission from Kiemeneij and Kloosterboer.[25] Copyright (1976) American Chemical Society.

Farey et al.[27] have discussed ultraviolet photochemical systems for the decomposition of organomercury compounds prior to analysis by cold vapour atomic fluorescence spectroscopy. These workers compared the effectiveness of a bromination treatment for the liberation of mercury from organomercury compounds and compared it with a pretreatment procedure involving oxidation with a permanganate–sulphuric acid mixture recommended by workers at the Water Research Centre, UK.[28] The basis of the bromination technique, in which a bromate–bromide reagent in hydrochloric acid reagent is used to generate bromine in the sample, is that the bromine quantitatively cleaves both alkyl and aryl mercury compounds to inorganic mercury bromide.

Method

Reagents

All reagents were of analytical reagent grade. Potassium bromate (0.1 M)–potassium bromide (1% m/v) solution: 2.784 g of anhydrous potassium bromate are dissolved in distilled water, 10 g of potassium bromide are added and the solution diluted to 1 litre with distilled water. Hydrochloric acid, sp. gr. 1.18.

Hydroxylammonium chloride (12% m/v)–sodium chloride (12% m/v) solution: 12 g of hydroxylammonium chloride and 12 g of sodium chloride are dissolved in distilled water and diluted to 100 ml.

Apparatus

For the inorganic mercury determinations a Shandon Southern spectrophotometer (A3400) in the emission mode was used, connected to a Baird-Atomic mercury-detection unit (A3460). The spectrophotometer was operated with the damping control set at position 1 (time constant 0.50 s) and the EHT at coarse 9 and fine 5. In the mercury-detection unit, argon gas was bubbled through the tin(II) chloride solution at 1 l min^{-1}. The mercury-signal peakheight measurement used in the determinations was monitored visually from an A3480 digital display unit (Shandon Southern).

For the bromination, 5 ml of hydrochloric acid (sp. gr. 1.18) were added to 50 ml of the water sample in a 100 ml calibrated flask, an aliquot of the bromate-bromide solution was added and the flask was shaken. After reaction, the bromination was terminated by removing the excess of bromine by the addition of 1–2 drops of hydroxylamine reagent. The solution was then made up to 100 ml with distilled water and the mercury was determined by cold vapour atomic fluorescence spectroscopy using the method of Thompson and Godden.[29]

Table 2 shows the recoveries of inorganic mercury from distilled water spiked with phenylmercury(II) chloride, thiomersal, ethylmercury(II) chloride, methylmercury(II) chloride, phenylmercury(II) acetate and *p*-tolylmercury(II) chloride. All recoveries are greater than 95%. When identical conditions of treatment were used in 50 ml samples of tap water and various river waters and sewage effluents, all with added methylmercury(II) chloride (Table 3) there were similar recoveries after 5 min, although the recoveries after a bromination period

Table 2 Recoveries of inorganic mercury from organomercurials in distilled water following bromination

Organomercurial	Organic Hg concentrations in distilled water (μg l^{-1})	Inorganic Hg found after bromination* μg l^{-1})	Recovery (%)
C_6H_5HgCl	15.1	14.9	99
$CH_3C_6H_4HgCl$	11.9	11.8	99
$C_2H_5HgSC_6H_4COONa$	9.7	9.9	102
$C_6H_5HgCOCCH_3$	10.0	10.0	100
C_2H_5HgCl	15.0	14.4	96
CH_3HgCl	14.0	14.2	101

*The figures quoted are the means of results obtained from duplicate experiments.
Bromination time 1 min using 1 ml of brominating reagent.
Reprinted with permission from Farey et al.[27] Copyright (1978) Royal Society of Chemistry.

Table 3 Recoveries of inorganic mercury from CH_3HgCl added to various waters and effluents

Spiked medium (14.8 µg l^{-1} of Hg as CH_3HgCl)	1 ml of $KBrO_3$–KBr added				2 ml of $KBrO_3$–KBr added			
	1 min reaction		5 min reaction		1 min reaction		5 min reaction	
	Inorganic Hg found (µg l^{-1})	Recovery (%)	Inorganic Hg found (µg l^{-1})	Recovery (%)	Inorganic Hg found (µg l^{-1})	Recovery (%)	Inorganic Hg found (µg l^{-1})	Recovery (%)
Tap water	13.6	92	14.5	98	14.1	95	13.9	94
River water*	6.8	46	13.2	89	13.8	93	13.6	92
River water*	9.9	67	14.4	97	13.9	94	14.4	97
River water*	12.6	85	14.1	95	14.1	95	13.3	90
Sewage effluent†	5.8	39	13.5	91	7.0	47	14.5	98
Sewage effluent‡	12.7	86	13.9	94	13.8	93	13.9	94

*Obtained from sampling points on the River Lee.
†From Rye Meads sewage works.
‡From East Hyde sewage works.
The figures quoted are the means of results obtained from duplicate experiments.
Reprinted with permission from Farey et al.[27] Copyright (1978) Royal Society of Chemistry.

of 1 min were lower. However, when 2 ml of bromate–bromide solution were added, higher recoveries were obtained after 1 min, indicating that the increase in the amount of brominating reagent accelerated the reaction. The best recoveries were obtained after treatment for 5 min with either 1 or 2 ml of the brominating reagent.

Table 4 shows a comparison of results obtained on a sewage effluent sample by the above method and a method involving digestion with acid permanganate reagent developed by the Water Research Centre.[30] It can be seen that the inorganic mercury value obtained after bromination for 5 min is similar to that found following the permanganate pretreatment. Moreover, the value obtained after a bromination period of 15 min was higher, showing a greater recovery of mercury in this instance.

Table 4 Total mercury determined by atomic fluorescence spectroscopy in Crossness sewage works final effluent

Pretreatment	Mercury found /(μg l^{-1})	
	Mean	Standard deviation
$KMnO_4$–H_2SO_4, 80 °C, 8 h	1.65	0.20
5 ml of HCl + 2 ml of $KBrO_3$–KBr		
(1) 5 min	1.56	0.14
(2) 15 min	1.91	0.13

Pretreatment and analysis performed on four separate aliquots of the sample.
Reprinted with permission from Farey et al.[27] Copyright (1978) Royal Society of Chemistry.

Farey et al.[27] claim that their treatment compares favourably with an established permanganate–sulphuric acid method. An advantage of the technique is that it can easily be carried out while sampling on-site. The sample is collected in glass bottles containing hydrochloric acid and the bromate–bromide solution is added. A bromination reaction time is then provided from the collection of the sample to the analysis in the laboratory and this is far in excess of that necessary to decompose the organic mercury. In addition, as aqueous mercury(II) solutions are stabilized by strong oxidizing agents, the oxidizing conditions so created will help to preserve the inorganic mercury formed.

Two methods,[30] both tentative have been described for the determination of mercury in water and effluents. The first method is designed for non-saline waters, effluents, and sludges and the second for saline waters. This method determines all forms of mercury provided they are first converted to inorganic mercury, while the second method is applicable to inorganic mercury and those organomercury compounds which form dithizonates. These methods are not described here in detail as they are readily available. All forms of mercury in non-saline waters, effluents, and sludges are converted to inorganic mercury using prolonged oxidation with potassium permanganate.[31] Solid samples require a more prolonged and vigorous oxidation to bring the mercury completely into solution in the inorganic form. A modification of the Uthé digestion procedure is used for such samples.[32] The inorganic mercury is determined by the flameless atomic absorption spectrophotometric technique using a method similar to that described by Osland.[33] Acid stannous chloride is added to the sample to produce elemental mercury:

$$Hg^{2+} + Sn^{2+} \rightarrow Hg^{\circ} + Sn^{4+}$$

The mercury vapour is carried by a stream of air or oxygen into a gas cuvette placed in the path of the radiation from a mercury hollow cathode lamp and the absorption of this radiation at 253.7 nm by the mercury vapour is measured (see Figure 6). Many of the potential interferences in the atomic absorption procedure are removed by the preliminary digestion/oxidation procedure. The most significant group of interfering substances is volatile organic compounds which absorb radiation in the ultraviolet. Most of these are removed by the pretreatment procedure used and the effect of any that remain are overcome by pre-aeration. Bromide and iodide ions may cause interference. Substances which are reduced to the elemental state by stannous chloride and then form a stable compound with mercury may cause interference; e.g. selenium, gold, palladium, and platinum. The effects of various anions, including bromide and iodide, were studied. These are not likely to be important interferers. Excellent performance characteristics are presented for this method (see Table 5).

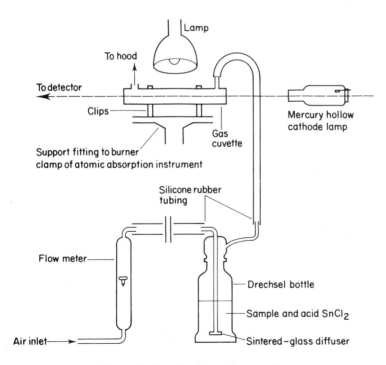

Figure 6 Detail of sample vaporizer

Table 5 Performance characteristics WRC method for mercury

Range of application	Up to 2.0 µg l^{-1} for liquid samples. Up to 2.0 µg g^{-1} for solid samples.	
Calibration curve	Linear to 20 µg l^{-1} for liquid samples. Linear to 20 µg g^{-1} for solid samples.	
Within-batch standard deviation for liquid samples	Mercury concentration (µg l^{-1})	Standard deviation (µg l^{-1})
	0.0	0.025
	0.2	0.022
	2.0	0.024
	0.0	0.078
	0.2	0.032
	2.0	0.104
	(All with 9 degrees of freedom)	
for solid samples	Not known	
Limit of detection	0.1–0.2 µg l^{-1} (with 9 degrees of freedom) for liquid samples depending on their nature. Not known for solid samples.	
Sensitivity	10 µg l^{-1} for liquid samples and 10 µg g^{-1} for solid samples are equivalent to an absorbance of approximately 0.3.	

Sampling techniques are described in detail including methods of cleaning sample bottles and fixing the sample with a solution of potassium dichromate in nitric acid.

The second method is suitable for determining in saline sea and estuary waters dissolved inorganic mercury and those organomercury compounds which form dithizonates. In this method inorganic mercury is extracted from the acidified saline water as its dithizonate into carbon tetrachloride, but not all these compounds form dithizonates and those which do not may not be determined by this method. In general, organomercury compounds of the type R-Hg-X, in which X is a simple anion, form dithizonates, whereas the type R_1-Hg-R_2 does not. Monomethylmercury ion is extracted though it only appears to have a transient existence in aerobic saline water. The dithizonates are decomposed by the addition of hydrochloric acid and sodium nitrite and the mercury or organomercury compound returned to the aqueous phase. Some organomercury compounds may not be completely re-extracted into the aqueous phase. The mercury in this aqueous phase is determined by the stannous chloride reduction–atomic absorption spectroscopic technique described earlier. The method is based on that used at the Department of Oceanography, University of Liverpool.[34]

The performance characteristics of this method are outlined in Table 6.

Abo-Rady[35] has described a method for the determination of total inorganic plus organic mercury in nanogram quantities in water, fish, plants, and sediments. This method is based on the decomposition of organic and inorganic mercury compounds with acid permanganate, removal of excess permanganate

Table 6

Range of application	Up to 100 ng l^{-1}
Calibration curve	Linear to 250 ng l^{-1}
Standard deviation	Mercury concentration (ng l^{-1}) Standard deviation (ng l^{-1}) 0.0 1.30 50.0 1.15 (each with 9 degrees of freedom)
Limit of detection	4 ng l^{-1} (with 9 degrees of freedom)
Sensitivity	100 ng l^{-1} is equivalent to an absorbance of approximately 0.1
Bias	None detected
Interferences	The combined effect of the commonly presently ions in estuarine and sea waters at the concentration normally encountered in these waters is less than 1 ng l^{-1} at a mercury concentration of 30 ng l^{-1}
Time required for analysis	For 6 samples the total analytical and operator times are approximately 140 minutes and 60 minutes respectively

Table 7 Reproducibility and accuracy of determinations in water, plants, fish, and sediments

		Concentration (μg l^{-1})			
Material	n (number of determinations)	Present	Found	x (mean)	±s.d. (standard deviation)
Water	10	0	0.1	0.10	0.02
	10	0.2	0.3	0.32	0.02
Fish	5	0	13	13	0.007
	5	40	53	57	0.002
	5	80	93	93	0.009
	5	100	113	103	0.011
Plants	5	0	29	29	0.02
	5	10	39	40	0.02
	5	30	59	65	0.02
	5	50	79	86	0.02
Sediment	4	0	43	43	0.02
	4	20	63	69	0.05
	4	40	83	93	0.07
	4	60	103	126	0.07

Reprinted with permission from Abo-Rady.[35] Copyright (1979) Springer-Verlag, Heidelberg.

with hydroxylamine hydrochloride, reduction to metallic mercury with tin and hydrochloric acid, and transfer of the liberated mercury in a stream of air to the spectrometer. Mercury was determined by using a closed, recirculating air stream. Sensitivity and reproducibility of the 'closed system' were better, it is

claimed, than those of the 'open system'. The coefficient of variation was 13.7%
for water, 1.9% for fish, 4.9% for plant, and 5.6% for sediment samples.

Lutze[36] has described a flameless atomic absorption method for determining mercury in surface waters and sediments and Grantham[37] has applied the method to waters, soil, foodstsuffs, and effluents.

Concentration of mercury prior to analysis

Fitzgerald et al.[38] reported a cold trap preconcentration technique for the determination of trace amounts of mercury in water. Krämer and Neidhart[39] determined ppb (μg l^{-1}) levels of mercury by using an aniline–sulphur resin for the selective enrichment of mercury from surface waters. Chan and Saitoh[40] reported a method for the determination of submicrogram amounts of mercury in lake water based on dithizone extraction. Preconcentration of mercury prior to the measurement has also been achieved by amalgamation with noble metals.[41-44]

Minagawa et al.[45] have described a technique employing chelating resins which has been applied to the determination of very low concentrations down to 0.2 ng l^{-1} of organic and inorganic mercury in natural waters including rivers, lakes, and rain waters.[38] The resin used contains dithiocarbamate groups which bind mercury but not alkali and alkaline earth metals. Both forms of mercury are collected at pH 1–11 and eluted with slightly acidic 4% thiourea in water. Large volumes of water can be concentrated to determine mercury by cold vapour atomic absorption spectrometry. Mercury vapour is generated from inorganic mercury with alkaline stannous chloride and from organic and inorganic mercury with a cadmium chloride–stannous chloride solution.

Method

Apparatus

The apparatus used for preconcentration consisted of a column (15 mm i.d. 5 cm long) for the resin and a 20 l high-density polyethylene bottle as reservoir for the samples. The 20–50 mesh wet dithiocarbamate-treated resin was packed in the column. The polyethylene bottles were cleaned by soaking in (1 + 9) nitric acid for 2 days and then rinsed thoroughly with distilled-deionized water before use.

The mercury vapour concentration meter, manufactured by Nippon Jarrell-Ash model AMD-F2, had a double-beam, dual-detector system. The inlet of the gas cell was connected by plastic tubing successively to a U-tube (10 × 200 mm Pyrex glass) containing calcium chloride (6–20 mesh) as a water absorbent, a Quickfit 30 ml test tube (as a reaction vessel) fitted with a Drechsel bottle head with a sintered-glass (porosity 2) bubbler, and an air pump. When recordings were made the output of the mercury vapour concentration meter was connected to a strip chart recorder. The operating conditions were as follows: air flow rate, 1.0 l min^{-1}; source, mercury lamp (253.7 nm); slit, 100 μm; gas cell, 20 × 200 mm quartz tube; scale expansion, × 10; recorder, full scale 10 mV, chart speed 5 mm min^{-1}.

Reagents

All chemicals used were reagent grade. All solutions were prepared with distilled-deionized water.

Aqueous 5% (w/v) solution of thiourea containing 5 ml of hydrochloric acid per litre was used.

Tin(II) chloride (10% w/v) and tin(II) chloride–cadmium chloride (10%w/v– 1%w/v) solutions were used as reductants, were heated to boiling in order to remove mercury and diluted to volume, as required, with distilled-deionized water.

A stock solution of mercury(II) chloride (1000 μg Hg ml^{-1}) was used for preparation of working standards (2–12 ng ml^{-1}) by appropriate dilutions with aqueous 5% solution of thiourea.

The chelating resin used was Sumichelate Q-10 (Sumitomo Chemicals),[46] a vinyl polymer containing dithiocarbamate groups. The resin was sieved to 20–50 mesh, and then suspended overnight in an aqueous saturated solution of thiourea containing concentrated hydrochloric acid (5 ml l^{-1}) in order to remove mercury. The mercury-free resin was rinsed with distilled water until the thiourea had been removed completely. To convert the resin to the hydrogen form, a portion was suspended in 0.2 M nitric acid.

Collection and treatment of water samples

River water and other fresh waters were sampled in a 20 l high-density polyethylene bottle which was rinsed three times with the water sampled before the sample was taken. The sample was adjusted to pH 2 with concentrated nitric acid, 1 mg of HAuCl$_4$ being added as preservative.[47] Samples of water should be analysed within one week of collection to avoid losses of mercury by adsorption and vaporization.

Procedures

For calibration, known concentrations of mercury standards and 10 ml 30% (w/v) potassium hydroxide were placed in the reaction vessel, and the volume was diluted to 20 ml with the aqueous 5% solution of thiourea. Air was passed through immediately after addition of reductant as described below, and calibration curves were obtained from peak absorption measurements. The resin column was not used in the calibration.

A 20 l polyethylene sample bottle containing the unfiltered fresh water was placed above the chelating resin column, and connected to it with polyethylene tubing. The sample was allowed to flow through the column at *ca.* 30 ml min^{-1}. Cleaned resin was used for each experiment.

After the collection, 30 ml of the acidic 5% thiourea solution served to elute total and inorganic mercury.[48] Inorganic and total mercury were determined separately in two 10 ml aliquots of this effluent.

For the determination of inorganic mercury, the 10 ml aliquot of well-mixed effluent was placed in the reaction vessel, 10 ml of 30% (w/v) potassium hydroxide was added followed by 2 ml of tin(II) chloride solution, and the air flow was started immediately. This mixture was allowed to react for 30 seconds, during which time the mercury vapour generated passed through the quartz gas cell. Peak height was used for measurements.

For the determination of total mercury, the same procedure was used, except for reduction with the tin(II) chloride–cadmium chloride mixture (10%–1%), instead of tin(II) chloride alone. The peak heights were again measured. The total mercury minus the inorganic mercury gives an estimate of the organic mercury.

The total blank was determined by carrying out the complete procedure of analysis with 20 l of distilled-deionized water. Five replicate measurements gave mean blanks of 0.07 ± 0.05 ng l^{-1} and 0.15 ± 0.06 ng l^{-1} for inorganic and total mercury, respectively.

The calibration graphs obtained for Hg^{2+} and CH_3Hg^+ were identical and are linear over the range 0–12 ng. Possible interferences are other ions, amino acids and naturally occurring chelating agents which could affect the preconcentration, desorption, and reduction steps. No interference was produced in the determination of 0.1 μg of mercury(II) by the presence of at least 1000 μg of each of the following ions or substances added to 5 l aliquots of river water: Cr^{3+}, Mg^{2+}, Na^+, K^+, Ca^{2+}, Ni^{2+}, Cu^{2+}, Pb^{2+}, Cd^{2+}, Au^{3+}, Fe^{3+}, Al^{3+}, Zn^{2+}, PO_4^{3-}, Cl^-, CO_3^{2-}, NO_3^-, SO_4^{2-}, silicate, cysteine, and humic acid. The accuracy of the method was tested by analysing river water samples spiked with known amounts of Hg^{2+} and CH_3Hg^+. As shown in Table 8, the accuracy of the method was satisfactory. The precision of the method was 9.5 ± 0.43 ng l^{-1} and 15.2 ± 0.36 ng l^{-1} for inorganic and organic mercury respectively.

Table 8 Results for river water spiked with mercury(II) chloride and methylmercury chloride*

Hg species	Hg added (ng)	Hg found (ng)†	Recovery (%)
Hg^{2+}	5	4.6	92
CH_2Hg^+	5	4.6	91
Hg^{2+}	10	9.1	91
CH_3Hg^+	10	9.3	93
Hg^{2+}	20	18.6	93
CH_3Hg^+	20	18.1	91
Hg^{2+}	100	95	95
CH_3Hg^+	100	91	91

*Each spike was added to 20 l of sample water.
†Each result represents the mean of three values after subtraction of the blank.
Reprinted with permission from Minagawa et al.[45] Copyright (1980) Elsevier Science Publishers.

Minagawa et al.[45] used this method to estimate inorganic and organic mercury in unfiltered Japanese river water samples (Table 9). Significantly, 35-60% of the mercury present in river and lake waters exists as organic compounds or in association with organic matter. In rain water only about 6% of the mercury was in this form. Figure 7 shows typical responses for standards, a sample, and a blank.

Table 9 Mercury fractions of river, lake, and rain waters in the Akita area of Japan

Location and type	No. of samples	Mean Hg concentration (ng l^{-1})		
		Total	Inorganic	Organic
Omono River	3	15.5	8.8	6.7
Taihei River	5	21.1	9.0	12.1
Tazawa Lake	3	12.4	6.5	5.9
	2	13.8	8.8	5.5
Akita city	2	10.8	10.1	0.7
Rain water	3	15.2	14.3	0.9

Samples were collected in June and July 1978.
Reprinted with permission from Minagawa et al.[45] Copyright (1980) Elsevier Science Publishers.

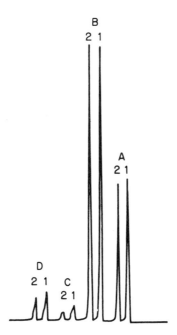

Figure 7 Typical responses for total and inorganic mercury. (A) 10 ng Hg; (B) 20 ng Hg; (C) blank; (D) river water: (1) total mercury; (2) inorganic mercury. Reprinted with permission from Minagawa et al.[45] Copyright (1980) Elsevier Science Publishers

Yamagami et al.[49] also applied chelating resins (dithiocarbamate type) to the determination of ppb of mercury in water. The samples are adjusted to pH 2.3 and passed through a column packed with 5 g of the resin, at a flow rate of 50 ml min^{-1}. The resin is then digested under reflux with concentrated nitric acid, and the mercury is determined by atomic absorption spectrophotometry, using the reduction–aeration technique. The method is relatively simple and inexpensive and the detection limit is 10 ng mercury in water samples as large as 10 litres.

Simpson and Nickless[50] have described a rapid dual channel method of cold vapour atomic absorption spectroscopy for determining mercury. A detection limit of 12.5 ng l^{-1} is claimed.

Gas chromatography

This technique has found limited applications in the determination of organomercury compounds in water and trade effluents.

Nishi and Horimoto[51,52] determined trace amounts of methyl-, ethyl-, and phenylmercury compounds in river waters and industrial effluents. In this procedure, the organomercury compound present at less than 0.4 ng l^{-1} in the sample (100–500 ml) is extracted with benzene (2 × 0.5 vol relative to that of the aqueous solution). The benzene layer is then back-extracted with 0.1% L-cysteine solution (5 ml), and recovered from the complex by extracting with benzene (1 ml) in the presence of hydrochloric acid (2 ml) and submitted to gas chromatography using a stainless steel column (197 cm × 3 mm) packed with 5% of diethylene glycol succinate on Chromosorb W (60–80 mesh) with nitrogen as carrier gas (60 ml min^{-1}) and an electron capture detector. The calibration graph is rectilinear for less than 0.1 μg of mercury compound per ml of the cysteine solution. This method is capable of determining mercury down to 0.4 μg l^{-1} for the methyl and ethyl derivatives and 0.86 μg l^{-1} for the phenyl derivative.

The above method has been modified[52] for the determination of methylmercury(II) compounds in aqueous media containing sulphur compounds that affect the extractions of mercury. The modified method is capable of handling samples containing up to 100 mg of various organic and inorganic sulphur compounds per 100 ml. The aqueous test solution (150 ml) containing 100 mg of methylmercury ions per 100 ml is treated with hydrochloric acid until the acid concentration is 0.4%, then 0.3–1 g of mercuric chloride is added (to displace methyl mercury groups bonded to sulphur), and the mixture is filtered. The filtrate is treated with aqueous ammonia in excess to precipitate the unconsumed inorganic mercury which is filtered off; this filtrate is made 0.4% in hydrochloric acid and extracted with benzene. The benzene solution is shaken with 0.1% L-cysteine solution, the aqueous phase is acidified with concentrated hydrochloric acid and then shaken with benzene for 5 minutes and this benzene solution is analysed by gas chromatography as described above.

Zarnegar and Mushak[53] have described a gas chromatographic procedure for the determination of organomercury compounds and inorganic mercury in water and biological materials. The sample is treated with an alkylating or arylating reagent and the organomercury chloride is extracted into benzene. Gas chromatography is carried out using electron capture detection. The best alkylating or arylating reagents were pentacyano(methyl)cobaltate(III) and tetraphenylborate. Inorganic and organic mercury could be determined sequentially by extracting and analysing two aliquots of sample, of which only one had been treated with alkylating reagent. The limits of detection achieved in the method were 10–20 ng.

Ealy et al.[54] have discussed the determination of methyl-, ethyl-, and methoxymercury(II) halides in water and fish. The mercury compounds were separated from the sample by leaching with M sodium iodide for 24 hours and then the alkylmercury iodides were extracted into benzene. These iodides were then determined by gas chromatography of the benzene extract on a glass column packed with 5% of cyclohexane-succinate on Anakron ABS (70–80 mesh) and operated at 200 °C with nitrogen (56 ml min^{-1}) as carrier gas and electron capture detection. Good separation of chromatographic peaks was obtained for the mercury compounds as either chlorides, bromides, or iodides. The extraction recoveries were monitored by the use of alkylmercury compounds labelled with ^{208}Hg.

Cappon and Crispin Smith[55] have described a method for the extraction, clean-up, and gas chromatographic determination of organic (alkyl and aryl) and inorganic mercury in biological materials. Methyl-, ethyl-, and phenylmercury are first extracted as the chloride derivatives. Inorganic mercury is then isolated as methylmercury upon reaction with tetramethyltin. The initial extracts are subjected to thiosulphate clean-up, and the organomercury species are isolated as the bromide derivatives. Total mercury recovery ranges between 75 and 90% for both forms of mercury, and is assessed by using appropriate ^{203}Hg-labelled compounds for liquid scintillation spectrometric assay. Specific gas chromatographic conditions allow detection of mercury concentrations of 1 ppb or lower. Mean deviation and relative accuracy average 3.2 and 2.2% respectively. These workers were concerned with the determination of different inorganic mercury and organomercury species (alkyl and aryl) in a variety of media including water, river sediments, and fish.

Another application of gas chromatography to natural water analysis is that of Longbottom[56] who uses a Coleman 50 Mercury Analyser System as a detector. A mixture of dimethyl-, dipropyl-, and dibutylmercury (1 mg of each) was separated on a 6 ft column packed with 5% of DC-200 and 3% of QF-1 on Gas-Chrom Q and temperature programmed from 60 to 180 °C at 20 °C min^{-1}. The mercury detector system was used after the column effluent had passed through a flame ionization detector; the heights of the resulting four peaks were related to the percentages of mercury in the compounds.

Dressman[57] also used the Coleman 50 system in his determination of dialkylmercury compounds in river waters. These compounds were separated in a glass column (1.86 m × 2 mm) packed with 5% of DC-200 plus 3% of

QF-1 on Gas Chrom Q (80-100 mesh) and temperature programmed from 70 to 180 °C at 20 °C min^{-1}, with nitrogen as carrier gas (50 ml min^{-1}). The mercury compound eluted from the column was burnt in a flame ionization detector, and the resulting free mercury was detected by a Coleman Mercury Analyser MAS-50 connected to the exit of the flame ionization instrument; down to 0.1 mg of mercury could be detected. River water (1 litre) was extracted with pentane–ethyl ether (4:1)(2 × 60 ml). The extract was dried over sodium sulphate, evaporated to 5 ml and analysed as above.

Zarnegar and Mushak[58] treated the water sample with an alkylating or arylating reagent and extracted. The benzene extract was examined by electron capture gas chromatography. The best alkylating or arylating reagents were pentacyano (methyl)cobaltate(III) and tetraphenylborate. Inorganic and organic mercury could be determined sequentially by extracting and analysing two aliquots of sample, of which only one had been treated with alkylating reagent. The limits of detection were about 10-30 ng.

Neutron activation analysis

Becknell et al.[59] first converted the organomercury to mercuric chloride using chlorine. The mercury was then concentrated by removal as $HgCl_4^{2-}$ by passing the sample solution ($\simeq$ 500 ml, adjusted to be 0.1 M in hydrochloric acid) through a paper filter disc loaded with SB-2 ion-exchange resin. The paper is then heat sealed in Mylar bags and irradiated for 2 hr in a thermal neutron flux of $\simeq 1.3 \times 10^{13}$ neutrons cm^{-2}s^{-1}, after which the concentration of mercury in the sample is determined by the comparison method using the 77-keV γ-ray photopeak from the decay of ^{197}Hg.

Miscellaneous

Radiochromatographic assay has been used[60] as the basis of a method for determining inorganic mercury and methylmercury in river water.

Ke and Thibert[61] have described a kinetic microdetermination of down to 0.05 μg ml^{-1} of inorganic and organic mercury in river water and sea water. Mercury is determined by use of the iodide-catalysed reaction between Ce(IV) and As(III), which is followed spectrophotometrically at 273 nm.

Van Ettekoven[62] has described a direct semi-automatic scheme based on ultraviolet light absorption for the determination of total mercury in water and sewage sludge. The full determination time is about 10 min. The lower limit of detection of mercury in water is 0.03 μg l^{-1} and 0.2 ppm (mg kg^{-1} dry matter) in sewage sludge.

Matsunaga et al.[63] have discussed possible errors caused prior to measurement of mercury in natural water and sea water.

Potentiometric titration with standard solutions of dithiooxamide at pH 5-6 has been used to estimate less than 100 μg mercury in water samples.[64] The precision in the range of 0.05-1.0 ppm mercury is about 4%. The first derivative should be used for end-point determination. A wide variety of ions can be tolerated but silver, copper, and chloride interfere, and must be separated in a preliminary step.

Mercury in potable waters

The proposed limit for total mercury in potable water has been set at 1 μg l^{-1}.[65,66] A method capable of rapid and reliable measurement of mercury at levels of one-tenth of this limit is required should this limit be adopted as a legal restriction.

Starý and Prášilová[67,68] have described a very selective radiochemical determination of phenylmercury and methylmercury.[69,70] These analytical methods are based in the isotope exchange reactions with the excess of inorganic mercury-203 or on the exchange reactions between phenylmercury and methylmercury chloride in the organic phase and sodium iodide-131 in the aqueous phase. The sensitivity of the methods (0.5–1 ppb in 5 ml sample) is not sufficient to determine organomercurials in natural waters. Subsequently Starý et al.[71] developed a preconcentration radioanalytical method for determining down to 0.01 ppm of methyl- and phenylmercury and inorganic mercury using 100–500 ml samples of potable or river water. Extraction chromatography and dithizone extraction were the most promising methods for the concentration of organomercurials in the concentration range 0.01–2 ppb. The dithizone extraction was used for the preconcentration of inorganic mercury.

Method

Apparatus and equipment

The scintillation counter with the well-type Na(I)(Ti) crystal was used for the radioactivity measurements.

A Mercury Vapour Meter (Hendrey) was applied for the inorganic mercury determination by the method of cold vapour atomic absorption spectrophotometry.

The extraction column was prepared as follows: Dry polyurethane foam in the form of a cylinder (diameter 15 mm, length 60 mm) was packed into the chromatographic column (i.d. 10 mm, length 80 mm) with a 100 ml liquid reservoir applying a gentle pressure with a glass rod. Two millilitres of xylene were pipetted into the column, followed by 10 ml of 1 M hydrochloric acid. The excess of xylene was removed from the column by suction with a syringe.

Reagents

Unless otherwise stated, all reagents were of analytical reagent grade purity.

Buffer-masking solution was prepared by the dissolution of 7.5 g of disodium salt of ethylenediaminetetraacetic acid, 20 g of sodium hydroxide and 57.2 ml of glacial acetic acid in 1000 ml of bidistilled water.

2×10^{-5} M Dithizone solution in distilled chloroform or isooctane was used.

Alkaline solution of tin chloride was prepared by the dissolution of 1 g of stannous chloride in 100 g of 30% sodium hydroxide.

Solutions of mercury-203 chloride or mercury-203 acetate were made 1 M in hydrochloric acid and purified by extraction with several portions of benzene. The solutions were diluted to the appropriate concentration with 0.5 M sulphuric acid (specific activity 100–500 mCi g^{-1}Hg. Carrier-free sodium iodide-131 in 0.01 M sodium hydroxide was used for labelling 2×10^{-5} sodium iodide in 1% ascorbic acid (specific activity 1000–2000 mCi per gram iodine. The stock solution was purified before each set of experiments by extraction with several portions of benzene.

Phenylmercury and methylmercury hydroxides were labelled with mercury-203 using isotope exchange method.

Extraction chromatography

One hundred millilitres of the aqueous sample are acidified with 10 ml of concentrated hydrochloric acid. This prepared solution is transferred immediately (phenylmercury chloride is partially decomposed by prolonged standing in acid solutions) into the liquid reservoir of the chromatographic column containing polyurethane foam loaded with xylene. Using a flow rate of about 5 ml min^{-1}, more than 97% of phenylmercury chloride is retained in the column. The column is washed with 10 ml of distilled water and phenylmercury are eluted from the column with 10 ml of 0.05 M potassium hydroxide. After the addition of 0.2 ml of concentrated sulphuric acid, 0.2 ml of 2 M silver nitrate, and 0.5 ml of 10^{-1} M mercury-203 sulphate, the mixture is kept standing at room temperature for 10 min. Then 2.0 ml of concentrated hydrochloric acid and 5.0 ml of benzene are added and the mixture shaken for 1 min to transfer phenylmercury chloride into the organic phase. Three millilitres of the centrifuged organic extract are measured using Na(I)(Ti) scintillation crystal. The total yield of the separation procedure described (75 ± 5%) was determined using labelled phenylmercury chloride added to potable and river waters. Inorganic mercury is not retained in the extraction column. Despite the fact that methylmercury chloride is partially extracted into xylene and eluted by potassium hydroxide, it does not interfere in the determination.

Dithizone extraction

Fifty millilitres of buffer-making solution are added to a 500 ml aqueous sample in a 1000 ml separation vessel and the prepared solution is shaken for 4 min with 50 ml of 2×10^{-5} M dithizone solution in isooctane. The separated organic phase is shaken for 2 min with a mixture of 9.5 ml of 0.5 M sulphuric acid and 0.5 ml of 2 M silver nitrate. To the separated aqueous phase, containing phenylmercury cations, 0.5 ml of 10^{-5} M mercury-203 sulphate is added and the mixture is kept standing for 10 min. The subsequent procedure is the same as described above.

The total yield of the separation of phenylmercury is about 70%. The calibration curve using synthetic samples is linear in the range 0.01–0.20 ppb of phenylmercury chloride. Methylmercury and inorganic mercury, which are also completely extracted into dithizone solutions, do not interfere in the determination even if present in a great excess.[68]

Determination of both phenylmercury and methylmercury

Benzene extraction

Ten millilitres of concentrated hydrochloric acid are added to a 100 ml aqueous sample. This solution is immediately shaken with 10 ml of benzene for 2 min. Four millilitres of the separated organic phase are transferred into a test tube containing 0.2–5 ml of 2×10^{-5} M sodium iodide-131 in 1% ascorbic acid. After 1 min of shaking 3.0 ml of the separated organic phase was measured using a Na(I)(Ti) scintillation crystal. Under the above conditions about 90% of phenylmercury and 50% of methylmercury are transferred into the organic phase.

Dithizone extraction

The extraction and back-extraction steps are identical with those described under dithizone extraction above. To the separated aqueous phase, containing both phenylmercury and methylmercury cations, 1 ml of concentrated hydrochloric acid and 5 ml of benzene are added. After 2 min of shaking 4 ml of the separated organic phase are treated as described under benzene extraction.

Using labelled phenylmercury and methylmercury chlorides (0.05–0.5 ppb) it was found that the total separation yield for both species is about 70%. The calibration curve for synthetic samples was linear in the concentration range 0.02–0.2 ppb.

Determination of inorganic mercury

Dithizone extraction

Fifty millilitres of buffer-masking solution are added to a 500 ml aqueous sample and the prepared solution is shaken for 4 min with two portions (30 ml and 20 ml) of 2×10^{-5} M dithizone in chloroform. Under these conditions more than 98–99% of mercury(II), methylmercury, and phenylmercury are transferred into the organic phase. The separated organic phase is shaken for 5 min with 4.5 ml of a mixture of 0.2 M hydrochloric acid and 0.5 M sulphuric acid to which 0.5 ml of 5% sodium nitrite solution was added. Inorganic mercury(II) is transferred into the aqueous phase (total yield about 95%) whereas more than 99% of methylmercury and phenylmercury remain as chlorides in the organic phase. The aqueous phase is filtered (Schleicher Schuell 8714; white strip) and transferred to a Mercury Vapour Meter. Immediately 2 ml of alkaline tin chloride solution and 2-3 drops of *n*-octanol (to prevent foam formation) are added. The calibration curve is constructed under the same conditions as for the analysed samples.

Using carbon tetrachloride as organic solvent, inorganic mercury(II), methylmercury, and phenylmercury are extracted quantitatively. Under the above conditions about 10% of methylmercury and 3% of phenylmercury are transferred into the aqueous phase due to the lower distribution ratio of methylmercury and phenylmercury chlorides between carbon tetrachloride and aqueous phase.

Jackson and Dellar[72] have described a photolysis–cold vapour flameless atomic absorption method for determining down to 0.1 mg l^{-1} total mercury in potable and natural waters and sewage effluents. Photolysis converts organomercury compounds to the inorganic state and these workers showed the advantages of this approach over chemical oxidation methods, for converting organic mercury to inorganic mercury.

Method

Materials and equipment

Hydrochloric acid, sulphuric acid, nitric acid, BDH Aristar or equivalent grade.

Mercury free water, deionized and aerated to remove traces of mercury.

Stannous chloride: 20% solution of hydrated AR reagent in hydrochloric acid (1:1), aerated to remove traces of mercury.

Methylmercury chloride solution: stock aqueous solution of 0.40 mg ml^{-1} prepared from standard material from the National Physical Laboratory. Dilute solutions are obtained by serial aqueous dilution.

Phenylmercury acetate solution: stock aqueous solution of 0.336 µg l^{-1} in 1% acetic acid. Dilute solutions obtained by serial aqueous dilution.

Mercuric nitrate solution: metallic mercury dissolved in nitric acid to give a 0.5 mg ml^{-1} solution. Dilutions are made in 2% sulphuric acid, 0.05% potassium dichromate medium to prevent loss of mercury.

Mercury meter: Laboratory Data Control monitor type 1235, cell length 300 mm, with chart recorder.

Mercury generations and aeration equipment: a non-recirculating system for manual use with sample capacity of 50 ml.

Photolysis unit: a Philips Cd-Zn-Hg quartz lamp, type 93146, 75 watts with associated switch gear, ballast transformer and light-tight cabinet.

Procedure

The irradiation is performed by immersing the ultraviolet lamp, protected from thermal shock by a silica sheath, in the water sample contained in a tall

Table 10 Recovery (%) of mercury from 150 ml of sample by photolytic–cold vapour AA method.

	Inorganic mercury (as nitrate)		Methylmercury (as chloride)		Phenylmercury (as acetate)	
mercury present	0.33 µg l^{-1}	2.00 µg l^{-1}	0.33 µg l^{-1}	2.00 µg l^{-1}	0.33 µg l^{-1}	2.00 µg l^{-1}
(a) Potable waters						
A Hardness 30 mg l^{-1}	107.106	102.99	101.85	99.98	100	91
B Hardness 150 mg l^{-1}	98.114	99.102	95.100	104.105	93	94
C Hardness 250 mg l^{-1}	111.96	99.94	100.97	97.101	100	95
D Harness 358 mg l^{-1}	100.103	99.100	102.102	96.102	97	88
Mean recovery (%)	104	99	98	100	98	92
Reproducibility (%)	6.3	2.5	5.7	3.3	—	—
Reproducibility (µg)	0.003	0.007	0.003	0.010	—	—
Reproducibility (µg l^{-1})	0.021	0.050	0.019	0.066	—	—
95% Confidence interval (µg l^{-1})	0.058	0.14	0.053	0.18	—	—
(b) Other waters						
River A suspended solids 10 mg l^{-1}	108.104	96.100	107.89	93.96	104	91
River B suspended solids 111 mg l^{-1}	126.97	98.97	104.124	100.98	97	96
Efluent A suspended solids 7 mg l^{-1}	97.102	93.93	105.86	99.98	—	—
Effluent B suspended solids 55 mg l^{-1}	127.125	89.87	82.66	94.76	—	—
Mean recovery (%)	111	94	95	94	101	94
Reproducibility (%)	13.1	4.5	18.1	7.8	—	—
Reproducibility (µg)	0.007	0.013	0.009	0.023	—	—
Reproducibility (µg l^{-1})	0.043	0.090	0.060	0.16	—	—
95% confidence interval (± µg l^{-1})	0.12	0.25	0.17	0.43	—	—

Reprinted with permission from Jackson and Dellar.[72] Copyright (1979) Pergamon Press.

250 ml beaker silvered on the outer wall and placed in an outer cold water bath to prevent overheating. The equipment is housed in a cabinet with a safety switch and sited under the fume hood to remove ozone formed during the irradiation.

One hundred and fifty millilitres of water sample is acidified to 0.25 M with hydrochloric acid, allowed to stand for 5 min, and then irradiated for 10 min in the 250 ml beaker. The sample (now at 30 °C) is cooled to room temperature. A 50 ml aliquot is placed in the reduction vessel of the mercury generation apparatus and 5 ml of stannous chloride solution are added. The released mercury is entrained with air flowing through the cell of the mercury meter. The response obtained is noted on the chart recorder and the mercury quantified by adding a suitable volume of standard mercury solution to the spent water sample, aerating once more and observing the response to the standard addition.

Table 10 shows results obtained by this procedure for four potable water samples spiked with 0.3 and 2 $\mu g\, l^{-1}$ mercury. When similar tests were carried out on surface waters and sewage effluents it was found that in the presence of high suspended solids, recoveries were reduced. For example an effluent with suspended solids of 36 mg l^{-1} gave an average recovery of 82%. The addition of nitric acid to 0.3 M, together with the hydrochloric acid improved the average recovery to 93%. The results obtained for effluents using this modification also appear in Table 10. The results in Table 10 show that for potable water, the reproducibilities for inorganic mercury and methylmercury are similar. At the 0.33 $\mu g\, l^{-1}$ level the 95% confidence interval is approximately 0.10 $\mu g\, l^{-1}$. The limit of detection based on the variation of results when estimating low levels is 0.1 $\mu g\, l^{-1}$ for a single estimation. Hence levels above 0.1 $\mu g\, l^{-1}$ may be estimated for 95% confidence intervals to the nearest 0.1 $\mu g\, l^{-1}$ up to 0.5 $\mu g\, l^{-1}$ and over this figure to the nearest 20%.

Similar considerations for effluents and surface waters give approximately 95% confidence intervals of 0.25 and 0.5 $\mu g\, l^{-1}$ at the 0.33 and 2.00 $\mu g\, l^{-1}$ levels respectively and the limit of detection of 0.25 $\mu g\, l^{-1}$. Thus the levels above 0.25 $\mu g\, l^{-1}$ may be estimated for 95% confidence to the nearest 0.25 $\mu g\, l^{-1}$ up to 1 $\mu g\, l^{-1}$ and over this level to the nearest 25%.

Jackson and Dellar[72] emphasize that to obtain results for concentrations of mercury present in water at sampling time the container and water itself must be stabilized to avoid loss or gain of mercury. Acidic potassium dichromate is believed to be the best preservative.[73] Any particulate matter present in water is likely to adsorb dissolved mercury. Before analysis a decision must be made whether total or dissolved mercury concentration is required.

Mercury in sea and coastal waters

Fitzgerald and Lyons[74] have described flameless atomic absorption methods for determining organic mercury compounds respectively in coastal and sea waters. Fitzgerald and Lyons[74] used ultraviolet light in the presence of nitric acid to decompose the organomercury compounds. In this method two sets of 100 ml samples of natural water are collected in glass bottles and then adjusted to pH 1.0 with nitric acid. One set of samples is analysed directly to give inorganically bound mercury, the other set is photo-oxidized by means of ultraviolet radiation for the destruction of organic material and then analysed to give total mercury. The element is determined by a flameless atomic absorption technique, after having been collected on a column of 1.5% of OV-17 and 1.95% of QF-1 on Chromosorb W-HP (80–100 mesh) cooled in a liquid nitrogen bath and then released by heating the column. The precision of analysis is 15%. It was found that up to about 50% of the mercury present in river and coastal waters is organically bound or associated with organic matter.

Millward and Bihan[75] studied the effect of humic material on the determination of mercury by flameless atomic absorption spectrometry. In both sea and fresh water association between inorganic and organic entities takes place within 90 min at pH values of 7 or above, and the organically bound mercury was not detected by an analytical method designed for inorganic mercury. The amount of detectable mercury was related to the amount of humic material added to the solutions. However, total mercury could be measured after exposure to ultraviolet radiation under strongly acid conditions.

Agemian and Chau[76] have described an automated method for the determination of total dissolved mercury in fresh and saline waters by ultraviolet digestion and cold vapour atomic absorption spectrocopy. A flow-through ultraviolet digester is used to carry out photo-oxidation in the automated cold vapour atomic adsorption spectrometric system. This removes the chloride interference. Work was carried out to check the ability of the technique to degrade seven particular organomercury compounds. The precision of the method at levels of $0.07\,\mu g\,l^{-1}$, $0.28\,\mu g\,l^{-1}$, and $0.55\,\mu g\,l^{-1}$ Hg was $\pm 6.0\%$, $\pm 3.8\%$, and $\pm 1.00\%$ respectively. The detection limit of the system is $0.02\,\mu g\,l^{-1}$.

Many of the standard methods for the reduction of organomercury compounds to metallic mercury such as acidic potassium permanganate and potassium period are inapplicable to saline waters as the large amount of chloride ion present reduces all of the oxidant used in the system, thus interfering with the oxidation of organomercurials. In addition, large amounts of chlorine are produced which unless reduced back to chloride, would absorb at the 253.7 nm line causing a positive interference. Because of these problems, high chloride samples could not be analysed in an automated system and required a manual predigestion step.

Agemian and Chau[76] showed that any organomercurials could be decomposed by ultraviolet radiation and that the rate of decomposition of organomercurials increased rapidly in the presence of sulphuric acid and with increased surface area of the u.v. irradiation. They developed a flow-through u.v. digestor which had a delay time of 3 min which was used to carry out the photo-oxidation in the automated system. The u.v. radiation has no effect on chloride. The method, described below, therefore, can be applied to both fresh and saline waters without the chloride interference.

Method and apparatus

The equipment used for the analysis is illustrated in Figures 8 and 9.

Manifold 1 (Figure 8)

(a) An automatic sampler (Technicon Auto Analyser II sampler with 30-2/1 cam). (b) Proportioning pump (Carlo Erba, Model 08-59-10202). (c) Technicon Auto Analyser tubing of specified dimensions and colour codes. (d) Ultraviolet digestor consisting of a 550-W photochemical lamp placed inside a quartz coil made of Puracil 453 quality fused silica tubing approximately 10 m long, 3 mm i.d., 0.6 mm wall thickness and a coil diameter of approximately 12 cm. (e) Gas separator. (f) Detector. Two systems were used for detection of mercury; at the 253.7 nm line: (1) Mercury Monitor (Pharmacia Fine Chemicals). This has a 30 cm long cell and has no background correction. (2) Model 603 Perkin Elmer atomic absorption spectrophotometer with automatic background correction and equipped with a home-made cell with quartz windows (10 mm diameter and 100 mm long). A mercury hollow cathode lamp was used in the instrument. (g) Strip chart recorder.

Manifold 2 (Figure 9)

As an alternative to manifold 1. It is the same as for manifold 1 (Figure 8) except the pump is replaced with a Technicon Auto Analyser II proportioning pump with the specified dimensions and colour codes for the tubing (*Note:* All tubing used in manifolds 1 and 2 was clear standard Technicon tubing except for the sulphuric acid line, in which case it was acidflex Technicon tubing. Manifolds 1 and 2 gave equivalent results.

Reagents

High-purity certified reagents were used.
Sulphuric acid, 36 N.
Stannous sulphate solution, 10% w/v in 2 N sulphuric acid.
Hydroxylamine sulphate (3% w/v–sodium chloride (3% w/v) solution.

Procedure

Preservation

Jenne and Avotins[77] have pointed out the requirement for a strong oxidizing agent together with a strong acid for the preservation of low levels of mercury. Both potassium permanganate and dichromate have been used as the oxidants. The former is inadequate for samples with high chloride levels since it would be readily consumed by chloride. Carron and Agemian[78] showed that 1% sulphuric acid containing 0.05% potassium dichromate make a very effective

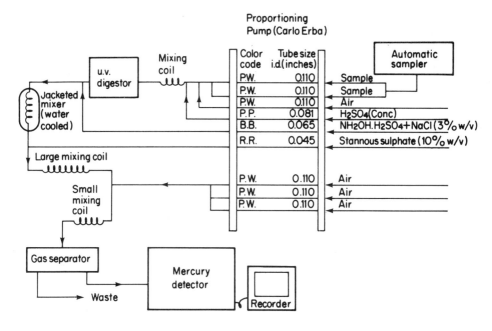

Figure 8 Mercury manifold No. 1. Reprinted with permission from Agemian and Chau.[76] Copyright (1978) American Chemical Society

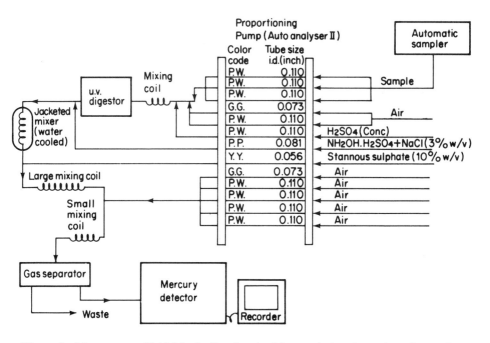

Figure 9 Mercury manifold No. 2. Reprinted with permission from Agemian and Chau.[76] Copyright (1978) American Chemical Society

preservative for sub-ppb levels of mercury in water, for extended periods of time especially when coupled with glass as the container. Agemian and Chau[76] preserved samples by adding 1 ml concentrated sulphuric acid and 1 ml of 5% potassium dichromate in glass containers at the start of the dilutions or sampling.

Analysis

Dilute mercuric chloride and methylmercuric chloride standards are prepared by serial dilution of stock solutions. The system has a detection limit of $0.02\,\mu g\,l^{-1}$ and is linear up to about $5\,\mu g\,l^{-1}$. A rate of 30 samples per hour was found to be the practical limit for the system.

Table 11 (column e) shows that complete recovery of seven organomercurials is obtained by using the u.v. digestor described above in the presence of sulphuric acid. The recoveries (Table 11) are for $5\,\mu g\,l^{-1}$ Hg solutions which is much higher than mercury levels in most natural waters. The systems provide similar recoveries throughout the working range of the calibration curve. The table shows that recoveries by u.v. oxidation and sulphuric acid (column e) are complete and comparable to the permanganate persulphate oxidation method (column c). Under the conditions of analysis, the u.v. radiation has no effect on chloride so that chloride ion behaves as an inert constituent of the sample. Analysis of synthetic mercury solutions of the seven compounds in distilled water and synthetic sea water (about 3% w/v sodium chloride) gave similar recoveries, proving that there was no chloride interference. Furthermore, the absence of any peaks when the u.v. method was used for sea water without the stannic sulphate line (Figures 8 and 9) proves that chloride is not oxidized to chlorine.

Levels of sulphide up to $100\,mg\,l^{-1}$ as S^{-2} did not when the u.v. digestor was used, have any interference effect on mercury determinations whilst when the digestor was not used levels of $1\,mg\,l^{-1}\,S^{2-}$ reduced mercury recoveries down to 50%.

Table 11 Recoveries of organomercury compounds for different oxidation methods in the automated system*

5 µg l⁻¹ of Hg as organic compound	Method (% recovery ± 5)†				
	a	b	c	d	e
(1) Phenylmercuric acetate	74	80	95	61	102
(2) Phenylmercuric nitrate	73	75	95	71	98
(3) Diphenylmercury	65	92	84	100	91
(4) Methylmercuric chloride	41	46	89	46	98
(5) Ethylmercuric chloride	81	88	88	98	95
(6) Methyoxyethylmercuric chloride	70	75	94	96	93
(7) Ethoxyethylmercuric chloride	85	91	91	85	95

*The manifolds in Figures 8 and 9 were used with the appropriate reagent lines given in each method.
†(a) H_2SO_4; (b) $H_2SO_4 + 4\%$ (w/v) $K_2Cr_2O_7$; (c) $H_2SO_4 + 0.5\%$ (w/v) $KMnO_4 + 0.5\%$ (w/v) $K_2S_2O_8$; (d) u.v. oxidation; (e) $H_2SO_4 +$ u.v. oxidation.
Reprinted with permission from Agemian and Chau.[76] Copyright (1978) American Chemical Society.

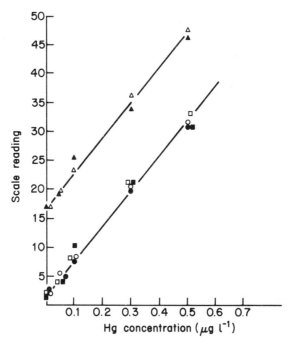

Figure 10 Calibration curves for Hg spiked into different matrix waters: —□— as $HgCl_2$ and —■— as CH_3HgCl in distilled water; —○— as $HgCl_2$ and —●— as CH_3HgCl, in Hamilton Harbour water, Hamilton, Canada (25 µg l^{-1} Cl$^-$); and —△— as $HgCL_2$ and —▲— as CH_3HgCl in sea water from New Brunswick, Canada (18,000 mg l^{-1} Cl$^-$). Reprinted with permission from Agemian and Chau.[76] Copyright (1978) American Chemical Society

Figure 10 shows that satisfactory calibration curves and standard additions curves are obtained for different waters using mercuric chloride and methylmercuric chloride in the range of normal analysis.

Sipos et al.[79] used subtractive differential pulse voltammetry at a twin gold electrode to determine total mercury levels in sea water samples taken from the North Sea.

Fish frequently have 80 100% of the total mercury in their bodies in the form of methylmercury regardless of whether the sites at which they were caught were polluted with mercury or not.[83] Methylmercury in the marine environment may originate from industrial discharges or be synthesized by natural methylation processes.[82] Fish do not themselves methylate inorganic mercury[87,89] but can accumulate methylmercury from both sea water[89] and food.[88] Methylmercury has been detected in sea water only from Minamata Bay, Japan,[80] an area with a history of gross mercury pollution from industrial discharge. It has been found in some sediments but at very low concentrations, mainly from areas of known mercury pollution. It represents usually less than 1% of the total mercury in the sediment, and frequently less than 0.1%.[86,91,92,94] Microorganisms within the sediments are considered to be responsible for the methylation[86,90] and it has been suggested that methylmercury may be released by the sediments to the sea water, either in dissolved form or attached to particulate material and thereafter rapidly taken up by organisms.[81,84,85,90] Davies et al.[93] set out to determine the concentrations of methylmercury in sea water samples much less polluted than Minamata Bay, viz. the Firth of Forth, Scotland. They described a tentative bioassay method for determining methylmercury at the 0.06 ng l^{-1} level. Mussels from a clean environment were suspended in cages at several locations in the Firth of Forth. A small number were removed periodically, homogenized, and analysed for methylmercury by solvent extraction-gas chromatography, as described by Westöö.[95] The rate of accumulation of methylmercury was determined, and by dividing this by mussel filtration rate, the total concentration of methylmercury in the seawater was calculated.

The methylmercury concentration in caged mussels increased from low levels (less than 0.01 µg g^{-1}) to $0.06-0.08$ µg g^{-1} in 150 days (Figure 11), giving a mean uptake rate of 0.4 ng g^{-1} day^{-1}, i.e. a 10 g mussel accumulated 4 ng day^{-1}. The average percentage of total mercury in the form of methylmercury increased from less than 10% after 20 days to 33% after 150 days. This may be compared with analyses of natural intertidal mussels from the area, in which the proportion of methylmercury was higher in mussels of lower (less than 10 µg g^{-1}) than of higher total mercury concentrations.

Davies et al.[93] calculated the total methylmercury concentration in the sea water as 0.06 µg l^{-1}, i.e. $0.1-0.3$% of the total mercury concentration as opposed to less than $5-32$ ng l^{-1} methylmercury found in Minamata Bay,

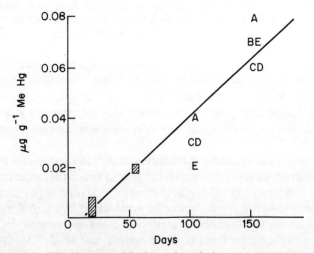

Figure 11 The increase with time of methylmercury concentration in cage mussels at positions A-E. Methylmercury was not detectable (<0.01 µg g^{-1}) after 20 days and animals from all five positions contained 0.02 µg g^{-1} after 55 days' exposure, as shown by the shaded rectangles. The cage at position B was not sampled at 106 days' exposure. Reprinted with permission from Davies et al.[93] Copyright (1979) Elsevier Science Publishers

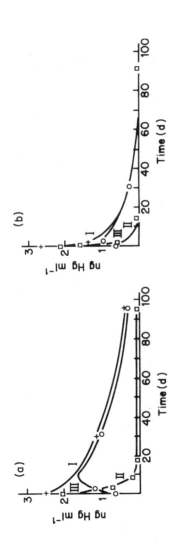

Figure 12 Storage of sea water at natural pH spiked with 2.05 ng Hg ml^{-1} as CH$_3$HgCl and 0.64 ng Hg ml^{-1} as HgCl$_2$. (a) Glass containers; (b) polyethylene containers; I, total Hg; II, CH$_3$HgCl; III, HgCl$_2$. Storage time 94 days. Reprinted with permission from Stoeppler and Matthes.[106] Copyright (1978) Elsevier Science Publishers

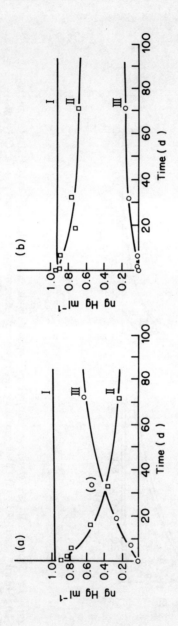

Figure 13 Storage of spiked (2.05 ng Hg ml^{-1} as Ch$_3$HgCl) sea water acidified to pH 2.5 in glass containers. (a) Nitric acid. (b) Hydrochloric acid. Curves I, II and III as in Figure 12. Reprinted with permission from Stoeppler and Matthes.[106] Copyright (1978) Elsevier Science Publishers

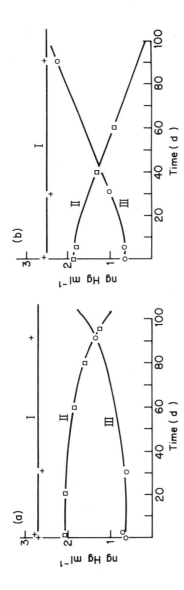

Figure 14 Storage of spiked (as in Figure 12) sea water at pH 2.5 (HCl). (a) Glass containers. (b) Polyethylene containers. Curves I, II and III as in Figure 12. Reprinted with permission from Stoeppler and Matthes.[106] Copyright (1978) Elsevier Science Publishers

Japan. These workers point out that a potentially valuable consequence of this type of bioassay is that it may be possible to obtain estimates of the relative abundance of methylmercury at different sites by the exposure of 'standardized' mussels as used in their experiment, in cages for controlled periods of time, and by the comparison of the resultant accumulations of methylmercury.

Various other workers 96–100 have reported on the levels of total mercury in sea waters. Generally, the levels are less than $0.2\,\mu\mathrm{g}\,\mathrm{l}^{-1}$ with the exception of some parts of the Mediterranean where additional contributions due to man-made pollution are found.[101-105]

Stoeppler and Matthes[106] have made a detailed study of the storage behaviour of methylmercury and mercuric chloride in sea water. They recommended that samples spiked with inorganic and/or methylmercury chloride be stored in carefully cleaned glass containers acidified with hydrochloric acid to pH 2.5. Brown glass bottles are preferred. Storage of methylmercury chloride should not exceed 10 days. Figures 12–14 show the effect of pH and preservation acid type on the long-term stability of such solutions.

Organic mercury in sediments

In lakes and streams, mercury can collect in the bottom sediments, where it may remain for long periods of time. It is difficult to release the mercury from these matrices for analysis. Several investigators have liberated mercury from soil and sediment samples by the application of heat to the samples and the collection of the released mercury on gold surfaces. The mercury was then released from the gold by application of heat or by absorption in a solution containing oxidizing agents.[107,108]

Bretthaur et al.[109] described a method in which samples were ignited in a high pressure oxygen-filled bomb. After ignition, the mercury was absorbed in a nitric acid solution. Pillay et al.[110] used a wet-ashing procedure with sulphuric acid and perchloric acid to digest samples. The released mercury was precipitated as the sulphide. The precipitate was then redigested using aqua regia.

Feldman digested solid samples with potassium dichromate, nitric acid, perchloric acid, and sulphuric acid.[111] Bishop et al.[112] used aqua regia and potassium permanganate for digestion. Jacobs and Keeney oxidized sediment samples using aqua regia, potassium permanganate, and potassium persulphate.[113] The approved US Environmental Protection Agency digestion procedure requires aqua regia and potassium permanganate and oxidants.[114]

These digestion procedures are slow and often hazardous because of the combination of strong oxidizing agents and high temperatures. In some of the methods, mercuric sulphide is not adequately recovered. The oxidizing reagents, especially the potassium permanganate, are commonly contaminated with mercury, which prevents accurate results at low concentrations.

Earlier work on the determination of total mercury in river sediments also includes that of Iskandar et al.[115] and Craig and Morton.[116] Iskandar applied flameless atomic absorption to a sulphuric acid–nitric acid digest of the sample following reduction with potassium permanganate, potassium persulphate, and stannous chloride. A detection limit of one part in 10^9 is claimed for this somewhat laborious method. Craig and Morton[116] found a 2.2 μg l^{-1} mean total mercury level in 136 samples of bottom deposits from the Mersey Estuary.

Ealy et al.[117] determined methyl-, ethyl-, and methoxyethylmercury compounds in sediments by leaching the sample with sodium iodide for 24 hours and then extracting the alkylmercury iodides into benzene. These iodides are then determined by gas chromatography of the benzene extract on a glass column packed with 5% of cyclohexylenedimethanolsuccinate on Anakrom ABS (70–80 mesh) and operated at 200 °C with nitrogen (56 ml min^{-1}) as carrier gas and electron capture detection (^{3}H foil). Good separation of chromatographic peaks is obtained for the mercury compounds as either chlorides, bromides, or iodides.

Batti et al.[118] determined methylmercury in river sediments from industrial and mining areas.

Gas chromatographic methods have been described for the determination of alkylmercury compounds in sediments.[119]

Bartlett et al.[120] observed unexpected behaviour of methylmercury containing river Mersey sediments during storage. They experienced difficulty in obtaining consistent methylmercury values; supposedly identical samples analysed at intervals of a few days gave markedly different results. They followed the levels of methylmercury in selected sediments over a period, to determine if any change was occurring on storage. They found that the amounts of methylmercury observed in the stored sediments did not remain constant; initially there was a rise in the amount of methylmercury observed, and then, after about 10 days, the amount present began to decline to levels which in general only approximate those originally present. They have observed this phenomenon in nearly all of the Mersey sediment samples they examined. It was noted that sediments sterilized, normally by autoclaving at approximately 120 °C, did not produce methylmercury on incubation with inorganic mercury, suggesting a microbiological origin for the methylmercury. A control experiment was carried out in which identical samples were collected and homogenized. Some of the samples were sterilized by treatment with an approximate 4% w/w solution of formaldehyde. Several samples of both sterilized and unsterilized sediments were analysed at intervals and all of the samples were stored at ambient room temperature (18 °C) in the laboratory. It can be seen from Figure 15 that there is a difference in behaviour between the sterilized and unsterilized samples. Some of the samples were separately inoculated into various growth media to test for microbiological activity.

This work suggests that the application of laboratory-derived results directly to natural conditions could, in these cases, be misleading: analytical results for day 10 if extrapolated directly might lead to the conclusion that natural methylmercury levels and rates of methylation are much greater than in fact they really are. Work in this area with model or laboratory systems needs to be interpreted with particular caution.

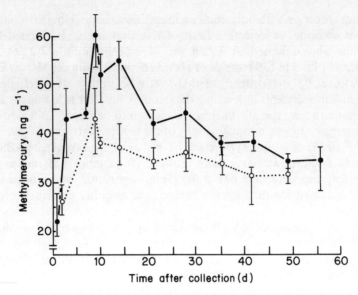

Figure 15 Analyses of sterilized and unsterilized sediments from Hale Point, for methylmercury. Total mercury is 7.24 µg g^{-1}. Results up to day 25 are the mean of eight determinations; results beyond day 25 are the mean of four determinations. Error bars represent range limits for each analysis series. The samples were stored at room temperature (18 °C), —■—, Untreated; ---O---, sterilized samples

Bartlett et al.[120] used the method of Uthe et al.[121] for determining methylmercury. Sediment samples of 2–5 g were extracted with toluene after treatment with copper sulphate and an acidic solution of potassium bromide. Methylmercury was then back extracted into aqueous sodium thiosulphate. This was then treated with acidic potassium bromide and copper sulphate following which the methylmercury was extracted into pesticide grade benzene containing approximately 100 µg dl^{-1} of ethyl mercuric chloride as an internal standard. The extract was analysed by electron capture gas chromatography using a Pye 104 chromatograph equipped with a nickel 65 detector. The glass column (1 m × 0.4 cm) was packed with 5% neopentyl glycol adipate on Chromosorb G (AW-DMCS). Methylmercury was measured by comparing the peak heights with standards of methyl mercuric chloride made up in the ethylmercury–benzene solution. The results were calculated as nanograms of methylmercury per gram of dry sediment. The detection limit was 1–2 ng g^{-1}.

A method[122] has been described for the determination of down to 2.5 ppb alkylmercury compounds and inorganic mercury in river sediments. This method uses steam distillation to separate methylmercury in the distillate and inorganic mercury in the residue. The methylmercury is then determined by flameless atomic absorption spectrophotometry and the inorganic mercury by the same technique after wet digestion with nitric acid and potassium permanganate.[123] These workers considered the possible interference effects of clay, humic acids, and sulphides, all possible components of river sediment samples on the determination of alkylmercury compounds and inorganic mercury.

Method

Methylmercury, ethylmercury, phenylmercury, and inorganic mercury standard solutions (0.1 ppm as Hg): methylmercury standard solution was freshly prepared from the stock solution (1000 ppm as Hg). The stock solution was kept in the dark.

40% $NH_2OH \cdot HCl$ solution: 40 g of $NH_2OH \cdot HCl$ was made up to 100 ml with distilled-deionized water and the solution was extracted with a suitable amount of a 1% dithizone chloroform solution.

HCl, HNO_3, and $KMnO_4^-$ analytical grade.

Determination of alkylmercury

Five to ten grams of the sediment were weighed accurately in a conical beaker, and 50 ml of 2 M hydrochloric acid solution and 10 g of sodium chloride were added. The mixture was stirred with a glass rod and was left for about one hour, and then transferred into a distilling flask with a small amount of distilled-deionized water. The steam distillation was continued to make the final amount of 200 ml of the distillate. The distillate was collected in the flask containing 10 ml of 2 M hydrochloric acid solution. Fifty millilitres of the distillate were mixed with 5 ml of saturated solution of sodium hydroxide, 3 ml of 1% cupric sulphate ($5H_2O$) and 2 ml of 10% stannous chloride solution. Then the vessel was closed with the lid and slightly shaken. The vessel was coupled to a measuring cell for the determination of mercury by flameless atomic absorption spectrophotometry.

Determination of inorganic mercury

After cooling the residue in the distillation flask distilled-deionized water was added to make the solution up to a known volume. The solution was transferred into a flat flask and attached to the closed type wet digestion apparatus. Then 20 ml of nitric acid were added to the solution; the mixture was stirred and heated at 130 °C for about 2 hours. After cooling the solution, 2 g of potassium permanganate were directly added in 0.5 g portions. In this process, the violet brown colour of potassium permanganate should remain in the solution. After potassium permanganate was added, the solution was heated for about 30 minutes, and after cooling and detaching the flask from the apparatus, 40% hydroxylamine hydrochloride solution was carefully added dropwise to decolorize the potassium permanganate solution. Distilled-deionized water was added to the digested solution to make it up to a known volume, and 50 ml of the solution were transferred to a reaction vessel, and 10 ml of sulphuric acid solution (sulphuric acid:water = 1:1) and 2 ml of 10% stannous chloride solution

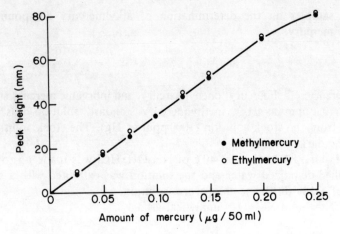

Figure 16 Calibration curves of methylmercury and ethylmercury

added. Mercury was measured in the same way as the alkylmercury. A few drops of tri-*n*-butyl phosphate was added as defoaming reagent.

The calibration curve was obtained by performing the analysis on standard solutions of methylmercury, ethylmercury, phenylmercury, and inorganic mercury. Figure 16 shows the linearity of the calibration curve for methylmercury and ethylmercury in the range 0.025–0.2 µg.

The well known absorptive properties of clays for alkylmercury compounds does not cause a problem in the above method. The presence of humic acid in the sediment did not depress the recovery of alkylmercury compounds by more than 20%. In the presence of metallic sulphides in the sediment sample the recovery of alkylmercury compounds decreased when more than 1 mg of sulphur was present in the distillate (Figure 17). The addition of 4 M

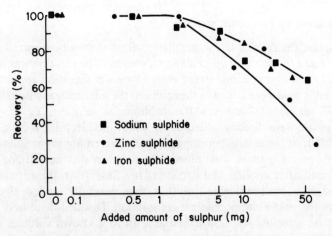

Figure 17 Effects of zinc sulphide, sodium sulphide, and iron sulphide on the determination of methylmercury

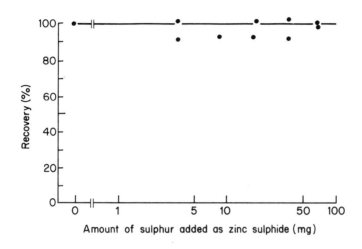

Figure 18 Elimination of the effects of sulphide on the determination of methylmercury by adding 4 M HCl instead of 2 M HCl

hydrochloric acid instead of 2 M hydrochloric acid before distillation completely eliminated this effect giving a recovery of 90–100% (Figure 18).

Figure 19 shows the results of applying the method described above to river sediment samples spiked with between zero and 0.06 μg g^{-1} methylmercury and 0–6 μg g^{-1} mercuric chloride. Smooth curves were obtained indicating the presence in the original sediment of about 0.02 μg g^{-1} methylmercury and 0 μg g^{-1} inorganic mercury.

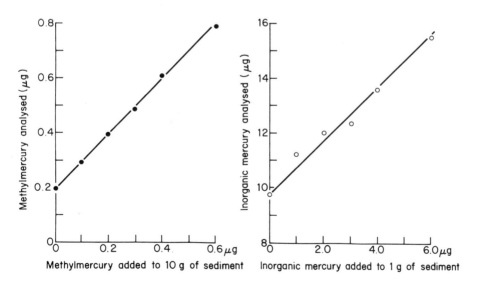

Figure 19 Inorganic mercury and methylmercury recovery test

Jurka and Carter[124] have described an automated determination of down to 0.1 µg l^{-1} mercury in river sediment samples. This method is based on the automated procedure of El-Awady[125] for the determination of total mercury in waters and wastewaters in which potassium persulphate and sulphuric acid were used to digest samples for analysis by the cold vapour technique. These workers proved that the use of potassium permanganate as an additional oxidizing agent was unnecessary.

Method and apparatus

A homogenizer was used to blend samples prior to analysis. All other apparatus used was described by El-Awady.[125] Additional air lines were added to the analytical system, and an air-bar was used for all airlines. The G0 fitting on the sample line was changed to G3. The analytical manifold is shown in Figure 20.

Reagents

A preservative solution was prepared by addition of 250 ml of concentrated nitric acid and 25 g of potassium dichromate to 500 ml of distilled water and dilution to 1000 ml. All other reagents were those described by El-Awady.[125]

Procedure

Sediment samples were passed through a No. 10 polypropylene sieve to remove large debris. If necessary, the samples were blended using a Waring blender. Approximately 1 g of wet sediment was accurately weighed into a 360 ml polyethylene bottle. Five ml of preservative solution were added to drive off or oxidize any free sulphides as well as to preserve the sample.[125,126] If the potassium dichromate was entirely reduced, as indicated by a green colour, additional preservative solution was added. Then 245 ml of distilled water was added, and the aqueous samples were blended. The samples were allowed to stand overnight. Additional preservative solution was added if the dichromate was entirely reduced after standing. The aqueous samples were then analysed, using the modified automated analytical system in the manner described by El-Awady.[125]

To convert the mercury concentrations in the aqueous samples to the concentrations in the original sediments, separate determinations of per cent solids were made, and the following formula was applied:

$$\text{mg Hg kg}^{-1}(\text{dry sample}) = \frac{\mu\text{g Hg l}^{-1}(\text{aqueous}) \times 25}{(\text{g sediment}) \times (\%\text{ solids})}$$

Table 12 contains the results of analyses for organic and inorganic mercury standards which were spiked with sulphide. There was no significant interference

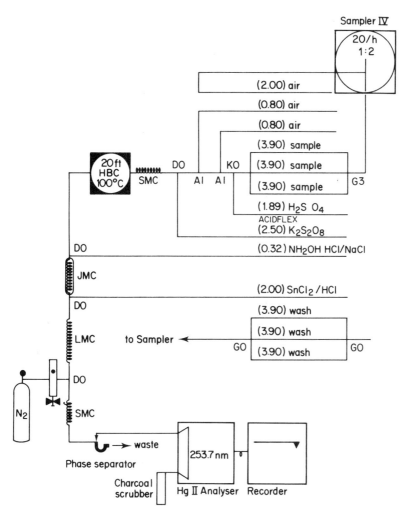

Figure 20 Modified automated total mercury manifold. Numbers in parentheses correspond to the flow rate of the pump tubes in ml min^{-1}. Numbers adjacent to glass coils and fittings are Technicon Corp. part numbers. Reprinted with permission from Jurka and Carter.[124] Copyright (1978) American Chemical Society

due to sulphide in the solutions containing 10 mg sulphide l^{-1}. However, a negative interference was observed for both organic and inorganic standards containing 100 mg sulphide l^{-1} which is equivalent to 25,000 mg sulphide kg^{-1} in the sediment. The spiked blank also resulted in a small negative interference.

It is interesting to note that exactly the same interference occurred for both organic and inorganic mercury standards, since methyl mercuric chloride does not directly react with sodium sulphide to form mercuric sulphide. Therefore the interference could not be the result of incomplete digestion of HgS or CH_3Hg^+.

Table 12 Recovery of mercury standards spiked with sulphide

Standard solution	S^{2-} spike (mg l^{-1})	observed Hg concentration (μg l^{-1})
Blank	—	0.0
1 mg Hg l^{-1}(HgCl$_2$)	—	1.0
1 μg Hg l^{-1}(CH$_3$HgCl)	—	1.0
1 μg Hg l^{-1}(HgCl$_2$)	10	1.0
1 μg Hg l^{-1}(CH$_3$HgCl)	10	1.0
Blank	10	0.0
1 μg Hg l^{-1}(HgCl$_2$)	100	0.4*
1 μg Hg l^{-1}(CH$_3$HgCl)	100	0.4*
Blank	100	−0.1
1 μg Hg l^{-1}(HgCl$_2$) + 4 times normal preservative	100	1.1
1 μg Hg l^{-1}(CH$_3$HgCl) + 4 times normal preservative	100	1.1
Blank + 4 times normal preservative	100	0.0

*Odd shaped peak.
Reprinted with permission from Jurka and Carter.[124] Copyright (1978) American Chemical Society.

Sulphur, ozone, and hydrogen sulphide were investigated as possible causes for the interference. Ozone and hydrogen sulphide were introduced directly into the mercury detector. No interference was observed. When the automated method was used, the interferences which were observed for the standards and blank spiked with 100 mg sulphide l^{-1}, occurred when an excess of dichromate did not exist in the solutions.

Aromatic organic compounds such as benzene, which are not oxidized in the digestion, absorb at the same wavelength as mercury. This represents a positive interference in all cold vapour methods for the determination of mercury. For samples containing aromatics, i.e. those contaminated by some industrial wastes, a blank analysis must be performed, and the blank results must be subtracted from the sample results. The blank analysis is accomplished by replacing the potassium persulphate reagent and the stannous chloride reagent with distilled water, and reanalysing the sample.

This automated procedure was estimated to have a precision of 0.13–0.21 mg Hg kg^{-1} at the 1 mg Hg kg^{-1} level with standard deviations varying from 0.011 to 0.02 mg Hg kg^{-1}, i.e. relative standard deviations of 8.4–12% at the 17.2–32.3 mg Hg kg^{-1} level in sediments. Recoveries in methyl mercuric chloride spiking studies were between 85 and 125%. The detection limit for the automated method is dependent upon the weight of sample taken for analysis. It is 0.1 μg Hg l^{-1} in the aqueous samples. The results for the automated method are routinely reported to a lower limit of 0.1 mg kg^{-1} which corresponds to a dry sample weight of 0.25 g.

Organic mercury in fish

Atomic absorption spectroscopy is the analytical finish most commonly used for the determination of organomercury compounds in fish, although gas chromatography has been used to some extent. Methylmercury compounds have been specifically dealt with by various workers.[126-130] Shum et al.[131] carry out a toluene extraction of the fish, then treat the extract with dithizone to form methylmercury dithizonate which is then determined in amounts down to $0.08\,\mu g\,Hg\,g^{-1}$ fish sample by graphite furnace atomic absorption spectroscopy.

Stuart[132] used ^{203}Hg-labelled methylmercuric chloride for in vivo labelling of fish to study the efficacy of various wet-ashing procedures.

Yamanaka and Ueda[133] determined ethylmercury, originating as man-made pollution, in fish found in Japanese waters.

Collett et al.[134] steam distilled the fish sample and determined alkylmercury compounds by cold vapour atomic absorption spectroscopy.

The applications of flameless atomic absorption spectroscopy to the determination of total mercury in fish have been dealt with by Shultz,[135] Stainton,[136] Kopp et al.,[137] the US Environmental Protection Agency,[138] and Hendzel and Jamieson.[139]

The determination of mercury in solid environmental samples such as fish requires low temperature preparation techniques to prevent loss of organomercury compounds. Armstrong and Uthe[140] used a sulphuric and nitric acid digestion at 58 °C followed by permanganate oxidation to extract mercury from fish tissue.

The Analytical Methods Committee[141,142] described the use of hydrogen peroxide and sulphuric acid for the destruction of organic matter, and it has been applied[143] to mercury in biological materials. This digestion system can be coupled with permanganate–persulphate oxidation to recover completely organomercury compounds from fish. This sample preparation technique has been adapted to automated reduction and determination of mercury by atomic absorption spectroscopy.[144]

Agemian and Cheam[145] developed a procedure for the simultaneous extraction of organomercury and organoarsenic compounds from fish tissues.

Gas chromatography

Kamps and McMahon[146] determined methylmercury in fish by gas chromatography. The method involves the partioning of methylmercury chloride in benzene and analysis with electron capture detection. Down to 0.02 ppm of methylmercury chloride were detected in a 10 g sample.

Longbottom[147] used the Westöö clean-up procedure to detect down to $0.01\,\mu g$ of methylmercury per g of fish, $0.001\,\mu g$ per g of sediment and $0.1\,\mu g$ per g of water. Longbottom et al. improved the Westöö clean-up procedure by replacing cystine with the more stable sodium thiosulphate when forming the methylmercury adduct. For the gas chromatography of methylmercury iodide, these workers recommend the use of a ^{63}Ni electron capture detector as it does not form an amalgam at 280 °C; the temperature at which it is used.

Uthe et al.[148] have described a rapid semi-micro method for determining methylmercury in fish, crustaceae, and aquatic mammal tissue. The procedure involves extracting the methylmercury into toluene as methylmercury(II) bromide, partitioning the bromide into aqueous ethanol as the thiosulphate complex, re-extracting methylmercury(II) iodide into benzene followed by gas chromatography on a glass column (4 ft × 0.25 in.) packed with 7% of Carbowax 20 M on Chromosorb W and operated at 170 °C with nitrogen as carrier gas (60 ml min^{-1}) and electron capture detection. Down to 0.01 ppm of methylmercury in a 2 g sample could be detected. A comparison of the results with those obtained by atomic absorption (total Hg content) indicated that all the fish samples examined contained more than 41% of the mercury as methylmercury.

Eye and Paus[149] determined alkylmercury compounds in fish tissues using an atomic absorption spectrometer tuned in at the mercury wavelength as a specific gas chromatographic detector.

Sewage and trade effluents

Takeshita[150] has used thin-layer chromatography to detect alkylmercury compounds and inorganic mercury in sewage. The dithizonates were prepared by mixing a benzene solution of the alkylmercury compounds and a 0.4% solution of dithizone. When a green coloration was obtained the solution was shaken with N sulphuric acid followed by aqueous ammonia and washed with water. The benzene solution was evaporated under reduced pressure, and the dithizonates, dissolved in benzene, were separated by reversed phase chromatography on layers of corn starch and Avicel SF containing various proportions of liquid paraffin. Solutions of ethanol and of 2-methoxyethanol were used as developing solvents. The spots were observed in daylight. The detection limit was from 5 to 57 ng (calculated as organomercury chloride) per spot.

Itsuki and Komuro[151] determined organomercury compounds in waste water by heating the sample with a 2:1 mixture of nitric and hydrochloric acids (10 ml) and 30% hydrogen peroxide (2 ml) at 90 °C for one hour followed by the addition of 50% ammonium citrate solution 5 ml, diaminocyclohexane-tetraacetone (5%v/v in 2% sodium hydroxide) (5 ml), and 10% hydroxylamine hydrochloride solution (1 ml), followed by pH adjustment to pH 3-4 with aqueous ammonia. The solution is shaken with 5 mg ml^{-1}1, 1, 1-trifluoro-4-(2-thienyl)-4-mercaptobuta-3-en-2-one in benzene (10 ml), the benzene layer washed with 0.1 M borate pH 11 (50 ml), and the solution evaluated spectrophotometrically at 365 nm.

Thin-layer chromatography has been used[152] to evaluate organomercury compounds in industrial wastewater. C_1-C_6 n-alkylmercury chlorides were separated on layers prepared with silica gel (27.75 g) plus sodium chloride (2.25 g in 60 ml water) using as development solvent cyclohexane–acetone–28% aq. NH$_3$ (60:40:1). The R_F values decrease with increasing C-chain length and phenylmercury acetate migrated between the C_1 and C_2 compounds. The spots are detected by spraying with dithizone solution in chloroform. Water samples (100–200 ml) were treated with hydrochloric acid (to produce a concentration of 0.1–0.2 M) and with potassium permanganate solution until a pink colour persists, then shaken ($\times$3) with chloroform (one-third the volume of the aqueous layer) for 3 min. The combined extracts are shaken with 0.1–0.2 M aqueous ammonia (3 $\times$ 20 ml); the aqueous solution is neutralized to p-nitrophenol with hydrochloric acid and adjusted to 2 M in hydrochloric acid and the organomercury compounds extracted with chloroform (4 ml). The chloroform extract usually recovered about 95% of the organomercury compounds and was in suitable form for thin-layer chromatography.

Murakami and Yoshinaga[153] determined organomercury compounds in industrial wastes by a spectrophotometric procedure using dithizone. The sample (100 ml) is neutralized to p-nitrophenol and hydrochloric acid added to give an acid concentration of 0.1–0.2 M. The solution is shaken with chloroform (one-third the volume of the sample), 6% potassium permanganate solution is added (until the mixture is pink, then 0.2 ml in excess), the mixture is shaken for 3 min, and the chloroform phase removed. Ten per cent hydroxylammonium chloride solution is added to decolorize the aqueous phase and the extraction repeated with chloroform twice. The combined chloroform extracts are washed with 0.1 M hydrochloric acid (3 $\times$ 50 ml), then the mercury is extracted with 0.1–0.5 M aqueous ammonia (2 $\times$ 20 ml), the aqueous solution is filtered and mercury determined.

Carpenter[154] has reviewed the application of flameless atomic absorption spectroscopy to the determination of mercury in paper mill effluents. Thiosulphate oxidation is recommended as a means of converting organomercury compounds to inorganic mercury. Carpenter concludes that the practical limit of detection of mercury in effluents using this technique is 1 part in 10^9.

Storage of mercury containing samples

The problems of preserving mercury in solution are well known. Although controversy still exists over which preservative is the best, agreement on several of the factors which affect the stability of mercury solutions seems to have been reached. For example, it is agreed that low pH values, high ionic strengths and oxidizing environments help in keeping mercury in solution. Acids such as sulphuric acid,[155] nitric acid,[156-162] and hydrochloric acid[161,163] have been widely used in different amounts. Oxidants such as permanganate[164-168] and dichromate[169-172] have been shown to prevent volatilization of mercury. Sodium chloride[163] and gold(III)[170,172] have also been used as preservatives. Various workers have commented on the instability of mercury solutions when stored in polyethylene or polypropylene containers.[173-176]

McFarland[174] concluded that polyethylene phials can be used for the neutron irradiation of mercury solutions without loss. Weiss and Chew[176] carried out neutron irradiations of aqueous and nitric acid solutions of mercury in polyethylene containers and showed that, whilst no losses occurred in the presence of nitric acid, both absorption and volatilization losses of up to 18% occurred in aqueous medium. Heiden and Aikens[173] studied the effect of differences in commercial polyethylene bottles on the stability of parts per billion mercury(II) solutions.

Coyne and Collins[175] showed that solutions containing 0.05-0.5 mg mercury in the presence of acetic acid formaldehyde preservative lost about 50% in 3 days. Loss of mercury appeared to be related to the original concentration of mercury and the presence or absence of a preservative. Nitric acid (added to give a final pH of 1.0) was moderately effective in preventing loss of mercury provided that the acid was placed in the container before the sample was added.

Other workers have compared the effectiveness of glass with various plastic containers for the safe storage of dilute mercury solutions.[177,178] Rosain and Wai[177] concluded that mercury losses in solutions pH 7 to nitric acid were more rapid from distilled water than from natural water and were most severe from poly(vinyl chloride) containers. It is recommended that sampling for mercury determination be done in glass or polyethylene containers, that the sample be acidified to pH of less than 0.5 with nitric acid and that analysis be carried out as soon as possible after sampling.

Feldman[178] stored 0.1-10 mg inorganic mercury solutions in glass and polyethylene containers in the presence of various reagents. He found that the solutions lost appreciable amounts of mercury even when solutions of 1-5% nitric acid, 0.5% sulphuric acid-0.01% potassium permanganate or nitric acid-0.01% dichromate were added. However, solutions could be stored in glass for up to 5 months in the presence of 5% nitric acid-0.01% dichromate (added as $K_2Cr_2O_7$ or CrO_3^-) and in polyethylene for at least 10 days in the presence of 5% nitric acid-0.05% dichromate.

Carron and Agemian[179] have recently pointed out that whilst the majority of fresh water samples rarely contain mercury at levels over 0.5 ppb and in most cases 0.2 ppb, most previous investigators of the stability of mercury solutions have carried out their tests at higher mercury levels. These workers studied preservation methods which provide both low pH values as well as oxidizing environments using both synthetic and natural samples in a variety of containers, in order to obtain a practical method which would be adaptable to routine analysis for mercury in natural waters at sub-ppb levels by the automated cold vapour atomic absorption technique. The essential requirement was that the preservation method should maintain mercury in waters of low salt content (low conductivity such as distilled water) and of high salt content (high conductivity). The outcome of this work was that Carron and Agemian[179] recommended glass containers washed with concentrated nitric or chromic acid and a preservative consisting of a mixture of 1% sulphuric acid and 0.05% potassium dichromate. This preservative gives good accuracy, precision, and low detection limits. It was also observed that the presence of methylmercury ions improves preservation efficiency.

Stoeppler and Matthes[180] have studied the storage behaviour of methylmercury chloride and inorganic mercury in sea water samples.

Carr and Wilkniss[181] showed if a water sample is acidified to pH 1 with nitric acid then of any mercury that is initially associated with particulate matter in the sample about 80% enters solution during storage for a period of about 1 week. There is no significant loss to the container. If the sample is not acidified, most of the mercury is retained by the particulate matter, 15% is adsorbed by the container and only 10% remains in solution.

ORGANOARSENIC COMPOUNDS

The knowledge and concern about the environmental impact and the ultimate fate of organoarsenicals applied to soils and other ecosystems is steadily growing. The extensive number of analytical procedures which have been reported in the literature would seem to indicate inherent difficulties in the analysis of arsenic. Most procedures intended to measure total arsenic incorporate some mode of wet or dry digestion to destroy any organically bound arsenic, in addition to any other organic constituent present in the sample.

Probably the most frequently used method of digestion incorporates the use of nitric and sulphuric acids. Kopp[182] used this digestion method and experienced 91–114% recovery of arsenic trioxide added to deionized water and 86–100% recovery of the compound added to river water. Evans and Bandemer[183] recovered 87% of the arsenic trioxide added to eggs. By modifying the above digestive method by the addition of perchloric acid, Caldwell et al.[184] observed 80–90% arsenic recovery with o-nitrobenzene arsenic acid, 85–94% arsenic recovery with o-arsanilic acid, and 76.7% arsenic recovery with disodium methylarsenate.

Two uncertainties seem to arise when reviewing the previous digestive methods using nitric and sulphuric acid. First, the addition of inorganic arsenic to an organic matrix and subsequent recovery of all the inorganic arsenic added is not definite proof of total recovery of any organoarsenicals present. Secondly, the choice of o-nitrobenzene arsenic acid and o-arsanilic acid seems unfortunate since both compounds represent arsenic attached to an aromatic ring which is atypical of cacodylic acid and disodium methylarsenate, two widely used organoarsenicals.

Aside from nitric and sulphuric acid, and relatively simple digestive method employing 30% hydrogen peroxide in the presence of sulphuric acid was reported by Kolthoff and Belcher[185] and subsequently used by Dean and Rues[186] to determine arsenic in triphenylarsine.

Atomic absorption spectroscopy

Edmunds and Francesconi[187] estimated methylated organoarsenic compounds by vapour generation atomic absorption spectroscopy. Fishman and Spencer[188] and Agemian and Cheam[189] have described automated atomic absorption spectrometric methods for the determination, respectively, of total arsenic in water and in fish tissues. Fishman and Spencer[188] used an ultraviolet radiation or an acid persulphate digestion procedure to decompose the organoarsenic compounds. The automated methods of Agemian and Cheam[189] uses hydrogen peroxide and sulphuric acid for the destruction of organic matter, combined with permanganate–persulphate oxidation for the complete recovery of organoarsenic compounds from fish. An automated system based on sodium borohydride reduction with atomization in a quartz tube is used for the determination of the inorganic arsenic thus produced.

Method

Sampling

Prior to digestion, the fish tissue samples must be homogenized. Efficient transfer of aliquots of such samples into long-necked digestion or extraction flasks requires a long delivery tube. The device shown in Figure 21 is satisfactory. The barrel and plunger are commercially available (Part 432, Oxford pipetter Model S-A, cat. No. 13-687-75, Fisher Scientific). The ground glass piston delivers a good vacuum so that thick samples can be used. The sample touches only the disposable glass tube; these are precut and prewashed suitably; a 6 in. length suits most analytical requirements.

Reagents

Certified analytical grade.

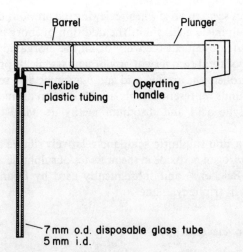

Figure 21 Transfer device. Reprinted with permission from Agemian and Cheam.[189] Copyright (1978) Elsevier Science Publishers

Digestion

To the digestion flask containing the sample is added 1 ml of 30% hydrogen peroxide solution. The solution is allowed to stand for *ca*. 10 min after which is added, slowly, 10 ml of 18 M sulphuric acid while cooling the flask efficiently in an ice bath. If, after 5 min the sample has not dissolved, the flask is placed in a shaking water bath at 60 °C for 30 min, removed, placed in an ice bath and 20 ml of 5% potassium permanganate solution added slowly to avoid frothing. The solution is allowed to stand for 60 min, then 60 ml of 5% (w/v) potassium persulphate solution are added and the solution left overnight.

The solution is cleared by adding 5 ml of 20% (w/v) hydroxylamine sulphate–20% (w/v) sodium chloride solution and adjusting to 100 ml.

Standards are prepared with the same reagents. The samples are analysed with the manifold shown in Figure 22. A 40-1/2 cam is used in the autosampler to obtain a sampling time of 30 s and a wash time of 60 s.

Apparatus

The equipment (Figure 22) consisted of a Technicon AutoAnalyser II sampler with 40-1/2 cam; proportioning pump (AutoAnalyserII); Technicon AutoAnalyser mixing coils and tubing of specified dimensions; flow meter

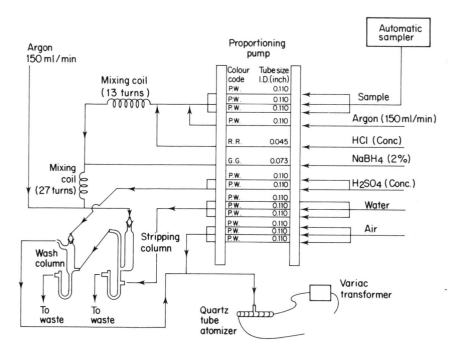

Figure 22 Arsenic manifold. Reprinted with permission from Agemian and Cheam.[189] Copyright (1978) Elsevier Science Publishers

for argon (150 ml min^{-1}); stripping column and wash column (8); furnace made from silica tubing (1 cm i.d. 10 cm long). At the centre of this tube a 'T' is made with silica tubing (2 mm i.d.) for the gas entry port. The furnace is wrapped with asbestos paper and chromel wire (26 gauge; resistance, 20 Ω). Silica tubing mounted close to the furnace contains a pyrometer; the furnace is heated to 850 °C through a variac transformer. The Perkin Elmer Model 503 atomic absorption spectrometer used was equipped with an arsenic EDL lamp and power supply and recorder (Hewlett Packard Model 7128A).

Agemian and Cheam[189] found that in the sodium borohydride reduction of inorganic arsenic to arsenic, concentrations from 0.5 to 1.5 M of hydrochloric acid gave the highest sensitivity; both As(III) and As(V) were equivalently detected. When the hydrochloric acid concentration was increased from 2 to 6 M, the sensitivity for both species decreased, particularly for As(V). Replacement of the hydrochloric acid line (Figure 22) with a sulphuric acid line, to give a common acid in the manifold and sample digests, reduced the sensitivity for As(III) by about 30% and As(V) gave a sensitivity of about 50% of As(III).

Figure 23 shows a standard additions curve obtained by this procedure for As(III) and As(V) for a typical fish sample. The recovery of arsenic was 10% from National Bureau of Standards SRM 1571 (Orchard Leaves), which contain 10 ± 2 µg As g^{-1}. Replicate determinations at a level of 0.10 µg As g^{-1} in a typical fish sample gave a relative standard deviation of 15%.

Stringer and Attrep[190] compared hydrogen peroxide–sulphuric acid digestion and ultraviolet photodecomposition methods for the decomposition of three organoarsenic compounds in water samples (triphenylarsine oxide, disodium-methane arsonate and dimethylarsinic acid) to inorganic arsenic prior to

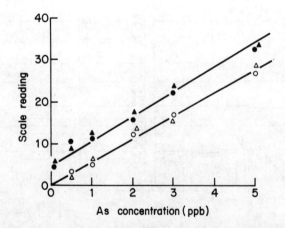

Figure 23 (○) Calibration curve for As(III). (●) Standard additions curve for As(III) in a fish sample; (△) calibration curve for As(V); (▲) standard additions for As(V) in a fish sample. Reprinted with permission from Agemian and Cheam.[189] Copyright (1978) Elsevier Science Publishers

reduction to arsine and determination by atomic absorption spectroscopy or by the silver diethyldithiocarbamate spectrophotometric method.[191]

Stringer and Attrep[190] carried out their ultraviolet radiation photodecompositions using a medium pressure 450 W mercury arc photochemical lamp (Hannovia lamp no. 67A0100) mounted vertically in the middle of an aluminium cylinder large enough in diameter to accommodate ten 1 inch diameter silica tubes. A blower fan was attached to the bottom of the lamp housing and served as an extractor fan to provide cooling of the lamp and samples. Excessive cooling of the lamp and resulting dimming of the lamp discharge was alleviated by enclosing the lamp in a 2 inch diameter quartz liner tapered at one end to retard air flow. The silica tubes which contained the samples were approximately 20 cm long and set 2 cm parallel to the lamp.

Prior to irradiation, each sample was acidified with 3 drops of nitric acid and also 3 drops of 30% hydrogen peroxide. Following irradiation, each sample was transferred to a 100 ml volumetric flask and made to volume. Then, 10 ml of the 100 ml were volumetrically transferred to a 125 ml Erlenmeyer flask, which also served as the arsine generator. The inorganic arsenic content was determined colorimetrically with silver diethyldithiocarbamate.

Sulphuric acid-hydrogen peroxide decompositions were carried out using the method of Kolthoff and Belcher.[185] Arsenate, resulting from the decomposition of the organoarsenicals, was reduced to arsenite with hydrazine sulphate and titrated with potassium bromate. The yellow colour for free bromine liberated by oxidation of bromide ion in the presence of excess bromate titrant served to distinguish the end-point of the titration.

Aliquots of each of the three organoarsenicals containing 5 µg of arsenic were added to 125 ml arsine generators. To each organic compound and a complete set of arsenic trioxide standards, 5 ml of concentrated sulphuric acid and 5 ml of 30% hydrogen peroxide were added. Samples and standards were boiled off and fumed for an additional 2 min. The samples were cooled and to each flask the following were added in succession: 25 ml of water, 7 ml of hydrochloric acid (concentrated), 5 ml of 15% potassium iodide, and after 10 min, 5 drops of 40% stannous chloride. The samples were refrigerated for 20 min to allow for reduction to occur. Arsine was generated into an absorption tube containing 4 ml of the silver diethyldithiocarbamate reagent for 1 hour. Alternatively, following hydrogen peroxide-sulphuric acid decomposition inorganic arsenic was determined by atomic absorption spectroscopy.

The percentage recoveries obtained by the hydrogen peroxide-sulphuric acid digestion followed by spectrophotometric evaluation using silver diethyldithiocarbamate by atomic absorption spectrometric determination are given in Table 13.

The recoveries obtained by the same method when applied to arsenic spiked water samples are shown in Table 14.

Figure 24 shows the effect of ultraviolet irradiation as a function of time for triphenylarsine oxide, disodium methanearsonate, and dimethylarsinic acid. The extent of arsenic recovery using photo-oxidation in conjunction with high

Table 13 Recovery of arsenic from triphenylarsine oxide, disodium methanearsonate, and dimethylarsinic acid employing wet digestion with 5 ml of 30% hydrogen peroxide and 5 ml of sulphuric acid with analyses by silver diethyldithiocarbamate and atomic absorption

Compound	Recovery (%)* AgDDC	Atomic absorption
Triphenylarsine oxide	101.0 ± 5.2 (4)	100.8 (2)
Disodium methanearsonate	103.5 ± 1.2 (4)	100.8 (2)
Dimethylarsinic acid	99.9 ± 2.3 (3)	99.2 (2)

*Each sample contained 5 µg arsenic. Number in parentheses indicates the number of samples run. Reprinted with permission from Stringer and Attrep.[190] Copyright (1979) American Chemical Society.

Table 14 Recovery of arsenic from triphenylarsine oxide, disodium methanearsonate, and dimethylarsinic acid spiked into water sample with wet digestion employing 15 ml of 30% hydrogen peroxide and 5 ml of sulphuric acid with analysis by silver diethyldithiocarbamate*

Sample	Recovery (%)† Sample 1	Sample 2
100 ml AF	(9.7 µg As l^{-1})	(6.9 µg As l^{-1})
100 ml AF + triphenylarsine oxide	92.4	94.1
100 ml AF + disodium methanearsonate	90.4	89.1
100 ml AF + dimethylarsinic acid	96.0	90.2
100 ml SBE	(22.6 µg As l^{-1})	(18.5 µg As l^{-1})
100 ml SBE + triphenylarsine oxide	101.1	100.6
100 ml SBE + disodium methanearsonate	98.1	104.4
100 ml SBE + dimethylarsinic acid	89.4	96.6

*All 5µg weights are as arsenic. †corrected for amount of arsenic initially present.
Reprinted with permission from Stringer and Attrep.[190] Copyright (1979) American Chemical Society.

sensitivity analysis when applied to water samples is shown in Table 15.

The conclusion to be drawn from this work is that the digestive method employing hydrogen peroxide and sulphuric acid with analysis by silver diethyldithiocarbamate gives good results with arsenic recoveries ranging from 98.5 to 104.9%. This same digestive method when applied to primary settled raw sewage gave arsenic recoveries ranging from 89.1 to 96.0%. Arsenic recoveries of 89.4–104.4% were experienced from an activated sludge effluent sample. The same digestive method employing the high sensitivity arsenic analysis by atomic absorption resulted in arsenic recoveries of 99.2–100.8%.

The hydrogen peroxide–sulphuric acid digestion seemed to provide the most consistent and complete recoveries of any of the wet digestive procedures previously examined. This digestion procedure coupled with the silver diethyldithiocarbamate colorimetric analysis resulted in quantitative arsenic recoveries from waste water samples. The hydrogen peroxide–sulphuric acid digestion combined with the high sensitivity arsenic analysis gave acceptable recoveries of arsenic and is a viable technique.

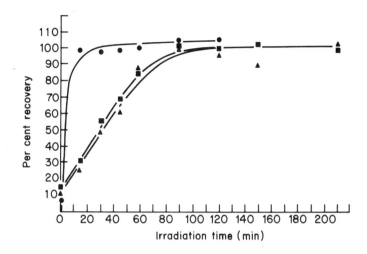

Figure 24 Recovery of arsenic after ultraviolet exposure as a function of time. ●, triphenylarsine oxide; ■ disodium methanearsonate; ▲ dimethylarsinic acid. Reprinted with permission from Stringer and Attrep.[190] Copyright (1979) American Chemical Society

Table 15 Recovery of arsenic from triphenylarsine oxide, disodium methanearsonate, and dimethylarsinic acid spiked into water samples with digestion by u.v. and analysis by high sensitivity atomic absorption

Compound	Sample 1	Sample 2	Recovery (%)* Sample 3
Triphenylarsine oxide	110.0	100.0	102.8
Disodium methanearsonate	100.0	102.4	102.8
Dimethylarsinic acid	100.0	109.8	88.6

*Corrections were made for arsenic present in the samples.
Reprinted with permission from Stringer and Attrep.[190] Copyright (1979) American Chemical Society.

Regarding the ultraviolet photodecomposition procedure, these workers showed that a 15 minute exposure of triphenylarsine oxide resulted in greater than 99% photodecomposition of the compound. The monoalkylated arsenic compound reacted much more slowly than triphenylarsine oxide, requiring 2 h for complete decomposition so, obviously, this method has to be used with caution. A 4 hour decomposition produced 100–110% conversion of three organoarsenic compounds.

Gas chromatography

The applications of this method are very limited. Andreae[192] has detected inorganic and methylated arsenic species at the $ng\,l^{-1}$ level in natural waters by volatilization and detection by atomic absorption, electron capture and flame ionization detectors.

Soderquist et al.[193] determined hydroxydimethyl arsine oxide in water and soil by converting it to iododimethylarsine using hydrogen iodide followed by determination at 105 °C on a column (15 ft × 0.125 in) packed with 10% of DC-200 on Gas-Chrom Q(60–80 mesh), with nitrogen as carrier gas (20–30 ml min^{-1}) and electron capture detection. The recovery of hydroxydimethylarsine oxide (0.15 ppm) added to pure water was 92.3% with a standard deviation of ±7.4%. The corresponding results from soil were 91.3 ± 5.1%.

Spectrophotometric methods

Haywood and Riley[194] have described procedures for the determination of arsenic in sea water, potable water, and waste waters. Whilst this method does not include organic arsenic species, these can be rendered reactive either by photolysis with ultraviolet radiation or by oxidation with potassium permanganate or a mixture of nitric acid and sulphuric acids. Arsenic(V) can be determined separately from total inorganic arsenic after extracting As(III) as its pyrrolidine dithiocarbamate into chloroform.

In the method for inorganic arsenic the sample is treated with sodium borohydride added at a controlled rate. The arsine evolved is absorbed in a solution of iodine and the resultant arsenate ion is determined photometrically by a molybdenum blue method. For sea water the range, standard deviation, and detection limit are 1–4 μg l^{-1}, 1.4%, and 0.14 μg l^{-1} respectively; for

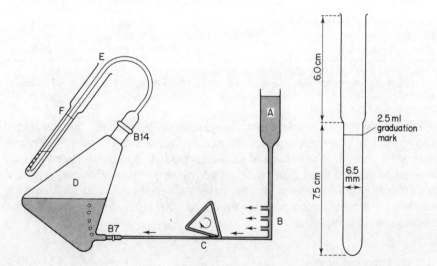

Figure 25 Apparatus for the evolution and trapping of arsenic. The dimensions of the absorption tube F are given in the inset. Reprinted from Haywood and Riley.[194] Copyright (1971) Elsevier Science Publishers

potable waters they are 0–800 μg l^{-1}, about 1% (at 20 μg l^{-1} level) and 0.5 μg l^{-1} respectively. Silver and copper cause serious interference at concentrations of a few tens of mg l^{-1}; however, these elements can be removed either by preliminary extraction with a solution of dithizone in chloroform or by ion-exchange.

Method

Apparatus

The apparatus used for the evolution of arsine is shown diagrammatically in Figure 25. Sodium borohydride solution was delivered from the reservoir (A) via a manifold (B) to a Watson-Marlow Type HR peristaltic pump (C) fitted with Acidflex tubing which served to inject the reagent at a controlled rate into the 250 ml evolution flask (D) mounted at an angle of about 35° on a stand. The evolved hydrogen, along with entrained arsine, was passed through the jet (E) (bore 0.05 mm) into the absorption tube (F) (calibrated at a volume of 2.5 ml) containing iodine solution. All glassware was cleaned initially by soaking in 18 M sulphuric acid overnight.

Reagents and standards

Sodium borohydride solution

Powdered sodium borohydride (25 g) are dissolved in 100 ml of distilled water, and filtered through a retentive filter paper. To purify the solution 2 g of calcium hydroxide are added, the flask is closed loosely with a bulb stopper and warmed to 75 °C in a water bath for 20 min; under these conditions a small proportion of the borohydride decomposes rapidly, thus removing traces of arsenic as arsine. The solution is cooled to room temperature, filtered through a retentive filter paper and diluted to 250 ml with water. This solution should be replaced after 2 days as it decomposes slowly.

Absorption solution

Iodine (0.25 g) is dissolved in a solution of 0.4 g of potassium iodide in 5 ml of water and diluted to 100 ml with water. The solution is stored in a well-stoppered bottle; it is stable indefinitely.

Ammonium molybdate solution, 4.8% (w/v)

Ammonium molybdate (4.8 g) is dissolved in 60 ml of water and diluted to 100 ml with distilled water. The solution is stored in a polyethylene bottle; it is rejected if it becomes discoloured, or if a precipitate forms.

Ascorbic acid solution, 1.76% (w/v)

This is prepared in redistilled water and stored at 0 °C. It is rejected when it becomes yellow.

Potassium antimonyl tartrate solution (1 mg Sb ml^{-1})

A 0.274% (w/v) solution in distilled water is used.

Mixed reagent

10 ml of 2.5 M sulphuric acid, 3 ml of ammonium molybdate solution, 1.0 ml of potassium antimonyl tartrate solution, and 6 ml of ascorbic acid solution are mixed. The reagent should be used within 1 h of preparation as it is unstable.

Cation-exchange column for removal of heavy metals

Zerolit 225 ion-exchange resin (52–100 mesh) is digested on a water bath at 80 °C with excess of 6 M hydrochloric acid. After 1 h the acid is poured off and the resin washed several times with distilled water. A wad of silica wool is placed at the bottom of a 5 cm × 0.8 cm^2 ion-exchange column and the column filled with about 5 ml of the resin. (Glass wool should not be used for supporting the resin since it tends to retain arsenic). The resin is converted to the sodium form by passing 50 ml of 1 M sodium chloride through it. Finally, the column is washed with 50 ml of distilled water. After use, the column can be regenerated by washing with 50 ml of 1 M hydrochloric acid and then reconverting to the sodium form.

Standard arsenic solution

This is prepared from reagent-grade arsenic(III) oxide a stock solution containing 100 μg As ml^{-1}. Working standards (0.1 μg As ml^{-1}) are prepared as required since they tend to be unstable.

Arsenic-free sea water

One litre of sea water is purified by passing it through an ion-exchange column (i.d. 1 cm) packed with a 5 cm bed of hydrous zirconium oxide (Bio-rad HZO-1; 100–200 mesh) treated previously with 50 ml of 2 M sodium hydroxide and then washed free from alkali.

Determination of total inorganic arsenic in sea water

The filtered (0.45 μm) sea water sample (150 ml) is placed in the evolution flask and 5 ml each of 2.5 M sulphuric acid and 2% (w/v) EDTA (disodium salt) solution are added. The flask is replaced on its stand and the head and delivery tube fitted. Then 1.2 ml of the iodine–potassium iodide absorption solution and 0.2 ml of 4.2% (w/v) sodium hydrogen carbonate solution are pipetted into an absorption tube. The delivery tube is inserted so that its jet is close to the bottom of the absorption tube. The absorption tube is secured to the flask with a rubber band. With the peristaltic pump sodium borohydride solution is added at a rate of 15 ml h^{-1}. After 1 h the absorption tube is lowered and the tip of the delivery tube rinsed into it with 0.2–0.3 ml of water. To this is added 0.5 ml of mixed reagent and the solution diluted to 2.5 ml with distilled water. The solution is thoroughly mixed, preferably in an ultrasonic bath, to release the carbon dioxide liberated from the sodium hydrogen carbonate in the absorption solution. After 30 min the absorbance of the solution at 866 nm is measured in a 40 mm microcuvette. The reagent blank is determined in the same way but substituting 150 ml of arsenic-free sea water for the sample; the absorbance of the blank is normally about 0.025. The method is calibrated in the same manner with 150 ml aliquots of arsenic-free seawater to which 1.0 and 2.0 ml increments of standard arsenic solution have been added (corresponding to 0.1 and 0.2 µg As, respectively).

Determination of arsenic(V) in sea water

The sample (150 ml) is placed in a separating funnel and 2 ml of 2.5 M sulphuric acid and 4 ml of freshly prepared 1% (w/v) ammonium pyrrolidine dithiocarbamate (APDC) solution are added. The APDC–As(III) complex is extracted by shaking for 2 min with two 5 ml aliquots of chloroform. The aqueous phase is transferred to a conical flask, covered with a bulb stopper and boiled gently for 20 min to decompose the excess of APDC. The solution is cooled and arsenic(V) determined as described above. A blank is determined in the same manner, substituting 150 ml of arsenic-free sea water for the sample. The method is calibrated similarly with 150 ml aliquots of arsenic-free sea water spiked with 1.0 and 2.0 ml increments of standard arsenic solution (corresponding to 0.1 and 0.2 µg As).

Determination of total arsenic in sea water

To determine total arsenic in sea water, the pH is adjusted to about 6, and the sample irradiated for 2 h in a fused silica tube at *ca.* 60 °C at a distance of 15 cm from a 1 kW, medium pressure, mercury lamp. After the irradiation, the sample is diluted to the original volume and the analysis contained as described above for total inorganic arsenic.

Determination of total inorganic arsenic in potable water

Fifty millilitres of the filtered water sample are placed in an evolution flask and 5 ml of 2.5 M sulphuric acid and 5 ml of 2% EDTA (disodium salt) solution are added. The analysis is continued as described for sea water with the exception that the sodium borohydride solution should be added at a rate of 10 ml h^{-1}. The blank is determined in the same way with 50 ml of distilled water instead of the sample; the absorbance of the blank is normally 0.025. The method is calibrated similarly with distilled water spiked with 2 and 4 ml increments of standard arsenic solution (corresponding to 0.2 and 0.4 µg As, respectively), and then diluted to a total volume of 50 ml.

Determination of total arsenic in potable and natural waters containing less than $0.1\,mg\,l^{-1}$ of heavy metals

Fifty millilitres of the sample are placed in a 250 ml conical flask and 5 ml of 2.5 M sulphuric acid and 25 ml of aqueous 1.0 M potassium permanganate solution (0.632% w/v) added. The flask is closed loosely with a bulb stopper and heated on a boiling water bath for 8 h. If necessary further permanganate solution is added to maintain an excess in the solution. After cooling, 10% hydroxylammonium chloride solution is slowly added, with shaking, until the permanganate colour is just discharged. The solution is transferred to an evolution flask, 5 ml of 2% EDTA solution are added, and the analysis continued as described in the previous paragraph. Blank determinations and standardizations are performed in the same manner with 50 ml aliquots of distilled water alone and spiked with 0.4 µg of arsenic respectively.

Determination of arsenic in effluents or natural waters containing more than $0.1\,mg\,l^{-1}$ of heavy metals.

Fifty millilitres of the sample are placed in a polytetrafluoroethylene beaker and evaporated to about 10 ml on a hot plate. The solution is cooled and 2 ml of concentrated nitric acid and 1 ml of concentrated sulphuric acid are cautiously added. The evaporation is continued until dense white fumes of sulphur trioxide begin to be evolved. After cooling, the residue is taken up in 25 ml of distilled water. When the concentration of heavy metals in the original sample exceeds $0.1\,mg\,l^{-1}$ these elements must be removed before proceeding in the evolution stage; this can be achieved in two ways as follows.

The solution of the residue is placed in a separatory funnel and adjusted to pH 7-8 by cautious addition of 2 M ammonia solution. The solution is extracted with 10 ml of a 0.05% solution of dithizone in chloroform. If necessary, the pH is readjusted to 7-8. The extraction is repeated with further aliquots of dithizone solution until the organic phase remains green. The aqueous phase is washed with 10 ml of chloroform and diluted to 50.0 ml with distilled water. The extracts and washings are rejected. This procedure is satisfactory up to a total heavy metal concentration of about $25\,mg\,l^{-1}$.

Alternately, the solution of the residue is adjusted to pH 3.0-3.5 by careful addition of 2 M ammonia solution. The solution is passed through a column of Zerolit 225 (Na^+ form) and the column washed with four 5 ml aliquots of distilled water. The percolate and washings are combined and diluted to 50.0 ml with distilled water. This procedure can be used for a total heavy metal concentration of up to $2\,g\,l^{-1}$.

A suitable volume of the solution ($\leqslant 1$ µg As) is transferred to an evolution flask, diluted to 50 ml with distilled water, and the analysis continued as described for the determination of inorganic arsenic in potable waters. Blank determinations and calibrations are performed through the whole procedure with 50 ml aliquots of distilled water and distilled water spiked with 0.4 µg As, respectively.

The precision of the method was tested by carrying out replicate analyses (10) on 150 ml aliquots of two sea water samples from the Irish Sea. Average arsenic concentrations of 2.63 ± 0.05 and $2.49 \pm 0.05\,\mu g\,l^{-1}$ were found; thus the average relative standard deviation was 1.4%. Replicate determinations of the reagent blank were performed with arsenic-free sea water and showed a standard deviation of $\pm 0.035\,\mu g\,As\,l^{-1}$; this implies a detection limit of $0.14\,\mu g\,As\,l^{-1}$ if this is assumed to be four times the standard deviation. The recovery of arsenic was checked by analysing 150 ml aliquots of arsenic-free sea water which had been spiked with known amounts of arsenic(V). The results of these experiments (Table 16) shows that there is a linear relationship between absorbance and arsenic concentration and that arsenic could be recovered from sea water with an average efficiency of 98.0% at levels of $1.3-6.6\,\mu g\,l^{-1}$. Analagous experiments in which arsenic(III) was used gave similar recoveries.

Table 16 Recovery of arsenic(V) from spiked samples of arsenic-free sea water (150 ml)

As added (μg)	0.200	0.300	0.400	0.600	0.800	1.000
Absorbance per $\mu g\,l^{-1}$	0.063_6	0.063_4	0.064_3	0.053_5	0.063_6	0.063_8
As found (μg)	0.196	0.294	0.396	0.586	0.778	0.985
Recovery (%)	98.1	97.8*	99.1	97.9	97.2	98.5

*average for 10 determinations; range 96.0–99.2%, standard deviation ± 1.35%
Reprinted with permission from Haywood and Riley.[194] Copyright (1971) Elsevier Science Publishers.

Determination of the As(III): As(V) ratio in sea water

Although purely thermodynamic considerations suggest that arsenic should exist in oxic sea waters practically entirely in the pentavalent state, equilibrium rarely appears to be attained, probably because of the existence of biologically mediated reduction processes. For this reason, the arsenic in most of these waters exists to an appreciable extent in the trivalent state, and indeed, As(III): As(V) ratios as high as 1:1 have been found in a number of instances.

Haywood and Riley[194] found that arsenic(III) can be separated from arsenic(V), even at levels of $2\,\mu g\,l^{-1}$, by extracting it as the pyrrolidine dithiocarbamate with chloroform. They applied this technique to samples of sea water spiked with As(V) and As(III) and found (Table 17) that arsenic(V) could be satisfactorily determined in the presence of arsenic(III).

Table 17 Determination of arsenic(V) in seawater*

As^{3+} added (μg)	1.000	1.000	0.000	0.000	1.000	1.000	1.000	1.000
As^{5+} added (μg)	0.000	0.000	1.000	1.000	1.000	1.000	1.000	1.000
As found (μg)	0.018	0.015	0.979	0.985	0.990	0.980	0.976	1.032

*150 ml aliquots extracted with APDC in chloroform before the evolution and determination of arsenic.
Reprinted with permission from Haywood and Riley.[194] Copyright (1971) Elsevier Science Publishers.

Potable water

The relationship between absorbance and arsenic(V) concentration in distilled water medium was investigated by the method described for potable waters. A linear relationship was closely adhered to for arsenic concentrations in the range 10–800 µg l^{-1}; for measurements of 50–800 µg As l^{-1}, the solutions were diluted to an appropriate final volume, and the absorbances normalized. Deviation from linearity at higher arsenic concentrations almost certainly resulted from the use of insufficient absorption reagent to trap the arsine completely. The reproducibility and precision of the method was tested by performing replicate analyses on 50 ml aliquots of distilled water spiked at three different levels with arsenic(V). Over the range 4–4 µg As l^{-1}, the relative standard deviation was 0.6–1.1% and the average arsenic recovery was 98.7% (Table 18). The detection limit of the method was assessed by making replicate measurements of the reagent blank. The standard deviation (s) of these was 0.13 µg l^{-1}; this corresponds to a detection limit (4 s) of 0.5 µg l^{-1}.

Table 18 Replicate (8) determinations of arsenic(V) in spiked samples of distilled water

Arsenic concentration (µg l^{-1})	4.0	20.0	40.0
Recovery range (%)	96.5–99.6	97.9–99.1	99.0–100.5
Average recovery (%)	98.1	98.6	99.5
Relative standard deviation	1.0$_5$	0.5$_8$	0.6$_1$

Reprinted with permission from Haywood and Riley.[194] Copyright (1971) Elsevier Science Publishers.

Destruction of organarsenic compounds

Haywood and Riley[194] showed that arsenic in the form of tetraphenylarsonium chloride 1-(o-arsono-phenylazo)-2-naphthol-3,6-disulphuric acid and o-arsonophenylazo-p-dimethylaminobenzene are quantitatively decomposed in sea water by ultraviolet radiation. This irradiation technique can also be applied to potable waters, provided that they do not contain significant amounts of substances which absorb ultraviolet radiation, e.g. humic acids. For strongly absorbing samples of effluents it is necessary to oxidize organic arsenic compounds chemically. For potable waters this can be achieved by digesting the acidified sample overnight at 100 °C with an excess of potassium permanganate. Practically quantitative recoveries (>98%) of arsenic were obtained from tetraphenylarsonium chloride, 1-(o-arsono-phenylazo)-2-naphthol-3,6-disulphuric acid and o-arsonophenylazo-p-dimethylaminobenzene. For effluents, it is preferable to oxidize the organic arsenic compounds by evaporating the sample to fuming with a mixture of nitric and sulphuric acids, as described by Caldwell et al.,[195] but reducing the amount of sulphuric acid used by a factor of 10 to minimize the reagent blank. Evaporations carried out in this way gave almost quantitative (>98%) recoveries of inorganic arsenic(III) and of the three organic arsenic compounds mentioned above.

Sandhu and Nelson[196] have also studied the interference effects of several metals on the determination of arsenic at the 0–100 μg l^{-1} range in water and waste water by the silver diethyldithiocarbamate method. They found that in general it is the total metal content that is important and not the specific metal. A total metal concentration greater than 7 mg l^{-1} interferes significantly. Antimony and mercury interfere specifically, forming complexes with silver diethyldithiocarbamate at absorbance maxima at 510 and 425 nm respectively. An absorbance peak at 410 nm has been assigned to a hydrogen–silver diethyldithiocarbamate complex. Recovery of arsenic released by digesting solutions was tested and shown to give about 90% recovery of organic arsenic.

Miscellaneous methods

Braman et al.[197] separated and determined nanogram amounts of inorganic arsenic and methylarsenic compounds by reducing them to the corresponding arsines which are then separated and detected by emission spectrometry.

A variety of techniques have been used to obtain speciation data for monomethyl arsonate and dimethyl arsinite. Spectrophotometry[198-201] and gas chromatography[202-205] have been useful for one or more of the arsenicals. The procedure developed by Braman and Foreback[215] for generation and selective volatilization of arsines resolved As(III), As(V), monomethyl arsonate, and dimethyl arsinite. However, molecular rearrangements[206] and incomplete recoveries at low concentrations[207,208] have been reported. Yamamoto[209] and Dietz and Perez[210] observed that dimethyl arsinite has a strong affinity for acid-charged cation-exchange resins. Elton and Geiger[211] used this fact to separate monomethyl arsonate and dimethyl arsinite prior to determinations of the organoarsenicals by differential pulse polarography. The authors reported detection limits of 0.1 μg ml^{-1} and 0.3 μg ml^{-1}, respectively. Henry et al.[212] reported a method for the determination of As(III), As(V), and total inorganic arsenic by differential pulse polarography. As(III) was measured directly in 1 M perchloric acid or 1 M hydrochloric acid.[213] Total inorganic arsenic was determined in either of these supporting electrolytes after the reduction of electroinactive As(V) with aqueous sulphur dioxide. As(V) was evaluated by difference. Sulphur dioxide was selected because it reduced As(V) rapidly and quantitatively, and excess reagent was readily removed from the reaction mixture.

Henry and Thorpe[214] separated monomethyl arsonate, dimethyl arsinite, As(III) and As(IV) on an ion-exchange column from samples of pond water receiving fly ash slurry from a coal-fired power station. They then determined these substances by differential pulse polarography. The above four arsenic species are present in natural water systems. Moreover, a dynamic relationship exists whereby oxidation–reduction and biological methylation–dimethylation reactions[215-219] provide pathways for the interconversions of the arsenicals. Analytical methods capable of distinguishing between the predominant species of arsenic are necessary if immediate and potential impacts are to be accurately assessed.

Method

Instrumentation

A Princeton Applied Research Corporation Model 174A polarographic analyser and a Hewlett Packard (Avondale, Pa) Model 7040A X-Y recorder were used for all determinations.

The flow rate of the DME was $0.845\,\text{mg s}^{-1}$. A 2.0 s drop time was employed for all measurements. Other instrumental parameters are as described by Henry et al.[212]

Reagents

Triply distilled, deionized water was used to prepare all solutions.

Preparation of ion-exchange columns

Cation, Dowex 50 W-X8 and anion, AG 1-X8 exchange resins were used. To remove potential interferents, the 50–100 mesh resins were extensively washed with alternating 0.5 M solutions of hydrochloric acid and sodium hydroxide at flow rates of $5-10\,\text{ml min}^{-1}$. The cation exchange resin, in the H^+ form, was slurry packed into a 1.0 cm i.d. glass column to a height of 16 cm. After use, the column was regenerated by reaction with 1.0 M hydrochloric acid. The anion-exchange resin was slurry packed into a 0.8 cm i.d. glass column to a height of 11 cm. The resin was converted to the acetate form by passing 250 ml of 0.5 M sodium hydroxide over the column followed by 100 ml of 1.0 M sodium acetate, both at flow rates of $5-10\,\text{ml min}^{-1}$. The column was rinsed with triply distilled water and a mobile phase that was 0.1 M in total acetate concentration and had a pH of 4.7 was passed over the column until the pH of the effluent was also 4.7.

Procedure

Figure 26 depicts the analytical procedure for separation and the subsequent determination of each of the arsenicals. The pH of the sample is adjusted to between 4 and 10, and the sample is divided into four aliquots. Two of these are used to determine As(III) and total inorganic arsenic as described by Henry et al.[212] The concentration of As(V) is calculated from the difference between these two values.

A third 200 ml aliquot is mixed with 2.0 ml of 1.74 M of 1.74 M acetic acid and passed through the cation-exchange resin at a flow rate of $5\,\text{ml min}^{-1}$ to isolate the dimethylarsinite from the matrix. The sample is followed by a mobile phase consisting of 0.02 M acetic acid. As(III), As(V), and monomethylarsinite elute within 70 ml. Dimethylarsinite is recovered by stripping the column with 1.0 M sodium hydroxide at $1.0\,\text{ml min}^{-1}$. Eluate collected from 31–42 ml after addition of the basic mobile phase contains the dimethylarsinite.

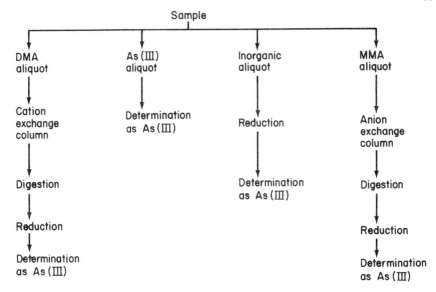

Figure 26 Flow chart for the determination of As(III), As(V), monomethylarsinic acid (MMA), and dimethylarsinic acid (CMA). Reprinted with permission from Henry and Thorpe.[214] Copyright (1980) American Chemical Society

The fraction of eluate containing dimethylarsinic acid is added to 7 ml of 70% perchloric acid contained in a flask fitted with an air condenser. The resulting solution is evaporated to fumes and heated to 200 °C for 2.5 h. The solution is cooled and diluted with 10 ml of triply distilled water. The As(V) produced in this digestion is reduced and determined As(III) according to the procedure of Henry et al.[212]

Monomethylarsonate is isolated using the anion-exchange column. A 50.0 ml aliquot of the sample is mixed with sodium acetate–acetic acid buffer to obtain a total acetate concentration of 0.01 M and a pH of 4.7. The resulting solution is loaded onto the column and followed with 50 ml of a mobile phase of similar composition. As(III) and dimethylarsinite will elute within this 100 ml (sample + eluant). Monomethylarsonate is retained on the resin. The flow rate is constant (1–2 ml min^{-1}), throughout the procedure.

To resolve monomethylarsonate from As(V), the total acetate concentration of the mobile phase is increased to 0.1 M while maintaining a constant pH. Eluate containing monomethylarsonate is collected between 20 and 40 ml from the introduction of the more concentrated phase. As(V) begins to appear after 75 ml from the same reference point.

The fraction containing monomethylarsonate is digested in perchloric acid under the same conditions as is dimethylarsinite. The reaction proceeds at a faster rate, however, and reaches completion within 30 min. The As(V) produced in this reaction is determined by the procedure of Henry et al.[212]

Henry and Thorpe[214] showed that digestion of dimethylarsinite completely decomposed this substance at the boiling point of the perchloric acid–water

Table 19 Efficiency of digestion of dimethylarsinite with perchloric acid

Added (ppb)	Determined (ppb)	Recovery (%)
400	380	95
400	400	100
400	370	92
400	380	95
290	310	107
270	260	104
150	150	100
120	120	100
110	110	100
59	61	103

Mean = 100%; s.d. = 4.6%; $n = 10$.

Recovery of dimethylarsinite from separation and digestion procedure

Arsenic added (ppb)	Arsenic found (ppb)	Final recovery (%)
396	376	95
99	107	108
89	86	97
89	86	97
69	68	99
49	51	104
25	25	100

Mean = 100%; s.d. = 4.5%; $n = 7$.
Reprinted with permission from Henry and Thorpe.[214] Copyright (1980) American Chemical Society.

azeotrope (203 °C) in a minimum of 2.5 hours (Table 19). The efficiency of the perchloric acid digestion and the recovery of dimethylarsinite from the entire procedure was measured and the data are presented in Table 19. The concentrations investigated ranged from low ppm to low ppb in both cases. The mean recoveries and standard deviations indicate the procedures are quantitative and reproducible. The detection limit for dimethylarsinite is 8 ppb.

Table 20 presents data obtained from studies of the efficiency of the perchloric acid digestion of monomethylarsonate and recovery results for monomethylarsonate standard solutions when subjected to the entire procedure. The digestion of monomethylarsonate occurred much more quickly than that of dimethylarsinite, reaching completion in less than 30 min. The mean efficiency and the standard deviation show the reaction was both quantitative and reproducible. Recovery of known amounts of monomethylarsonate is also seen to have been complete. The detection limit for monomethylarsonate was 18 ppb.

Tables 21 and 22 respectively, show results obtained in determinations of the four arsenic species in a spiked water sample and a fly ash basin slurry sample. The recoveries of the known additions of the four arsenic species substantiate the accuracy of the procedure and these results demonstrate the utility of the method for the speciation of arsenicals in complex aqueous matrices.

Table 20 Efficiency of digestion of monomethylarsonate with perchloric acid

Added (ppb)	Determined (ppb)	Recovery (%)
1640	1680	102
1640	1670	102
1640	1640	100
1640	1670	102
821	783	95
821	843	103
737	790	107
125	113	90
85	80	94
43	43	100

Mean = 100%; s.d. = 5.1%; $n = 10$.

Recovery of monomethylarsonate from separation and digestion procedure

Added (ppb)	Determined (ppb)	Recovery (%)
1640	1560	95
737	729	99
266	277	104
177	168	95
88	86	98
88	85	97

Mean = 98%; s.d. = 3.0%; $n = 6$.
Reprinted with permission from Henry and Thorpe.[214] Copyright (1980) American Chemical Society.

Table 21 Results of the analysis of an Environmental Protection Agency standard reference sample*

Species	Concentration present	Determination 1	2	3	4	Average concentration	Standard deviation
As(III)	26	25	33	26	34	30	3.7
As(V)	82	85	81	76	76	80	4.4
MMA	83	72	86	85	79	80	6.5
DMA	69	69	71	63	65	67	3.7

*Values in ppb arsenic. MMA, monomethylarsonate acid; DMA, dimethylarsinic acid.
Reprinted with permission from Henry and Thorpe.[214] Copyright (1980) American Chemical Society.

Table 22 Ash basin slurry sample

Species	Initially present (ppb)	Spiked (ppb)	Total present (ppb)	Found (ppb)	Recovered (%)
As(III)	63	76	139	134	96
As(V)	46	73	119	121	102
DMA	24	62	86	83	97
MMA	n.d.*	85	85	93	109

*Not detectable. MMA, monomethylarsonate; DMA, dimethylarsinite.
Reprinted with permission from Henry and Thorpe.[214] Copyright (1980) American Chemical Society.

ORGANOTIN COMPOUNDS

Luijten[220] has published a bibliography of organotin analysis including the analysis of water samples.

Methyltin compounds

Braman and Tompkins[221] have developed methods for the determination of trace amounts (ppb) of inorganic tin and methyltin compounds in rain, river, esturial and tap waters and, indeed, these workers were the first to confirm the presence of methyltin componds in natural water. Tin compounds are converted to the corresponding volatile hydride (SnH_4, CH_3SnH_3, $(CH_3)_2 SnH_2$, and $(CH_3)_3SnH$) by reaction with sodium borohydride at pH 6.5 followed by separation of the hydrides and then atomic absorption spectroscopy using a hydrogen-rich hydrogen–air flame emission type detector (Sn-H band).

The technique described has a detection limit of 0.01 ng as tin and hence parts per trillion of organotin species can be determined in water samples.

Braman and Tompkins[221] found that stannane (SnH_4) and methylstannanes (Ch_3SnH_3, $(CH_3)_2SnH_2$, and $(CH_3)_3SnH$) could be separated very well on a column comprising silicone oil OV-3 (20% w/w) supported Chromosorb W. A typical separation achieved on a water sample is shown in Figure 27.

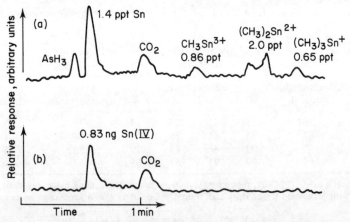

Figure 27 Environmental sample analysis and blank: (A) environmental analysis, Old Tampa Bay; (b) typical blank. Reprinted with permission from Braman and Tompkins.[221] Copyright (1979) American Chemical Society

The procedure is given below.

Samples containing 0–5 mg of tin in up to 100 ml of solution are adjusted to pH 6.5 by addition of 4 ml of 2 M tri (hydroxymethyl) amino methane hydrochloride buffer solution. The volatilization chamber is sealed and degassed with helium at a flow rate of 183 ml min^{-1} for approximately 1 min. At the end of the degassing period, the apparatus (Figures 28 and 29) is connected to

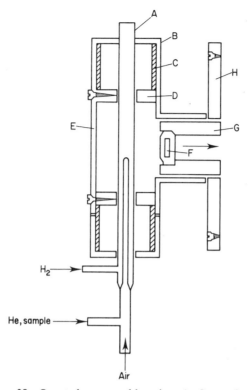

Figure 28 Quartz burner and housing. A, Quartz burner; B, PVC cap; C, PVC tubing; D, mounting ring; E, PVC T-joint, 1.25 inch; F, filter and holder; G, PVC coupling; H, connection to photomultiplier housing (PM). Reprinted with permission from Braman and Tompkins.[221] Copyright (1979) American Chemical Society

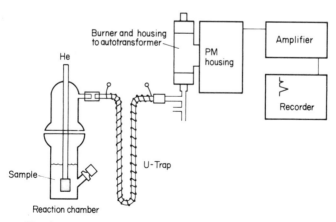

Figure 29 Apparatus arrangement for tin analysis. Reprinted with permission from Braman and Tompkins.[221] Copyright (1979) American Chemical Society

the detector. The U-trap is then immersed in the liquid nitrogen bath. Addition of two 1 ml portions of a 1% sodium borohydride (nitrogen degassed) solution (1 ml every 15 s) generates the stannanes which are swept by the carrier gas into the cold U-trap. After a total reduction time of 6–8 min, the air–hydrogen flame is ignited and the photometric system placed in operation. Upon removal of the liquid nitrogen bath, the strip chart recorder is started and after approximately 15 s the U-trap is slowly warmed. The stannanes and carbon dioxide are separated as they emerge from the U-trap and enter the detector. After analysis, the U-trap is disconnected from the system and heated while passing carrier gas through it to remove water prior to reuse. Integration of peak areas is carried out by one of several available methods and the response compared to a calibration curve. Peak area responses are linear with sample size.

No interferences were observed in this procedure except that which occurs when As(III) is present in sea water samples. This is reduced to arsine at pH 6.5 and as arsine is not separated from stannanes on the OV-3 column it does in effect comprise an interference in the method. The arsenic(III) peak can be eliminated by oxidation of arsenic(III) to arsenic(V), by addition of a few drops of dilute iodine solution to samples followed by addition of a few drops of sodium thiosulphate solution to dispel excess iodine. Neither arsenic(V) nor the methylarsenic acids are reduced to corresponding arsines at pH 6.5.

Certain metal ions, Ag^+, Cu^{2+}, Hg^{2+}, Ni^{2+}, MoO_4^{2-}, and Pb^{2+} at 2 ppm in analysed solutions were found to reduce the complete removal of stannane. Sea water did not inhibit recovery of stannane or methylstannanes. The ions Al^{3+}, CrO_4^{2-}, F^-, I^-, Mn^{2+}, PO_4^{3-}, and Sb^{3+} did not interfere at 20 ppm while Fe^{3+}, BiO_3^-, Cd^{2+}, S^{2-}, VO_3^{2-} and Zn^{2+} did not interfere at 2 ppm.

The generation and recovery of stannane, methyl-, dimethyl,- and trimethylstannane were studied in sea water. Average tin recoveries ranged from 96 to 109% for six samples analysed to which were added 0.4–1.6 µg of methyltin compounds and 3 ng inorganic tin. Reanalysis of analysed samples shows that all methyltin and inorganic tin is removed in one analysis procedure.

Analysis of environmental samples

A number of natural waters, from in and around the Tampa Bay, Florida, area were analysed for tin content. All samples were analysed without pretreatment. Samples which were not analysed immediately, were frozen until analysis was possible. Polyethylene bottles, 500 ml volume, were used for sample acquisition and storage. The results of the analyses appear in Tables 23 and 24. The average total tin content of fresh, saline, and estuarine waters are 9.1, 4.2, and 12 ng l^{-1}, respectively. Approximately 17–60% of the total tin present was found to be in the methylated forms. The saline waters appear to have the highest percentage of methylated tin compounds, 60% of the total tin present was found to be in the methylated forms, the dimethyltin form contributes approximately half of this value.

A number of rain waters collected in a Plexiglas collector were analysed over

Table 23 Analysis of saline and estuarine water samples*

Sample	Tin(IV) ng l^{-1}	%	Methyltin ng l^{-1}	%	Dimethyltin ng l^{-1}	%	Trimethyltin ng l^{-1}	%	Total tin ng l^{-1}
Saline waters									
Gulf of Mexico Sarasota	62	73	15	18	7.0	8.3	0.98	1.2	85
Gulf of Mexico Fort Desoto	2.2	6.0	n.d.†		0.74	2.0	0.71	20	3.6
Gulf of Mexico St Petersburg	4.5	54	0.62	7.4	3.2	39	n.d.		8.3
Old Tampa Bay Oldsmar	0.3	9.7	0.86	33	0.88	34	0.61	24	2.6
Old Tampa Bay Safety Harbour	1.4	29	0.86	17	2.0	40	0.65	13	5.0
Old Tampa Bay Philipee Park	0.8	32	1.1	44	0.60	24	n.d.		2.5
Old Tampa Bay Davis Municipal	n.d.		0.98	35	0.91	32	0.95	34	2.8
Old Tampa Bay Courtney Campbell	2.7	54	n.d.		1.7	34	0.61	12	5.0
Average	1.7	40	0.63	15	1.4	33	0.50	12	4.2
Estuarine surface waters									
Sarasota Bay	5.7	47	3.3	27	2.0	16	1.1	9.1	12
Tampa Bay	3.3	27	8.0	66	0.79	6.5	n.d.		12
McKay Bay	20	88	n.d.		2.2	9.6	0.45	2.0	23
Hillsborough Bay	n.d.		d.d.		1.8	71	0.71	29	2.5
Hillsborough Bay, Seddon Channel North	12	86	0.74	5.3	0.91	6.6	0.35	2.5	14
Hillsborough Bay, Seddon Channel South	13	83	n.d.		2.4	15	0.31	1.9	16
Manatee River	4.8	61	1.4	1.7	1.1	14	0.65	8.2	7.9
Alafia River	3.4	73	n.d.		0.75	16	0.55	12	4.7
Palm River‡	567	98	n.d.		4.6	0.80	4.0	0.69	576
Bowes' Creek	8.6	42	8.5	42	3.3	16	n.d.		20
Average	7.9	63	2.4	19	1.7	14	0.46	3.7	12

*Data are average of duplicates.
†n.d. less than 0.01 ng l^{-1} for methyltin compounds and 0.3 ng l^{-1} for inorganic tin.
‡This set of values was not used in computing the average.
Reprinted with permission from Braman and Tompkins.[221] Copyright (1979) American Chemical Society.

a period of several days. These results, and several tap water analyses, appear in Table 25. The average total tin content of rain is near 25 ng l^{-1} and the methyltin form comprises 24% of that total. It is not clear why methyltin is predominantly found in rain waters, but it may be a stable form resulting from demethylation of tetramethyltin. No tin blanks were observed from the Plexiglas rain collector.

Nelson[222] has used gas chromatography to determine the solubility of tin tetraalkyls in water.

Table 24 Analysis of fresh water samples*

Fresh waters	Tin(IV) ng l^{-1}	%	Methyltin ng l^{-1}	%	Dimethyltin ng l^{-1}	%	Trimethyltin ng l^{-1}	%	Total tin ng l^{-1}
Lake Eckles, Tampa	10	52	0.99	4.9	1.2	5.8	7.6	38	20
Lake Carroll, Tampa Inlet	7.7	65	n.d.†		0.96	8.1	3.3	28	12
Lake Carroll, Tampa South	6.0	58	n.d.		0.80	7.7	3.6	34	10
Lake Carroll, Tampa North	3.3	46	n.d.		0.56	7.8	3.3	46	7.2
15th Fairway Pond, USF	2.7	44	1.4	24	0.67	11	1.3	21	6.0
Hillsborough River, Morris Bridge Road	730‡	100	n.d.		n.d.		n.d.		730
Hillsborough River Fletcher	n.d.		1.8	71	0.74	29	n.d.		2.5
Hillsborough River, Rowlett	4.1	52	1.5	19	0.39	5.0	1.6	23	7.8
Lake Alice, Gainesville	3.5	47	4.5	31	5.5	37	1.3	8.9	15
Withlacoochee River Nobleton	1.0	17	2.8	46	2.2	37	n.d.		6.1
Little Manatee River	17	47	12	32	7.5	21	n.d.		37
Half-Moon Lake, Tampa	2.4	48	2.6	52	Trace	—	n.d.		Trace
Cowhouse Slough, Tampa	n.d.		n.d.		n.d.		Trace	—	Trace
Hillsborough River, State Park	0.4	32	0.49	39	0.36	29	n.d.		1.3
Hillsborough River, Fowler	n.d.		1.5	76	0.48	25	n.d.		2.0
Average	4.2	46	2.0	22	1.4	15	1.5	16	9.1

*Data are average of duplicate analyses.
†n.d., less than 0.01 ng l^{-1} for methyltin compounds of 0.3 ng l^{-1} for inorganic tin.
‡This set of values not used in computing the average.
Reprinted with permission from Braman and Tompkins.[221] Copyright (1979) American Chemical Society.

Butyltin compounds

Meinema et al.[223] have described a sensitive and interference-free method for the simultaneous determination of tri-, di-, and monobutyltin species in aqueous systems at tin concentrations of 0.01–5 ppm. The species are concentrated from hydrobromic acid solutions into an organic solvent by solvent extraction in the presence of a metal coordinating ligand. The butyltin species in the organic extract are transformed into butylmethyltin compounds and analysed by a gas chromatography–mass spectrometry method. The inorganic tin IV species in the organic extract are butylated to tetrabutyltin which is detected by the same technique.

Table 25 Analysis of rain water and tap water samples*

Sample	Tin(IV) ng l^{-1}	%	Methyltin ng l^{-1}	%	Dimethyltin ng l^{-1}	%	Trimethyltin ng l^{-1}	%	Total tin ng l^{-1}
Rain (8/21/77,pm) after a steady rain	6.4	65	1.2	12	1.7	17	0.61	6.1	9.9
Rain (8/22/77,am) after all night rain	2.7	44	2.1	34	0.83	14	0.45	7.4	6.1
Rain (8/31/77,pm) last rain	6.7	6.0	3.7	33	0.88	7.9	n. d.†		11
Rain (9/3/77,pm) first rain	7.7	50	3.3	22	3.1	20	1.1	7.5	15
Rain (9/29/77,pm) first rain	6.5	1.9	22	66	4.8	14	n.d.		34
Rain (9/12/77)	41	91	4.2	9.3	n.d.		n.d.		45
Rain (9/16/77)	12	92	0.74	5.7	n.d.		n.d.		13
Rain (9/16/77)	5.9	81	1.4	19	n.d.		n.d.		7.3
Rain (9/17/77)	3.0	73	1.1	27	n.d.		n.d.		4.1
Rain (9/25/77)	23	74	0.65	2.1	n.d.		n.d.		31
Average	11	44	5.9	24	7.4	30	0.22	0.88	25
Tap water (9/6/77)	2.2	17	8.1	62	2.2	17	0.55	4.2	13
Tap water (9/7/77)	2.2	40	0.49	8.9	0.40	7.3	2.4	44	5.5
Average	2.2	24	4.3	47	1.3	14	1.5	16	9.2

*Data are averages of duplicate analyses.
†n.d., less than 0.01 ng l^{-1} for methyltin compounds.
Reprinted with permission from Braman and Tompkins.[221] Copyright (1979) American Chemical Society.

Method

Apparatus

A Finnigan Model 9500 gas chromatograph, coupled with a Finnigan Model 3100 D quadrupole mass spectrometer, was used for the analysis. A glass jet separator forms the interface between the gas chromatograph and the mass spectrometer. The g.c.–m.s. combination is provided with a vent-off system between the end of the column and the separator to prevent large amounts of solvent from entering the mass spectrometer.

Reagents

Standard solutions of inorganic Sn(IV) (100 ppm Sn) were prepared[224,225] by dissolution of metallic tin in a small amount of concentrated sulphuric acid that is subsequently diluted with water to a fixed volume. Aqueous solutions containing 1 ppm in Sn(IV) were obtained by dilution of 5 ml of the standard solution to 500 ml with tap water.

Tropolone was obtained from EGA-Chemie. Zinc *bis*(diethyldithiocarbamate) was prepared as described by Wyttenbach and Bajo.[226] Cyclohexane was purified by distillation after removal of contaminating unsaturated products by washing with concentrated sulphuric acid.

Benzene thiophene-free standard solutions of MeMgBr and BuMgBr in diethylether (concentration 1.5–2.5 mol l^{-1}) were prepared by reaction of methyl bromide and n-butylbromide, respectively, with magnesium metal. The internal standard was prepared by the reaction of dibutylmethyltin bromide with n-hexylmagnesium bromide in diethylether.[227]

Quantitative detection of butyltin species present in dilute aqueous solutions by g.c.–m.s.

Bu_3Sn, Bu_2Sn and Bu_1Sn species, either individually or simultaneously present in such a solution (concentration range 0.01–5.0 ppm as Sn), were extracted twice with 25 ml of an extractant. Mixtures kept in a 1 l round bottom flask were shaken twice for 15 min on a mechanical shaker. Optimal results were obtained from aqueous solutions that were acidified with hydrobromic acid 48% (5–20 ml) prior to extraction. After leaving the solution for 15 min at room temperature, extraction with benzene or chloroform resulted in almost complete migration of Bu_3Sn and Bu_2Sn species into the organic solvent. With 0.05% of tropolone dissolved in the organic solvent, not only Bu_3Sn and Bu_2Sn but also Bu_1Sn and inorganic Sn(IV) species are extracted from the aqueous solution.

Benzene extracts can be submitted to a Grignard methylation procedure without any further precautions. The presence of traces of water in these extracts does not interfere with the methylation of butyltin species, Bu_n, Sn to butylmethyltin compounds. $Bu_nSnMe_{(4-n)}$ ($n = 1-3$), whenever an excess of MeMgBr is added (3 ml of a 2.5 N solution of MeMgBr in diethylether). The reaction mixture is stirred for 0.5 h at room temperature and subsequently treated with 25 ml of benzene. The combined organic layers are concentrated at reduced pressure to a volume of about 25 ml.

Chloroform and methylene chloride extracts are concentrated almost to dryness prior to the methylation procedure since these solvents react with the Grignard reagent. Diethylether (25 ml) or benzene (25 ml) is added to the residue prior to the methylation procedure. The amounts of $Bu_nSnMe_{(4-n)}$ species present in the methylated extracts were determined by mass fragmentography. Aqueous solutions (0.5 l) of $(Bu_3Sn)_2O$, Bu_2SnCl_2, and Bu_1SnCl_3 (Bu_n concentration 1 ppm as Sn) acidified with hydrobromic acid (10 ml, 48%) were extracted with benzene to give butyltin recoveries of 89% for $(Bu_3Sn)_2O$, 86% for Bu_2SnCl_2, and 0% for Bu_1SnCl_3. Upon extraction with benzene containing 0.05% of tropolone, these figures were 93, 90, and 84% respectively.

The sensitivity of the latter procedure was demonstrated in the almost quantitative recovery of butyltin species, $(Bu_3Sn)_2$, $(Bu_3Sn)_2O (n = 1-3)$ from aqueous solutions containing Bu_2SnCl_2 and Bu_1SnCl_3 at a concentration of 0.01 ppm as Sn. One litre of this aqueous solution, acidified with 20 ml of hydrobromic acid (48%) was extracted twice with 25 ml of benzene containing 0.05% tropolone. The methylated benzene extract concentrated to a volume of 10 ml was submitted to the g.c.-m.s. detection procedure. Butylmethyltin compounds were detected in 95% (Bu_3SnMe), 90% (Bu_2SnMe_2), and 84% ($BuSnMe_3$) – yields calculated on the amounts of $(Bu_3Sn)_2O$, Bu_2SnCl_2, and $BuSnCl_3$ originally present.

Tetramethyltin formed by the methylation of inorganic tin(IV) cannot be determined by gas chromatography–mass spectrometry since it has the same retention time as the solvent. Butylation of an organic extract (50 ml) that contains inorganic Sn(IV) species by the addition of an excess of butyl magnesium bromide in diethylether (4 ml of a 2.0 N solution) results in the formation of tetrabutyltin. The reaction mixture is stirred for 0.5 h at room temperature and subsequently treated with 25 ml of a 1 N sulphuric acid solution. The aqueous layer is separated and extracted with 25 ml of benzene, and the combined organic layers are concentrated at reduced pressure to a volume of about 25 ml. The tetrabutyltin content of this sample can be determined quantitatively by gas chromatography–mass spectrometry.

Mass Fragmentography

Alternatively, 1 µl of the standard test solution of methylated butyltin compounds ($BuSnMe_3$, Bu_2SnMe_2, and Bu_3SnMe and the internal standard $HexBu_2SnMe$) are prepared by solution of these compounds in cyclohexane to a concentration of about 50 ppm each. $HexBu_2SnMe$ in cyclohexane is prepared by accurately weighing a sample of this compound in a volumetric flask to give a final solution with a precisely known concentration of about 2% (wt/vol). To act as an internal standard 50 µl of this standard solution of $HexBu_2SnMe$ was added to a sample to be analysed) and 1 µl of an actual sample were injected into the gas chromatograph–mass spectrometer combination. The gas chromatographic separation was performend on a 1.2 m glass column (o.d. 6.2 mm; i.d. 2mm) filled with 3% OV-17 on Supelcoport 80–100 mesh or filled with 3% OV-1 on Chromosorb WHP 100–120 mesh. Column temperature was: initial 70 °C; program rate: 15° min^{-1}, maximum 170 °C. Duration of the analyses was 11 min. Helium was used as carrier gas; inlet pressure, 50 psi; the flow rate was 20 ml min^{-1}, which resulted in an analyser pressure of 10^{-5} torr. The temperatures of the injection port, separator, and transfer line were: 225, 200, and 225 °C respectively. The vent-off system was opened during the first 54 s after injection to prevent the solvent from entering the mass spectrometer. Mass spectrometer conditions were: electron energy, 70 eV; beam current, 14 µA, preamplifier, 10^{-8}. Mass fragments $m/e = 135$ and $m/e = 193$ were recorded as shown in Figure 30. $HexBu_2SnMe$ cannot be used as an internal standard in the detection of tetrabutyltin formed by butylation of inorganic Sn(IV) species because both have almost the same retention time. Moreover, tetrabutyltin has practically no fragment ions $m/e = 135$ and $m/e = 193$. Instead the fragment ions $m/e = 121$, 177, 178, and 179 are recorded, and no internal standard is used. Concentrations were estimated by comparison with a solution of a known concentration of tetrabutyltin.

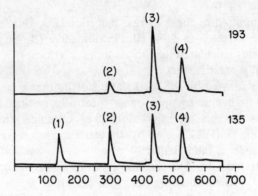

Figure 30 Mass fragmentogram of standard test solution of BuSnMe$_3$(1), Bu$_2$SnMe$_2$(2), Bu$_3$SnMe(3), and HexBu$_2$SnMe(4) at $m/e = 135$ and $m/e = 193$. Reprinted with permission from Meinima et al.[223] Copyright (1978) American Chemical Society

Qualitative detection of Bu$_3$Sn and Bu$_2$Sn species by thin-layer chromatography

The presence of Bu$_3$Sn and Bu$_2$Sn species in chloroform or benzene extracts from hydrobromic acid acidified aqueous solutions can be demonstrated qualitatively by thin-layer chromatography on Eastman chromatogram sheets using hexene–acetone–acetic acid 40:4:1 as an eluant. After spraying with a 0.1% dithizone solution in chloroform, Bu$_3$Sn species are visualized as a yellow spot (R_f value 0.75) and Bu$_2$Sn species as a red spot (R_f value 0.50). The detection limit is improved by keeping the thin-layer strip for 10 s in bromine vapour after elution. Bromine breaks carbon–tin bonds, and both Bu$_3$Sn and Bu$_2$Sn species are now detected as red spots after spraying. The detection limit after bromination is about 0.5 μg tin per spot.

A comparison of the mass spectra of Bu$_n$SnMe$_{(4-n)}$($n = 1$-3) compounds reveals that they all show a common fragment ion $m/e = 135((\text{Me}^{120}\text{Sn})^+)$. Moreover, the mass spectra of Bu$_2$SnMe$_2$ and Bu$_3$SnMe show a common fragment ion $m/e = 193$ ((MeBuH^{120}Sn)$^+$) (see Figure 31). This implies that gas chromatographic–mass spectrometric analysis of these compounds can be performed by multiple peak scanning of these two fragments (mass fragmentography). This procedure has advantages over the procedure of detecting the total ion current in that it is considerably more sensitive, whereas no or substantially less disturbance by impurities with the same retention time as the Bu$_n$SnMe$_{(4-n)}$ species does occur. A contaminant with the same retention time as one of the Bu$_n$SnMe$_{(4-n)}$ compounds is not detected at all unless it presents the same mass fragments, $m/e = 135$ and also $m/e = 193$. The presence of such a contaminant is easily observed because the mutual ratio of these fragments will be disturbed. When this is the case, it is possible to monitor fragment ions containing one of the other isotopes of tin. However, this will affect the sensitivity as can be seen from the data presented in Table 26 and from the spectra presented in Figure 31.

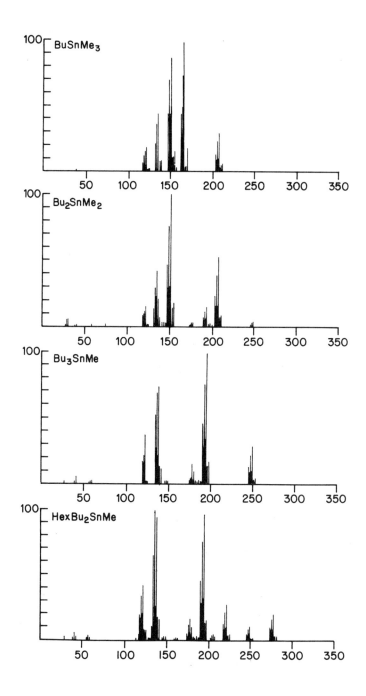

Figure 31 Mass spectra of $BuSnMe_3$, Bu_2SnMe_2, Bu_3SnMe, and $HexBu_2SnMe$. Reprinted with permission from Meinima et al.[223] Copyright (1978) American Chemical Society

Table 26 Natural abundances of main isotopes of tin

Mass no.	Abundance (%)
116	14.30
117	7.61
118	24.03
119	8.58
120	32.85
122	4.72
124	5.94

Reprinted with permission from Meinima et al.[223] Copyright (1978) American Chemical Society

Quantitative determination is made possible by addition to the sample of an internal standard. Such an internal standard should have mass fragments at $m/e = 135$ and $m/e = 193$. Moreover, it should display a retention time in the gas chromatograph different from those of the $Bu_nSnMe_{(4-n)}$ ($n = 1-3$) compounds. As shown in Figure 31 HexBu$_2$SnMe meets these requirements.

The standard deviations in the ratios of the peak areas of the butylmethyltin compounds to the peak areas of the internal standards, HexBu$_2$SnMe, have been determined. At seven injections on the same day, each time 1 µl of a test solution containing methylated butyltin species and internal standard, at a concentration of about 35 ppm in tin each, standard deviation appeared to be:

	$m/e = 135$	$m/e = 193$
BuSnMe$_3$	3.4%	—
Bu$_2$SnMe$_2$	2.6%	3.4%
Bu$_3$SnMe	1.6%	1.9%

Upon reducing the concentration of the butylmethyltin species present to 3.5 ppm in tin with the internal standard present at a concentration of 35 ppm in tin the same accuracy was achieved. Samples containing butylmethyltin species in the concentration range of 0.35 ppm in tin have to be analysed by injecting 5 µl of sample solution.

As a general conclusion, for the quantitative analysis of butylmethyltin compounds, $Bu_nSnMe_{(4-n)}$ ($n = 1-3$) by gas chromatography–mass spectrometry at least about 0.2 ng of each compound must be injected.

In Table 27 are shown recoveries of butyltin species at concentrations of 1 ppm in tap water obtained by the above procedures.

Detection of Bu$_3$Sn and Bu$_2$Sn species

Extraction of aqueous solutions of butyltin species (concentrate of tetrabutyltin 1 ppm as Sn) with benzene or chloroform by the above procedures resulted in

Table 27 Recovery of butyltin species, Bu_nSn ($n = 1-3$) from dilute aqueous solutions of $BuSnCl_3$, Bu_2SnCl_2, and $(Bu_3Sn)_2O$ (Bu_nSn concentration 1 ppm as Sn) in tap water (0.5 l)

Extractant Organic solvent/complexing agent[†] (2 × 25 ml)	HBr (10 mL)	Recovery*(%)		
		Bu_1Sn	Bu_2Sn	Bu_3Sn
C_6H_6	−	...	...	80–95
$CHCl_3$	−	...	20–25	80–95
C_6H_6	+	...	80–90	80–95
$CHCl_3$	+	...	80–90	80–95
$C_6H_6/0.1\%Zn(DDC)_2$	+	80–90	n.d.[‡]	n.d.
$C_6H_6/0.1\%Zn(DDC)_2$	−	0–5	n.d.	n.d
$C_6H_6/0.05\%HOx$	+	40–50	n.d.	n.d.
$C_6H_6/0.05\%HOx$	−	80–90	n.d.	n.d.
$C_6H_6/0.05\%HTrop$	+	70–90	80–90	80–95
$C_6H_6/0.05\%HTrop$	−	50–90	80–90	80–95
$CHCl_3/0.05\%HTrop$	+	70–90	80–90	80–95
$CHCl_3/0.05\%HTrop$	−	40–80	60–70	70–90
$CH_2Cl_2/0.05\%HTrop$	+	70–90	80–90	80–95
$CH_2Cl_2/0.05\%HTrop$	−	60–70	80–90	80–95

*In each experiment at least three identical samples were extracted. The methylated extracts were analysed in duplicate by g.c.-m.s.
[†]Complexing agents: $Zn(DDC)_2$, zinc bis (diethyldithiocarbamate); HOx, 8-hydroxyquinoline; HTrop, tropolone.
[‡]Not determined.
Reprinted with permission from Meinima et al.[223] Copyright (1978) American Chemical Society.

the almost complete recovery of Bu_3Sn species in the organic extract. Under the same conditions Bu_2Sn and Bu_1Sn species, originating from Bu_2SnCl_2 and Bu_1SnCl_3, respectively, appear to be rather reluctant to extract. Addition of hydrobromic acid 48% prior to extraction markedly influences the recovery of butyltin species. Now both Bu_3Sn and Bu_2Sn species are extracted into the organic solvent in over 80% yield, whereas Bu_1Sn species reside in the aqueous solution. This procedure offers a reliable quantitative detection technique for Bu_3Sn and Bu_2Sn species either individually or simultaneously present in trace amounts in aqueous solutions. On a qualitative basis the presence of Bu_3Sn and Bu_2Sn species in these chloroform or benzene extracts can easily be demonstrated by thin-layer chromatography.

Detection of Bu_1Sn species

The procedure outlined above, although valuable in itself, is incomplete in that any Bu_1Sn species present escape detection. Excellent recoveries of Bu_1Sn species are achieved with tropolone.

Meinima et al.[223] studied the effect of combinations of various solvents with 0.05% tropolone on the recoveries of mono-, di-, and tributyltin species either individually or simultaneously present in aqueous solutions. The results obtained by gas chromatography–mass spectrometry after methylation are presented in Table 27. Bu_3Sn and Bu_2Sn recoveries appear to be almost quantitative both from neutral and from hydrobromic acid-acidified aqueous solutions. Bu_1Sn recovery appears to be influenced by the presence of hydrobromic acid in that, in general, recovery rates are higher from solutions acidified with hydrobromic acid than from non-acidified solutions. Bu_3Sn and Bu_2Sn recoveries remain fairly constant with aging of an aqueous solution of these species over a period of several weeks. Bu_1Sn recoveries, however, do decrease with time to a notable extent (20–40%), most likely as a result of adsorption/deposition of Bu_1Sn species to the glass wall of the vessel. Addition of hydrobromic acid obviously affects the desorption of these species as recovery of Bu_1Sn species increases to almost the same values as obtained from hydrobromic acid-acidified freshly prepared aqueous solutions of Bu_1Sn species.

These workers[223] tested out their procedures on canal water and sea water spiked with 1 ppm (as Sn) of the following butyltin compounds; $(Bu_3Sn)_2O$, $Bu_2Sn_2Cl_2$, and $BuSnCl_3$. One litre samples were acidified with hydrobromic acid prior to extraction to disrupt any interaction of butyltin species with inorganic or organic material present in natural water. Recovery data obtained upon extraction with benzene and benzene–0.05% tropolone solutions did not notably differ from those obtained in experiments with tap water solutions.

To test out the recovery of inorganic tin(IV) from aqueous solutions, 1 ppm tin solutions were extracted with chloroform–0.05% tropolone or benzene–0.05% tropolone. The tin content of these extracts was determined both by atomic absorption spectrometry after destruction and by gas chromatography–mass spectrometry after butylation to tetrabutyltin. Both detection procedures revealed the presence of over 80% of the inorganic tin species originally present in water in the organic extract irrespective of whether the aqueous solution had been acidified with hydrobromic acid prior to extraction. The presence of tropolone as a metal complexing reagent was found to be essential. In the absence of tropolone, by use of pure benzene or chloroform as an extractant, not even a trace of tin-containing species was found present in the organic solvent.

A strong tendency of tin(IV) ions to adsorb to glass surfaces was demonstrated. Aqueous solutions of Sn(IV) ions (1 ppm Sn) kept in glass beakers for a period of 3 days showed only 15–20% recovery of tin(IV) species upon extraction with benzene–0.05% tropolone, whereas duplicate experiments in polyethylene beakers showed recovery rates of 70–80%.

The results obtained in the recovery of inorganic tin(IV) species from aqueous solutions are compiled in Table 28.

Hodge et al.[228] have described an atomic absorption spectroscopic method for the determination of butyltin chlorides and inorganic tin in natural waters, coastal sediments, and macro algae in amounts down to 0.4 ng.

Waggon and Jehle[229,230] have reported on the quantitative detection of triphenyl and tri-, di-, and monobutyltin species in aqueous solution by a combination of liquid-liquid extraction, thin-layer chromatography, and anodic stripping voltametry. Neubert and Wirth[231] reported on the quantitative determination of mono-, di-, tri-, and tetraalkyltin compounds, present in a

Table 28 Recovery of inorganic tin(IV) from dilute aqueous solutions (0.5 l) containing 1 ppm tin

Experiment*	Extractant		
	Organic solvent/complexing agent (2 × 25 ml)	HBr 48% 10 ml	Recovery†(%)
A	C_6H_6/0.05%HTrop‡	+	80-90
B	C_6H_6/0.05%HTrop	+	15-20
C	C_6H_6/0.05%HTrop	+	70-80
D	C_6H_6	+	0
E	$CHCl_3$/0.05%HTrop	+	80-90
F	$CHCl_3$/0.05%HTrop	−	80-90
G	$CHCl_3$	+	0
H	$CHCl_3$	−	0

*A,D,E,F,G,H: freshly prepared solutions of Sn(IV). B: stored for 3 days in a glass beaker prior to extraction. C: stored for 3 days in a polyethylene beaker prior to extraction. C: stored for 3 days in a polyethylene beaker prior to extraction.
†In each experiment two identical samples were extracted. The methylated extracts were analysed in duplicate by g.c.-m.s.
‡Trop = tropolone.
Reprinted with permission from Meinima et al.[223] Copyright (1978) American Chemical Society.

mixture by gas chromatography after alkylation to be mixed tetraalkyltins. This technique was applied by Neubert and Andreas[232] to the quantitative detection of tributyl- and dibutyltin species present in dilute aqueous solution. Butyltin species were concentrated on a cation-exchange column, desorbed into diethylether–hydrogen chloride, and determined by gas chromatography after methylation.

Triphenyltin compounds

Soderquist and Crosby[233] have described a gas chromatographic procedure for the determination of down to 0.01 μg l^{-1} triphenyltin hydroxide and its possible degradation products (tetraphenyltin, diphenyltin oxide, benzene stannoic acid, and inorganic tin) in water. The phenyltin compounds are detected by gas-liquid chromatography using an electron capture detector after conversion to their hydride derivatives, while inorganic tin is determined by a novel procedure which responds to tin dioxide and aqueous solutions of tin salts. The basis for the method involves extraction of the phenyltin species from water followed by their quantitation as phenyltin hydrides by electron capture gas chromatography and analysis of the remaining aqueous phase for inorganic tin (Sn^{4+} plus SnO_2) by colorimetry.

$Ph_3Sn^+_{aq} \rightarrow Ph_2Sn^{+2}_{aq} \rightarrow PhSn^{+3}_{aq}$ Sn^{+4}_{aq} Possible degradation

↓ SnO_{2aq}

$Ph_3SnH, Ph_2SnH_2, PhSnH_3$ Sn–PCV complex
e.c.-g.l.c.   colorimetry   Analysis

Method

Gas-liquid chromatography was performed on a dual column/dual detector Varian Model 2400 instrument equipped, on one side, with a flame ionization detector and a 0.7 by 2 mm (i.d.) glass column containing 3% OV-17 on 60–80 mesh Gas Chrom Q. Column, injector and detector temperatures were 265, 275, and 300°C, respectively; carrier gas (nitrogen) flow rate was 25Z ml min^{-1}. Tetraphenyltin eluted within 8 min under those conditions. The second side of the chromatograph was equipped with a tritium electron capture detector and a 1.1 m by 2 mm (i.d.) glass column containing 4% SE-30 on 60–80 mesh Gas Chrom Q. The injector and detector temperatures were 210 °C, the carrier gas (nitrogen) flow rate was 20 ml min^{-1}, and column temperatures which eluted the following compounds within 6 min were: Ph_3SnH, 190 °C; Ph_2SnH_2, 135 °C; $PhSnH_3$, 45 °C. Combined gas-liquid chromatography–mass spectrometry was performed on a Finnigan Model 1015 utilizing a 1.0 m by 2 mm (i.d.) glass column containing 3% OV-17 on 60–80 mesh Gas Chrom Q. Infrared spectra were obained in hexane solution.

Reagents

Unless otherwise noted, water was distilled and passed through a column of Amberlite XAD-4 resin. Hexane and dichloromethane were Nanograde quality or equivalent. All glassware was cleaned by soaking in 2.0 M hydrochloric acid followed by rinsing with copious volumes of water.

Lithium aluminium hydride solution

Lithium tetrahydridoaluminate (Ventron) (100 ± 10 mg) was added to 25 ml of dry, reagent grade diethylether in a glass-stoppered flask, the mixture shaken for 2 min, and the grey precipitate allowed to settle before use. The solution was prepared fresh daily.

Sensitized PCV solution

Pyrocatechol violet (Eastman Organic Chemicals) (12 mg) and 11 mg of cetyltrimethyl ammonium bromide were added to 100 ml of water. The solution was prepared fresh daily.

Ascorbic acid solution

Ascorbic acid (2.5 g) was added to 50 ml of water. The solution was prepared fresh daily.

Acetate buffer (2.0 M, pH4.7)

Sodium acetate trihydrate (68 g) and 28 ml of glacial acetic acid were combined and diluted to 500 ml with water.

Citric acid solution

Citric acid monohydrate (10 g) was added to 100 ml of water.

Sulphuric/citric acid solution

Analytical reagent grade sulphuric acid (25 g) and 13 g of citric acid monohydrate were combined and diluted to 500 ml with water.

Fortification standards

A mixed standard was prepared by dissolving about 20 mg of each phenyltin chloride and tetraphenyltin in separate 10 ml volumetric flasks. Aliquots of each were combined, in dichloromethane, to yield Ph_4Sn, Ph_3Sn^+, Ph_2Sn^{2+}, and $PhSn^{3+}$ at 0.20 $\mu g/ul^{-1}$ each. The inorganic tin (Sn^{4+}) standard was prepared by dissolving 0.250 g of pure tin in 150 ml of concentrated hydrochloric acid and diluting to 500 ml with water. A 5.0 $\mu g\, ml^{-1}$ working standard was prepared by dilution with sulphuric–citric acid solution.

Procedure (Figure 32)

Organotins

The 200 ml sample in a 250 ml separatory funnel was mixed with 5 ml of acetate buffer, the mixture extracted with two 15 ml portions of dichloromethane, and the pooled extract divided into three equal portions, each of which was concentrated to about 0.1 ml in a screw-capped test tube at less than 40 °C under a gentle stream of nitrogen. To one of the concentrates (EX-1, Figure 32) was added 5 ml of hexane followed by 0.5 ml of lithium aluminium hydride solution. After 2–3 min, the mixture was diluted with hexane, about 0.5 ml of water carefully added, the phases were mixed, and the hexane phase was analysed by electron capture gas chromatography for Ph_3SnH, Ph_2SnH_2, and $PhSnH_3$. A standard curve was prepared for each of the hydrides with the ng μl^{-1} hexane standard solutions, generally in the 0.2–2.0 ng range. Quantitation was done by comparison of sample peak heights to the standard curve.

To the second dichloromethane concentrate (EX-2) was added 0.50 ml of sulphuric acid and the dichloromethane removed from the mixture with a vigorous stream of nitrogen. The tube was sealed and heated at 100 °C in a water bath for 20 min. After cooling, 4.0 ml of citric acid solution was added and the sample treated as described below under colour development.

To the third dichloromethane concentrate (EX-3) was added 1 ml of hexane, the contents were concentrated under nitrogen to about 0.1 ml, and then diluted back to 1.0 ml with hexane. The sample was transferred to a Florisil microcolumn (prepared by packing a disposable Pasteur pipette with 0.35 g of 60–100 mesh Florisil held with a small glass wool plug and rinsing with two 5 ml portions of hexane before use) and eluted with hexane. The first 2.5 ml of eluate was

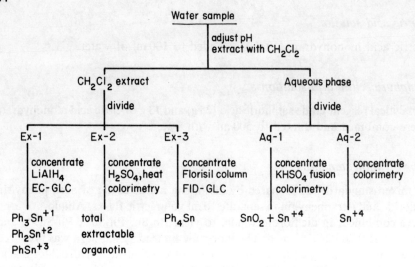

Figure 32 Flow diagram for analytical procedure. Reprinted with permission from Soderquist and Crosby.[233] Copyright (1978) American Chemical Society

collected, concentrated to 0.1–0.5 ml and analysed by flame ionization gas chromatography for tetraphenyltin. Quantitation was accomplished by comparison of sample peak heights to the tetraphenyltin standard curve in the 10–50 ng range.

Inorganic tin

The extracted aqueous sample was divided equally between two 125 ml Erlenmeyer flasks and the samples were concentrated by boiling just to dryness on a hot plate. To one of the dry aqueous concentrates (AQ-l, Figure 32) was added 2.0 g of potassium hydrogen sulphate and the flask heated at 300–350 °C for 30 min. After cooling, 4.0 ml of citric acid solution was added and the solids were dissolved with gentle heating if necessary. Any insoluble particulate matter which would interfere with subsequent colorimetric measurements was removed at this point by filtering through tightly packed glass wool. To the second aqueous concentrate (AQ-2) was added 4.0 ml of sulphuric acid–citric acid solution. Both samples were then treated as described below under colour development.

Colour development

The hydrolysed dichloromethane extract (EX-2) and the two aqueous samples (AQ-1 and AQ-2) were analysed for tin as follows: ascorbic acid solution (2.0 ml) and 4.0 ml of sensitized PCV solution were added and, after 30 min, the absorbance was read at 660 nm. Developed samples with absorbance values exceeding those of the standard curve were diluted with water. The concentrations were obtained from standard curves prepared in a manner consistent with the individual sample work-up. That is, a curve for the aqueous tin subsamples (AQ-1 and AQ-2) was prepared by the addition of 0, 1.0, 3.0, and 5.0 μg of tin to 3.5 ml of sulphuric–citric acid solution followed by 2.0 ml of ascorbic acid solution, 4.0 ml of sensitized PCV solution and enough water to give 11.5 ml total volume. A separate standard curve for the hydrolysed dichloromethane extracts was prepared when the final sample volume, after colour development, was less than about 15 ml; in these cases, addition of the same tin standards was to 0.5 ml of sulphuric acid plus 4.0 ml of citric acid, again followed by the usual amounts of ascorbic acid and sensitized PCV.

Soderquist and Crosby[233] point out that the phenyltin hydrides, either neat or in solution, should not be brought into contact with readily reducible materials. Conversion of microgram amounts of the phenyltins to their hydrides is instantaneous and appears to be quantitative as long as lithium aluminium hydride was present in excess of other reducible materials (e.g. water). When the derivatization proceeded in the presence of only residual dichloromethane and added reagent, the yields of the hydrides decreased with time, particularly for the mono- and diphenyl compounds. Dilution with hexane before the addition of lithium aluminium hydride reagent alleviated this difficulty (Figure 33), perhaps because of the increased stability of the products in dilute hexane solution. In fact, dilute solutions (ng μl^{-1}) of the synthetic hydride standards were stable indefinitely in hexane.

The hydride derivatives differ widely in volatility and could not be simultaneously detected with any gas chromatographic columns tried by these workers. Since chromatography at different temperatures was required it was found most efficient to analyse all of a series of derivatized samples, blanks,

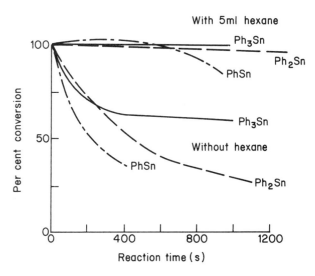

Figure 33 Conversion of phenyltin chlorides to their hydride derivatives. Reprinted with permission from Soderquist and Crosby.[233] Copyright (1978) American Chemical Society

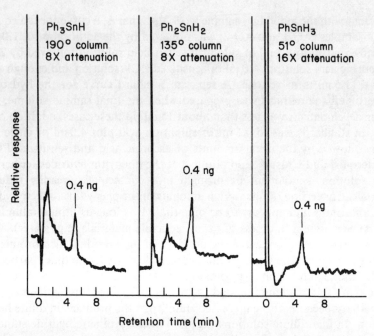

Figure 34 Typical chromatograms of the hydride derivatives. Reprinted with permission from Soderquist and Crosby.[233] Copyright (1978) American Chemical Society

and standards for one of the hydrides before equilibration of the chromatograph at new temperature. Typical chromatograms of the hydride standards near the limit of detectability are shown in Figure 34.

The chemistry of phenyltin compounds is dominated by the lability of the bond associated with the accompanying anionic group as compared to the relative stability of the carbon–tin bond. For example, both Ph_3SnCl and $(Ph_3Sn)O_2$ exist in water as Ph_3SnOH, which, in the presence of excess acetate, partitions into dichloromethane as Ph_3SnOAc. Ph_3SnCl, Ph_3SnOH, $(Ph_3Sn)_2O$ gave identical recoveries using the above procedure.

The true identity and behaviour of the diphenyltin (IV) compounds is less clear. It is expected that neutral aqueous solutions of Ph_2SnCl_2 rapidly produce $Ph_2Sn(OH)_2$ or the corresponding hydrated oxide Ph_2SnOH_2O. While recoveries of Ph_2SnO equalled those for Ph_2SnCl_2, solutions of Ph_2SnO in dichloromethane containing 5% acetic acid, prepared as fortification standards, always decomposed within a day to yield an unidentified possibly polymeric material unresponsive to lithium aluminium hydride.

The monophenyltin compounds are the least well characterized. While solutions of $PhSnCl_3$ in dischloromethane were stable, pure $PhSnCl_3$ fumes upon exposure to moist air and presumably exists in aqueous solution as $PhSn(OH)_3$ (or the hydrated benzenestannoic acid, $PhSnO_2H.H_2O$). In spite of the quantitative conversion of microgram amounts of $PhSnCl_3$ (or $PhSnO_2H$) to $PhSnH_3$, fortification of samples at equivalent levels resulted in consistently low recoveries. Even at relatively high levels (1-2 μg ml^{-1}), less than 35% of the PhSn^{3+} was recovered. The rapid initial loss was independent of sample pH, volume or nature of extractant, or level of fortification. As with the diphenyltin species, the possibility of formation of a non-extractable polymeric material seems reasonable. While the mode of this rapid loss of PhSn^{3+} in aqueous solution under mild laboratory conditions remains unexplained, it is surely indicative of the transitory existence which this compound must have under environmental conditions.

Table 29 Recovery of tin species from water

Species	Procedure*	μg ml^{-1} added	Fortified samples μg ml^{-1} found	Blank sample μg ml^{-1} found
Ph$_4$Sn	EX-3	0.050	0.035 $\pm$ 0.002†	<0.003
Ph$_3$Sn^{1+}	EX-1	0.050	0.054 $\pm$ 0.001†	<0.002
PH$_2$Sn^{2+}	EX-1	0.050	0.031 $\pm$ 0.003†	<0.006
PhSn^{3+}	EX-1	0.050	<0.005†	<0.005
Total extractable organotins (as Sn)	EX-2	0.082	0.050 $\pm$ 0.005†	<0.008
Sn^{4+}	AQ-2	0.050	0.050 $\pm$ 0.001(0.019)‡,§	
Sn^{4+} + SnO$_2$	AQ-1	...	0.072 $\pm$ 0.001(0.028)‡,§	<0.008

*Refers to subanalyses given in procedure and Figure 32.
†Average and average diviation of four replicates.
‡Average and average deviation of three replicates.
§Parenthetic values are from single samples to which no inorganic tin was added.
Reprinted with permission from Soderquist and Crosby.[233] Copyright (1978) American Chemical Society.

Distilled water samples (200 ml) spiked with 50 μg ml^{-1} of various organotin compounds and inorganic tin were put through the procedure described above to check on recoveries. All the organotins except PhSn^{3+} were successfully recovered (Table 29). Analysis for total extractable organotin, on a mole basis, resulted in slightly more tin than the sum of the specific hydride analyses and in the absence of any intentionally added inorganic tin, some inorganic tin was still found (parenthetic values in Table 29); both observations support the notion that PhSn^{3+} is unstable in water. There was no cross-reference between any of the phenyltin compounds when analysis was made via the hydride procedure for any one organotin in a sample containing a tenfold excess of each of the other organotins. While some of the added tin species were not recovered intact, the sum of the recoveries of procedures EX-2 plus AQ-1(0.207 μmol Sn^{4+}) was 92% of the total added tin (0.224 μmol) – an acceptable overall accountability.

It is significant that the bisulphate fusion procedure consistently accounted for more organotin than did the simpler procedure.

The sensitivity of the method for each of the tin species examined is summarized in Table 30. These limits could be decreased either by increasing the sample size, avoiding some of the subanalyses, or both. None of the natural

Table 30 Method sensitivity

Species	Method	Minimum detectable amount	Method sensitivity ($\mu g\ ml^{-1}$)*
Ph_4Sn	FID-g.l.c.	5.0 ng as Ph_4Sn	0.015
Ph_3Sn^{1+}	e.c.-g.l.c.	0.2 ng as Ph_3SnH	0.003
Ph_3Sn^{2+}	e.c.-g.l.c.	0.2 ng as Ph_2SnH_2	0.003
$PhSn^{3+}$	e.c.-g.l.c.	0.2 ng as $PhSnH_3$	0.003
Total extractable organotins	colorimetry	1.0 μg as Sn	0.01
Sn^{4+}	colorimetry	1.0 μg as Sn	0.007
$SnO_3 + Sn^{4+}$	colorimetry	1.0 μg as Sn	0.007

*For 200 ml samples.
Reprinted with permission from Soderquist and Crosby.[233] Copyright (1978) American Chemical Society.

water samples analysed by Soderquist and Crosby[233] contained materials which interfered with the determination of any of the tin compounds of interest.

Fluorimetry has been used to determine triphenyltin compounds in distilled, tap, canal, and sea water.[234] These toxic compounds are extensively used as fungicides and therefore possible contaminants of run-off. Triphenyltin compounds in water at concentrations of 0.004-2 ppm are readily extracted into toluene and can be determined by spectrofluorimetric measurements of the triphenyltin-3-hydroxyflavone complex.

Method

Excitation and emission wavelengths

The observed excitation and emission maxima for the compounds used by Bluden and Chapman[234] are given in Table 31. (Perkin Elmer Model 1000 fluorescence spectrometer).

Table 31 Excitation and emission maxima

Compound	Approximate wavelength (nm)	
	Excitation	Emission
$(C_6H_5)_3SnCl$	415	495
$(C_6H_5)_3SnOCOCH_3$	415	495
$(C_6H_5)_2SnCl_2$	415	450
$(C_6H_5)SnCl_3$	415	450
$(C_6H_3)_3SnCl$	415	510
$SnCl_4$	395	450
3-Hydroxyflavone	380	525

Reprinted with permission from Blunden and Chapman.[234] Copyright (1978) Royal Society of Chemistry.

Procedure

Fifty millilitres of the water sample are shaken with 10 ml of toluene in a separating funnel for approximately 30 min and the two layers allowed to separate. For the determination of triphenyltin compounds in the range of 0.2–2.0 ppm 1.0 ml of toluene is added to 5.0 ml of a 0.01% solution of 3-hydroxyflavone in toluene and 1 ml of a saturated aqueous solution of sodium acetate in a stoppered 10 ml flask covered with black paper and the contents shaken for approximately 10 min. The fluorescence emission of the organic layer at approximately 495 nm is measured, using an excitation wavelength of approximately 415 nm. For the 0.004–0.2 ppm range of triphenyltin compounds 5.0 ml of toluene are added to 1.0 ml of 3-hydroxyflavone solution and 1 ml of saturated sodium acetate solution and the procedure continued as above. A reagent blank is prepared by shaking 50 ml of water free from triphenyltin with 10 ml of toluene and continuing as above.

Calibration

Calibration graphs are prepared by mixing known volumes of triphenyltin chloride solutions (5×10^{-7} to 5×10^{-6} M) in toluene with either 1.0 or 5.0 ml (as appropriate) of 0.01% solution of 3-hydroxyflavone in toluene and the final volume made up to 6 ml. The solution is shaken with 1 ml of saturated aqueous sodium acetate solution in a darkened container for approximately 10 min and the fluorescence measured as before.

It has been shown that the triphenyltin-3-hydroxyflavone complex was unstable even in subdued light. It was essential to keep the solutions in flasks that were covered with black paper both before and after the addition of the reagent to the triphenyltin solution. Even when the solution was kept in the dark, the fluorescence from the triphenyltin chloride–3-hydroxyflavone complex was unstable, whereas the emission from the complex with triphenyltin acetate was stable over a period of hours. However, if the triphenyltin chloride–3-hydroxyflavone complex in toluene was shaken with a saturated aqueous solution of sodium acetate, a stable emission was obtained. It was also found that contamination of the fluorescent species by chloride ions after separation from the aqueous sodium acetate layer gave rise to quenching of the fluorescence emission (e.g. contamination of the fluorescence cells arising from the use of hydrochloric acid as a cleaning solution). It was concluded that chloride ions quenched the fluorescence but that on shaking with aqueous sodium acetate solution a stable complex was formed. Provided that these precautions were observed the emissions, of the triphenyl-3-hydroxyflavone complex were stable for several hours.

Shaking the triphenyltin chloride–3-hydroxyflavone solutions with a saturated aqueous solution of sodium acetate quenched the fluorescence from di- and monophenyltin compounds and at least a 10 molar excess of these compounds had no effect on the determination of triphenyltin. Inorganic tin in at least a 50 molar excess was also found not to interfere with the determination.

Tri-, di-, and monobutyl and di- and monomethyltin compounds did not fluoresce under the conditions used for the determination of triphenyltin. However, trimethyltin compounds react in a similar manner with 3-hydroxyflavone, and although the emission maximum is at approximately 510 nm, this is not sufficiently different from the emission maximum of triphenyltin compounds (approximately 495 nm) for these compounds to be determined in the presence of each other.

Spiking recoveries by the above procedure carried out on standard solutions of triphenyltin chloride in various types of water ranged from 74% at the 0.004 ppm tin level (rel. s.d. 8.9%) to 93.6% at the 2 ppm tin level (rel. s.d. 4.2%).

Luskima and Syavtsillo[235] have described a spectrophotometric procedure utilizing phenylfluorone for the determination of organotin compounds in water. They also used gas chromatography to separate tetraethyltin and tetrabutyltin. Smith[236] discussed the determination of tin in water and in organisms. In the determination of low concentrations of the order of 40 μg of trialkyltin chlorides in sea water it has been observed that these compounds are very volatile and are easily lost upon evaporation with acid. Quantitative recovery of tin is, however, obtained in the absence of chloride ion during evaporation with acid. Preliminary removal of chlorides from sea water by passage down a column of IRA 400 resin before digestion with acid completely overcame loss of tin on subsequent evaporation with acid giving a tin recovery of 90%. Work[237–240] has been carried out in the determination of tin in aqueous leachates from organotin containing antifouling paint compositions. This includes, tributyltin compounds[237] and *bis*(tributyltinoxide).[238,239]

ORGANOLEAD COMPOUNDS

In water

Potter *et al.*,[241] as described below, have applied gas chromatography and thin-layer chromatography to the detection and determination of alkyllead compounds and alkyllead salts in natural waters.

Analysis for tetraalkyllead compounds

Gas-liquid chromatography

Analyses were carried out on a Pye Model 104 gas-liquid chromatograph coupled to a Pye Unicam electron capture detector operated in the pulsed mode at 300 °C with a ^{63}Ni source. Samples were injected onto a 150 cm glass column packed with Chromosorb W (60–85 mesh) coated with 10% SE-30. Column temperature and flow rate of nitrogen carrier gas were adjusted to give the optimum retention for the tetraalkyllead present: samples suspected of containing tetraethyllead were chromatographed at 100 °C with a flow rate of 60 ml min^{-1}; those suspected of containing tetramethyllead, at 64 °C and 40 ml min^{-1}. Confirmation of the identity of the suspected tetraalkyllead was obtained by rechromatographing with a genuine sample and, for all except tetramethyllead, by rechromatographing the sample on a 300 cm glass column packed with a 10% 1,2,3-*tris*(2-cyanoethoxy)propane on Chromosorb W at 64 °C and 40 ml min^{-1} flow rate. Peak areas were integrated electronically and quantified by comparison with those of standards of tetraalkyllead in benzene.

Thin-layer chromatography

For higher concentrations of tetraalkyllead, identity was confirmed by thin-layer chromatography. A sample of sediment was extracted into 40:60 petroleum ether (50 ml) and iodine monochloride (0.3 ml) was added to the separated petroleum ether extract. The solution was shaken, allowed to stand for 10 minutes and extracted into water (10 ml). The water from the aqueous extract was evaporated under reduced pressure and the residue taken up in acetone. After evaporation of the bulk of the acetone, the sample was run by thin-layer chromatography against samples of Et_2PbCl_2 and Me_2PbCl_2. Tetraalkyllead added to tetraalkyllead-free samples showed the presence of R_2Pb^{2+} by thin-layer chromatography at concentrations down to 2 m l^{-1} in the petroleum ether extract.

Analysis for alkyllead salts

Samples (10 ml) of the aqueous extract suspected of containing alkyllead salts were analysed by a colorimetric technique using pyridylazo-resorcinol, the absorption of the resulting solution being measured at 515 mm on a spectrophotometer. Confirmation of the presence of dialkyllead salts was attempted by evaporation of the water from the aqueous extract at reduced pressure, dissolution of the residue in acetone, and analysis by thin-layer chromatography. For trialkyllead salts the aqueous extract was saturated with sodium chloride and extracted into benzene; the benzene extract was then concentrated and analysed by thin-layer chromatography. These techniques also identified the alkyl groups present, but the sensitivity was restricted by the limits of detection of the thin-layer chromatographic analysis.

Thin-layer chromatography

Samples suspected of containing alkyllead salts were applied from an appropriate solution to plates coated with a 0.25 mm layer of MN Aluminoxid G (Camlab, Cambridge) equilibrated with the atmosphere. Chloroform was found to be a suitable solvent for R_3PbX and water or acetone for R_2PbX_2. Plates were eluted with acetic acid–toluene (1:19, v/v) and the spots were developed by spraying with a solution of dithizone in chloroform (0.1% w/v). R_3PbX gave a yellow spot and R_2PbX_2 gave a salmon red spot; inorganic lead gave a crimson spot on the baseline and tetraalkyllead was not detected. Under these conditions the R_f values for Et_3PbCl and Me_3PbCl were 0.5 and 0.2, respectively, and for Et_2PbCl_2 and Me_2PbCl_2 were 0.3 and 0.1 respectively. The mixed methyllead salts gave distinct spots with intermediate R_f values. The limit of detection of this method was 0.5-1.0 µg alkyllead salt.

The total concentration of lead in natural waters is generally low owing to absorption onto sediment unless pH is exceptionally low. This explains the observation that the levels of lead in rain water are twice those in ground water. Tetraalkyllead is insoluble in, and denser than, water and would be expected to accumulate with the sediment. Alkyllead salts are generally much more soluble in water than the corresponding inorganic lead salts and would remain in solution in the absence of suspended solids. However, alkyllead salts at low concentrations are totally absorbed onto a variety of natural sediments which are nearly always present in natural waters.

Potter et al.[241] took samples of sediment in drains and natural water courses from the top 50-100 cm, such that a portion of the supernatant water was included, and stored in polypropylene bottles until analysed. Portions (30 ml) of the sediment were shaken for 1 min with petroleum ether (10 ml) and centrifuged. The petroleum ether extract was transferred to a small glass-stoppered flask and stored at 0 °C. The extraction procedure was repeated. The extracted sediment was acidified to pH-5-6 with dilute nitric acid, allowed to stand until evolution of hydrogen sulphide had ceased, and lead nitrate (5 g in 20 ml water) was added and mixed. After standing for 5 min the sediment was filtered, the solids were dried and weighed and the filtrate was neutralized with aqueous sodium hydroxide (2 M), filtered and made up to 100 ml. Samples (10 ml) of this filtrate were analysed for alkyllead salts. The above operations were carried out on the day of sampling. The petroleum ether extract was analysed for tetraalkyllead compounds.

Potter et al.[241] showed that the recovery of alkyllead salts obtained by this procedure from sediments, was 90% for Et_3PbCl and Me_3PbCl, 75% for Et_2PbCl_2, and 40% for Me_2PbCl_2. Extraction of Et_2PbCl_2 added to sediment, containing no alkylleads, from a clean and polluted river, from a clean and polluted canal, and from road drainage grids gave recoveries of between 65 and 75%. The lowest detectable concentration of alkyllead salts was 2 mg kg^{-1} dry weight of sediment. For direct analysis of a sample of filtered water and lowest detectable concentration was 0.1 mg l^{-1}. Extraction of tetraalkyllead (0.01 µl) from sediment (30 ml) gave recoveries of 36-41% for the initial extraction with petroleum ether and 13-15% for the second. The lowest detectable concentration of tetraalkyllead was 0.02 mg kg^{-1} dry weight of sediment. By extracting samples of filtered water with one-tenth their volume of petroleum ether, tetraalkyllead could be detected down to concentrations of 0.002 mg l^{-1} in the water.

Using these procedures Potter et al.[241] found appreciable quantities of alkyllead compounds in some samples of drainage grid sediments (Table 32). No alkyllead compounds were detected in the filtered water from any of these sediment samples.

Table 32 Concentration of alkyllead compounds in drainage grid sediments

Site	Concentration (mg kg^{-1} dry weight of sediment) of			
	Tetraethyl lead^a	Tetraethyl lead^b	Tetramethyl lead	Alkyllead salts
(1) Garage A, near to pumps	96.^{e}	93^e	5.3	6.1
(2) Garage A, near to entrance	4.7^e	5.1^e	0.11	
(3) Garage B, near to pumps	54^e	58^e	1.8	4.0
(4) Garage B, near to entrance	25^e	24^e	0.36	
(5) 10 m downhill from garage A	0.33	0.33		
(6) 10 m downhill from garage A	0.51	0.53		
(7) 100 m downhill from garage A	0.20	0.18		
(8) 30 m downhill from garage B	0.38	0.35		
(9) 50 m downhill from garage B	0.24	0.24		
(10) Aston Expressway	0.25	0.22		
(10)^c Aston Expressway	0.09	0.07		
(11) Residential area	0.34	0.38		
(12) Residential area	0.36	0.35		
(13)^d Residential area	0.05	0.05		

^{a}Determined on SE-30 column (see Experimental)
^{b}Determined on TCEP column (see Experimental)
^{c}Site 10 resampled after (14)^d
^{d}Solids settlement tank into which water from site 12 passed.
^{e}Confirmed by t.l.c.
Reprinted with permission from Potter et al.[241] Copyright (1977) Institute of Water Pollution Control.

Imura et al.[242] have described a spectrophotometric procedure employing 1-hydroxy-4-(4-nitrophenylazo)-2-naphthoate as chromogenic reagent for the determination of triethyllead ions in industrial waste waters. The coloured adduct is extracted with chloroform. The absorption maximum is at 440 nm and the optimum pH for the extraction from 1% aqueous sodium chloride is 8.1–8.3. In the determinatin of about 60 μg of triethyllead ions, dimethyl- and diethyllead ions (100 μg) and Pb^{2+} (1.8 mg) are masked with 0.01 M ethylenediamine NN^--bis-(2-hydroxyphenylacetic acid) (disodium salt) (2 ml) and copper and ferrous iron are masked with 0.01 M 1,2-diaminocyclohexane-$NNN'N'$-tetraacetate (2 ml). Several other ions do not interfere. The limit of determination is 0.2 ppm of triethyllead.

Organolead compounds in water, fish, and sediments

Chau et al.[243] have described a simple and rapid extraction procedure to extract the five tetraalkyllead compounds (Me_4Pb, Me_3EtPb, Me_2Et_2Pb, $MeEt_3Pb$, and Et_4Pb) from water, sediment, and fish samples. The extracted compounds are analysed in their authentic forms by a gas chromatographic–atomic adsorption spectrometry system. Other forms of inorganic lead do not interfere. The detection limits for water (200 ml), sediment (5 g) and fish (2 g) are 0.50 μg l^{-1}, 0.01 μg g^{-1}, and 0.025 μg g^{-1}, respectively. Whilst this method would be applicable to the determination of tetraalkyllead compounds originating from automobile exhausts in water, fish, and sediment samples, the main interest of Chau et al. was in the determination of organically bound lead produced by biological methylation of inorganic and organic lead compounds in the aquatic environment by microorganisms.[243] The gas chromatographic-atomic absorption system used by Chau et al. (used without a sample injection trap) for this procedure has been described.[244] The extract was injected directly into the column injection port of the chromatograph. Instrumental parameters were identical. A Perkin Elmer electrodeless discharge lead lamp was used; peak areas were integrated. The procedures used by Chau et al. are outlined below.

Water analysis

Water (200 ml) and 5 ml of high purity hexane were placed in a 250 ml separatory funnel and shaken rigorously for 30 min in a reciprocating shaker. The solution was allowed to stand for about 20 min for phase separation. Approximately 195 ml of the water was drained off and the remaining mixture transferred into a 25 ml tube with a Teflon-lined cap. Without separating the phases, a suitable aliquot, 5–10 μl of the hexane, was injected into the g.c.–a.s. system.

Sediment analysis

Five grams of wet sediment, 5 ml of EDTA reagent (0.1 M, 37 g Na_2 EDTA $2H_2O\, l^{-1}$ disperses fish and sediment homogenates in a suspension to provide better extraction and produce clarified organic phase), and 5 ml of hexane were placed in a 25 ml test tube with a Teflon-lined screw cap and the contents shaken rigorously in a reciprocating shaker for 2 h. The sample was centrifuged for 10 min at $2000 \times g$. A suitable aliquot, 5–10 μl, of the hexane extract was injected into the g.c.–a.a.s. system.

Fish analysis

The frozen fish tissue was homogenized in a Hobart grinder and a Polytron homogenizer and 2 g of the fish homogenated with 5 ml of EDTA reagent, and 5 ml of hexane were immediately placed in a 25 ml test tube with a Teflon lined screw cap. The contents were shaken rigorously for 2 h in a reciprocating shaker and centrifuged to facilitate phase separation. A suitable aliquot, 5–10 μl of the hexane phase was withdrawn and injected into the g.c.–a.a.s. system.

Calibration

A known amount of standard tetramethyllead, 5 μg, was added to the hexane layer after injection of the sample, mixed gently, and centrifuged again if necessary. The same volume as used in sample analysis was injected. The increase in peak area due to the standard added was used to calculate the amount of tetraalkyllead in the sample. It is not necessary to separate the phases or to know the volume of hexane after extraction.

The calibration curves for each of the five tetraalkyllead compounds expressed as lead were identical and linear up to at least 200 ng, above which overlapping peaks occurred. If only one compound was present (e.g. tetramethyllead), the curve was linear up to at least 2000 ng.

Chau et al.[243] pointed out that as the authenticity of the compounds to be analysed must be preserved, any of the digestion methods with acids or alkalis is not suitable, and that extraction seemed to be the method of choice for removing these compounds from samples. For this extraction, they adopted benzene as recommended by Sirota and Uthe[245] for the quantitative extraction of tetramethyllead and tetramethyllead from fish homogenates suspended in aqueous EDTA solution. Although ionic forms of lead such as Pb(II), diethyllead dichloride, and trimethyllead acetate do not extract in the benzene phase, any lead compounds that distribute into the benzene phase as tetraalkyllead will be determined. Chau et al.[243] found that environmental samples can contain other forms of organolead compounds that are extractable into benzene but which are not volatile enough to be analysed by the gas chromatographic–atomic absorption spectroscopic technique, hence the need for a speciation specific analytical system.

Recovery experiments carried out by Chau et al.[243] showed that benzene, hexane, and octanol gave the most satisfactory recovery of tetraalkyllead compounds (Table 33).

Table 33 Extraction of tetraalkyllead compounds from fish tissue by different solvents*

Solvent	Averaged recovery (%)
Hexane	80.0
Cyclohexane	54.0
Octanol	90.0
Butyl acetate	55.0
Methylisobutyl ketone	30.0
Chloroform	57.0
Benzene	78.0

*Fish homogenate 2 g; EDTA, 5 ml; solvent, 5 ml.
Reprinted with permission from Chau et al.[243] Copyright (1979) American Chemical Society.

Chau et al.[243] found that tetraalkyllead compounds have high vapour pressures and are not stable in water. It was observed that water containing 4.2 µg l^{-1} Me$_4$Pb decreased to 2.8 and 3.9 µg l^{-1} when stored respectively at room temperature and at 4 °C overnight.

For this reason, water samples should not be filtered by suction but should be extracted with hexane immediately after collection. It was found convenient to add 5 ml hexane to the water sample (200 ml) and to shake the mixture briefly for 5 min. The sample can then be stored for at least one week. Similar practice is recommended for sediment samples. After collection, the sediment is weighed (5 g) and shaken with 5 ml of EDTA and 5 ml hexane for 5 min in a 25 ml stoppered tube. The treated samples can be stored for at least up to one week for further analysis.

Table 34 Recovery of tetraalkyllead compounds from water, sediment, and fish samples*

Compound	Water			Sediment		Fish	
	added μg	found μg	recovery %	found μg	recovery %	found μg	recovery %
Me_4Pb	10.00	8.78	87.8±3	8.27	82.7±9	7.22	72.2±8
Me_3EtPb	13.15	11.80	89.7±4	10.65	81.0±5	9.15	72.3±5
Me_2Et_2Pb	14.30	12.50	87.4±3	11.68	81.0±7	10.26	76.2±5
$MeEt_3Pb$	10.15	9.08	89.5±4	8.32	82.0±2	7.22	75.2±9
Et_4Pb	14.20	12.82	90.3±7	12.09	85.2±6	10.69	75.3±8
Average			88.9±7		83.7±9		74.2±9

*Four determinations for each sample.
Reprinted with permission from Chau et al.[243] Copyright (1979) American Chemical Society.

In Table 34 are tabulated recoveries obtained by the procedure for five tetraalkyllead compounds from lake water, sediment, and fish samples. The five tetraalkyllead compounds were added to respectively 200 ml of lake water, 5 g of sediment and 2 g of fish homogenate.

The recovery of the five alkylated lead compounds from fish tissue averaged 74%. The reproducibility of the procedure was evaluated by analysing 11 replicates of fish homogenate, 1 g, each spiked with 5 μg of tetramethyllead. The coefficient of variation was 7.3%.

The detection limits were $0.50 \mu g \, l^{-1}$, $0.01 \mu g \, g^{-1}$ and $0.025 \mu g \, g^{-1}$, respectively, for water, sediment, and fish.

In Table 35 are shown results obtained in measurements of the accumulation of tetramethyllead in rainbow trout. The trout after exposure for different periods of time to water containing $3.5 \mu g \, l^{-1}$ tetramethyllead were found to contain tetramethyllead Preliminary results showed that this compound was mainly concentrated in the lipid layer of the tissues.

Table 35 Accumulation of tetramethyllead in rainbow trout

Exposure (day)	Weight of fish (g)	Fish, alive or dead	Conc. of Me_4Pb in		Conc. factors*
			water (averaged μg l^{-1})	fish (μg g^{-1} wet wt.)	
1	0.1211	Dead	3.46	0.43	124
2	0.3661	Dead		1.08	312
	0.7982	Dead		2.00	578
3	0.4116	Dead		1.32	382
	0.6300	Dead		2.09	604
7	1.3045	Alive		2.94	850
	1.5466	Alive		3.23	934
	0.8100	Alive		2.25	650
	0.4926	Alive		1.73	500

*Concentration factor = concentration of Me_4Pb in fish/concentration of Me_4Pb in water.
Reprinted with permission from Chau et al.[243] Copyright (1979) American Chemical Society.

The application of a combination of gas chromatography and atomic absorption spectrometry to the determination of tetraalkyllead compounds has also been studied by Chau et al.[246] and by Segar.[247] In these methods the gas chromatography flame combination showed a detection limit of about 0.1 μg Pb. Chau et al. have applied the silica furnace in the atomic absorption unit and have also shown that the sensitivity limit for the detection of lead can be enhanced by three orders of magnitude. They applied the method to the determination of tetramethyllead in sediment systems and in the atmosphere.

The system used by these workers consisted of a Microtek 220 gas chromatograph and a Perkin Elmer 403 atomic absorption spectrophotometer. These instruments were connected by means of stainless steel tubing (2 mm o.d.) connected from the column outlet of the gas chromatograph to the silica furnace of the a.a.s. A four-way valve was installed between the carrier gas inlet and the column injection port so that a sample trap could be mounted, and the sample could be swept into the g.c. column by the carrier gas. The recorder (10 mV) was equipped with an electronic integrator to measure the peak areas, and was simultaneously actuated with the sample introduction so that the retention time of each component could be used for identification of peaks.

The furnace was constructed from silica tubing (7 mm i.d., 6 cm long) with open ends. The lead compounds separated by gas chromatography were introduced to the centre of the furnace through a side-arm. Hydrogen gas was introduced at the same point at a flow rate of 1.35 ml min^{-1}; the burning of hydrogen improved the sensitivity. The furnace was wound with 26-gauge Chromel wire to give a resistance of about 5 ohms. The voltage applied to the furnace was about 20 V a.c. regulated by a variable transformer so that the furnace temperature with the hydrogen burning was about 1000 °C. The silica furnace was mounted on top of the a.a.s. burner and aligned to the light path.

The sample trap was a glass U-tube (6 mm dia., 26 cm long) packed with 3% OV-1 on Chromosorb W, which was immersed in a dry ice–methanol bath at ca. −70 °C as described by Chau et al.[246] A known amount of gaseous sample was drawn through the trap by a peristaltic pump operated at 130–150 ml min^{-1}. After sampling, the trap was mounted to the four-way valve and heated to ca. 80–100 C by a beaker of hot water, and the adsorbed compounds were swept into the gas chromatographic column.

Liquid samples can be directly injected into the column through the injection port, without a sample trap.

Chau et al.[244] have applied gas chromatography atomic absorption to the determination of tetraalkyllead compounds in water, sediments, and fish samples in high lead areas. Of some 50 fish samples analysed, only one sample was found to contain detectable amounts of tetramethyllead in the fillet. Since there was no known tetraalkyllead industry the area, the possibility that it comes from *in vivo* lead methylation in the sediment or in the fish cannot be totally disregarded.

Another procedure[248] for determining tetraalkyllead compounds in fish samples employs vacuum extraction of the tetraalkyllead into a cold trap under liquid nitrogen, followed by solvent extraction of the condensate for gas chromatographic determination. In this method, tetraalkyllead compounds have been found in fish and mussels. The presence of tetraethyllead in aquatic organisms may indicate that the alkyllead compounds are not immediately metabolized by living organisms and may remain in their authentic forms in the living tissues for a long time.

The solvent extraction-atomic absorption extraction procedure employed by Sirota and Uthe[245] gave concentrations of tetraalkyllead compounds in fisheries products between 0.01 and 4.8 ppm, accounting for 30-90% of the total lead present in the tissues.

ORGANOGERMANIUM COMPOUNDS

Braman and Tompkins[249] have described an atomic emission spectrometric method for the determination of inorganic germanium and methylgermanium (and inorganic antimony) in amounts down to 0.4 ng in environmental samples. These compounds are reduced to hydrides using sodium borohydride, then separated prior to atomic emission spectrography.

ORGANOANTIMONY COMPOUNDS

Andreae *et al.*[250] determined methylantimony species and antimony(III) and antimony(V) in natural waters using atomic absorption spectrometry with hydride generation. The limit of detection was $0.3-0.6\,\mu g\,l^{-1}$ for a 100 ml water sample.

SILICONES

Van der Post[251] has described a method for the determination of silanols in water based on their ability to reduce nitrite or nitrate to ammonia at normal temperature. Individual silanols are identified by mass spectrometry.

Pellenberg[252] analysed river sediment for silicone content by nitrous oxide-acetylene flame atomic absorption spectrophotometry. He showed that total carbon and total carbohydrates both correlate well with silicone content and the correlation between sedimentary silicone and presumed sewage material is good enough to suggest silicone as a totally synthetic, specific tracer for sewage in the aquatic environment.

REFERENCES

1 Magos, L. *Analyst (London)*, **96**, 847 (1981).
2 Kimura, Y. and Miller, V. L. *Anal. Chim. Acta*, **27** 325 (1962).
3 Rains, T. C. and Menis, O. *J. Assoc. Off. Anal. Chem.*, **55**, 1339 (1972).
4 Environmental Protection Agency. *Methods for Chemical Analysis of Water and Waste Water*. EPA publication No. EPA-625/6-74-003, p. 118. U.S. E.P.A. office of technology transfer. Washington DC 20460 (1972).

5. Goulden, P. D. and Afghan, B. K. *Tech. Bull.* Inland Waters Branch Department of Energy, Mines and Resources, Ottawa, Canada.
6. Armstrong, F. A. J., Williams, P. M. and Strickland, J. D. *Nature (London)*, **211**, 481 (1966).
7. Bennett, T. B., McDaniel, W. H. and Hemphill, R. N. *Advances in Automated Analysis. Technical International Congress*, Vol. 8, Mediad, Inc., Tarrytown, N.Y. (1972).
8. El-Awady, A. A., Miller, R. B. and Carter, M. J. *Anal. Chem.*, **48**, 110 (1976).
9. Kalb, G. W. *Atomic Absorption Newsletter*, **9**, 84 (1970).
10. Umezaki, U. and Iwamoto, K. *Japan Analyst*, **20**, 173 (1971).
11. Kimura, O. and Miller, O. *Anal. Abstr.*, **10**, 2943 (1963).
12. Doherty, P. E. and Dorsett, R. S. *Anal. Chem.*, **43**, 1887 (1971).
13. Brandenberger, O. and Bader, O. *Anal. Abstr.*, **15**, 5883 (1968).
14. Brandenberger, O. and Bader, O. *Anal. Abstr.*, **17**, 2617 (1969).
15. Graf, E., Polos, L., Bezur, L. and Pungor, E. *Magy. kem. Fely*, **79**, 471 (1973).
16. Baltisberger, R. J. and Knudson, C. L. *Anal. Chim. Acta*, **73**, 265 (1974).
17. Bisagni, J. J. and Lawrence, A. W. *Environ. Sci. Technol.*, **8**, 850 (1974).
18. Frimmel, F. and Winckler, H. A. *Zeitschrift für Wasser und Abwasser*, **8**, 67 (1975).
19. Watling, R. J. and Watling, H. R. *Water S. A.*, **1**, 113 (1975).
20. Chan, Y.-K. and Saitoh, H. *Environ. Sci. Technol*, **4**, 839 (1970).
21. Umezaki, Y. and Iwamoto, K. *Japan. Analyst*, **20**, 173 (1971).
22. Stainton, M. P. *Anal. Chem.*, **43**, 625 (1971).
23. Carr, R. A., Hoover, J. B. and Wilkniss, P. W. *Deep-Sea Res.*, **19**, 747 (1972).
24. Fitzgerald, W. F., Lyons, W. B. and Hunt, C. D. *Anal. Chem.* **46**, 1882 (1974).
25. Kiemeneij, A. M. and Kloosterboer, J. G. *Anal. Chem.*, **48**, 575 (1976).
26. Hatch, W. R. and Ott, W. L. *Anal. Chem.*, **40**, 2005 (1968).
27. Farey, B. J., Nelson, L. A. and Rolph, M. J. *Analyst (London)*, **103**, 656 (1978).
28. Report Water Pollution Research Laboratory, Stevenage U.K. *Report No. 1272* (1972).
29. Thompson, K. C. and Godden, R. C. *Analyst (London)*, **100**, 544 (1975).
30. Department of the Environment and National Water Council (UK) HM Stationary Office, London, 23pp (Pf 22Ab ENV) (1978).
31. Omang, S. H. *Anal. Chim. Acta*, **52**, 415 (1972).
32. Uthé, J. E., Armstrong, F. A. J. and Stainton, M. P. *J. Fish. Res. Board, Can.*, **27**, 805 (1970).
33. Osland R. Pye Unicam Spectrovision, *No. 24*, 11 (1970).
34. Gardner, D. and Riley, J. P. *J. Cons. Int. Explor. Mer.*, **35**, 202 (1974).
35. Abo-Rady, M. D. K. *Fresenius Zeitschrift für Analytiche Chemie*, **299**, 187 (1979).
36. Lutze, R. L. *Analyst (London)*, **104**, 979 (1979).
37. Grantham, D. L. *Laboratory Practice*, **27**, 294 (1978).
38. Fitzgerald, W. F., Lyons, W. B. and Hunt, C. D. *Anal. Chem.*, **46**, 1882 (1974).
39. Krämer, H. J. and Neidhart, J. *Radio. Anal. Chem.*, **37**, 835 (1977).
40. Chan, Y. K. and Saitoh, H. *Environ. Sci. Technol.* **4**, 839 (1970).
41. Fishman, M. J. *Anal. Chem.*, **42**, 1462 (1972).
42. Muscat, V. I., Vickers, T. J. and Andrery, A. *Anal. Chem.*, **44**, 218 (1972).
43. Matsunaga, K. *Mizushori-gijutsu*, **15**, 431 (1974).
44. Olafsson, J. *Anal. Chim. Acta*, **68**, 207 (1974).
45. Minagawa, K., Takizawa, Y. and Kifune, I. *Anal. Chim. Acta*, **115**, 103 (1980).
46. Nahoa, S., Sekine, S. and Matsuda, K. *Sumitomo Kagaku, Toku* **1974I**, 19 (1974).
47. Lo, J. M. and Wai, C. M. *Anal. Chem.*, **47**, 1869 (1975).
48. Law, S. L. *Science*, 174, 285 (1971).
49. Yamagami, E., Tateishi, S. and Hashimoto, A. *Analyst (London)*, **105**, 491 (1980).
50. Simpson, W. R. and Nickless, G. *Analyst (London)*, **102**, 86 (1977).
51. Nishi, S. and Horimoto, H. *Japan Analyst*, **17**, 1247 (1968).

52. Nishi, S. and Horimoto, H. *Japan Analyst*, **19**, 1646 (1970).
53. Zarnegar, P. and Mushak, P. *Anal. Chim. Acta*, **69**, 389 (1974).
54. Ealy, J., Schultz, W. D. and Dean, D. A. *Anal. Chim. Acta*, **64**, 235 (1974).
55. Capon, C. J., Crispin Smith, V. *J. Anal. Chem.*, **49**, 365 (1977).
56. Longbottom, J. E. *Anal. Chem.*, **44**, 1111 (1972).
57. Dressman, R. C. *J. Chromat. Sci.*, **10**, 472 (1972).
58. Zarnegar, P. and Mushak, P. *Anal. Chim. Acta*, **69**, 389 (1974).
59. Becknell, D. E., Marsh, R. H. and Allie, W. *Anal. Chem.*, **43**, 1230 (1971).
60. Cullen, M. C. and McGuinness, E. T. *Analyt. Biochem.*, **42**, 455 (1971).
61. Ke, P. J. and Thibert, R. T. *Mikrochim. Acta*, **3**, 417 (1973).
62. Van Ettekoven, K. G. H_2O, **13**, 326 (1980).
63. Matsunaga, K., Konishi, S. and Nishimura, M. *Environ. Sci. Technol.*, **13**, 63 (1979).
64. Rubel, S. *Anal. Chim. Acta*, **115**, 343 (1980).
65. *World Health Organization International Standards for Drinking Water*, WHO, Geneva (1971).
66. *Official Journal of the European Community* Proposal for a Council directive relating to the quality of water for human consumption. **18**, C214, 2–17. (1975).
67. Starý, J. and Prášilová, J. *Radiochem. Radioanal. Letters*, **24**, 143 (1976).
68. Starý, J. and Prášilová, J. *Radiochem. Radioanal. Letters*, **26**, 33 (1976).
69. Starý, J. and Prášilová, J. *Radiochem. Radioanal. Letters*, **26**, 193 (1976).
70. Starý, J. and Prášilová, J. *Radiochem. Radioanal. Letters*, **27**, 51 (1976).
71. Starý, J., Havlik, B., Prášilová, J., Kratzer, K. and Hanušová, J., *Int. J. Environ. Chem.* **5**, 89 (1978).
72. Jackson, F. and Dellar, D. *Water Research*, **13**, 381 (1979).
73. Feldman, C. *Anal. Chem.*, **46**, 99 (1974).
74. Fitzgerald, W. F. and Lyons, W. B. *Nature (London)*, **242**, 452 (1973).
75. Millward, G. E. and Bihan, A. I. *Water Research*, **12**, 979 (1978).
76. Agemian, H. and Chau, A. S. Y. *Anal. Chem.*, **50**, 13 (1978).
77. Jenne, E. A. and Avotins, P. *J. Environ. Qual.*, **4**, 427 (1975).
78. Carron, J. and Agemian, H. *Anal. Chem. Acta*, **92**, 61 (1977).
79. Sipos, L., Nurnberg, H. W., Valenta, P. and Brancia, M. *Anal. Chim. Acta*, **115**, 25 (1980).
80. Egawa, H. and Tajima, S. *Proc. 2nd U.S.-Japan Experts Meeting, Oct. 1976, Tokyo, Japan* (1977).
81. Gillespie, D. C. *J. Fish. Res. Board Can.*, **29**, 1035 (1972).
82. HMSO Dept. of the Environment, Central Unit of Environmental Pollution. *Pollut. Pap.*, **10**, 92pp (1976).
83. Holden, A. V. *J. Food Technol.*, **8**, 1 (1973).
84. Jernelöv, A. *Limnol. Oceanogr.*, **15**, 958 (1970).
85. Langley, D. G. *J. Water Poll. Cont. Fed.*, **49**, 44 (1973).
86. Olsen, B. H. and Cooper, R. C. *Water Research*, **10**, 113 (1976).
87. Pennacchioni, A., Marchetti, R. and Gaggino, G. F. *J. Environ. Qual.*, **5**, 451 (1976).
88. Pentreath, R. J. *J. Exp. Mar. Biol. Ecol.*, **25**, 51 (1976).
89. Pentreath, R. J. *J. Exp. Mar. Biol. Ecol.*, **25**, 103 (1976).
90. Shin, E. and Krenkel, P. A. *J. Water Poll. Cont. Fed.*, **48**, 473 (1976).
91. Andren, A. W. and Harriss, R. C. *Nature*, **245**, 256 (1973).
92. Bartlett, P. O., Craig, P. J. and Morton, S. F. *Nature (London)*, **267**, 606 (1977).
93. Davies, I. M., Graham, W. C. and Pirie, S. M. *Marine Chemistry*, **7**, III (1979).
94. Windom, H., Gardner, W., Stephens, J. and Taylor, F. *Est. Coast. Mar. Sci.*, **4**, 579 (1976).
95. Westöö, G. *Acta Chem. Scand.*, **22**, 2277 (1978).
96. Chester, R., Gardner, D., Riley, J. P. and Stoner, J. *Mar. Poll. Bull.*, **2**, 28 (1973).
97. Renzoni, A., Bacci, E. and Falciai, L. *Rev. Intern. Océanogr. Méd.*, **32**, 31 (1973).
98. Olafsson, J. *Anal. Chim. Acta*, **68**, 207 (1974).

99. Fitzgerald, W. F., Lyons, W. B. and Hunt, C. D. *Anal. Chem.*, **46**, 1882 (1974).
100. Fitzgerald, R. A., Gordon, Jr., D. C. and Cranston, R. E. *Deep-Sea Res.*, **21**, 139 (1974).
101. Thibaud, Y. *Science et Pêche, Bull. Inst. Pêche Marit.*, **209**, 1 (1971).
102. Cumont, G., Viallex, G., Lelievre, H. and Bobenrieth, P. *Rev. Intern. Océanogr. Méd.*, **26**, 95 (1972).
103. Renzoni, A. and Baldi, F. *Accua and Aria*, 597 (1975).
104. Stoeppler, M., Backhaus, F., Matthes, W., Bernhard, M. and Schulte, E. *Proc. Verb. XXVth. Congress and Plenary Assembly of ICSEM, Split* (1976).
105. Stoeppler, M., Bernhard, M., Backhaus, F. and Schulte, E. *Mar. Poll. Bull.*, in press.
106. Stoeppler, M. and Matthes, W. *Anal. Chim. Acta*, **98**, 389 (1978).
107. Leong, P. C. and Ong, H. P. *Anal. Chem.*, **43**, 940 (1971).
108. Anderson, D. H., Evans, J. H., Murphy, J. J. and White, W. W. *Anal. Chem.*, **43**, 1511 (1971).
109. Bretthaur, E. W., Moghissi, A. A., Snyder, S. S. and Matthews, N. W., *Anal. Chem.*, **46**, 445 (1974).
110. Pillay, K. K. S., Thomas, C. C., Sondel, J. A. and Hyche, C. M. *Anal. Chem.*, **43**, 1419 (1971).
111. Feldman, C. *Anal. Chem.*, **46**, 1606 (1974).
112. Bishop, J. N., Taylor, L. A. and Neary, B. P. *The Determination of Mercury in Environment Samples*, Ministry of the Environment, Ontario, Canada, 1973.
113. Jacobs, L. W. and Keeney, D. R. *Environ. Sci. Technol.*, **8**, 267 (1976).
114. *Methods for Chemical Analysis of Water and Wastes*, U.S. Environmental Protection Agency, Cincinnati, Ohio, p. 134 (1974).
115. Iskander, I. K., Syens, J. K., Jababs, L. W., Keeney, D. R. and Gilmour, J. T. *Analyst (London)*, **97**, 388 (1972).
116. Craig, P. J. and Morton, S. F. *Nature (London)*, **261**, 125 (1976).
117. Ealy, J. A., Shultz, W. D. and Dean, J. A. *Anal. Chim. Acta*, **64**, 235 (1973).
118. Batti, R., Magnaval, R. and Lanzola, E. *Chemosphere*, **4** 13 (1975).
119. Longbottom, J. E., Dressman, R. C. and Lichtenberg, J. J. *J. Ass. Off. Anal. Chem.*, **56**, 1297 (1973).
120. Bartlett, P. D., Craig, P. J. and Morton, S. F. *Nature*, **267**, 606 (1977).
121. Uthe, J. F., Solomon, J. and Grift, B. *J. Ass. Off. Anal. Chem.*, **55**, 583 (1972).
122. *Official Methods of Analysis of AOAC* (11th Edition) 418 (1970).
123. Nagase, H., Sato, T., Ishikawa, T. and Mitani, K. *Int., J. Environ. Anal. Chem.*, **7**, 261 (1980).
124. Jurka, A. M. and Carter, M. J. *Anal. Chem.*, **50**, 91 (1978).
125. El-Awady, A. A., Miller, R. B. and Carter, M. J. *Anal. Chem.*, **48**, 110 (1976).
126. Aspila, K. I. and Carron, J. M. *Inter-Laboratory Quality Control Study No. 1'-Total Mercury in Sediments*, Report Series, Inland Waters Directorate Water Quality Branch, Special Services Section, Department of Fisheries and Environment, Burlington, Ontario, Canada.
127. Analytical Methods Committee Chemical Society, London, *Analyst (London)*, **102**, 769 (1977).
128. Davies, I. M. *Anal. Chim. Acta*, **102**, 189 (1978).
129. Jones, P. and Nickless, G. *Analyst (london)*, **103**, 1120 (1978).
130. Capelli, R., Fezia, C., and Franchi, A. Zanicchi *Analyst (London)*, **104**, 1197 (1979).
131. Shum, G. T. C., Freeman, H. C. and Uthe, J. F. *Anal. Chem.*, **51**, 414 (1979).
132. Stuart, D. C. *Anal. Chem.*, **96**, 83 (1978).
133. Yamanaka, S. and Ueda, K. *Bull. Environ. Contam. Toxicol.* **14**, 409 (1975).
134. Collett, D. L., Fleming, D. E. and Taylor, G. E. *Analyst*, **105**, 897 (1980).
135. Shultz, C. D., Clear, D., Pearson, J. E., Rivers, J. B. and Hyliu, J. W. *Bull. Environ. Contam. Toxicol.*, **15**, 230 (1976).
136. Stainton, M. P. *Anal. Chem.*, **43**, 625 (1971).

137. Kopp, J. F., Longbottom M. C. and Lobring, L. B. *J. Am. Water Works Ass.*, **64**, 20 (1972).
138. Environmental Protection Agency. *Mercury in Water – Provisional Method.* Analytical Quality Control Laboratory, Cincinnati, Ohio (1972). Environmental Protection Agency. 1972b. *Mercury in Fish – Provisional Method.* Analytical Quality Control Laboratory, Cincinnati, Ohio (1972).
139. Hendzel, M. R. and Jamieson, D. M. *Anal. Chem.*, **48**, 926 (1976).
140. Armstrong, F. A. J. and Uthe, J. F. *Atomic Absorption Newsletter*, **10**, 101 (1971).
141. Analytical Methods Committee, *Analyst*, **92**, 403 (1967).
142. Analytical Methods Committee, *Analyst*, **101**, 62 (1976).
143. Friend, M. T., Smith, C. A. and Wishart, D. *Atomic Absorption Newsletter*, **16**, 46 (1977).
144. Agemian, H. and Chau, A. S. Y. *Anal. Chim. Acta*, **75**, 297 (1975).
145. Agemian, H. and Cheam, V. *Anal. Chim. Acta*, **101**, 193 (1978).
146. Kamps, L. R. and McMahon, B. *J. Ass. Off Anal. Chem.*, **18**, 351 (1970).
147. Longbottom, J. E., Dressman, R. C. and Lichtenberg, J. J. *J. Ass. Off Anal. Chem.*, **56**, 1297 (1973).
148. Uthe, J. F., Solomon, J. and Grift, B. *J. Ass. Off. Anal. Chem.*, **55**, 583 (1972).
149. Bye, R. and Paus, P. E. *Anal. Chim. Acta*, **107**, 169 (1979).
150. Takeshita, R., Akagi, H., Fujita, M. and Sakegami, Y. *J. Chromat.*, **51**, 283 (1970).
151. Itsuki, K. and Komoro, H. *Japan Analyst*, **19**, 1214 (1970).
152. Murakami, T. and Yoshinaga, T. *Japan Analyst*, **20**, 1145 (1971).
153. Murakami, T. and Yoshinaga, T. *Japan Analyst*, **20**, 878 (1971).
154. Carpenter, W. L. *NCASI Stream Improvement Tech. Bull. No. 263* (1972).
155. Goulden, P. D. and Afghan, B. K. *Technicon International Congress* Vol. II, Nov. 2-4, Futura, New York, p. 317 (1970).
156. Rosain, R. M. and Wai, C. M. *Anal. Chim. Acta*, **65**, 279 (1973).
157. Carr, R. A. and Wilkniss, P. E. *Environ. Sci. Technol.*, **7**, 63 (1973).
158. Gaston, G. N. and Lee, A. K. *J. Am. Water Works Ass.*, **66**, 495 (1974).
159. Kopp, J. F., Longbottom, M. C. and Lobring, L. B. *J. Am. Water Works Ass.*, **20**, 64 (1972).
160. Coyne, R. V. and Collins, J. A. *Anal. Chem.*, **44**, 1093 (1972).
161. Bothner, M. H. and Robertson, D. E. *Anal. Chem.*, **47**, 592 (1975).
162. Weiss, H. V. and Chew, K. *Anal. Chim. Acta*, **67**, 444 (1973).
163. Masri, M. S. and Friedman. *Environ. Sci. Technol.*, **7**, 951 (1973).
164. Newton, D. W. and Ellis, Jr., R. *J. Environ. Qual.*, **3** 20 (1974).
165. Jonasson, I. R., Lynch, J. J. and Trip, L. J. *Geol. Surv. Can. Paper*, 73-21 (1973).
166. Toribara, T. Y., Shields, C. P. and Koval, L. *Talanta*, **17**, 1025 (1970).
167. Shimomura, S. and Kise, A. *Bunseki Kagaku*, **18**, 1412 (1969).
168. Avotins, P. and Jenne, E. A. *J. Environ. Qual.*, **4**, 515 (1975).
169. Feldman, C. *Anal. Chem.*, **46**, 99 (1974).
170. Lo, J. M. and Wai, C. M. *Anal. Chem.*, **4**, 1869 (1975).
171. El-Awady, A. A., Miller, R. B. and Carter, M. J. *Anal. Chem.*, **48**, 110 (1976).
172. Christman, D. R. and Ingle, Jr., J. D. *Anal. Chim. Acta*, **86**, 53 (1976).
173. Heiden, R. W. and Aikens, D. A. *Anal. Chem.*, **49**, 668 (1977).
174. McFarland, R. C. *Radiochem. Radioanal. Lett.*, **16**, 47 (1973).
175. Coyne, R. V. and Collins, J. A. *Anal. Chem.*, **44**, 1093 (1972).
176. Weiss, H. V. and Chew, K. *Anal. Chim. Acta*, **67**, 444 (1973).
177. Rosain, R. M. and Wai, C. M. *Anal. Chim. Acta*, **65**, 279 (1973).
178. Feldman, C. *Anal. Chem.*, **46**, 99 (1974).
179. Carron, J. and Agemian, H. *Anal. Chim. Acta*, **92**, 61 (1977).
180. Stoeppler, M. and Matthes, M. *Progress in Water Technology*, **9** 389 (1978).
181. Carr, R. A. and Wilkniss, P. E. *Environ. Sci. Technol.*, **7**, 62 (1973).
182. Kopp, J. F. *Anal. Chem.*, **45**, 1789 (1973).

183. Evans, R. J. and Bandemer, S. L. *Anal. Chem.*, **26**, 595 (1954).
184. Caldwell, J. S., Lishka, R. L. and McFarren, E. F. *J. Am. Water Works Ass.*, **65**, 731 (1973).
185. Kolthoff, I. M. and Belcher, R. *Volumetric Analysis*, Vol. 3, Interscience Publishers, New York, pp. 511–513 (1967).
186. Dean, J. A. and Rues, R. E. *Anal. Lett.*, **2**, 105 (1969).
187. Edmunds, J. S. and Francesconi, K. A. *Anal. Chem.*, **48**, 2019 (1976).
188. Fishman, M. and Spencer, R. *Anal. Chem.*, **49**, 1599 (1977).
189. Agemian, H. and Cheam, V. *Anal. Chim. Acta*, **101**, 193 (1978).
190. Stringer, C. E. and Attrep, M. *Anal. Chem.*, **51**, 731 (1979).
191. Manning, D. C. *Atomic Absorption Newsletter*, **10**, 6 (1971).
192. Andreae, M. O. *Anal. Chem.*, **49**, 820 (1977).
193. Soderquist, C. J., Crosby, D. G. and Bowers, J. B. *Anal. Chem.*, **46**, 155 (1974).
194. Haywood, M. G. and Riley, J. P. *Anal. Chim. Acta*, **85**, 219 (1976).
195. Caldwell, J. S., Lishka, R. J. and McFarren, E. F. *J. Am. Water Works Ass.*, **65**, 731 (1973).
196. Sandhu, S. S. and Nelson, P. *Anal. Chem.*, **50**, 322 (1978).
197. Braman, R. S., Johnson, D. L., Craig, C., Foreback, C. C., Ammons, J. M. and Bricker, J. L. *Anal. Chem.*, **49**, 621 (1977).
198. Peoples, S. A., Lakso, J. and Lais, T. *Proc. West. Pharmacol. Soc.*, **14**, 178 (1971).
199. Haywood, M. G. and Riley, J. P. *Anal. Chim. Acta*, **85**, 219 (1976).
200. Kamada, T. *Talanta*, **23**, 835 (1976).
201. Sandhu, S. S. *Analyst (London)*, **101**, 856 (1976).
202. Fickett, A. W., Daughtrey, E. H. and Mushak, P. *Anal. Chim. Acta*, **79**, 93 (1975).
203. Lodmell, J. D., PhD Thesis, University of Tennessee, Knoxville, Tenn., (1973).
204. Johnson, L. D., Gerhart, K. O. and Aue, W. A. *Sci. Total Environ.*, **1**, 108 (1972).
205. Soderquist, C. J., Crosby, D. G. and Bowers, J. B. *Anal. Chem.*, **46**, 155 (1974).
206. Taimi, Y. and Bostik, D. T. *Anal. Chem.*, **47**, 2145 (1975).
207. Portman, J. E. and Riley, J. P. *Anal. Chim. Acta*, **31**, 509 (1964).
208. Casvalho, M. B. and Hercules, D. M. *Anal. Chem.*, **50**, 2030 (1978).
209. Yamamoto, M. *Soil Sci. Soc. Am. Proc.*, **39**, 859 (1975).
210. Dietz, E. A. and Perez, M. E. *Anal. Chem.*, **48**, 1088 (1976).
211. Elton, R. K. and Geiger, Jr., W. E. *Anal. Chem.*, **50**, 712 (1978).
212. Henry, F. T., Kirch, T. O. and Thorpe, T. M. *Anal. Chem.*, **51**, 215 (1979).
213. Myers, D. J. and Osteryoung, J. *Anal. Chem.*, **45**, 267 (1973).
214. Henry, F. T. and Thorpe, T. M. *Anal. Chem.*, **52**, 80 (1980).
215. Braman, R. S. and Foreback, C. C. *Science*, **182**, 1247 (1973).
216. Challenger, F. *Chem. Rev.*, **36**, 315 (1945).
217. McBride, B. C. and Wolfe, R. S. *Biochemistry*, **10**, 4312 (1971).
218. VonEndt, D. W., Kearny, P. C. and Kaufman, D. D. *J. Agric. Food Chem.*, **16**, 17 (1968).
219. Woolson, E. A. and Kearny, P. C. *Environ. Sci. Technol.*, **7**, 47 (1973).
220. Luijten, J. G. A. *Organic Chemistry* Institute T. N. O. Utrecht, Netherlands. Publication No. 417. Tin Research Institute, Greenford, U.K. 48pp (1970).
221. Braman, R. S. and Tompkins, M. A. *Anal. Chem.*, **51**, 12 (1979).
222. Nelson, H. D. Doctorial Dissertation, University of Utrecht. 29 (1967).
223. Meinema, H. A., Burger Wiersina, T., Verslins-Dehaan, G. and Geners, E. C. *Environ. Sci. Technol.*, **12**, 288 (1978).
224. Adcock, L. H. and Hope, W. G. *Analyst (London)*, **95**, 868 (1970).
225. Ashton, A., Fogg, A. G. and Thorburn Burns, D. *Analysis (London)*, **98**, 202 (1973).
226. Wyttenbach, A. and Bajo, S. *Anal. Chem.*, **47**, 2 (1975).
227. Anderson, H. H. *Inorg. Chem.*, **1**, 647 (1962).
228. Hodge, V. F., Seidel, S. L. and Goldberg, E. D. *Anal. Chem.*, **51**, 1256 (1979).
229. Woggon, H. and Jehle, D. *Die Nahrung*, **17**, 739 (1973).

230. Woggon, H. and Jehle, D. *Die Nahrung*, **19**, 271 (1975).
231. Neubert, G. and Wirth, H. O. *Z. Anal. Chem.*, **273**, 19 (1975).
232. Neubert, G. and Andreas, H. *Z. Anal. Chem.*, **280**, 31 (1976).
233. Soderquist, C. J. and Crosby, D. C. *Anal. Chem.*, **50**, 1435 (1978).
234. Blunden, S. J. and Chapman, A. H. *Analyst (London)*, **103**, 1266 (1978).
235. Luskina, B. M. and Syavtsillo, S. V. *Nov. Obl. Prom. Sauit, Khim* 186 (1969).
236. Smith, J. D. *Nature (London)*, **225**, 103 (1970).
237. Rivett, P. *J. Appl. Chem.*, **15**, 469 (1965).
238. Chromý, L. and Uhaez, Z. *J. Oil Colour Chem. Ass.*, **51**, 494 (1968).
239. Warchol, R. *J. Oil Colour Chem. Ass.*, **53**, 121 (1970).
240. McCallum, I. R. *J. Oil Colour Chem. Ass.*, **52**, 434 (1969).
241. Potter, H. R., Jarview, A. W. P. and Markell, R. N. *Water Pollution Control*, **76**, 123 (1977).
242. Imura, S., Fukutako, K., Aoki, H. and Sakai, T. *Japan Analyst*, **20**, 704 (1971).
243. Chau, Y. K., Wong, P. T. S., Beugeut, G. A. and Kramer, O. *Anal. Chem.*, **186**, 51 (1979).
244. Chau, Y. K., Wong, P. T. S. and Goulden, P. D. *Anal. Chim. Acta*, **421**, 85 (1976).
245. Sirota, G. R. and Uthe, J. F. *Anal. Chem.*, **49**, 823 (1977).
246. Chau, Y. K., Wong, P. T. S. and Saitoh, H. *J. Chromatogr. Sci.*, **162**, 14 (1976).
247. Segar, D. A. *Anal. Lett.*, **7**, 89 (1974).
248. Epstein, M. S. and O'Haver, T. C. *Spectrochim. Acta, Part B*, **30**, 135 (1975).
249. Braman, R. S. and Tompkins, M. A. *Anal. Chem.*, **50**, 1088 (1978).
250. Andreae, M. O., Asmodé, J. F., Foster, P. and Van't dack, L. *Anal. Chem.*, **53**, 1766 (1981).
251. Van Der Post, D. C. *Water Pollution Control*, **77**, 520 (1978).
252. Pellenberg, R. *Mar. Poll. Bull.*, **10**, 267 (1979).

Chapter 2
Oxygen Compounds

FATTY ACIDS

River water

There are two sources of fatty acids in river water, man-made contamination and naturally occurring. Regarding the latter the determination of free and bound fatty acids present in aquatic systems is important first because fatty acids are sufficiently diverse in structure that they can be used to determine the source and cycling of organic carbon[1-5] and second because fatty acids *in vivo* function primarily as structural components of membranes and energy storage products. In the reduced state the amount and type of fatty acids may be indicative of the trophic status of the ecosystem at the time of fatty acid formation.[2,6] Unsaturated, short chain ($\leqslant C_{20}$) and microbial fatty acid are indicative of productive systems, whereas long chain acids dominate in oligotrophic systems. Fatty acids have been shown to comprise 5-10% of the weight of humic/fulvic acid structure.[7] As an integral portion of the structure of these refractory materials, fatty acids can be used to determine the source of organic carbon and the physical/chemical characteristics of these materials. The behaviour of humic/fulvic acids may determine the transport of toxic trace metals and anthropogenic organics.

Gas chromatography has been used for the determination of acetic acid in industrial waste waters,[8] short chain C_1-C_4 fatty acids in anaerobic digester samples[9] and dilute aqueous solutions.[10] Van Huyssteen[9] completely separated normal and iso acids on glass columns (2.13 m × 3 mm i.d.) packed with 3% of FFAP on Chromosorb 101 (80-100 mesh) at 180 °C in an instrument equipped with a dual flame ionization detector; nitrogen was used as carrier gas (77 ml min^{-1}).

Bethge and Lindstroem[10] first removed metal cations from a 10 ml sample of water by elution with water from a Dowex 50W-X8 ion-exchange column and the eluate was titrated to pH 8 with standard tetrabutylammonium hydroxide. A calculated amount (as determined from the titration) of hexanoic acid was added as internal standard, the solution was concentrated to a syrup, the syrup was dissolved in acetone, and α-bromotoluene is added in slight excess. After 2 h to ensure complete reaction, 1 μl of the acetone solution was injected into a stainless steel column (2 m × 2 mm) packed with 3% of butane-1,4-diol succinate with nitrogen (30 ml min^{-1}), as carrier gas and flame ionization detection. The column was kept at 120 °C for 17 min, then temperature programmed to 150 °C at 2.5 °C min^{-1}. Down to 50 μM concentration of the acids could be determined by this procedure.

Other procedures for determining fatty acids in water, sediments, and biota involve liquid-liquid extraction, liquid-solid adsorption chromatography followed by gas-liquid chromatographic analysis.[11-13] Liquid extractions have been performed with methanol–chloroform,[2,14] methylene chloride,[15] and benzene-methanol.[16,17] Typical liquid-solid adsorbents are silicic acid.[2] Standard gas chromatographic separations for complex mixtures employ non-polar columns packed with OV-1, OV-17, OV-101, SE-30, or glass capillary columns containing similar phases.

Khomenko et al.[18] used thin-layer chromatography for determining non-volatile organic acids dissolved in natural water. The organic acids extracted from the water and concentrated are separated on a silica gel column into four groups[19] which are concentrated to 0.1–0.2 ml and thin-layer chromatography is carried out on layers of Silica gel KSK previously air-dried for 20 min and activated for 30 min at 105 °C. The acids in the first, second, and third fractions are developed in butanol–benzene-acetic acid (10:20:3, 15:85:2, and 15:35:8 respectively) and in the fourth fraction in ethyl acetate–water formic acid (9:1:1). After drying the chromatograms for 1.5 h at 120 °C, the organic acids are detected by spraying with 0.4% solution of bromocresol green in 20% ethanolic alkali and the spot areas are measured for a semi-quantitative determination of the acids.

Richard and Fritz[20] employed macroreticular XAD-4 resin aminated with trimethylamine for the concentration, isolation, and determination of acidic material from aqueous solutions. Acidic material is separated from other organic material by passing the water sample through a resin column in hydroxide form; other organic compounds are removed with methanol and diethylether. The acids are eluted with diethylether saturated with hydrogen chloride gas. After concentration, the eluate is treated with diazomethane and the esters formed are separated by gas chromatography.

Application of high performance liquid chromatography to the resolution of complex mixtures of fatty acids[21,22] has provided an alternative to the high temperature separation obtained by gas chromatography. Both techniques have similar limits of detection, but lack the ability to analyse directly environmental samples. Analysis requires that the fatty acids be separated from the organic and inorganic carbon matrices followed by concentration. Typically, these processes can be accomplished simultaneously by the appropriate choice of methods. Initial isolation of the fatty acids is based on the relative solubility of the material of interest in an organic phase compared to the aqueous phase. Secondary separation is determined by the functional group content and affinity for a solid support.

Hullett and Eisenreich[23] used high performance liquid chromatography for the determination of free and bound fatty acids in river water samples. The technique involves sequential liquid-liquid extraction of the water sample by 0.1 M hydrochloric acid, benzene–methanol (7:3), and hexane–ether (1:1). The resultant extract was concentrated and the fatty acids were separated as a class on Florisil using an ether–methanol 1:1 and 1:3 elution. Final determination of individual fatty acids was accomplished by forming the chromophoric phenacyl ester and separating by high performance liquid chromatography.[21] Bound fatty acids were released by base saponification or acid hydrolysis of a water sample from which the free fatty acids had been removed by solvent extraction.

These workers[23] defined free fatty acids as those acids which are readily extractable into organic solvents without sample pretreatment with strong acid or base. Bound fatty acids are those acids associated with mineral and humic/fulvic acid surfaces, esterified with mineral and humic/fulvic acid surfaces, esterified to humic/fulvic acids, or entrapped in natural macromolecules and released with strong acid or base.[24]

Method

Reagents

All solvents were analytical reagent grade and were redistilled in glass before use. Granular anhydrous sodium sulphate was extracted for 24 h with hexane–acetone (1:1) and dried at 140 °C overnight before use. Reagent grade water was obtained from a Milli-Q water purification system.

Florisil, 60–100 mesh was cleaned by sequential 24 h extractions with methanol, acetonitrile, and diethylether, oven-dried, activated at 600 °C for 4 h, and deactivated with 9% water (w/w). Solvents used to elute organic compounds from the Florisil column were equilibrated with water before use by allowing the dry solvent to pass through 30% (w/w) water-deactivated Florisil by gravity flow. The 9% Florisil column was dry-packed and conditioned by passing 50 ml of ether–methanol (1:1) through the column, followed by 50 ml of hexane. Glass fibre filters of 1.0 μm diameter were cleaned by extraction with acetone. Phenacyl bromide was recrystallized two times from pentane before use; BF_3–methanol (14%) was used.

Procedure

Samples were placed in 20 l glass carboys which were previously rinsed with hexane–ether (1:1) and three times with river water. Samples were immediately subjected to continuous flow centrifugation to remove suspended matter. Sodium azide was added to the centrifuged samples at a final concentration of 1 mg l^{-1} to prevent bacterial growth. Centrifugation and subsequent storage at 4 °C was carried out within 6 h of sampling. A flow diagram for extraction,

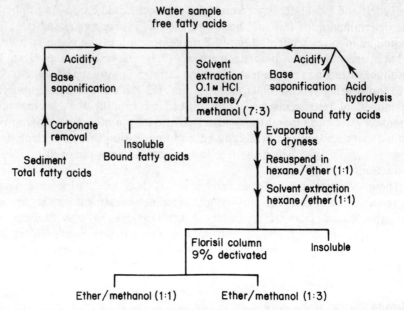

Figure 35 Flow diagram for the isolation and concentration scheme. Reprinted with permission from Hullett and Eisenreich[23]. Copyright (1979) American Chemical Society.

concentration, and isolation of fatty acids from river water is shown in Figure 35. River water (1500 ml) was extracted three times with 100 ml of 70:30 benzene:methanol, made 0.1M with respect to hydrochloric acid. The organic layer from each extraction was pooled; the glassware was rinsed with hexane-ether (1:1) and added to the pooled extractions. The sample was then dried over anhydrous sodium sulphate, decanted, and brought to dryness by rotary evaporation at room temperature. The extracted matter was resuspended in 50 ml of hexane-ether (1:1) to which 50 ml of Milli-Q (Millipore) was added. The organic layer was separated and the aqueous layer re-extracted twice with 50 ml of hexane-ether (1:1). The pooled organic layers were dried over anhydrous sodium sulphate and the volume reduced to 0.5–1.0 ml by rotary evaporation at room temperature. The sample was then transferred to the top of a preconditioned Florisil column. The Florisil column (1.5 cm i.d. × 30 cm) was dry packed first with 18.0 g of anhydrous sodium sulphate followed by 7.0 g of 9% deactivated Florisil separated by glass wool plug. Elution from the Florisil column was accomplished with an eluotrophic solvent series consisting of 5–50 ml fractions (∼2.4 bed volumes) of hexane-ether (1:3) ether-methanol (9:1), ether-methanol (3:1), ether-methanol (1:1), and ether-methanol (1:3). The fractions were dried over anhydrous sodium sulphate and filtered through 1.0 μm glass fibre filters to remove any particulate matter. Each fraction was then evaporated to approximately 0.5 ml via rotary evaporation, which was accomplished at elevated temperatures in the case of the last three fractions of the series. The samples were then transferred to sample vials with acetone and

Table 36 Retention index for fatty acid separation by h.p.l.c. and g.l.c.*†

	H.p.l.c.				G.l.c.	
Fatty acid	Relative‡ retention	Time	Relative§ retention	Time	Relative retention	Time
10.0	0.78	1830	0.78	1210	0.26	212
12.0	0.87	2034	0.87	1332	0.49	371
14.0	0.92	2154	0.94	1434	0.74	565
16.0	1.00	2334	1.00	1524	1.00	760
18.0	1.10	2562	1.07	1638	1.25	950
20.0	1.25	2922	1.16	1768	1.49	1133
22.0	1.46	3402	1.27	1966	1.70	1289
24.0	1.77	4122	1.47	2244	1.99	1514
16.1	0.92	2158	0.24	1434	0.97	734
18.1	1.00	2340	1.00	1524	1.22	924
18.2	0.94	2190	0.96	1452	1.19	906
20.4	0.91	2130	0.93	1422	1.39	1056
24.1	1.42	3318	1.26	1922	1.94	1472
$aC_{13.0}$	0.89	2082	0.92	1368	0.59	451
$aC_{15.0}$	0.94	2220	0.96	1470	0.85	643
$aC_{17.0}$	1.02	2382	1.04	1578	—	—
$aC_{19.0}$	1.15	2682	1.11	1687	1.34	1020
$aC_{21.0}$	1.30	3942	1.20	1834	1.56	1198
$iC_{12.0}$	0.86	1998	0.86	1334	0.46	348
$iC_{16.0}$	0.99	2304	0.99	1512	0.95	726
$iC_{18.0}$	1.08	2526	1.06	1620	1.20	912
$iC_{22.0}$	1.42	3318	1.27	1934	1.66	1260
13.0	0.90	2106	0.91	1392	0.61	466
15.0	0.96	2244	0.98	1488	0.86	654
17.0	1.04	2436	1.04	1584	1.12	874
19.0	1.17	2742	1.11	1692	1.36	1032

*All time values are given in seconds.
†Relative index is calculated relative to $C_{14.0}$.
‡Two μBondapak C_{18} columns in series.
§Single μBondapak C_{16} column.
Reprinted with permission from Hullett and Eisenreich.[23] Copyright (1979) American Chemical Society.

the volume was adjusted to 3.0 ml. At this point 100 μl of sample was removed and placed in another vial for high performance liquid chromatography and gas chromatographic analysis without formation of the phenacyl esters. The remaining sample was split for derivatization.

High Performance Liquid Chromatography

Each of the five fractions, consisting of a total volume of 1.45 ml, was derivatized to the phenacyl ester according to the procedure of Borch.[21]

High performance liquid chromatography analysis was performed with a Waters Associates, Inc. liquid chromatograph, Model 202, with an absorbance detector set at 254 nm. The solvent programme was a 30 min linear gradient (curve 6) ranging from 40% acetonitrile–60% water to 100% acetonitrile followed by 30 min of isocratic operation at 100% acetonitrile. A flow rate of 1 ml min^{-1} was employed which gave operating pressures of about 900 psi.

Peak identification was based upon retention data generated with analytical standards under identical conditions (Table 36). Confirmation of peak identification was obtained by mass spectral fragmentation patterns using direct inlet injection of collected peaks on a LKB Model 9000 mass spectrometer combined with an on-line computer.

Gas chromatography

Samples for gas chromatographic analysis were methylated according to the procedure of Metcalfe and Schmitz.[25] Analyses were performed on a gas chromatograph equipped with a flame ionization detector. The separation was performed on a stainless steel column (3.2 mm i.d. × 3 m) packed with 5% OV-101 on Chromosorb W. The fatty acids were eluted with a temperature programme from 150 to 280 °C at a rate of 6 °C min^{-1}.

Base saponification

Methanol (500 ml) and 72.0 g of sodium hydroxide (1.75 M) were added to 500 ml of extracted river water. Saponification was allowed to proceed for 16 h at 70 °C. After complete reaction, 300 ml of Milli-Q water and 200 ml of concentrated hydrochloric acid were added to the sample (final volume 1500 ml and 0.4 M hydrochloric acid) and cooled at room temperature. Methanol was then removed by rotary evaporation at 30 °C to facilitate the formation of an organic phase in the 0.1 M hydrochloric acid, benzene-methanol (70:30) extraction. The resulting sample was treated as shown in the analytical isolation and concentration scheme (Figure 35). Bound fatty acids were determined via high performance liquid chromatographic analysis using the procedures described above.

Acid hydrolysis

Hydrolysis (6 M hydrochloric acid) was carried out on 500 ml of extracted river water by adding 500 ml of concentrated hydrochloric acid to the sample, sealing the sample under vacuum and heating for 20 h at 70 °C. The resulting sample was isolated and concentrated as shown in Figure 35.

Table 36 gives the net retention times and relative retention of analytical standards used to characterize the high performance liquid chromatographic and gas-liquid chromatographic behaviour. The values indicate the baseline resolution is obtained from a homologous series of fatty acids. However, when a complex mixture containing branched and unsaturated fatty acids was chromatographed, identification and quantification became more difficult. The addition of methyl branching decreased the retention relative to the straight chain saturated homologue. Alkyl substitutions introduce dipole moments which interact with the permanent dipole moments of the mobile phase, allowing greater solubility of the solute. Unsaturation decreases the retention of the solute by allowing specific interactions to occur between the polar mobile phase and the π electrons. This increases the solubility of the fatty acids in the mobile phase while decreasing the solubility in the non-polar stationery phase, resulting in a net decrease in retention of the solute by the stationary phase. In comparing high performance liquid chromatography and gas chromatographic separation, the greater resolution of unsaturated and branched chain fatty acids obtained by high performance liquid chromatography is important in the determination of the source of organic carbon. These fatty acids were separated by high performance liquid chromatography (Figure 36(a), peaks 4,7,8;$C_{18.2}$, $C_{18.1}$, and $iC_{18.0}$, respectively), whereas under the conditions used for gas-liquid chromatography, resolution of these acids was poorer (Figure 36, peaks 4,7,8; $C_{18.2}$, $C_{18.1}$ and $iC_{18.0}$). This is a major experimental problem since branched acids are indicative of bacterial sources while polyunsaturated materials are usually found in the protists. Differentiation of $C_{18.1}$ and $C_{18.2}$ is of interest when determining specific sources, such as diatoms, which contain no significant amounts of $C_{18.2}$ but do contain $C_{18.1}$. Another problem encountered with separations by gas-liquid chromatography as compared to high performance liquid chromatography was the greater variability in gas-liquid chromatography retention times, necessitating the use of secondary identification tools such as mass spectrometry.

Figure 36 shows the resolution obtainable for a complex mixture containing odd and even saturated and unsaturated fatty acids by high performance liquid chromatography and gas-liquid chromatography. In neither case is baseline resolution obtained for all peaks; however, the high performance liquid chromatography appears to better resolve the compounds of particular interest. In most cases adequate resolution is attained for qualitative identification of individual fatty acids; however, neither procedure was entirely adequate for quantification.

Hullett and Eisenreich[23] evaluated their fatty acid isolation and concentration procedure using Milli-Q water spiked with acetone solutions of C_{12}, C_{16} and C_{18} fatty acids (400 mg of each). The sample was carried through the analytical fractionation scheme (Figure 35) and the fatty acids identified and quantitated from the ether–methanol (1:1) and ether–methanol (1:3) fractions.

It was found that the amount recovered was dependent on the method used to clean and activate the Florisil, and the preferred method is described above. A precision of ± 7.8 µg was obtained on river water samples spiked with C_{17}–C_{19} fatty acids. Total recoveries for the entire procedure were 68% of the fatty acid applied and were found to be independent of the concentration of organic carbon present in the river water samples since identical recoveries were obtained from Milli-Q water and environmental samples. The recovery was also shown to be independent of fatty acid concentration and chain lengths. Fatty acids

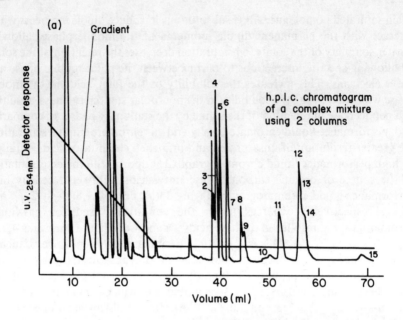
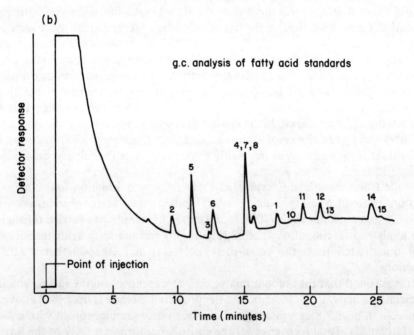

Figure 36 H.p.l.c. (a) and g.l.c. (b) chromatograms of a complex mixture of standard fatty acids. Flow rate 1.0 ml min^{-1} 0.05 AUFS. Peak identification: (1) $C_{20.4}$; (2) $C_{14.0}$; (3) $C_{16.1}$; (4) $C_{18.2}$; (5) $C_{15.1}$; (6) $C_{16.0}$; (7) $C_{18.1}$; (9) $C_{18.0}$; (10) $C_{20.0}$; (11) $aC_{21.0}$; (12) $iC_{22.0}$; (13) $iC_{22.0}$; (14) $C_{24.1}$; (15) $C_{24.0}$. Reprinted with permission from Hullett and Eisenreich[23].
Copyright (1979) American Chemical Society

Table 37 Recovery of standard fatty acids*

Fatty acid	Added	Found	%	Added	Found	%	Added	Found	%	Added	Found	%
$C_{12.0}$	400	272.1	68	400	249.7	62						
$C_{16.0}$	400	274.0	69	400	280.3	70	400	287.1	72			
$C_{18.0}$	400	357.2	89	400	261.9	66	400	285.4	71			
$iC_{18.0}$										142	68	48
$C_{20.0}$										11.2	7.6	68
$C_{22.0}$										20.7	12.4	60
$C_{22.0}$										220	132	60
$C_{24.0}$										38.1	23.8	63

*All values are given as µg.
Reprinted with permission from Hullett and Eisenreich[23]. Copyright (1979) American Chemical Society.

ranging from $C_{12.0}$ to $C_{24.0}$ were used to spike a 1 l sample of Milli-Q water in which the range of concentration varied from 20.7 to 400 µg L^{-1} (Table 37). Hullett and Eisenreich[23] concluded that the low recoveries were due to losses occurring at the liquid–liquid extraction step of the analyses.

Figure 37 shows high performance liquid chromatograms obtained for the two Florisil fractions (ether–methanol (1:1) and (1.3)) of a river water sample analysed for free fatty acids. The chromatograms show good separation of the fatty acid material from other compounds. The major fatty acids (Figure 38) determined by high performance liquid chromatography were $C_{16.0} > C_{18.2} > C_{14.0} > aC_{17.0} > C_{18.0}$ while the major fatty acids determined by gas-liquid chromatography were $C_{16.0} > C_{16.1} > C_{18.0} > C_{17.0} > aC_{17.0}$. Microbial fatty acids, which include the branched and odd chain length acids, were a significant proportion of the total fatty acids (19.3% by gas-liquid chromatography and 16.9% by high performance liquid chromatography). Underivatized samples were analysed by high performance liquid chromatography to determine first the ability of extracted material to absorb at 254 nm, indicating the aromatic or conjugated character of the sample and second to determine the presence of solutes which may coelute from Florisil and high performance liquid chromatography with fatty acids and contribute to these peaks. No such materials were found indicating that all identified peaks were due to derivatization only and contained acid functional groups.

Direct comparison between the high performance liquid chromatography and gas-liquid chromatography for individual fatty acids revealed a number of differences (Table 38). Greater detection and subsequent qualitative identification of a larger number of minor fatty acids is possible with high performance liquid chromatography, permitting characterization of the organic matter in the system. In some cases a wide discrepancy existed between the amount of material determined by high performance liquid chromatography and the amount determined by gas-liquid chromatography. However, application of the carbon distance index $(DI)_{26}$ which quantifies the resemblance of the two

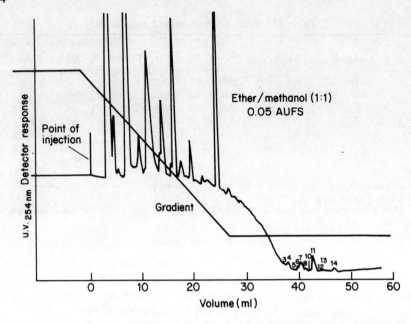

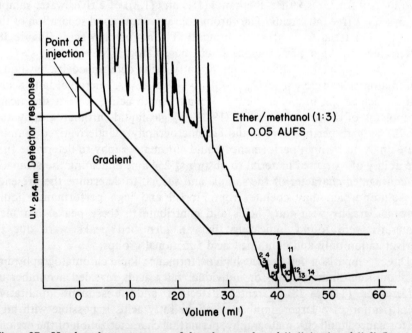

Figure 37 H.p.l.c. chromatograms of the ether–methanol (1:1) (top) and (1:3) (bottom) fractions. Flow rate 1.0 ml min^{-1} attenuation 0.05 AUFS. Peak identification: (1) phenacyl bromide; (2) iC$_{12.0}$; (3) C$_{12.0}$; (3) C$_{12.0}$; (4) aC$_{13.0}$; (5) C$_{20.4}$; (6) C$_{14.0}$; (7) C$_{18.2}$; (8) aC$_{15.0}$; (10) iC$_{16.0}$; (11) C$_{16.0}$; (12) aC$_{17.0}$; (13) C$_{17.0}$; (14) C$_{18.0}$. Reprinted with permission from Hullett and Eisenreich[23]. Copyright (1979) American Chemical Society

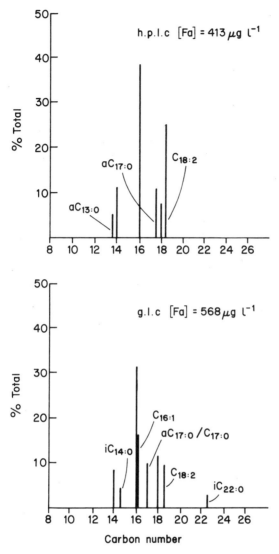

Figure 38 Fatty acid composition in river water determined by H.p.l.c. Reprinted with permission from Hullett and Eisenreich[23] Copyright (1979) American Chemical Society

populations, shows that the samples are similar (DI = 25.1). Both techniques tend to identify the same major fatty acids but differ in their relative abundances. The systems showed an obvious lack of long chain ($>C_{20}$) fatty acids in the free fraction, indicating that allochthonous inputs to the labile organic carbon fraction are minor with the majority derived from autochthonous sources. The fatty acids were generally equally distributed between the ether–methanol (1:1) and (1:3) fractions, with the longer non-polar fatty acids showing a tendency to appear in the less polar ether–methanol (1:1) fraction

Table 38 Fatty acid composition of Mississippi River water by h.p.l.c. and g.c.

Fatty acid	H.p.l.c. $\mu g\ l^{-1}$	H.p.l.c. % total	G.c. $\mu g\ l^{-1}$	G.c. % total
10	tr	tr		
12	tr	tr		
14	47	11	49	9
16	161	39	180	34
18	31	8	66	12
20				
22				
24				
16.1			77	14
18.1				
18.2	104	25	54	10
20.4	tr	tr		
24.1				
iC_{12}	tr	tr		
iC_{14}			26	5
iC_{16}	tr	tr		
iC_{18}				
iC_{22}			17	3
aC_{13}	23	6		
aC_{15}	tr	tr		
aC_{17}	47	11	58*	11
aC_{19}				
aC_{21}				
13				
15				
17	tr	tr		
19				
Total	413		527	
% DOC (dissolved organic carbon)	3.3		4.6	

*$aC_{17.0} + C_{17.0}$.
Reprinted with permission from Hullett and Eisenreich[23]. Copyright (1979) American Chemical Society.

while unsaturated acids were equally distributed between the two fractions.

Mass spectral identification of the major peaks was performed and confirmed the results indicated by the retention index. The primary fragments (Figure 39) used for structural determination were the M$^+$ ion, the molecular weight of the derivative; M-119, the molecular weight of the fatty acid, and M-135 the molecular weight of the derivative minus the phenacyl moiety. Secondary mass spectrometric identification of structure was based on the ability to follow the methylene fragmentation pattern to C_{n-1}. Double bond positions were determined by shifts in the methylene fragmentation

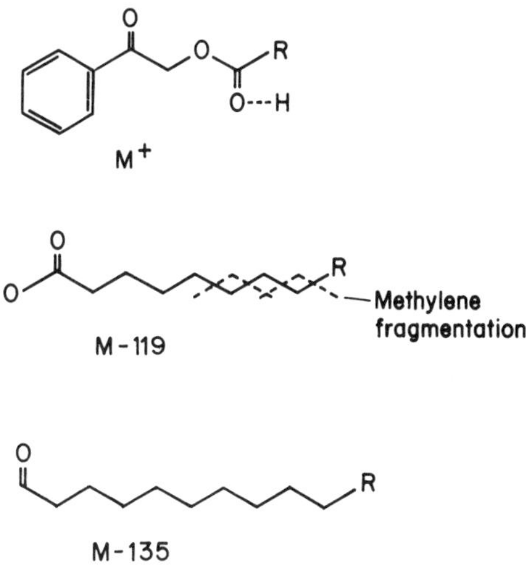

Figure 39 Mass spectral fragmentation patterns. Reprinted with permission from Hillett and Eisenreich[23]. Copyright (1979) American Chemical Society

pattern. The baseline hump present in the chromatograms is thought to be due to a larger number of unresolved solutes.

Results of saponification and 6 M hydrochloric acid hydrolysis procedures indicated that bound fatty acid exceeded the free fatty acid by a factor of approximately 3. Differences in the relative abundance of individual fatty acids were seen. The distribution indicated the occurrence of maturation of the fatty acids. As with the free fatty acids, no long chain ($>C_{20}$) fatty acids were found, suggesting that the source of organic carbon for refractory matter of the water column is derived primarily from autochthonous sources. Microbial fatty acids constituted a greater fraction of the fatty acids present and indicated the importance of microbes in the formation of natural organic matter in water. Saponification procedures allowed extraction of the colour into the organic phase which did not occur in the liquid-liquid extraction of free fatty acids or water subjected to 6 M-hydrochloric acid hydrolysis. This implies the presence of a number of linkages in the backbone structure of coloured humic/fulvic materials which are susceptible to base saponification but not to acid hydrolysis.

Fatty acids in sea water

Gorcharova and Khomenko[27] have described a column chromatographic method for the determination of acetic, propionic, and butyric acids in sea water and thin-layer chromatographic methods for determining lactic, aconitic, malonic, oxalic, tartaric, citric, and malic acids. The pH of the sample is adjusted to 8–9 with sodium hydroxide solution. It is then evaporated almost to dryness at 50–60 °C and the residue washed on a filter paper with water acidified with hydrochloric acid. The pH of the resulting solution is adjusted to 2–3 with

hydrochloric acid (1:1), the organic acids are extracted into butanol, then backextracted into sodium hydroxide solution; this solution is concentrated to 0.5-0.7 ml, acidified and the acids separated on a chromatographic column.
Treguer et al.[28] determined total dissolved free fatty acids in sea water. The sample (1 litre) was shaken with chloroform (2 × 20 ml) to remove the free fatty acids and the extract evaporated to dryness under reduced pressure at 50 °C. Chloroform-heptane (29:21) (2 ml) and fresh copper reagent (M triethanolamine-M acetic acetic acid-6.8% $CuSo_4 5H_2O$ solution (9:1:10) (0.5 ml) was added to the residue. The solution was shaken vigorously for 3 min and centrifuged at 3000 rpm for 5 min. A portion (1.6 ml) of the organic phase was evaporated to dryness and 1% ammonium diethyldithiocarbamate solution in isobutylmethyl ketone (2 ml) was added to the residue to form a yellow copper complex. The copper in the solution was determined by atomic absorption spectrophotometry at 324.8 nm (air-acetylene flame). Palmitic acid was used to prepare a calibration graph. The standard deviation for samples containing 30 μg 1^{-1} of free fatty acids (as palmitic acid) was ±1 μg per 1^{-1}.

Quinn and Meyers[29] discuss a gas-liquid chromatographic method for the determination of dissolved organic acids in sea water.

Fatty acids in sewage

Methods have been described for the determination of total fatty acids in raw sewage sludge. These methods[30-32] require a concentration steps such as simple distillation, steam distillation, evaporation, or extraction[33-35] which resulted in great losses of the volatile matter.[36,37]

Straight distillation or steam distillation of volatile acids and the chromatographic separation have been proposed in Standard Methods[38] for the organic acids in sludge. In this method an acidified aqueous sample, containing relatively high concentrations of organic acids, is adsorbed on a column of silicic acid and the acids are eluted with n-butabol in chloroform. The eluate is collected and titrated with standard base. All short chain (1-6 carbon) organic acids are eluted, but so are crotonic, adipic, pyruvic, phthalic, fumaric, lactic, succinic, malonic, aconitic, and oxalic acids, as well as alkyl sulphates and alkyl-aryl sulphonates. No information on the individual volatile acids is obtained by this method and the results are reported collectively as total organic acids. Various chromatographic methods, such as paper[39,40] and gel chromatography,[32] have been used for the analysis of sludge digester liquor. Mueller et al.[37] have modified the indirect chromatographic method for samples of raw sewage and river water involving tedious concentration steps, leading to losses. In paper chromatography individual volatile acid concentrations should be higher than 600 mg 1^{-1}, while in other methods the minimum detectable level is 1000 mg 1^{-1}.[39,40]

Gas chromatography is a very attractive possibility for volatile acids determination since it makes the separation of the individual acids for qualitative and quantitative determination *in situ* possible. In practice, many difficulties in analysing volatile acids in aqueous systems, resulting mainly from the presence of water, have been reported.[41] The volatile acids' high polarity as well as their tendency to associate and to be adsorbed firmly on the column require esterification prior to gas chromatographic determination. The presence of water interferes in esterification so that complex drying techniques and isolation of the acids by extraction, liquid-solid chromatography, distillation, and even ion exchangers had to be used.[42-45]

The introduction of the more sensitive hydrogen flame ionization detector has made possible the analysis of dilute aqueous solutions of organic acids by gas-liquid chromatography. Problems, such as 'ghosting' at high acid concentrations and an excessive tailing effect of the water in dilute solutions, masking the components, have been reported for aqueous solutions.[46] Subsequently phosphoric[47] or metaphosphoric acids[48] were added to the liquid phase, resulting in more reproducible column performance and reduced 'ghosting'. Addition of formic acid to the carrier gas was recommended by Cochrane[49] to overcome all the problems normally associated with analysing free fatty acids by gas chromatography.

Baker[50] used FFAP column for direct injection of dilute aqueous solutions of acids (FFAP, a reaction product of polyethylene glycol 20,000 and 2-nitrophthalic acid developed by Varian Aerograph). The acetic acid peak was not clear and the ability of this column to separate normal and iso fatty acids was not reported. Van Huyssteen[41] successfully used a Chromosorb 101 column coated with 3% FFAP for separation of volatile acids by direct injection of synthetic aqueous solutions and anaerobic digesters samples, which were first centrifuged and acidified with hydrochloric acid. He injected 1 μl at acid concentration greater than 50 mg l^{-1} and 2 μl below 50 mg l^{-1}. Ghosting was observed upon injecting 2 μl 25 mg l^{-1} C_2-C_6 acid solutions. Van Huyssteen did not try to inject volumes greater than 2 μl. His column affected complete separation of the C_2-C_6 straight and branched short chain fatty acids from synthetic aqueous solutions, but less sharpened peaks were obtained from anaerobic digester samples. The response with acetic acid approximated that of the other acids.

Hindin[36] made sewerage samples alkaline with sodium hydroxide prior to evaporation of water for at least ten-fold concentration; 20% metaphosphoric acid was then added and the samples centrifuged. Hindin[36] used a Carbowax 20 M plus phosphoric acid on Chromosorb W column.

The Van Huyssteen[41] procedure is described below. An isothermal oven temperature of 180 °C was used. The inlet temperature was 240 °C, the inlet line 260 °C, the detector line 280 °C, and the detector 330 °C. The columns were conditioned for 48 h or more at 200 °C with a flow of carrier gas. During this period, about 40 injections of 10 μl distilled water and 10 mg l^{-1} C_2-C_6 fatty acid solutions were made.

Chromatograms of low (2 μl of 25 mg l^{-1} of each acid) and high (1 μl of 750 mg l^{-1} of each acid) concentrations of the synthetic acid mixtures are presented in Figures 40 and 41 respectively. These chromatograms illustrate the complete separation of the acids, even at low concentrations.

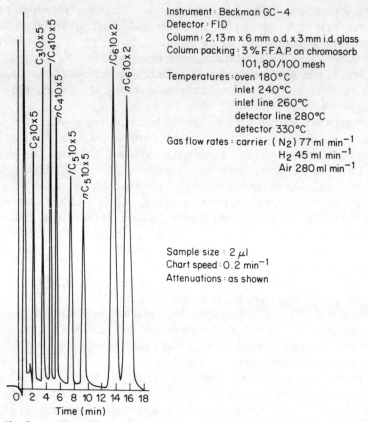

Figure 40 Separation of low concentrations of C_2-C_6 acids in aqueous solution at a concentration of 25 mg l^{-1} each. Reprinted with permission from Van Huyssteen.[41] Copyright (1970) Pergamon Press

Anaerobic digester samples were centrifuged at 105,000 g for 20 min at 2 °C, about 5 ml of the clear supernatant liquid decanted and acidified to pH 1-2 with hydrochloric acid. Figures 42 and 43 represent the chromatograms obtained from a balanced and unbalanced digester, respectively. The acids were identified by their retention times. For the low level fatty acid content, high sensitivity was used and additional peaks, probably alcohol, appeared between the acid peaks. To eliminate ghosting, 10 μl distilled water was injected between samples. Ghosting was observed by injecting 2 μl distilled water after 2 μl 25 mg l^{-1} C_2-C_6 solution or 1 μl distilled water after 1 μl 750 mg l^{-1} C_2-C_6 solution.

Peak height, measured from the point of rise of the leading edge to the peak, was taken as a measure of response. At very low concentrations of the acids in digester samples, peak heights were measured from the peak to the point of fall of the tail edge, where the rise of the leading edge apparently occurred before the previous peak had fully emerged (cf. C_3, i- and nC_4, i- and nC_5, i- and nC_6, Figure 42.

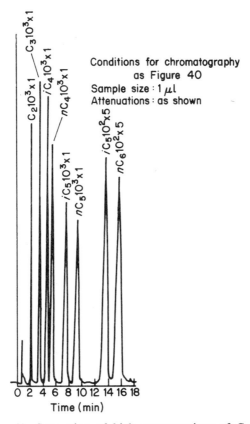

Figure 41 Separation of high concentrations of C_2–C_6 acids in aqueous solution at a concentration of 750 ml l^{-1} each. Reprinted with permission from Van Huyssteen.[41] Copyright (1970) Pergamon Press

Narkis and Henfeld-Furie have described a direct method for the identification and determination of volatile water-soluble C_1–C_5 acids in municipal waste water and raw sewage. The method involves direct injection of the sewage into a gas chromatograph equipped with a Carbowax 20M on acid-washed Chromosorb W column and a flame ionization detector. Preliminary preparation of the sample is limited to the addition of solid metaphosphoric acid to the sewage and removal of precipitated proteins and suspended solids by centrifuging.

Method

Gas chromatography

A gas chromatograph equipped with a hydrogen flame ionization detector was used. The following operational conditions were used: a glass column, 8ft long

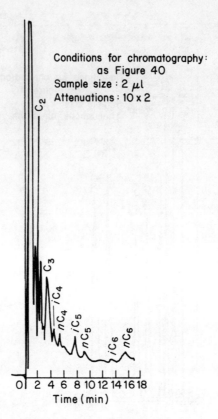

Figure 42 Separation of short chain fatty acids (C_2–C_6) in a sample from a balanced digester. Reprinted with permission from Van Huyssteen.[41] Copyright (1970) Pergamon Press

and ⅛ in. in diameter, packed with 60–80 mesh acid-washed Chromosorb W coated with 20% Carbowax 20M and 3% Phosphoric acid; nitrogen carrier gas at 55–60 ml min^{-1} flow rate, 350–400 ml min^{-1} air and 30–60 ml min^{-1} hydrogen gas. The air was passed through a silica gel column for drying and dust removal. Column temperature was 100–140 °C and gas temperature was 140–180 °C at the entrance to the column and 200–240 °C at the exit. Detector temperature was 180–210 °C. Sensitivity was kept constant at 3×10^{-11}.

Raw sewage samples were prepared for analyses by either of the procedures described below.

Evaporation to dryness and redissolution

The pH of 600 ml raw sewage was raised to 11 with sodium hydroxide. The sample was then centrifuged at 8000 rpm for 20 min and 500 ml of the supernatant were evaporated to dryness on steam bath. The residue was

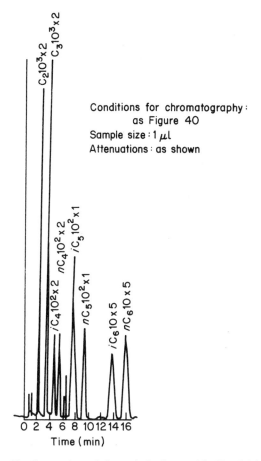

Figure 43 Separation of short chain fatty acids (C_2–C_6) in a sample from an unbalanced digester. Reprinted with permission from Van Huyssteen.[41] Copyright (1970) Pergamon Press

redissolved in 30% metaphosphoric acid in a 50 ml volumetric flask, adjusting pH to 2–3. The precipitated proteins were then removed by centrifuging at 14,000 rpm for 20 min. Due to heat evolution, losses of volatile acids can occur in this procedure (see below). The supernatant was then injected directly into a gas chromatograph. Total organic acids were determined by chromatographic separation followed by titration according to the method described by the American Public Health Association.[38]

The following procedure is recommended by Narkis and Henfield-Furie[51] as being superior to the method described above. Practically no losses of volatile acids occur during handling. To 10 ml of raw sewage, in a volumetric flask, less than 1 g of solid metaphosphoric acid was added to lower the pH to 2–3 without changing the volume of the solution. Suspended solids and precipitated proteins were then removed by centrifugation

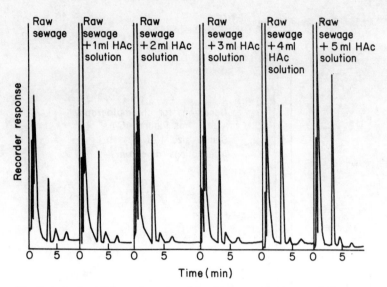

Figure 44 Standard addition of acetic acid solution to Haifa municiple raw sewage. Reprinted with permission from Narkis and Henfield-Furie.[51] Copyright (1978) Pergamon Press

at 14,000 rpm for 20 min at 5 °C. The supernatant was used for direct injection into a gas chromatograph.

Narkis and Henfeld-Furie[51] found that acid-washed Chromosorb W coated with 20% Carbowax 20M plus 3% phosphoric acid column successfully resolved both normal and iso volatile acids, upon direct injection of raw sewage and standard aqueous solutions pretreated only by acidification with metaphosphoric acid and centrifugation, avoiding any other preparations steps. An unidentified peak appeared immediately after injection in all the raw sewage and chemically treated raw sewage chromatograms. The various volatile acids appeared in the following order: acetic, propionic, isobutyric, butyric, isovaleric, valeric, and hexanoic. Formic acid is not detected by flame ionization. Excellent recoveries were obtained by this procedure when raw sewage spiked with increasing amounts of acetic acid to hexanoic acid were put through the procedure. Figure 44 shows the chromatograms obtained by standard addition of acetic acid to raw sewage sample. This figure shows the linear growth of area under the peak with increasing concentrations of acetic acids in a raw sewage sample. The curve passes through the point representing the original acid concentration in raw sewage. The standard deviation was ±0.17 for acetic acid, ±0.20 for propionic acid, ±0.18 for butyric acid, ±0.20 for isovaleric acid and ±0.13 for valeric acid.

Table 39 summarizes the individual volatile acid concentrations in raw sewage determined by the direct injection procedure of Narkis and Henfield-Furie[51] and by that of Standard Methods.[38] The results were also expressed as acetic acid for comparison with the collective total amount of organic acids determined

Table 39 Volatile acids concentrations in raw sewage and in lime-treated raw sewage determined by the direct injection procedure and by that of Standard Methods[38]

Exp. no.	Date	Source of raw sewage	Treatment	Lime dose (mg l⁻¹)	pH	Soluble COD (mg l⁻¹)	Acetic acid as HA_c (mg l⁻¹)	Gas-liquid chromatography							
								Priopionic acid (mg l⁻¹)	as HA_c (mg l⁻¹)	Butyric acid (mg l⁻¹)	as HA_c (mg l⁻¹)	Isovaleric acid (mg l⁻¹)	as HA_c (mg l⁻¹)	Valeric acid (mg l⁻¹)	as HA_c (mg l⁻¹)
5	16.8.73	NS*	Ca(OH)₂	1000	11.3	260	31.05	9.09	7.36	0	0	0	0	0	0
7	20.9.73	NS	Ca(OH)₂	900	11.0	168.5	29.4	4.28	3.48	0.82	0.56	0.44	0.26	0	0
8	24.9.73	NS	Raw	0	—	—	38.91	7.21	5.78	11.9	0.81	0.95	0.56	0	0
8	24.9.73	NS	Ca(OH)₂	950	11.0	189	35.4	5.72	4.58	1.19	0.81	0.50	0.29	0	0
9	3.10.73	HMSTP†	Raw	0	—	696	121.4	33.01	26.76	6.6	4.49	2.35	1.39	0.56	0.33
9	3.10.73	HMSTP	Ca(OH)₂	950	11.2	620	104.5	27.1	22.05	5.32	3.63	1.24	0.73	0.50	0.29
10	4.10.73	NS	Raw	0	—	236.1	36.5	6.84	5.54	1.51	1.03	0.63	0.37	0.22	0.13
10	4.10.73	NS	Ca(OH)₂	1000	11.3	174.3	31.2	6.0	4.86	1.43	0.97	0.32	1.89	0.198	0.114

Exp. no.	Date	Total volatile acids (mg l⁻¹) as HA₂	Standard methods analysis total organic acids (mg l⁻¹) as HA_c	Comparison of g.l.c. and Standard Method Difference (mg l⁻¹) as HA_c	% of volatile acids
5	16.8.73	38.41	45.2	6.79	85
7	20.9.73	33.70	48	14.3	70
8	24.9.73	46.06	50.4	4.34	91.4
8	24.9.73	41.08	48.3	7.21	85
9	3.10.73	154.36	158.4	4.04	97.5
9	3.10.73	131.2	142.5	11.3	92
10	4.10.73	43.57	57.6	14.03	75.6
10	4.10.73	37.33	45.2	3.96	83

*NS = raw sewage.
†HMSTP = municipal sewage treatment plant raw sewage.
Reprinted with permission from Narkis and Henfield-Furie[51]. Copyright (1978) Pergamon Press.

by the Standard Method.[38] The total amount of organic acids determined according to the Standard Method is higher than that found by the Narkis and Henfield-Furie[51] method. On average between 85 and 98% of the organic acids determined by the Standard Methods procedure were found to be volatile acids by the direct injection method.

Fatty acids in trade effluents

McKaveney and Byrnes[52] have described an apparatus incorporating semiconductor electrodes which they have used for the measurement of the concentration of organic acids in picking bath effluents. Genkin and Zel'manova[53] determined volatile fatty acids and their salts in effluents from aniline dye manufacture. To determine acids the sample was mixed with an equal volume of acetone and titrated potentiometrically (glass electrode and SCE). If mineral acids are absent the inflection in the curve corresponds to the organic acids, but if mineral acids are present this inflection is preceded by one corresponding to these acids. To determine the salts of organic acids, the acids are freed by addition of hydrochloric acid and the solution is titrated similarly.

Fatty acids in silage

Hueni and Uebersax[54] have used gas chromatography to determine low fatty acids in silages. In this method, fresh silage (100 g) is mixed with 100–200 ml of water and allowed to stand for 30 min. The juice is expressed, and 1.0 ml of 25% hydrochloric acid is added per 10 ml of fluid. Insoluble material is removed by centrifugation and an aliquot (1 µl) of the supernatant liquid is injected directly into a gas chromatograph equipped with glass columns containing Porapak Q and temperature programmed from 150 to 230 °C at 8 °C min^{-1} and with a flame ionization detector; nitrogen is used as carrier gas (30 ml min^{-1}). A standard solution containing 0.5% of acetic acid, 0.25% of propionic acid, 0.25% of butyric acid, and 2.0% of lactic acid is also analysed for comparison and peak identification.

Fatty acids in river sediments

Farrington and Quinn[55] give details of procedure involving sponification and extraction. Between 32 and 65% of the fatty acids was not released from sediments by organic solvent extraction.

Miscellaneous (non-fatty) acids

Kawahara[56,57] carried out the microdetermination of pentafluorobenzylester derivatives of benzoic, oleic, linoleic and undecanoic acids by electron capture gas chromatography. The method was applied to chloroform extracts of natural water.

Khomenkô and Goncharova[58] separated and concentrated the following acids dissolved in natural water: oxalic, malonic, succinic, glutaric, adipic, fumaric, aconitic, lactic, malic, tartaric, citric, pyruvic, and gallic acids. Ethyl ether–butanol (1:1) and isobutyl alcohol and ethyl acetate were used as solvents. The preferred method was extraction (3 × 5 min) at pH 2 into an equal volume of butanol.

Citric acid and citrate

Bustin and West[59] determined traces of citrate in aqueous systems. The citrate is converted into ammonium 2,6-dihydroxy-isonicotinate and the resulting blue fluorescence gives a measure of the amount of citrate. The calibration graph is rectilinear for 0.01–10 μg of citrate ml^{-1}, the optimum working range is 0.1–10 μg ml^{-1}. There was no interference from species normally encountered in water samples.

Malic acid

Smotrakov and Stradomskaya[60,61] determined this acid spectrophotometrically in natural water. Malic acid is oxidized to carbonyl compounds with potassium permanganate in sulphuric acid medium and then the carbonyl compounds are reacted with 2,4-dinitrophenylhydrazine. The 2,4-dinitrophenylhydrazones are then dissolved in potassium hydroxide in 50% ethyl alcohol and evaluated at 554 nm.

Lactic acid

Stradomskaya and Goncharova[62] described a spectrophotometric method for the determination of lactic acid in amounts down to 5 μg in natural water. The method is based on oxidation of lactic acid to acetaldehyde and reaction of the latter with l-naphthol to form a coloured product. Amino acids, aldehydes (other than acetaldehyde), organic acids, and inorganic ions do not interfere.

Phenylglyoxylic and mandelic acids

Hatsue and Masayuki[63] determined these acids spectrophotometrically as follows: The sample (0.5 ml) is acidified with 0.05 ml of M hydrochloric acid and shaken with 5 ml of ethyl ether for 10 min. A portion of the ether layer is evaporated to dryness on a water bath at 70 °C. To the residue is added 4 ml of concentrated sulphuric acid–40% aqueous formaldehyde (100:1) reagent, and after 15–60 min the extinctions at 350 and 450 nm are measured. The extinctions due to phenylglyoxylic acid has a maximum at 350 nm and that due to mandelic acid a maximum at 450 nm, and the extinctions are proportional to the amounts of each acid and are additive when mixtures of the acids are analysed.

Salicyclic and chlorophenoxyisobutyric acids

These acids have been determined[64] in primary treated sewage works effluent and river waters.

Benzoic acid

Goncharova and Khomenkô[65] determined benzoic acid in ether extracts of natural and contaminated waters. The benzoic acid was back extracted from the ether into 0.1 M aqueous sodium hydrochloric acid and evaluated spectrophotometrically at 230 nm. Beer's law is obeyed with 20–140 μg of benzoic acid in 20 ml. Phenol, benzene, and carboxylic acids (succinic, citric, lactic, oxalic, malic, glutaric, propionic, acetic, and formic) do not interfere. The presence of humic and fulvic acids has little effect since they are not extracted into ether.

Abietic-type resin acids

These have been determined in hexane extracts of kraft-mill effluents.[66] The effluent was adjusted to pH 2.4–2.7 prior to spectrophotometric evaluation at 490 nm of the levoprimaric acid—1-amino-2-naphthol-4-sulphonic acid—potassium ferricyamide reaction product. Down to 1 ppm of abietic-type resin acids could be determined in the water sample.

Acrylic acid

Brown[67] has described a technique for the routine determination of acrylic acid monomer in natural and polluted waters which uses high performance liquid chromatography for separation for interferences and ultraviolet detection for quantification.

Methacrylic acid

Krotova[68] determined this acid spectrophotometrically. A portion of the effluent was extracted with toluene or benzene to remove N-phenyl-2-naphthylamine, $\mu\alpha$-dimethylbenzyl alcohol and acetophenone and a 50 ml sample was adjusted to pH 8 (to prevent polymerization) and subjected to distillation in a stream of nitrogen to remove benzene, toluene, butadiene, acetone, or other volatile substances. The solution was diluted to 50 ml with water and adjusted to pH 4–5 with 0.1 M hydrochloric acid. The extinction is measured at 208 nm.

Formic acid

Jordan[69] has described a procedure for the determination of traces of formic acid and formaldehyde in river water. Formic acid plus formaldehyde are determined after reducing the acid with nascent hydrogen; formaldehyde alone is determined without reduction. The chromotropic acid method is used in 6–7.5 M sulphuric acid. The limit of determination is 0.05 μg ml^{-1} without preconcentration and 0.05 ng ml^{-1} with preconcentration by extraction with diethylether.

ALCOHOLS

Methanol

Igarashi[70] has described a spectrophotometric method for the determination of methanol using chromotrophic acid. The test solution containing 1–15 μg of methanol, is treated with 0.2 ml of 0.3% (v/v) propionaldehyde solution and 0.4 ml of potassium permanganate solution (prepared by dissolving 1.5 g of potassium permanganate in 30 ml of water and 7.5 ml of 85% phosphoric acid, diluting to 50 ml). After 5 min unconsumed permanganate is reduced with 0.2 ml of 20% sodium sulphite solution. Then 0.3 ml of 2% aqueous chromotrophic acid and 4 ml of 75% (v/v) sulphuric acid are added the mixture is heated at 80–85 °C for 10 min and cooled to room temperature and the extinction is measured at 575 nm against a reagent blank.

Fox[71] has described a rapid gas chromatographic method for the determination of residual methanol in sewage. In this method the filtered sample is adjusted to pH 2 and injected directly into a U-shaped stainless steel column packed with Tenax GC (60–80 mesh) and pretreated with 85% phosphoric acid for 4 h. For gas chromatography the column was operated at 70 °C, with nitrogen as carrier gas (25 ml min^{-1}) and nitrogen flame ionization detection. The detector response was rectilinear in the range 0.5–100 ppm of methanol, and for a sample containing 50 ppm of methanol, the coefficient of variation were 2.4%.

An automated method has been described to determine methanol in dinitrified effluents.[72] The method is based on oxidation of methanol to formaldehyde which is determined colorimetrically after reaction with acetylacetone.

Cyclohexanol

Romantsova[73] has described a spectrophotometric determination of small amounts of cyclohexanol (and cyclohexanone) in river water samples and effluents. The sample (0.2 ml containing down to 10 ppm of cyclohexanol and 3 ppm of cyclohexanone) is mixed with 0.4 ml of a 5% solution of 4-dimethylaminobenzaldehyde in 83% sulphuric acid and 4.6 ml of 83% sulphuric acid, and heated on a boiling-water bath for 15 min. Extinctions are measured against a reagent blank at 435 nm (max. for cyclohexanone) and 520 nm (max. for cyclohexanol).

GLYCOLS

Gas chromatography has been used[74] to determine parts per million of glycols in water. Ethylene glycol and di- and triethylene glycols in amounts down to 0.5 mg l^{-1} have been determined in surface waters by spectrophotometry.[75] To a 5 ml sample is added 0.5 ml 2 M sulphuric acid and 1 ml of 0.0126 M potassium permanganate. After heating for one minute at 100 °C 0.07 M sodium arsenite (1 ml) and 1 ml of 2% 3-methylbenzothiazolin-2-one hydrazone hydrochloride are added and the solution again heated for 6 minutes at 100 °C. After cooling 1 ml of 2% ferric chloride (6H$_2$O)–3% sulphuric acid is added and the solution diluted to 10 ml. After 20 minutes the extinction is measured at 630 nm against a reagent blank. These workers also describe a potassium periodate method for the determination of ethylene glycol alone.

ALDEHYDES

Formaldehyde

Musselwhite and Petts[76] have described an automated procedure for the determination of formaldehyde in sewage and sewage effluents. The method is based on the reaction of formaldehyde with acetylacetone in the presence of excess of an ammonium salt to form a yellow compound, diacetyldihydrolutidine which is determined colorimetrically.

Furfuraldehyde

This has been determined by spectrophotometry in water and gas chromatographically[77] in industrial effluents. Kelus and Waviernia[223] have reviewed spectrophotometric and colorimetric methods for determining 2-furfuraldehyde in water used as an anticorrosive lining in water tanks. The sample of water is mixed with an equal volume of freshly prepared aniline–anhydrous acetic acid (1:9) and the extinction measured at 520 nm in a spectrophotometer. Down to 0.04 mg of furfuraldehyde per litre of water can be determined. Iron in concentrations of about 0.1 mg l^{-1} interferes and must be removed by distilling the sample.

Karmil'chik et al.[77] found that the most suitable gas chromatographic column packing for determining 2-furfuraldehyde was 10% of Polysorbate 80 on Chromosorb W operated at 130 °C with argon as carrier gas (24 ml min^{-1}), a 5 μl sample and flame ionization detection. The relative error was less than 7%. The range of the method is 0.004–8% of 2-furfuraldehyde. Methanol and acetic acid do not interfere.

Acrolein

This substance has been determined in water samples in amounts down to 0.05 mg l^{-1} by differential pulse polarography.[78]

Syringaldehyde

Syringaldehyde has been determined gas chromatographically in oxidized neutral sulphite pulping effluents.[79] A copper column (2 m × 2 mm i.d.) containing Apiezon N on granulated Teflon was used. It was operated at 220 °C using nitrogen as carrier gas and hydrogen-flame ionization detection. Preconditioning of the column by repeated injection of vanillin and syringaldehyde is necessary.

KETONES

Acetone

Vajta et al.[80,81] investigated the determination of down to 5 µg l^{-1} acetone in aqueous petroleum refinery effluents. The effluent (1 ml) containing acetone (or ethylmethyl ketone) is mixed with 1 ml of saturated methanolic 2,4-dinitrophenylhydrazine and 1 drop of concentrated hydrochloric acid. The mixture is maintained at 50 °C for 30 min and after being set aside for 1 h is made alkaline with 5 ml of a 10% solution of potassium hydroxide in aqueous methanol (1:4). After 10 min the extinction is measured at 490 and 540 nm. The concentration of acetone is determined by reference to a calibration graph.

Karyakin and Chirkova[82] described a method for determining down to 10 ppm acetone in water based on the photochemical reaction of acetone with fluorescein sodium. The latter is concerted into a non-fluorescent compound. The acetone is determined from a graph of the log of the fluorescence intensity at 515 nm of a solution of fluorescein sodium at about 20 °C against the log of the acetone concentration after exposure for 10 min to u.v. radiation. Hydrocarbons and ethers do not interfere but organic acids do.

ESTERS

In an early reference to the determination of phthalic acid esters in river water Hites[83-85] used computerized gas chromatography–mass spectrometry, high resolution mass spectrometry, and high pressure liquid chromatography to determine the plasticizers dibutyl phthalate-*bis*(2-2-butoxyethoxy)methane, *bis*(2-ethylhexyl) adipate, and various isomers of dioctyl phthalate and di-isodecyl phthalate in river water from 1 to 30 parts per 10^9. They also identified tricholorobenzene biphenyl, and butyl benzoate in concentrations from 0.01 to 0.5 parts per 10^9, together with diethyl, dibutyl, and *bis*(2-ethylhexyl) phthalates.

Mori[86] has identified and determined very low levels of phthalate esters in river water using reversed phase high performance liquid chromatography using an ultraviolet detector. Phthalates were extracted with *n*-hexane and the uncleaned or concentrated extracts were injected into three chromatographic systems, these being cross-linked porous polymer beads (Shodex HP-225, Showa Penko Co.) and *n*-hexane; porous polymer beads and methanol; and polystyrene

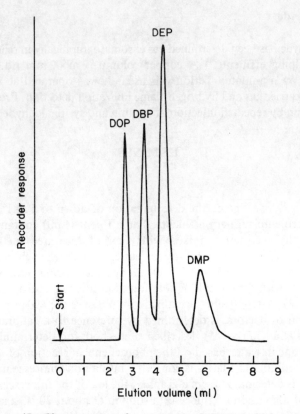

Figure 45 Chromatogram for system A. Column 1, 50 cm × 2 mm i.d., packed with Shodex Polymer Beads HP-225. Eluant, n-hexane. Sample volume injected, 10 μl. Flow rate, 0.6 ml min^{-1}. Sample concentration, $ca.$ 0.1%. Detector, u.v. at 254 nm. Attenuation, 0.16 AUFS. DMP, dimethyl phthalate; DEP, diethyl phthalate; DBP, di-n-butyl phthalate; DOP, di-2-ethylhexyl phthalate. Reprinted with permission from Mori.[86] Copyright (1976) Elsevier Science Publishers

GPC gel and chloroform. The eluants used were n-hexane (system A) and methanol (system B), and chloroform (system C).

Procedure

Suspended particles were removed from all types of river water by filtration using filter paper rinsed in n-hexane for 24 h before use. A 100 ml volume of this filtrate was then added to a separating funnel, followed by 10 ml of n-hexane. After shaking for 10 min, the organic layer was made up to 10 ml with n-hexane. This solution was filtered with 0.5 μm Millipore filter paper prior to chromatography.

Typical chromatograms for systems A, B, and C are shown in Figure 45–47 respectively. Elution volumes and apparent and corrected theoretical plate numbers (N and N^1) were calculated as follows:

$$N = 5.54 \frac{(V_e)^2}{W_{1/2}}$$

$$N' = 5.54 \frac{(V_e - V_o)^2}{W_{1/2}}$$

where V_e is the elution volume, V_o is the void volume or interstitial volume plus dead volume and $W_{1/2}$ is the peak width at half-height, all measured in mm.

An example of a chromatogram of the extract of a river water is shown in Figure 48. The presence of n-dibutyl phthalate and di-2-ethylhexyl phthalate was observed. The concentrations of phthalates in the extract were 450 ppb of n-dibutyl phthalate and 100 ppb of di-2-ethylhexyl phthalate and their

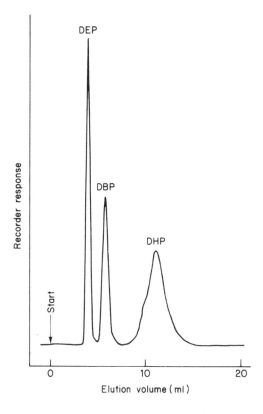

Figure 46 Chromatogram for system B. Column 1, 50 cm × 2 mm i.d., packed with Shodex Polymer Beads HP-255, Eluant, methanol. Sample volume injected, 10 μl. Flow rate, 1.0 ml min^{-1}. Sample concentration, ca. 0.1% (DHP 0.2%). Detector, u.v. at 254 nm. Attenuation, 0.16 AUFS. DHP, diheptyl phthalate; DBP, di-n-butyl phthalate; DEP, diethyl phthalate. Reprinted with permission from Mori.[86] Copyright (1976) Elsevier Science Publishers

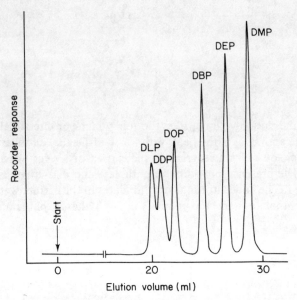

Figure 47 Chromatogram for system C. Column II, 50 cm × 8 mm i.d., packed with Shodex A801. Eluant, chloroform. Sample volume injected, 50 μl. Flow rate, 1.5 ml min^{-1}. Sample concentration, *ca*. 0.1%. Detector, u.v. at 254 nm. Attenuation, 0.64 AUFS. DMP, dimethyl phthalate; DEP, diethyl phthalate; DBP, di-*n*-butyl phthalate; DOP, di-2-ethylhexyl phthalate; DDP, didecyl phthalate; DLP, dilauryl phthalate. Reprinted with permission from Mori.[86] Copyright (1976) Elsevier Science Publishers

concentrations in river water were 45 and 10 ppb, respectively. The first peak in Figure 48 is contaminant(s) in *n*-hexane.

Chlorinated hydrocarbons and other pesticides that might be present in river water interfere in the gas chromatographic analysis of the phthalates. Hence a separation of these classes of compounds is required prior to gas chromatographic analysis, whereas no pretreatment was required for the liquid chromatographic analysis of the phthalates.

Mori[86] investigated the limits of quantitative determination of phthalate esters by this procedure. Phthalate esters have λ_{max} values near 224 nm and the molar absorption coefficient is about ten times that at 254 nm. Calibration graphs for di-*n*-butyl phthalate and di-2-ethylhexyl phthalate at 22 nm for solvent system A were linear over the range 20 ppb–6 ppb. The operating sensitivities were about 13 ng for di-*n*-butyl phthalate and 12 ng for di-2-ethylhexyl phthalate giving 10% of full-scale deflection, if the absorbance was measured at attentuation × 0.01 (0.01 AUFS) and 100 μl of sample solution were injected. The limit of the determination in the extract was assumed to be 20 ppb (2 ng in this instance) by considering the noise levels. As the volume of *n*-hexane used for extraction was one-tenth that of the river water, the limit of determination could be lowered to 2 ppb of phthalates in river water. A concentration procedure

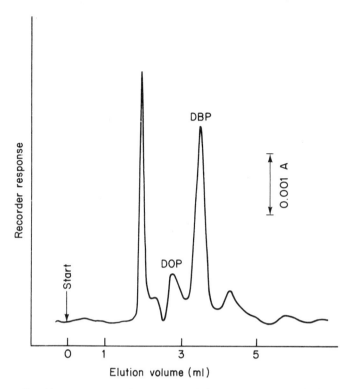

Figure 48 Chromatogram of an extract of river water measured with system A. Sample volume injected, 100 μl. Detector, u.v. at 224 nm. Attenuation, 0.01 AUFS. DPB, di-n-butyl phthalate, DOP, di-2-ethylhexyl phthalate. Reprinted with permission from Mori.[86] Copyright (1976) Elsevier Science Publishers

can be used in order to decrease this limit further. Similarly, limits of determination in the extract for solvent system B were 20 ppb of di-n-butyl phthalate and 600 ppb of di-2-ethylhexyl phthalate and for system C 200 ppb of both di-n-butyl phthalate and di-2-ethylhexyl phthalate. Solvent extraction recoveries at the 0.1 ppm ester level were between 88 and 100%. Mori[86] identified phthalate esters in river water samples by comparing the chromatograms with those of standard mixtures of known esters.

Schouten[87,88] *et al.* used high performance liquid chromatography to determine very low levels of di-2-ethylhexyl phthalate and di-n-butyl phthalate in Dutch river waters, and compared results with those obtained by gas-liquid chromatography. Good agreement was obtained between the two techniques, although high performance liquid chromatography was shown to be the less time-consuming technique. These two esters account for about 95% of the total phthalate production in Western Europe, used mainly in plastic production. Hence, they are the phthalate esters most commonly found as contaminants in river water.

Method

Sampling and extraction procedure

One litre glass bottles having screw caps with a Teflon inlay, were filled with 50 ml of hexane, tightly closed, and transported to the sampling stations. Here 500 ml samples of river water were poured into the bottles which were shaken thoroughly for about 5 min. The contents of the bottles were transferred to a 1 l glass separatory funnel provided with a Teflon stopper and stopcock. The bottles were rinsed with 2×20 ml hexane which were added to the separatory funnel. After phase separation the organic layer was collected, the funnel rinsed with 2×20 ml hexane and the total extract evaporated to a volume of about 1 ml in an all-glass still heated to a temperature of approximately 90 °C. The concentrated extract was carefully weighed and aliquot portions were used for analysis by means of high performance liquid chromatography and/or gas chromatography.

To avoid contamination of the samples and extracts which will generally contain extremely low concentrations of phthalates, all glassware was soaked for 16 h and/or washed thoroughly with 1% Extran (Alkalisch; Merck, Darmstad, GFR) in tapwater and then rinsed three times with hot tapwater. Drying was done by rinsing twice with acetone. Immediately prior to use all glassware was rinsed three times with hexane.

Procedure

High performance analysis was carried out on a Siemens S 100 liquid chromatograph equipped with a six-port injection valve (100 μl loop) and a u.v. detector. The stainless steel separation column (25 cm $\times$ 3 mm i.d.) was prepacked with 5 μm LiChrosorb S160 silica gel. Hexane–dichloromethane (1:2, v/v) containing 0.1–0.2% of ethanol at a flow rate of about 0.5 ml min^{-1}, was used as mobile phase. Detection was done at 223 nm, which is close to the wavelength of maximum adsorption of di-2-ethylhexyl phthalate and di-n-butyl phthalate (224 nm). Phthalate concentrations were determined by measuring peak heights at 0.04 AUFS, the detection limits for both compounds being 5–10 ng. Chromatograms were run at a temperature of 25 ± 2 °C.

Gas chromatographic analysis was performed on a gas chromatograph equipped with a flame ionization detector, using a 7 ft $\times$ 2 mm i.d. glass column packed with 4% OV-101 on Chromsorb W (HP). Four microlitre injections were made using an all-glass solid injector. The injector, column, and detector temperature were 240, 240 and 300 °C, respectively. The flow rate of nitrogen, hydrogen, and air were 25, 30, and 300 ml min^{-1}, respectively. Quantitative analysis was done by means of peak height measurements. The detection limits for the di-2-ethylhexyl phthalate and di-n-butyl phthalate were about 0.2 and 002 ng, respectively.

The results of high performance liquid chromatographic analysis of series of water samples taken from the river Meuse are listed in Table 40. For reasons

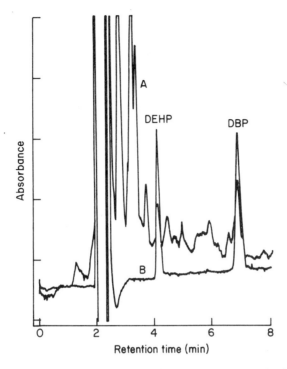

Figure 49 H.p.l.c. chromatogram of (A) hexane extract of a river water sample taken at Station 12, and (B) a standard solution of DEHP and DBP; detection at 0.04 AUFS. DEHP, di-2-ethylhexyl phthalate; DBP, di-n-butyl phthalate

Table 40 Di-2-ethylhexyl phthalate (DEHP) and di-n-butyl phthalate (DBP) content (ppb) of water samples taken from the river Meuse

Sampling area	DEHP		DBP	
	h.p.l.c.	g.c.	h.p.l.c.	g.c.
13	0.7	0.5	0.4	0.3
14	0.7	0.5	0.2	0.1
15	1.6	1.5	0.4	0.2
16	0.4	0.2	0.1	0.1
17	0.5	0.4	0.2	0.1
18	1.1	0.9	0.2	0.1
19	1.1	0.8	1.0	0.9
20	4.0	4.2	0.4	0.2
21	0.8	0.8	0.1	<0.1

of comparison, data on analysis by gas chromatography are included. In all cases blank values (cf. below) have already been subtracted. Figure 49 shows an high performance liquid chromatogram of a hexane extract of a water sample taken from the River Ijssel.

Blank values were determined by subjecting 50 ml of hexane to the procedure, calculating the ultimate result as $(x/500)$ppb, x being the number of nanograms of phthalates found. For di-n-butyl phthalate blank values invariably were below the detection limit (0.1 ppb) while for di-2-ethylhexyl phthalate blank values were from 0.3 to 1.0 ppb (average 0.5 ppb); analysis by both techniques yielded identical results.

The detection limits for the two esters were found to be 0.1 ppb by high performance liquid chromatography and 0.1–0.2 ppb by gas chromatography.

Schouten et al.[87] improved the accuracy and precision of the above method by the use of a trace-enriched technique, in which 500 ml samples are pumped through a short precolumn filled with 5 μm LiChrosorb RP-18 at high flow rates. During this step, the phthalates are concentrated into a small zone on the precolumn, from which they are subsequently eluted using a simple methanol–water gradient. With this technique, blank values for the two phthalates investigated of typically less than 0.1 and 0.05 ppb were obtained.

These workers also observed that the di-2-ethylhexyl phthalate and even more so, the di-n-butyl phthalate content of river water samples rapidly decreased with time. At the 50 ppb level, a 50% loss of di-2-ethylhexyl phthalate typically occurred within 5 days in the dark while over 90% of di-n-butyl phthalate was lost in about 3 days. These losses were not caused by adsorption of the phthalates to the glass wall of the container. The absence of degradation in a sample spiked with phthalates and sterilized immediately afterwards (less than 10% in 2 weeks; Figure 50(b)) may suggest that breakdown was caused by biological factors.

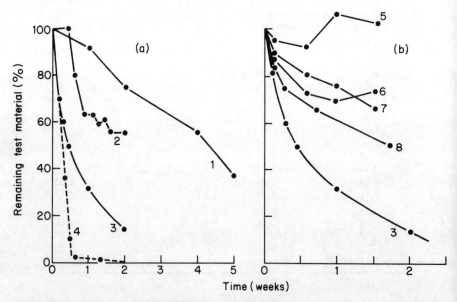

Figure 50 (a) Breakdown of DEHP (———) and DBP (----) in river water. (1) 1 ppm, Mississippi; (2) 25 ppm, Okawa; and (3) and (4) 50 ppb, Rhine. (b) Breakdown of DEHP (50 ppb, Rhine) in the absence of microbial inhibitors (3), after sterilization (5), or in the presence of 3000 ppm Halamid (6), 1500 ppm NaN_3 (7), or 500 ppm $HgCl_2$ (8).

Their attempts to prevent biodegradation of the phthalates by adding microbial inhibitors such as mercuric chloride, sodium azide, and Halamid were only partly successful. Figure 50(b) demonstrates that, despite the large excess of inhibitor used, degradation — although at a slower rate — continued. To offset the effects of degradation, extraction of the phthalate was carried out immediately after sampling.

Several workers have reviewed methods for the determination of phthalic acid esters[89,90] in water.

Phthalate esters in river sediments

Schwartz et al.[91] have also described a high performance liquid chromatographic method for determining di-2-ethylhexyl and di-n-butyl phthalate in river sediments. This method requires no sample clean-up and consists of a single extraction step followed by quantitative analysis using high performance liquid chromatography.

Method

High performance liquid chromatograph equipped with an Orlita Model Mk 00 pump and a multiple-wavelength detector was used to quantitate the phthalates. The injection port was a Siemens Model F 10–34 six-port valve to which a 100 μl loop was attached. The Chrompack separation column was a stainless steel tube (25 cm × 3 mm i.d.) prepacked with 5 μm LiChrosorb SI 60 silica gel. Mixtures of n-hexane–dichloromethane (containing approximately 0.2% of ethanol) (1:2, v/v) — which were thoroughly mixed before use — were used as mobile phase. All chromatograms were run at a temperature of approximately 25 °C.

Reagents

Nanograde n-hexane reagent grade, acetone, dichloromethane analysed-grade, methanol, and diethyl ether were used as solvents without further purification.

Folding filters, cellulose Soxhlet thimbles and Florisil adsorbent (60–100 mesh) were washed by means of extraction in a Soxhlet apparatus for at least 12 h to remove phthalate impurities. The filters were stored under n-hexane. Prior to use, all glassware was washed with a detergent solution and thoroughly rinsed with hot tap water, acetone, and n-hexane. Soxhlet and distillation apparatus were rinsed with n-hexane. These precautions eliminated background contamination to a level of 10 ng of phthalate, i.e. to below the detection limit of the method.

Procedure

After arrival in the laboratory, 100–200 g samples were immediately frozen at a temperature of −30 °C and freeze-dried. Subsequently, the samples were homogenized and 10 g were placed in a Soxhlet thimble. Extraction was carried out for 17 h, using 200 ml of a n-hexane–acetone–methanol (8:1:1, v/v) mixture. The extract was carefully evaporated to a volume of 1–2 ml, dissolved in 30 ml of n-hexane, and filtered over a paper filter; 100 μl aliquots from a known weight of the sample solution were injected into the high performance liquid chromatographic system which was run at a flow rate of approximately 0.5 ml min^{-1}. Analysis was done at a wavelength of 223 nm, di-2-ethylhexyl phthalate and di-n-butyl phthalate contents being determined by measuring peak heights. The calibration curve was linear over the concentration range 0.1–15 μg of phthalates per ml of n-hexane solution. Recovery was found to be between 90 and 100% for both esters.

Some samples were also examined by gas chromatography.

Gas chromatographic method

Apparatus

Gas chromatography was carried out on a Pye Unicam Model 104 gas chromatograph equipped with an electron capture and a flame ionization detector, using a 7 ft × 2 mm i.d. glass column packed with 4% OV-101 on Chromosorb WHP (80–100 mesh). The injector, detector, and column temperatures were 250, 300, and 220 °C, respectively. Nitrogen was used as carrier gas at a flow rate of 30 ml min^{-1}.

Procedure

Samples were precleaned by applying an aliquot of the hexane extract to a 0.9 × 20 cm column filled with 7 g of Florisil. After elution of the column with 100 ml of n-hexane–diethylether (50:3, v/v) (eluate discarded) percolation of the column was continued with n-hexane–diethylether (17:3, v/v) to yield two further 100 ml fractions, which contained di-2-ethylhexyl phthalate and di-n-butyl phthalate respectively. These fractions were carefully evaporated to dryness, dissolved in n-hexane, and subjected to gas chromatography.

In the high performance liquid chromatographic method the retention times of di-2-ethylhexyl phthalate and di-n-butyl phthalate showed unexpectedly large variations. This turned out to be affected by the presence of varying amounts of ethanol in different brands, or even lots of dichloromethane. With decreasing percentage of ethanol, the retention times of the phthalates sharply increase. The relationship between percentage of ethanol or n-butanol added to the mobile phase and the value of the capacity ratio, k', of the phthalates is shown in Figure 51. Small capacity ratios and sufficient resolution were obtained by using n-hexane–dichloromethane (1:2, v/v) containing approximately 0.1% of ethanol (or 0.15% of n-butanol) as mobile phase. Under these conditions, reproducibility of the chromatographic system was excellent.

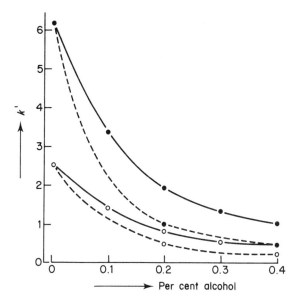

Figure 51 Dependence of k' values of DEHP (○) and DBP (●) on the amount of ethanol (----) or n-butanol (———) present in the mobile phase (n-hexane–dichloromethane, 1:2, v/v)

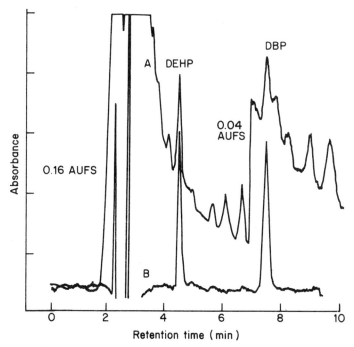

Figure 52 H.p.l.c. chromatogram of (A) hexane extract of a sediment and (b) a standard solution of DEHP and DBP. Retention times: DEHP, 4.5 min; DBP, 7.5 min

Table 41 DEHP and DBP content (ppm) and composition of sediments taken from the river Rhine in 1977

Sample station	Date	DEHP	DBP	% dry weight	Fraction with $d_p < 16\,\mu m$
01	10–05	15.0	5.0	—	—
	22–08	70.5	15.5	34	—
	03–10	9.5	ND	57	0.12
02	10–05	20.5	7.5	—	—
	22–08	25.0	4.5	43	—
	03–10	19.0	ND	45	0.40
03	22–08	26.0		34	—
	03–10	27.0	ND	33	0.40
04	22–08	10.0	ND	58	—
	03–10	17.0	ND	48	0.36
05	10–05	13.0	4.0	—	—
	22–08	21.0	3.0	37	—
	03–10	27.0	ND	24	0.51
07	22–08	10.5	ND	46	—
	03–10	6.5	ND	59	0.11
08	10–05	0.1*	ND	77	—
09	22–08	14.5	ND	45	—
10	10–05	9.0	2.0	-	—
	22–08	28.5	0.5	32	—
	03–10	25.0	ND	25	—
11	22–08	45.0	0.5	34	—

ND not detectable.
*Procedure: 25 instead of 10 g dry sample; 5 instead of 30 ml n-hexane.

Table 42 DEHP and DBP content (ppm) and composition of sediments taken from the river Ijssel in 1977

Sample station	Date	DEHP	DBP	% dry weight	Fraction with $d_p < 16\,\mu m$
12	21–06	21.5	2.0	—	—
	15–08	24.0	ND	28	—
	03–10	8.0	ND	56	0.12
13	21–06	12.0	0.5	—	—
	15–08	11.5	ND	45	—
	03–10	4.0	ND	60	0.16
14	21–06	27.0	1.0	—	—
	15–08	12.0	ND	52	—
	03–10	6.5	ND	56	0.18
15	21–06	14.5	1.0	—	—
	15–08	23.0	ND	40	—
	03–10	36.0	ND	35	0.42
16	21–06	28.0	ND	—	—
	15–08	25.5	ND	23	—
17	22–08	52.5	7.5	29	—
18	22–08	34.0	6.5	35	—
19	22–08	2.5	0.5	73	—

ND not detectable

Table 43 DEHP and DBP content (ppm) and composition of sediments taken from the river Meuse in 1977

Sample station	Date	DEHP	DBP	% dry weight	Fraction with $d_p < 16 \mu m$
20	31-08	13.0	1.5	38	0.37
21	31-08	3.5	0.5	67	0.11
22	31-08	11.5	1.5	42	0.32
23	31-08	9.5	1.5	44	0.32
24	31-08	16.0	1.0	46	0.29
25	31-08	2.5	1.0	50	0.45
26	31-08	17.0	0.5	47	0.27
	31-10	11.0	ND	52	0.26
27	31-10	7.0	ND	50	0.32
28	31-10	6.0	ND	52	0.30
29	31-10	4.5	ND	50	0.37
30	31-10	1.0	ND	59	0.38
31	31-10	1.0	ND	79	0.07
32	31-10	11.5	ND	32	0.48

ND not detectable

Following the procedure described above it is possible to detect down to 10 ng of both esters, i.e. 0.5 ppm in river sediments. Figure 52 shows a typical chromatogram of a hexane extract of sediment taken from the river Rhine.

Schwartz et al.[91] investigated the biodegradation of phthalic acid esters adsorbed in river sediments by repeated analysis of the sediment over a 2-week period. The di-2-ethylhexyl phthalate content turned out to remain essentially constant (s.d., 5%; $n = 10$), irrespective of the absence or presence of a microbial inhibitor (500 ppm of sodium azide or mercuric chloride added immediately after sampling); i.e. no marked biodegradation occurred.

The results of analyses of sediment samples taken from the rivers Rhine, Ijssel and Meuse are listed in Tables 41-43 respectively. From Tables 41 and 42 it is seen that for the sediments of the rivers Rhine and Ijssel di-2-ethylhexyl phthalate and di-n-butyl phthalate concentrations generally are between 2 and 50 and 0 and 7 ppm respectively. For the river Meuse, distinctly lower values are observed i.e. 1-17 and 0-2 ppm respectively. For all cases the di-2-ethylhexyl phthalate content is higher than the di-n-butyl phthalate content.

The phthalate levels of the sediment samples should actually be regarded as minimum values, since only the amount of extractable phthalates has been determined. Eglinton et al.[92] report that some organic pollutants in sediments may be converted into insoluble complexes, such as humates. On the other hand, data by Cifrulak[93] suggest that the use of a methanol-containing solvent mixture, rather similar to the one employed by Schwartz et al.,[91] effectively removes all phthalates from soil samples.

Generally low phthalate levels were found when the percentage dry weight of the sediment was high; this was especially true with samples containing course particles, i.e. sand. Similarly, in soil samples collected at a sanitary landfill, which displayed high dry weight values of 70–80%, low levels of di-2-ethylhexyl phthalate (0.2–0.8 ppm) and di-n-butyl phthalate (0.1–0.4 ppm) were determined.

It has been observed[94,95] that a relationship exists between the heavy metal level of sediments in polluted areas and the particle size of the sediment, measured as the fraction of particles having a diameter d_p smaller than 16 μm. Schwartz[91] determined for 25 samples the di-2-ethylhexyl phthalate content, the percentage dry weight, and the fraction having $d_p < 16\,\mu$m. The correlation between the di-2-ethylhexyl phthalate content and the percentage dry weight displayed a correlation coefficient r_s, of -0.77 and -0.91 (no significant difference) for the river Meuse and the rivers Rhine and Ijssel, respectively, while the correlation coefficients between the di-2-ethylhexyl phthalate level and the fraction having $d_p < 16\,\mu$m were $r_s = 0.03$ and 0.85, respectively. That is, with the rivers Rhine and Ijssel more sorption of phthalates occurs onto particles having $d_p < 16\,\mu$m than onto larger particles; in the case of the river Meuse, no such difference is observed.

Phthalates in biota samples

Giam et al.[96] used electron capture gas chromatography to detect phthalate ester plasticizers in open ocean biota samples. Chlorinated hydrocarbons and polychlorinated biphenyls, which are present in almost all marine biota samples, interfere with the analysis and must first be separated by column chromatography on deactivated Florisil. The very low levels of phthalates anticipated in these samples required the reduction of background contamination to levels much lower than had previously been reported. With careful decontamination of all reagents and equipment and precautions to avoid recontamination during the procedure, background levels as low as 25 ng of dibutyl phthalate and 50 ng of di-2-ethylhexyl phthalate were attained.

CARBOHYDRATES

Steinle[97] has reviewed methods for the determination of carbohydrates in water samples.

Shaova and Kaplin[98] have described a spectrophotometric method for the determination of down to 0.1 mg l^{-1} of reducing sugars in natural water. To a 200 ml portion of sample is added 20 ml of zinc sulphate 7H$_2$O (5%) to precipitate proteins, then 2.5 ml sodium hydroxide (5%) is added until zinc is completely precipitated. The precipitate is filtered off and the filtrate passed through an ion-exchange column (KU-2 and AN-22 resins) discarding the first 100 ml of percolate; to 50 ml of subsequent percolate is added 0.5 ml water. The mixture is heated at 100 °C for 20 minutes and 0.5 ml cupric chloride 2H$_2$O (2%) and 1 ml oxalic acid (1 N) is added to the hot solution. After one minute the colour of the solution is compared with that of similarly prepared standards.

Cavari and Phelps[99] have described a sensitive enzymic assay for glucose in natural waters in amounts down to 0.1 ppm without prior concentration or extraction. The method is based on reaction with adenosine 5′-triphosphate catalysed with hexakinase to form glucose-6-phosphate. Adenosine 5′-triphosphate consumed is measured by the luciferin–luciferase assay.

Josefsson[100] and Stabel[101] determined monosaccharides in natural water by gas chromatography of the trimethylsilyl derivatives.

Ochiai[102] used gas chromatography to determine dissolved carbohydrates in natural water. The carbohydrates are first hydrolysed to alditol acetates of monosaccharides by refluxing for 7 hours with 1 M hydrochloric acid under nitrogen. The hydrolysate is then reduced with sodium borohydride at 60 °C.

Method

Apparatus

The gas chromatograph used was equipped with a flame ionization detector. A glass column (2 m × 3 mm i.d.) packed with 5% OV-275 on Chromosorb W was employed at a nitrogen flow rate of 40 ml min^{-1}. A temperature programmed analysis from 160 °C to 240 °C at 2 °C min^{-1} required 40 min to elute the acetyl derivatives of the monosaccharides and the internal standard inositol.

Procedure

The water samples were filtered through Whatman GF/C glass fibre filter, previously baked in a furnace at 450 °C for 2 h. A 100 or 200 ml volume of filtered water sample was dried in a freeze dryer. Inositol was added to the dried sample as internal standard and dissolved in 1 M hydrochloric acid. The sample was hydrolysed under nitrogen in 1 M hydrochloric acid for 7 hours (critical) at 100 °C. The hydrolysate was again dried and reduced with sodium borohydride for 1 h at 60 °C. The reduced sample was applied to the top of a column (150 × 8 mm) of Dowex 50W X8 cation exchange resin which had previously been cleaned with 1 M sodium chloride solution, regenerated (to H$^+$) with 1 M hydrochloric acid and rinsed to neutrality with distilled water. The monosaccharides were eluted with 25 ml of distilled water. The solution was evaporated to dryness, 20 ml of methanol added and re-evaporated to dryness. The dehydrated residue was transferred to a 1 ml glass ampoule with methanol and evaporated to dryness under vacuum. The residue was acetylated with 100 μl of acetic anhydride–pyridine (1:1) for 2 h at 100 °C. The acetylation mixture was evaporated and the residue dissolved in 30 μl of chloroform. An aliquot of this solution was injected into the gas chromatograph for analysis.

Ochiai[102] pointed out that although natural water contains lipids, amino acids, humic substances etc., in addition to carbohydrates, the monosaccharides of the dissolved carbohydrates could be measured without clean-up. A typical

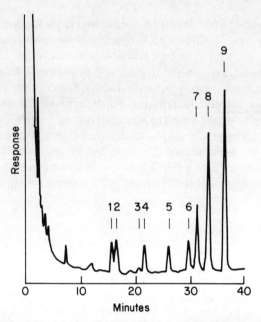

Figure 53 Gas-liquid chromatography of alditol acetates of monosaccharides, formed by hydrolysis of DCHO in lake water (taken from the surface of Lake Nakanuma, Japan, on May 26th 1978). Peaks: 1, rhamnose; 2, fucose; 3, ribose; 4, arabinose; 5, xylose; 6, mannose; 7, galactose; 8, glucose; 9, inositol internal standard. Reprinted with permission from Ochiai[102.] Copyright (1980) Elsevier Science Publishers

chromatogram of monosaccharides of the dissolved carbohydrates in natural water is shown in Figure 53.

Carbohydrates in sea water

LeB Williams[103] has studied the rate of oxidation of ^{14}C-labelled glucose in sea water by persulphate. After the oxidation, carbon dioxide was blown off and residual activity was measured. For glucose concentration of 2000, 200, and 20 μg l^{-1} residual radioactivities (as percentage of total original radioactivity) were 0.04, 0.05, and 0.025 respectively, showing that biochemical compounds are extensively oxidized by persulphate. With the exception of change of temperature, modifications of conditions had little or no effect. Oxidation for 2.5 h at 100 °C was the most efficient.

Josefsson[104] determined soluble carbohydrates in sea water by partition chromatography after desalting by ion-exchange membranes. The electrodialysis cell used had a sample volume of 430 ml and an effective membrane-surface area of 52 cm^2. Perinaplex A-20 and C-20 ion-exchange membranes were used. The water-cooled carbon electrodes were operated at up to 250 mA at 500 V. The desalting procedure normally took less than 30 h. After the desalting, the samples were evaporated nearly to dryness at 40 °C *in vacuo*, then taken up in 2 ml of 85% ethanol and the solution was subjected to chromatography on anion-exchange resins (sulphate form) with 85% ethanol as mobile phase. By this procedure, it was possible to determine eight monosaccharides in the range 0.15–46.5 µg l^{-1} with errors of less than 10% and to detect traces of sorbose, fucose, sucrose, diethylene glycol, and glycerol in sea water.

Eklund et al.[105] developed a method for sensitive gas chromatographic analysis of monosaccharides in sea water, using trifluoracetyl derivatization and electron capture detection. It is difficult to determine accurately the monosaccharide concentrations by this method because a number of chromatographic peaks result from each monosaccharide.

Carbohydrates in lake sediments

Mopper[106] determined monosaccharides with a sensitivity of 0.1 nmole. using an automated chromatographic sugar analyser. The test solution is forced by nitrogen pressure into a nylon column (110 cm × 2.8 mm) packed with Echnicon type S resin (sulphate form) and maintained at 76 °C. The sugars are separated by pumping 89% ethanol through the column and the eluate is mixed with an alkaline solution of tetrazolium blue which is considerably more sensitive and less corrosive than other dyes. The extinction is monitored at 520 nm.

Carbohydrates in effluents

Bark et al.[107] determined mono- and disaccharides in aqueous effluents by gas chromatography of the polytrimethylsilyl derivatives. Difficulties due to the low solubility of sucrose (and other disaccharides) in the silylation reagent were overcome by hydrolysing the sucrose to glucose and fructose under reflux in 0.01 M hydrochloric acid medium for 10 min, then immediately neutralizing the solution before the evaporation stage. The increase in fructose concentration effected by the hydrolysis was used as a measure of the original sucrose concentration. The derivatives of the saccharides were determined by using dual columns (270 cm × 3 mm o.d.) packed with 10% of FFAP on AW-DMCS chromosorb W (85–100 mesh) and operated with temperature programming from 150 to 220 °C 2 °C min^{-1}, using nitrogen as carrier gas (50 ml min^{-1}) and dual flame ionization detectors. Down to 0.05 mg l^{-1} of each saccharide could be determined.

Collins and Webb[108] used gel chromatography to detect and determine carbohydrates in pulp mill effluents. The phenol–sulphuric acid method was used to monitor column effluents.

The applicability of the phenol–sulphuric acid reagent to carbohydrate analysis in waste water and biological sludges has been examined by Benefield and Randall.[109] The test is based on the formation of a yellow colour when concentrated sulphuric acid is added to the sample mixed with phenol solution. This method includes sugars such as heptoses, pentoses, uronic acids, and the sugar components of nucleic acids. Details of the test are given below:

Method

Reagents

5% (w/v) solution of phenol in water.
Concentrated sulphuric acid (analytical reagent grade).

Procedure

On millilitre of sample containing the equivalent of 20–100 µg of glucose is placed in a tube. To each tube is added 1.0 ml of 5% phenol solution and the solution mixed. To each tube is rapidly added 5.0 ml of concentrated sulphuric acid, directing the liquid stream at the sample surface while simultaneously shaking the tube. The tubes are allowed to stand for 10 minutes at room temperature. The tubes are shaken and placed in a water bath at 25–30 °C for 10–20 minutes. A yellow colour is produced the absorbance of which is measured at 488 nm for hexoses and 480 nm for pentoses. The colour is stable for several hours. A reagent blank containing 1.0 ml of water and a set of glucose standards (20–100 µg glucose) is carried through the analysis with the samples.

LACTAMS

ε-Caprolactam

Eremin and Kopylova[110] have described a polarographic method for the determination of ε-caprolactam in synthetic fibre production effluents. The sample (500 ml) is adjusted to pH 7 with sulphuric acid and shaken with light petroleum (3 × 25 ml), the combined extracts are washed with water (3 × 50 ml) and then discarded and the main aqueous phase plus washings is made 0.5 M in sulphuric acid and heated under reflux for 1 h to hydrolyse ε-caprolactam to 6-aminohexanoic acid. The cooled solution is adjusted to pH 2 with 50 ml of 10.6 M sodium hydroxide and passed through a column of KU-1. The 6-aminohexanoic acid is eluted with M sodium bicarbonate, the eluate is mixed with 0.5 ml of 40% aqueous formaldehyde and diluted to 25 ml with M sodium bicarbonate and the polargram of the deoxygenated solution is recorded from −0.9 to 1.4 V.

DIOXANS

Dimethyldioxan

Kavelenko et al.[111] have described an absorptiometric method for determining this substance in effluents. The sample to which has been added 0.5 g of potassium sulphate is extracted with benzene (3 × 5 ml) and to the separated benzene layer is added 5 ml of a solution of dimethylaminobenzaldehyde (0.5 g) in ethyl ether (85 ml) plus sulphuric acid (15 ml). After 20 min, with occasional shaking the organic phase is extracted with water (2 × 20 ml), the combined aqueous layers are diluted, and the extinction measured at 530 nm within 15 minutes.

QUINONES

Thielemann[112] used 4-aminopyrine as a spray reagent for the thin-layer chromatographic identification of p-benzoquinone and anthraquinone in extracts of coal processing plant effluents. The separation is achieved on Kieselgel G with benzene–acetone (9:1) as solvent, and the spots are revealed by spraying with an aqueous solution of the reagent. Red-brown and violet colours are obtained with benzoquinone and anthraquinone respectively. The intensity of the colour may be increased by spraying the plates with 0-1 M hydrochloric acid before development of the chromatogram.

More recently Thielemann[113] discussed the determination of hydroquinone and its oxidation product 1,4-benzoquinone, both of which are toxic constituents of coal industry waste water. He reviews methods for the qualitative detection of benzoquinone and describes a semi-quantitative method for its determination by thin-layer chromatography on Kieselgel G using a 2% ethanolic solution of 4-aminoantipyrine (1-phenyl-2,3-dimethyl-4-aminopyrazol-5-one) as spray reagent.

Suslov and Stom[114] have described a polarographic method for determining quinones formed during the oxidation of phenols in waste water. These workers showed that a change of half-wave potentials with quinone concentration can be used for determining quinones in the presence of excess phenols in waste water.

PHENOLS

Trace amounts (<1 mg l^{-1}) of phenolic compounds can have significant detrimental effects on water quality. Phenols are toxic to aquatic life[115] and mammals[116,117] and can impart objectional tastes and odours to water and fish.[115,118,119] The US Environmental Protection Agency[120] recommends a maximum of 1 μg l^{-1} for total phenolic compounds in domestic water supplies. Chlorophenols are discussed in Chapter 4. Various methods have been used to determine traces of phenols in water samples. The earlier methods were usually spectrophotometric or colorimetric. A number of sensitive colorimetric methods have been developed for determining phenols in water based on the reactions with various chromogenic reagents including 4-aminoantipyrine (4-aminophenazone) in the presence of potassium ferricyanide (at various pHs) p-amino-N,N-diethylaniline in the presence of potassium ferricyanide (indophenol blue method), 2,6-dibromoquinone clorimide (Gibbs reagent), pyramidone (i.e. dimethylaminoantipyrine or aminopyrine) in the presence of potassium ferricyanide, nitroaniline diazotization, and 3-methyl-2-benzothiazoline hydrazone. Although these methods are sensitive they cannot differentiate between substituted phenols and are used primarily for determining total phenol concentrations. Also, the colour-forming reagents will not react completely with most para-substituted phenols and thus results obtained for total phenol concentrations may be significantly in error. For this reason, alternate, more specific techniques such as gas chromatography have been investigated in recent years.

Spectrophotometric methods

The basic chromogenic reactions in the methods mentioned above as as follows.

4-Aminoantipyrine (4-aminophenazone)

Phenols react with 4-aminopyrine at a pH of 10 ± 0.2 in the presence of potassium ferricyanide as an oxidizing agent to form a coloured antipyrine dye. The dye is extracted from aqueous solution into chloroform and evaluated at 460 nm.

In an alternate version of this method the phenols are extracted with diethylether from the water sample, then back extracted with sodium hydroxide from the extract, followed by colour development in the aqueous phase at pH 7.9 with 4-aminoantipyrine and potassium ferricyanide. This method is much more sensitive than the method carried out at pH 10 for the determination of di- and trichlorophenols in chlorinated water samples.

p-Amino-N,N-diethylaniline (indophenol blue method)

Tentative reaction

An alkaline solution of phenol reacts with *p*-amino-*N,N*-dethylaniline sulphate in the presence of potassium ferricyanide to form a coloured indophenol dye. The dye is extracted into carbon tetrachloride and evaluated spectrophotometrically.

2,6-dibromoquinone

$$\underset{\underset{Br}{\overset{Br}{|}}}{O=\bigcirc=N-Cl} + \bigcirc-OH \xrightarrow{\text{alkaline buffer}} \underset{\underset{Br}{\overset{Br}{|}}}{O=\bigcirc-N=\bigcirc-OH} + HCl$$

Phenolic compounds, excluding p-substituted compounds react with 2,6-dibromoquinone at pH 9.4 ± 0.2 to form a coloured indophenol dye which is extracted into butanol-1 and evaluated at 670 nm.

Pyramidone (dimethylaminoantipyrine or aminopyrine)

[Reaction scheme: pyramidone + phenol with $K_4Fe(CN)_6$ oxidation yields coloured product + $2CH_3OH$]

Phenolic compounds react with pyramidone in the presence of potassium ferricyanide to produce orange coloured compounds which are extracted into a mixture of chloroform and isoamyl alcohol and evaluated spectrophotometrically.

p-Nitroaniline

$$O_2N-\bigcirc-N_2Cl + \bigcirc-OH \longrightarrow O_2N-\bigcirc-N=N-\bigcirc-OH + HCl$$

An alkaline solution of the phenol is treated with diazolized p-nitroaniline and the resulting dye extracted into butanol-1 prior to spectrophotometric evaluation at 530 nm.

Workers at the Water Research Association[121] have investigated the applicability of these various methods to the determination of phenols in water at levels down to 0.001 mg l^{-1}. None of the methods was completely satisfactory in that they were less sensitive than desirable and did not respond to all types of phenolic compounds.

These workers did not investigate the 3-methyl-2-benzothiazoline hydrazone method which had only just been introduced at the time their work was carried out.

As a result of their work Cheeseman and Wilson[121] recommended that the best spectrophotometric method for the determination of phenol in unchlorinated water samples was the 4-aminoantipyrine method operated at pH 10 as described in the American Public Health Association method.[122-124] For chlorinated water supplied they recommended the 4-aminoantipyrine method operated at pH 7.9, as described below, as this method has an appreciably greater sensitivity towards chlorinated phenols. This reagent reacts with the phenolic group on each molecule but, due to steric considerations, not with equal sensitivity. It is, for example, relatively insensitive to some para-substituted forms.[125-127] Lack of selectivity and variable sensitivity is a serious drawback to this method because the many diverse sources with different types of phenols can cause the total phenol concentration of natural waters to appear to vary considerably.[119]

4-Aminoantipyrine method for determination of 1–50 ng l^{-1} phenol in water (pH 7.9)

Reagents

Analytical reagent grade issued except when otherwise stated.

Water

Water of very low phenol content is used for the preparation of reagents, blanks, and standards. This water should be prepared by passing distilled water through a column of activated carbon. This column should be about 450 nm long by 50 mm diameter. When prepared the water should be stored in a glass aspirator.

Ammonia solution 0.5 N

This is prepared by diluting 27.5 ml (± 1 ml) of ammonia solution (sp.gr. 0.880) with distilled water to the mark in a 1 litre volumetric flask. This solution is stored in a glass bottle with a ground-glass stopper.

Phosphate buffer solution

This solution is prepared by dissolving 104.5 g of anhydrous dipotassium hydrogen phosphate, K_2HPO_4, and 72.3 g of anhydrous potassium dihydrogen orthophosphate, KH_2PO_4, in distilled water and diluting with distilled water to the mark in a 1 litre volumetric flask. The pH of this buffer solution should be 6.8. It is stored in a glass bottle with a ground-glass stopper.

4-Aminoantipyrine solution (2% w/v)

Two grams of 4-aminoantipyrine (laboratory reagent grade) are dissolved in distilled water and diluted with distilled water to 100 ml. This must be prepared freshly on the day of use and stored in the dark in a dark glass bottle when not in use.

Potassium ferricyanide solution (8% w/v)

Eight grams of potassium ferricyanide are dissolved in distilled water and diluted with distilled water to 100 ml. This is prepared freshly each week and stored in the dark in a dark glass bottle when not in use.

Chloroform

Analytical reagent grade chloroform is used.

Standard solutions of phenol

Solution A 1 ml = 1 mg of phenol: 1.000 g ($\pm$ 0.001 g) of phenol is dissolved in distilled water and diluted with distilled water to 1 litre. This is prepared freshly each month and stored in the dark in a dark glass bottle.

Solution B 1 ml = 5 μg of phenol: 5 ml of the phenol Solution A are pipetted into a 1 litre volumetric flask and diluted to 1000 ml with distilled water. This is prepared freshly on the day of use.

Solution C 1 ml = 0.5 μg of phenol: 25 ml of the phenol Solution B are pipetted into a 250 ml volumetric flask and diluted to 250 ml with distilled water. This is prepared immediately before use.

Cupric sulphate solution

This is prepared by dissolving 100 g of cupric sulphate, $CuSO_4 5H_2O$ in distilled water and diluting with distilled water to 1 litre. The solution is stored in a glass bottle.

Phosphoric acid solution

One hundred millilitres of 88% w/v orthophosphoric acid (sp.gr.1.75) are diluted with distilled water to 1 litre. The solution is stored in a glass bottle.

Sodium arsenite solution (0.04% w/v)

This solution is prepared by dissolving 0.4 g ($\pm$ 0.001 g) of sodium arsenite in distilled water and diluting with distilled water to 1 litre. The solution is stored in a glass bottle.

Apparatus

One litre separating funnels are required and should first be cleaned by standing in chromic acid cleaning solution overnight and then washing sequentially with tap water and distilled water. After each determination the funnels should be washed with three portions of about 300 ml of distilled water. Any yellow stains due to 4-aminoantipyrine complex may be removed with acetone before rinsing with distilled water. One litre distillation flasks, splash heads, condensers, and 1 litre beakers for collecting distillation should first be cleaned as described above and rinsed with distilled water after each determination.

Sample collection and preservation

Biochemical and chemical processes in the sample may occur between sampling and analysis and affect the concentration of phenols. Thus, unless samples are to be analysed within 4 hours of collection, the addition of preserving reagents is necessary.

The samples should be collected and stored in glass bottles with ground-glass stoppers. If oxidizing agents (such as chlorine) are present, 10 ml of the sodium arsenite solution per litre of sample are added immediately after sampling. Then sufficient phosphoric acid is added to the sample that a pH between 4 and 4.5 is achieved. Finally, 10 ml of the cupric sulphate solution are added.

The procedure described below, for analysing the sample, assumes that no preservatives have been added to the sample. If this is not the case, stages 1, 2 and 3 of the procedure should be omitted and the pH of the sample should be checked and adjusted to 4.0 (± 0.2) with phosphoric acid solution. A volume equivalent to 500 ml of sample (before the addition of preservatives and phosphoric acid solution) is transferred to a distillation flask and stages 4–17 are carried out.

Procedure

(1) Add 500 ml (± 5 ml) of sample to a 1 litre beaker.
(2) Add phosphoric acid solution to the beaker until the pH of the solution is 4.0 (± 0.2). Pour the solution into a distillation flask.
(3) Add in sequence to the distillation flask, carefully swirling the contents after each addition: 5 ml (± 0.1 ml) of 0.4% sodium arsenite solution and 5 ml (± 0.1 ml) of 10% w/v cupric sulphate solution.
(4) Distil 450 ml (± 1 ml) of the sample and add 50 ml (± 1 ml) of distilled water to the distillate.
(5) Add the distilled sample (from stage 4) to a 1 litre separating funnel.
(6) Add in sequence to the separating funnel, swirling the contents after each addition: 5 ml (± 0.1 ml) of 0.5 N ammonia solution, the appropriate volume of phosphate buffer (± 0.1 ml), 1 ml (± 0.05 ml) of 2% w/v 4-aminoantipyrine solution, and 3 ml (± 0.1 ml) of 8% w/v potassium ferricyanide solution. (For each batch of 0.5 N ammonia prepared, determine the appropriate volume of phosphate buffer as follows. Add 500 ml (± 5 ml) of distilled water and 5.0 ml (± 0.1 ml) of 0.5 N ammonia solution to a 1 litre beaker. Titrate the solution with the phosphate buffer

solution and note the volume required to give a pH of 7.9. Use that volume of buffer solution for all determinations using the same batch of 0.5 N ammonia solution.)

(7) Allow the solution to stand for 3 minutes (± 30 seconds).
(8) Add 25 ml (± 0.1 ml) of chloroform from a burette to the separating funnel. Shake the funnel gently for a few seconds and then release the excess pressure in the funnel by opening the tap or stopper.
(9) Shake the funnel vigorously (200 shakes/minute) for 1 minute (± 10 seconds) and allow the two phases to separate for at least 5 minutes.
(10) At the beginning of each batch of measurements measure the optical density at 460 nm of the sample cuvette (40 mm) against the reference cuvette when both are filled with pure chloroform. If the optical density differs from that in the previous batch of analyses by more than 0.005, attempt to reduce this difference by cleaning and/or polishing one or both of the cuvettes.
(11) Pass the chloroform extract from stage 9 above through a 90 mm Whatman No. 4 filter paper into the clean, dry sample cuvette. Discard the filter paper.
(12) Within 30 minutes measure the optical density of the sample as in stage 10 above against the reference cuvette filled with pure chloroform. Let the optical density $= A_S$.
(13) After measurement, discard the contents of the sample cuvette, rinse it with pure chloroform, and allow to dry before the next extract to be measured is placed in the cuvette.

Blank determination

(14) Add 500 ml (± 5 ml) of water of low phenol content to a 1 litre beaker.
(15) Repeat stages 2-4 and 5-13 above. Let the optical density $= A_B$.

Calculation

(16) Calculate the optical density A_R due to phenolic compounds in the sample:

$$A_R = A_S - A_B$$

(17) Determine the apparent phenol concentration in the sample from A_R and the calibration curve.

Calibration curve

500, 495, 490, 480, 470, 460, and 450 ml of distilled water (all ± 5 ml) are added to a series of separating funnels, then to these funnels are added 0, 5, 10, 20, 30, 40, and 50 ml (0–50 μg l^{-1}) phenol standard solution C. Stages 2-4 and 5-13 inclusive are repeated on these samples. The average optical density of the blank is subtracted from the average optical densities for each of the phenol concentrations and the corrected optical densities plotted against the concentration of phenol added to the water.

Caution has to be exercised when using this method for the determination of certain types of phenols. Thus recovery of phenols in the distillation stage ranged from near 100% (phenol, 2-chlorophenol) to below 30% (catechol, resorcinol, and other polyhydric phenols). Precision in the pH 10 method as measured within-batch standard deviation is 0.5 µg phenol l^{-1} at the zero phenol concentration level and 1.1 µg l^{-1} at the 50 ng l^{-1} level (without distillation decreasing to 0.6 µg l^{-1} at the 50 µg l^{-1} level with distillation). The precision of results obtained using the pH 7.9 method is shown in Table 44. The analyses of surface waters gives less precise results than for distilled water solutions. In the range 3–63 µg l^{-1} the standard deviation for surface waters was not dependent on concentration and had an estimated value of 2.5 µg l^{-1}, i.e. a relative s.d. of 10% at a concentration of 25 µg phenol l^{-1}.

Table 44 Precision of results using 4-aminoanytipyrine (pH 7.9) method

Distilled water solutions without distillation		Surface water solutions with distillation	
Concentration (µg phenol l^{-1})	Within-batch* standard deviation (µg phenol l^{-1})	Concentration (µg phenol l^{-1})	Within-batch* standard deviation (µg phenol l^{-1})
Blank	1.0 (9)	3–13	2.1‡ (8)
5.0	1.4† (9)	8–18	3.4‡ (8)
50.0	1.2† (9)	53–63	2.1‡ (8)
0–50	1.4§ (27)	3–63	2.6‡§ (24)

*The number of degrees of freedom for each result is given in brackets.
†Calculation from the results—phenol standard minus distilled water blank.
‡Calculated from the results—surface water minus distilled water blank.
§Results obtained by pooling estimates for all concentrations.
Reprinted with permission from Cheeseman and Wilson.[121] Copyright (1972) Water Research Centre.

As the method involves a distillation stage, no interference is to be expected from inorganic impurities in the sample. Distillable organic impurities in the sample, are, of course, always a potential source of interference.

Other workers have investigated the 4-aminoantipyrine (4-aminophenazone) spectrophotometric methods for the determination of phenols, in water[128-131] and effluent.[129,134-137] In the procedure described by Stroehl[129] a 500 ml sample of effluent or water is adjusted to pH 2–3 with phosphoric acid and distilled to 450 ml to separate the volatile phenols. The remaining non-volatile components comprise only a small proportion of the phenol content, but may if desired be separated by liquid-liquid extraction). The distillate is treated with 4-aminoantipyrine in the presence of potassium ferricyanide to give an orange-red complex, which is measured photometrically against a standard. The use of steam distillation was found to give results 10% lower than normal distillation and even the normal method gave a low result when 450–500 ml were distilled from a 600 ml sample; addition of a further 50 ml of water and subsequent distillation brought the result to 98% of the expected phenol content.

Vinson[130] used adsorption of phenols on a column followed by desorption prior to spectrophotometric estimation with 4-aminoantipyrine to improve the sensitivity of the determination of phenols.

Automation of the 4-aminoantipyrine method using an autoanalyser has been investigated.[132,133] Gales and Booth[132] used this method to estimate phenols in waste waters, sewage, and industrial wastes.

Method

Apparatus

Technicon AutoAnalyzer (I or II) or equivalent, consisting of sampler, manifold, proportioning pump II or III, heating bath with distillation coil, distillation head, colorimeter equipped with 50 mm flow cell and 505 nm filter, and recorder.

Reagents

The distillation reagent is produced by adding 100 ml of concentrated phosphoric acid (85% H_3PO_4) to 800 ml of distilled water, cooling, and diluting to 1 l. Sodium hydroxide (1 M) is used.

Buffered potassium ferricyanide, 2.0 g potassium ferricyanide, 3.10 g boric acid, and 3.76 g potassium chloride are dissolved in 800 ml of distilled water. The pH is adjusted to 10.3 with 1 M sodium hydroxide and the solution is diluted to 1 l. Then 0.5 ml of Brij-35 is added.

4-Aminoantipyrine, 0.065% w/v is used.

Ferrous ammonium sulphate, 1.1 g is dissolved in 500 ml distilled water containing 1 ml of sulphuric acid diluted to 1 l with freshly boiled and cooled distilled water.

Phenol, 1 mg ml^{-1} in distilled water, copper sulphate (1 g), and concentrated phosphoric acid (0.5 ml) are added as preservatives.

Procedure

The manifold is set up as shown in Figure 54 or 55. Polyethylene tubing should be used for the sample line and glass test tubes for standards and samples.

The precision of this method was determined at various concentration levels for two working ranges (2–100 μg l^{-1} and 5–500 μg l^{-1}). For phenol concentrations of 4, 15, 43, and 89 μg l^{-1}, the standard deviations were ± 0.5, ± 0.6, ± 0.6, and ± 1.0 μg l^{-1} respectively. At concentrations of 73, 146, 299, and 447 μg l^{-1} the standard deviations were ± 1.0, ± 1.8, ± 4.2, and ± 5.3 μg l^{-1} respectively.

The percentage recovery was determined at two different levels for each working range by spiking the waste sample with four concentrations of phenol.

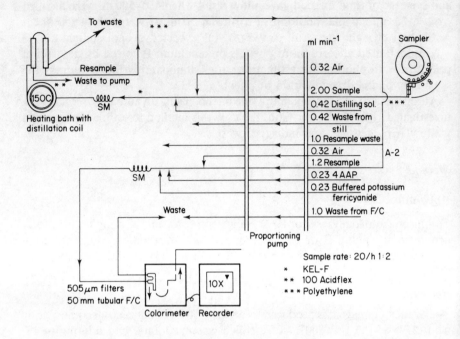

Figure 54 Flow setup for Analyzer I

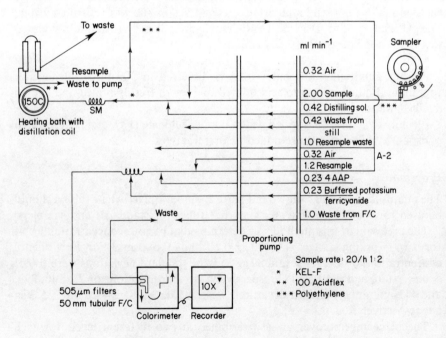

Figure 55 Flow setup for Analyzer II

At concentrations of 5, 83, 168, and 490 µg l^{-1}, the recoveries were 72, 98, 97, and 98% respectively. Results obtained by the automated method were 3–4% lower than those obtained by the manual method.

The earliest references found to the use of 3-methyl-2-benzothiazolinone hydrazone as a colorimetric reagent for phenols are those of Pays and Bourden[138] and Freistad et al.[139,140] However, the first workers to apply this reagent to water samples were Freistad,[140] followed by Goulden et al.,[141] Lei and Pang,[142] and Gales.[143] Although the method had a limit of detection of about 10 µg l^{-1} for phenol, this limit could possibly be improved if the coloured reaction products of the reagent and phenols could be concentrated by solvent extraction (as in the aminoantipyrine procedures). This new reagent generally reacts with parasubstituted phenols to a greater extent than aminoantipyrine, though the latter reagent offers rather greater sensitivity for some compounds, e.g. 2,4-dichlorophenol.

Goulden et al.[141] described an automatic continuous flow apparatus for this spectrophotometric determination, which is based on reaction of the phenols with 3-methylbenzothiazolin-2-one hydrazone. Up to 80% of the sample flow is distilled and condensed and after the colour-formation step, the dye is extracted into chloroform before measurement. At a concentration of 5 µg of phenol l^{-1} the coefficient of variation was 3.6% and 2.4%.

Lei and Pang[142] determined phenols in water in amounts down to 0.01 ppm using this reagent. The sample (40 ml) is adjusted to pH 4 with aqueous sulphuric acid, 10% aqueous cupric sulphate solution (1.2 ml) is added, and the solution is steam distilled. To this distillate (40 ml) are added 0.5% aqueous 3-methyl-benzothiazolin-2-one hydrazone hydrochloride (1 ml), 1% 2(NH$_4$)$_2$SO$_4$–Ce(SO$_4$)$_2$ in 4% sulphuric acid (1 ml), NaOH–EDTA–H$_3$BO$_3$ buffer solution (2 ml) and water (6 ml) and the extinction is measured at 520 nm after 15 min. Recoveries of 2–10 µg of phenol added to water were 95–110%. The method is relatively free from interference by various cations and anions, which are separated at the distillation stage. The effect of aldehydes is avoided by heating the final solution at 60 °C for a few minutes before the measurement of colour.

Gales[143] investigated the use of 3-methyl-2-benzothiazolinone hydrazone for the determination of down to 1 µg l^{-1} of phenols in water and waste waters. Phenolic compounds were determined by manual and automated methods based on coupling with 3-methyl-2-benzothiazolinone hydrazone in an acidic medium using ammonium cerium(IV) sulphate as an oxidant.

Ultraviolet and fluorescence methods

Fountaine et al.[144] used an ultraviolet spectroscopic method for the determination of traces of phenolic compounds using two conventional sealed hollow cathode lamps to monitor the ultraviolet bathochromic shift which occurs when phenolic compounds are made basic and its use in the determination of traces of several phenols, including phenol, o-cresol, p-cresol, resorcinol, thymol, p-methoxyphenol, and tyrosine, is outlined, compared to the 4-aminoantipyrine method and results for both methods are presented. The ultraviolet ratio spectrometric method has the advantage that it can detect para-substituted phenols.

Afghan et al.[145] mentions the use of fluorescence methods for determining phenols in natural waters. This method uses n-butyl acetate or isoamyl acetate extraction followed by back extraction with sodium hydroxide solution to extract the phenol from the water sample.

Gas Chromatography

One of the earliest gas chromatographic methods for the determination of phenols in water involved absorption of phenols on carbon, extraction of the carbon with chloroform to remove the phenols, concentration of the phenols in the chloroform by evaporation, and determination of phenols in the chloroform extract by gas chromatography.[146] Since then various workers have studied the application of gas chromatography to the determination of traces of mono- and dihydroxy phenols in natural and polluted waters, including phenol, cresols, xylenols, and the resorcinol.[147-165] Some gas chromatographic methods are applied direct to the sample[166-168] whilst others use a prior concentration technique, frequently solvent extraction.[169-174]

Semenchenko and Kaplin[152] acidified the sample with hydrochloric acid, then saturated it with sodium chloride. The phenols were extracted with diethylether. The extract was shaken with aqueous sodium bicarbonate (5%) to remove any organic acids, then the phenols extracted into aqueous sodium hydroxide and methylated with dimethyl sulphate. The methyl derivatives were analysed in a glass column (1.5 m × 3.5 mm i.d.) packed with 10% of tritolylphosphate–Apiezon L (1:1) supported on Chromosorb W; the column was operated at 120 °C, with nitrogen (40 ml min^{-1}) as carrier gas and a flame ionization detector. Determination is by means of peak-area measurement. Down to 1 μg of phenols l^{-1} could be determined (in 200 ml water samples) with a relative error less than ±15%.

Eichelberger et al.[153] collected the phenols from water samples on columns of activated carbon. The phenols were stripped from the carbon with chloroform prior to gas chromatography of the cleaned up extract. Clean-up was achieved by a treble extraction of the chloroform phase with aqueous sodium hydroxide. The aqueous extracts were combined, acidified with concentrated hydrochloric acid to pH 2 and extracted three times with ethyl ether, and the combined ether extracts are passed through a 10 cm Florisil column topped with anydrous sodium sulphate (2 cm). The phenols are eluted with ether, and the eluate is concentrated by evaporation and analysed by gas chromatography. An aluminium column (10 ft × 0.125 in o.d.) packed with 10% of Carbowax 20 M terephthalic acid on HMDS-treated Chromsorb W (60–80 mesh) was used operated at 210 °C for the majority of phenols with nitrogen (50 ml min^{-1}) as carrier gas and flame ionization detection. Recoveries were on average, 85% for alkylphenols and 94% for chlorophenols.

Cooper[155] converted phenols to their pentafluorobenzyl derivatives prior to gas chromatography.

More recent gas chromatographic methods for the determination of traces of phenols in water samples, include those of Murray,[175] the Yorkshire Water Authority,[176] Voznakova and Popl,[177] and Goldberg and Weiner.[178]

Murray[175] determined low concentrations of phenols, cresols, and xylenols in chloroform extracts of water. The trimethylsilyl derivatives of the phenols were formed and analysis completed by gas chromatography. The method was rapid and required a minimum of sample manipulation. The lower limit of detection was 0.100 mg l^{-1} for phenol, 0.025 mg l^{-1} for cresols, and 0.050 mg l^{-1} for xylenols.

In this method an internal standard is added to the water samples, followed by a single extraction with chloroform. Since a ratio is established when the internal standard is added, and since a constant proportion of each organic compound is removed by a solvent extraction, only a portion of the total organic material present need be removed for analysis. The extract is a concentrated derivative formed and analysed by gas chromatography.

Method

A dual column gas chromatograph with flame ionization detectors was used in a single column mode in conjunction with a digital integrator. The columns used were 8 ft × 1/16 inch o.d. stainless steel and were packed with 5% Imol coated on Chromosorb W, AW, and DMCS treated. The columns were used isothermally at 120 °C, or programmed from 80 to 120 °C at 2 °C min^{-1} to provide a better separation.

Operating conditions

Injector temp	150 °C
Manifold and detector temp	150 °C
Carrier gas (nitrogen)	25 ml min^{-1}
Hydrogen	35 ml min^{-1}
Air	400 ml min^{-1}

A 1 litre sample of water calibration containing known weights of phenols was placed in a 1 litre separatory funnel and 5 µl of o-xylene carefully added by syringe as an internal standard. The sample was shaken and a single solvent extraction made by shaking with 50 ml of chloroform for 2 min. When the solvent layer had separated it was transferred to a 100 ml pear-shaped flask with a ground-glass joint and reduced to about 1 ml on a rotary evaporator at 25 °C under vacuum. Care must be taken to avoid losses of volatile material at this stage by preventing the flask from drying out. The sample was treated with 250 µl of Tri-Sil concentrate (Pierce Chemical Co.[179]) at room temperature for 30 min to form trimethylsilyl derivatives. It was necessary to leave the xylenols overnight to react, possibly because of steric hindrance. The solution of the derivatives was injected into the gas chromatograph, the peak areas measured, and the ratios of the phenols to the internal standard calculated. The ratios were plotted against the original amount of phenol, cresol, or xylenol added and a linear relationship was found in all cases over the range 0–10 mg l^{-1}. The calibration factors were calculated as in Table 45.

Table 45 Calibration factors in $R = KC$ where R is ratio to a constant amount of internal standard and C is concentration in mg l^{-1}

	K
Phenol	0.027
m-Cresol	0.100
p—Cresol	0.104
3,5-Dimethylphenol	0.135
o-Cresol	0.136
3,4-Dimethylphenol	0.143
2,3-Dimethylphenol/2,5 dimethylphenol	0.178
2,4-Dimethylphenol/2,6 dimethylphenol	0.192
2-Ethylphenol	0.230

Reproduced with permission from Murray.[175] Copyright (1975) Fisheries Research Board, Canada.

In Figure 56 is shown a gas chromatogram of a mixture of phenols obtained by this procedure. The percentage extraction from a 1 litre sample was: phenol, 20%; m-cresol, 39%; p-cresol, 40%; o-cresol, 48%; 3,4-xylenol, 8%; 2,6-xylenol, 83%; o-xylene, 95%. Phenol is the most hydrophilic and its extraction was the lowest. To achieve a 90% recovery of phenol, more than 10 extractions would be required. Once a ratio to the internal standard had been established only a portion of the total organic material need be extracted and only one solvent extraction is necessary. When the chloroform is evaporated to a small volume, the net effect is a 200 times concentration for phenol. Recoveries of phenols obtained in spiking experiments at the 5–10 ml l^{-1} level on hard water, soft water, river water, and sewage effluent were in the range 98–100%.

The method used by the Yorkshire Water Authority[176] for determining phenols in potable waters has a criterion of detection of 3–70 μg depending on the phenol. It is capable of determining phenols and cresols, xylenos, dihyric phenols, monochlorophenols, dichlorophenols, and trichlorophenols. The phenols are extracted from the water sample with ethyl acetate using a liquid-liquid extractor. After drying the extract is concentrated to a small volume then treated with *bis*(trimethylsilyl)trifluoroacetamide to produce the phenol trimethylsilyl ethers. Retention times of various phenols on these gas chromatographic columns are tabulated in Tables 46–48. Unfortunately this method is rather lengthy, requiring about 2 days per sample.

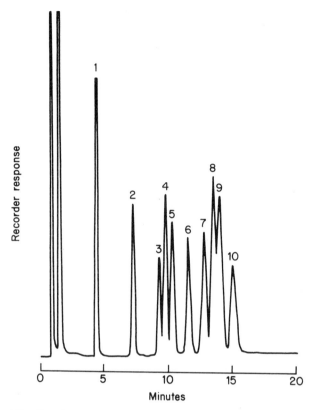

Figure 56 Gas-chromatographic temperature programme, 40 min^{-1}. 80–120 °C. 1, o-xylene; 2, phenol; 3, o-cresol; 4, m-cresol; 5, p-cresol; 6, 2,5-xylenol; 7, 2,4-xylenol, 3,5-xylenol; 8, 2,6-xylenol; 9, 2,3-xylenol; 10, 3,4-xylenol. Peak 7 contains both compounds listed. Reproduced with permission from Murray.[175]. Copyright (1975) Fisheries Research Board, Canada

Voznakova and Popl[180] departed from the usual approach to phenol analysis by the gas chromatography made by earlier workers, i.e. direct injection gas chromatography or solvent extraction followed by gas chromatography. These workers described a method for determining phenol, o-cresol, m-cresol, and 2,6-xylenol in water in which the phenols are sorbed on a macroporous polymer sorbent and then undergo thermal desorption for gas chromatographic analysis. This method is claimed to be quick and reproducible and have a detection limit of between 1 and 10 ppb, depending on the phenol.

Method

Apparatus

For the sorption of the phenols from aqueous solutions a stainless steel column was used, 100 mm × 5 mm i.d., with a wall thickness of 0.4 mm, equipped with a

Table 46 Retention times of phenol–TMS ethers on 7% Carbowax 20M at (a) 130 °C, (b) 180 °C

Phenol–TMS ether of	Retention time (a) (min from injection)	Relative retention time (compared to phenol–TMS ether)
Phenol	3.22	1.00
o-Cresol	4.07	1.26
p-Cresol	4.37	1.36
p-Cresol	4.65	1.45
2-Ethylphenol	4.95	1.54
2,4-Xylenol	5.97	1.85
3,5-Xylenol	5.97	1.85
4-Ethylphenol	6.50	2.02
2,6-Xylenol	6.75	2.10
Catechol	6.93	2.16
2-Chlorophenol	7.77	2.41
4-Chlorophenol	9.22	2.87
Resorcinol	10.37	3.22
Quinol	11.80	3.67
4-Chloro-2-methylphenol	12.42	3.86
5-Indanol	18.37	5.71
2,4-Dichlorophenol	>30	—
2,6-Dichlorophenol	>30	
2,4,6-Trichlorophenol	>30	
	Retention time (b)	
Pyrogallol	3.27	
-octylphenol	6.03	
-cyclohexylphenol	6.83	
-phenylphenol	11.30	

Table 47 Retention times of phenol–TMS ethers on 5% Trixylenlyl phosphate at 135 °C

Phenol–TMS ether of	Retention time (min from injection)	Relative retention time (compared to phenol–TMS ether)
Phenol	3.54	1.00
	Interfering peak	
o-Cresol	4.72	1.33
mCresol	5.10	1.44
p-Cresol	5.54	1.56
2,4-Xylenol	7.63	2.16
3,5-Xylenol	7.63	2.16
2,6-Xylenol	8.57	2.42
Catechol	10.50	2.97
Resorcinol	15.87	4.48
Quinol	17.52	4.95

heating mantle coupled with a timing voltage programmer. The column was packed with 0.4 or 0.8 g of the sorbent, Separon SE.[181]

A micropump was employed for the pumping of aqueous solutions of the phenols through the sorption column. A glass column was used, 3 mm i.d. with

Table 48 Retention times of phenol–TMS ethers on 5% E301 at 140 °C

Phenol–TMS ethers of	Retention time (min from injection)	Relative retention time (compared to phenol–TMS ether)
Phenol	2.13	1.00
2-Chlorophenol	3.80	1.78
4-Chlorophenol	4.25	1.99
4-Chloro-2-methylphenol	6.08	2.85
2,6-Dichlorophenol	6.95	3.26
2,4-Dichlorophenol	7.37	3.45
2,3,6-Trichlorophenol	12.50	5.86

a length of 120 cm packed with Separon SDA. The operating temperature limit was 190 °C, the injection temperature 200 °C, carrier gas nitrogen, flow rate 80 ml min^{-1}, and a flame ionization detector was used.

Materials

The phenols were isolated from the aqueous solutions by means of Separon SE, a styrene–ethylenedimethacrylate copolymer (particle size 32–40 μm), which is thermally stable up to 280 °C with a sorption capacity for *m*-cresol of 10 mg g^{-1} of the sorbent for the cresol concentration in water, 7.7 mg l^{-1}.[182] Separon SDA, a styrene–divinyl benzene–acrylonitrile copolymer, served as the column packing (particle size 100–160 μm stable up to 230 °C).[183] Separon SDA remains unaffected even with large quantities of water in the column.

Procedure

The procedure of preconcentration and determination of phenols comprises four steps: sorption of the phenols from the aqueous solution on the solid sorbent; partial drying of the sorbent prior to the phenol desorption; thermal desorption of the phenols and their transfer into the chromatographic column, and the determination of the phenols by gas chromatography.

The apparatus is shown in Figure 57. In the first step, the six-way valve, 4, was adjusted in the position, a–f, and the sample was pumped with a given flow rate through the sorption column, 5; the inlet, a, of the six-way valve was connected to the micropump, 2, output. From the outlet, b, of the valve the sample freed from the pollutants was drawn off. After this step, the sorption column was washed with distilled water, the micropump was disconnected and the nitrogen stream was fed from the chromatograph carrier gas supply system into the inlet a. The other branch of the gas supply system was employed for the chromatographic determination itself. The nitrogen displaced most of the water remaining on the sorption column after the washing. Thereafter, the temperature programme for the sorption column drying was switched in on the voltage programmer, and the column packing was partly dried in the nitrogen stream. The moisture was passed into the atmosphere through the six-way valve.

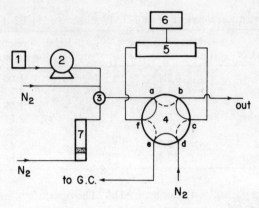

Figure 57 Simplified flow diagram of the apparatus for the phenols determination. 1, sample; 2, micropumps; 3, three-way valve; 4, six-way valve; 5, sorption column; 6, timing voltage programmer; 7, stripping vessel. Reproduction from the *Journal of Chromatographic Science* by permission of Preston Publications, Inc.

After the drying, the six-way valve was set in the position, a–b; the sorption column was connected to the carrier gas supply and to the analytical column of the gas chromatograph; the carrier gas flow rate was adjusted for the desorption, and the desorption temperature programme was switched in. In this step the thermally released phenols were carried to the cold chromatographic column and trapped at the inlet. After the completed desorption the six-way valve was set back to the position, a–f, the carrier gas flow rate was adjusted to the value appropriate for the chromatographic determination, the column thermostat was heated up to 190 °C at a rate 20 °C min^{-1}, and the chromatogram was recorded. When the chromatographic analysis was finished, the thermostat was cooled down. Simultaneously with this step the sorption of the phenols from the next sample proceeded.

The quantitative evaluation relied on the comparison of the peak area of the phenol obtained from the above procedure with that of the same phenol obtained by direct injection. During the analysis of the real water samples the internal standard method was applied.

Depending on the phenol Voznakova *et al.*[181] found that phenol recovery was dependent on desorption time, nitrogen purge volume, and desorption temperature (Table 49). When a nitrogen flow rate of 480 ml min^{-1} is used, 20 min is sufficient for desorption. Other components which exhibited a greater affinity towards the sorbent than phenol, e.g. the two cresols and 2,6-xylenol, were also examined. The application of higher temperature is limited by the thermal stability of Separon SE.

Phenol and its derivatives occur in waters frequently together with petroleum products. The two substance types differ considerably from one another as to their chemical natures and physical properties. While hydrocarbons are typical

Table 49 Dependence of phenol recovery on the temperature and time of desorption and on the volume of nitrogen passed

Desorption time (min)	Nitrogen volume (l)	Recovery (%) Desorption temperature		
		170 °C	180 °C	200 °C
3.5	1.68	24	40	71
10.0	4.80	30	48	80
15.0	7.20	38	56	86
30.0	15.40	42	55	84
45.0	21.50	43	59	83

Sample volume: 25–50 ml.
Sorption: amount of the sorbent, 0.8 g; temperature, 20 °C; sample flow rate 2 ml min^{-1}.
Drying: temperature, 70 °C; nitrogen flow rate; 170 ml min^{-1}, time, 15 min.
Desorption: temperature; 200 °C; nitrogen flow rate; 480 ml min^{-1}, time, 20 min.
Reprinted from the *Journal of Chromatographic Science* by permission of Preston Publications, Inc.

representatives of inert substances, only partly miscible with water, interaction can occur between the functional hydrogen bond formation accompanied by the creation of associates. This can be conveniently utilized for a separation of the two types of substances: while volatile hydrocarbons can be well stripped from the aqueous solutions with an inert gas, phenols can be expected to be retained by the aqueous phase dueto the interaction with the water molecules. In order to verify this assumption, the conditions used for the stripping of volatile aromatic hydrocarbons from water, were applied by Voznakova and Popl[180] to aqueous solutions of phenols. No phenol stripping was observed for neutral or acidic solutions even after the passage of 9000 ml nitrogen, at temperatures up to 85 °C. This was also true in acidified solutions (pH = 1).

Based on these results, the following procedure is suggested for the determination of trace quantities of volatile aromatic hydrocarbons and phenols in waters. The water sample (100 ml) is placed in the stripping vessel. Volatile aromatic compounds and other possible volatile substances are stripped with nitrogen, which is trapped from the gaseous phase on a sorption column of Separon SE, thermally desorbed, and swept into the chromatographic column and determined as described by Voznakova et al.[181] Of the remaining solution containing phenols and possible other higher-boiling hydrocarbons, a 50 ml aliquot is then pumped through the dried, cold (after the desorption of the volatile substances) sorption column. After a partial drying and thermal desorption the phenols, along with the other possible desorbed pollutants, are swept into the cold chromatographic column and determined.

The presence of both types of substances—hydrocarbons and phenols—simultaneously in the sample does not appreciably affect their recovery, as can be seen for various concentration ratios in Table 50. The limit of determination of phenols for 50 ml batches is a few ppb. The technique ensures efficient preconcentration of the phenols and reproducible results, and owing to the relatively small sample volume analysed it is not too time-consuming.

Table 50 Recovery of alkylbenzenes and phenols for different concentration ranges during simultaneous determinations

Compound	Concentration (ppb)	Recovery (%)	Concentration (ppb)	Recovery (%)	Concentration (ppb)	Recovery (%)
Toluene	10	98	25	99	125	100
Ethylbenzene	10	95	25	96	125	96
n-Butylbenzene	10	95	25	93	125	92
Phenol	100	86	50	89	20	84
m-Cresol	100	98	50	100	20	96
2,6-Xylenol	100	98	50	100	20	96

Conditions for recovery of alkylbenzenes, see reference 181.
Reproduced from the *Journal of Chromatographic Science* by permission of Preston Publications, Inc.

Goldberg and Weiner[224] used solvent-heavier-than-water, two-cycle, liquid extractors to concentrate phenols at the μg l^{-1} level from water into dichloromethane. The non-aqueous solution containing the extract was concentrated by Kuderna-Danish concentrators, and this was followed by gas chromatography. Overall concentration factors were around 1000 with efficiencies 23.1–87.1%. Sediment concentration efficiencies were 18.9–73.8% using a batch method. Determinations could be made with accuracy of only 15–20% because of solute losses during concentration. The range of concentration used was of the order of 1 μg l^{-1}.

Method

Apparatus

The specially designed solvent-heavier-than-water, two-cycle, liquid-liquid extractors have been described by Goldberg *et al.*[184] The non-aqueous solution containing the extract collected by this apparatus was concentrated by evaporation in Kuderna-Danish concentrators with 250 ml sample flasks.[185] The Kuderna-Danish concentrators employ a three-ball Snyder column furnishing about 2.7 plates. Gas chromatographic separations on the concentrated solutions utilized a dual-column instrument with flame ionization detector. The column (6 ft long with a ⅛ in stainless steel wall) was packed with silanized 5% SP 2401 on Chromosorb W.

Procedure

The extraction efficiencies and concentration factors obtainable were determined as follows. Four extractors were connected in series such that the effluent from extractor 1 was the influent for extractor 2, and so forth. A test sample was made by dissolving approximately 5 mg each of three different phenols in 18 l of water. The flow rate through the extractors was 7–8 l h^{-1}. After the dichloromethane extraction (technical-grade dichloromethane distilled through a 50 plate, programmed-head glass still) the non-aqueous solvents containing the extracted solutes from each extractor (ca. 500 ml each) were evaporated in the Kuderna-Danish concentrators to a final total volume of about 6 ml. The final solute concentrate was silanized by refluxing with hexamethyldisilazane and analysed by gas chromatography. To calibrate the system, standard phenol samples were dissolved in dichloromethane and silanized by the same reflux procedure. Phenolic concentrations were determined from the peak area. Overall extraction and concentration efficiencies were calculated from $E = (C_c V_c / m_s) \times 100$, where E is the extraction and concentration efficiency, C_c the concentration of a particular phenol in the final concentrate, V_c the volume of the final Kuderna-Danish concentrate, and m_s the mass of the particular phenol present in the water sample that was introduced into the liquid-liquid extractors. The concentration factor was determined only for the liquid-liquid extractor system. It was taken as the ratio of the concentration of a particular phenol in the final Kuderna-Danish concentrate to its concentration in the initial sample. The values are reported in Table 51.

Table 51 Concentration factors and overall extraction and concentration efficiencies for various phenolic compounds

Phenolic compounds	Continuous liquid-liquid extractor-concentrator system*		Soxhlet extractor-concentrator system[†]
	Overall extraction and concentration efficiency (%)[‡]	Overall concentration factor[§]	Overall concentration efficiency (%)[‡]
Phenol	24.8	910	18.9
o-Chlorophenol	39.4	1200	—
m-Chlorophenol	49.5	1500	—
p-Chlorophenol	23.1	690	—
2,4-Dichlorophenol	87.1	3750	64.7
2,4,5-Trichlorophenol	—	—	73.8
1-Naphthol	—	—	68.6
2-Naphthol	—	—	63.8
o-Methoxyphenol(guaiacol)	36.5	1250	31.6
o-Aminophenol	—	—	42.7
o-Nitrophenol	43.5	1320	46.6
o-Cresol	29.4	1080	65.9
m-Cresol	29.9	1100	—
2,6-Dimethylphenol	61.5	1890	51.0

*Four continuous liquid-liquid extractors connected in series and Kuderna-Danish concentrators.
[†]Soxhlet extractor and Kuderna-Danish concentrator.
[‡](Mass of phenol in final concentrate) (mass of phenol in original sample) × 100.
[§]Concentration of phenol in final concentrate/concentration of phenol in original sample.
Reprinted with permission from Goldberg and Weiner.[224] Copyright (1980) Elsevier Science Publishers.

The results obtained by this method are compared with those obtained by conventional Soxhlet extraction in Table 51. The overall extraction and concentration efficiencies of the liquid-liquid extractor system are of more significance for monitoring trace phenolic levels in water than are the overall concentration sample and final volume of concentrate. The ability of the liquid-liquid extractor system to provide large overall concentration factors is the result of its ability to handle large sample volumes in conveniently short times. The principal limiting factor in obtaining arbitrarily large concentration factors is the loss of solute during the evaporation in the Kuderna-Danish concentrators. Goldberg et al.[184] showed that it is not unusual to lose 50% or more of the extracted solute during the evaporation from 500 ml to around 5 ml. There is still an increase in solute concentration, however, because the solvent volume decrease factor is approximately 50 times greater than the solute loss factor. The solute loss rate increases as the solvent volume decreases. In the Kuderna-Danish concentrator, this factor becomes increasingly important as the solvent volume is reduced below about 5 ml and a limit is reached where the solute loss and solvent volume decrease during evaporative concentration results in no net increase in solute concentration.

Various other workers have discussed the application of gas chromatography to the determination of phenols in tapwater,[186] gas condensates,[187] and coke work effluents.[188-190]

Thin-layer chromatography

Thielemann[191] has described a thin-layer chromatographic method for the identification of phenolic impurities in diethylether or active carbon extracts of water. The separation and identification of o, m-, and p-cresols, phenol, 2-naphthol, and 2,5-, 3,4-, and 3,5-xylenols, after coupling with Fast red salt AL (C.I. Azoic Diazo Component 36), was carried out on Kieselgel G impregnated with potassium carbonate with the use of one of three solvents. Polyhydric phenols were not identifiable by this method, but were separated by direct thin-layer chromatography on Kieselgel G or D with dioxan–benzene–anhydrous acetic acid (25:9:4) or benzene–acetone (9:10) as solvent, and detected by spraying with diazotized sulphanilic acid or with molybdophosphoric acid.

Thielemann[192] also described a thin-layer method for the determination of phenol, cresols, xylenols, and naphthols in active carbon extracts of natural and potable water. Although the direct separation of these phenols was not possible, products of their coupling with Fast blue salt BB (C.I. Azoic Diazo Component 20) could be separated. The R_f values are reported for the dye complexes for phenol o-, m-, and p-cresols, 1- and 2-naphthols, and 3,4-, 3,5-, and 2,5-xylenols.

Thielemann[193] also achieved a thin-layer chromatographic separation of phenol, three xylenols, and 1- and 2-naphthols after coupling water and effluents with diazolized Fast Blue Salt BB (C.I. Azoic Diazo Component 20). The separation of the coupled phenolic compounds was achieved with three solvent systems on plates coated with Kieselgel G–potassium carbonate (1:2). Rf values are reported. The separation of phenol and the cresols by paper chromatography is also reported.

Edeline et al.[194] described a method for determining phenols in environmental samples based on separation on thin-layers of activated carbon.

Ragazzi and Giovanni[195] determined phenols in waste water for processing olives by a thin-layer chromatographic method. The phenols were separated in amounts from 10 to 100 µg on silica gel or cellulose with various solvent systems. The compounds were detected by spraying the plate with 20% sodium carbonate solution and then with Folin-Ciocalteu reagent. The material from each spot was removed from the plate and suspended in water or methanol then Folin-Ciocalteu reagent and 20% sodium carbonate solution were added. The solution was diluted to appropriate volume and centrifuged. The extinctions of the blue solution were measured at 725 nm against blank solution obtained by similarly treating blank areas of the plate.

Rump[196] has described a cellulose thin-layer method for the detection of phenolic acids such as m-hydroxybenzoic acid, m-hydroxyphenylacetic acid and m-hydroxyphenylpropionic acid, in water samples suspected to be contaminated with liquid manure. The phenolic acid is extracted with ethyl acetate from a volume of acidified sample equalling 1 mg of oxygen consumed (measured with potassium permanganate). The ethyl acetate is evaporated and the residue dissolved in ethanol. After spotting of a 1 µl aliquot on a cellulose plate the chromatogram is developed by capillary ascent with the solvent n-propanol: n-butanol: 25% NH_3: water (4:4:1:1 by vol). The solvent front is allowed to advance 10 cm. The air-dried plate is sprayed with a diazotized p-nitroaniline reagent to make the phenolic acids visible.

Location reagent

Diazotized p-nitroaniline

Solution A: 0.1 g p-nitroaniline is dissolved in 50 ml of 2 N hydrochloric acid and diluted to 100 ml with water. Solution B is a 0.2% solution of sodium nitrite in water. Solution C is a 10% solution of potassium carbonate in water.

The solutions are mixed in the following ratio: solution A: solution B: solution C (1:1:2.5, by vol). Before mixing, the required volume of each solution is cooled separately to 5 °C in an ice bath. Solution A is cooled in the spray bottle. Solution B is added, and the mixture is kept cool for 5 min. Afterwards solution C is added and the plate sprayed at once.

Preparation of samples for chromatography

A volume of sample is pipetted into a 100 ml separatory funnel and then diluted to 50 ml. The pH is adjusted to 2 with 2 ml of 1 M hydrochloric acid and 13 g of sodium chloride is added. The solution is extracted with 20 ml of ethyl acetate. The lower, aqueous layer is discarded. The ethyl acetate layer is transferred to a beaker and allowed to evaporate over an electric hot plate down to 3 ml. If precipitation occurs, the remaining amount of ethyl acetate phase is decanted to another beaker and the precipitate washed with a small amount of ethyl acetate. This is transferred to the other beaker while the precipitate is discarded. Evaporation on the heating plate continues until approximately 1 ml of liquid remains. The residue is quantitatively made up to 0.5 ml with 96% ethanol.

Chromatography

One μl of the ethanol solution and 1 μl of the working standard solution (0.01 μg μl^{-1} are spotted separately on the plate. The plate is transferred to the developing tank containing the developing solvent. The solvent front is allowed to ascend 10 cm from the origin. This takes about 2.5 h. The plate is removed from the tank and air-dried for 30 min. The dry plate is sprayed with the location reagent to make the phenolic acids visible (Table 52).

Table 52 R_f values and qualitative colour reactions of the phenolic acids

Phenolic acid	R_f	Colour
m-Hydroxybenzoic acid	0.32	Reddish brown
m-Hydroxyphenylacetic acid	0.37	Purple
m-Hydroxyphenylpropionic acid	0.47	Purple

Reprinted with permission from Rump.[196] Copyright (1974) Pergamon Press.

Thielemann[197] has used thin-layer chromatography to study the effect of chlorine dioxide on 1- and 2-naphthols in potable water. The coloured products obtained are thought to be condensation products of chloroderivatives of 1,2- or 2,6-naphthaquinone. Thielemann applied paper chromatography to a study of the reaction products of polyhydric phenols with chlorine dioxide.

High performance liquid chromatography

Conventional column chromatography has been applied to the determination of traces of phenols in industrial effluent and surface waters.[198,201-204] More recently, high performance liquid chromatography has been applied to this problem.[199] Hashimoto *et al.*[200] carried out a high performance liquid chromatography study on the radiolysis of phenol in aqueous solution. This showed that in the oxygenated solution, hydroquinone, pyrocatechol, hydroxyhydroquinone, and trace amounts of resorcinol and phloroglucinol are produced. In the deaerated solution there were hydroquinone, pyrocatechol, and small amounts of resorcinol and hydroxyhydroquinone. The decomposition rates of phenol, hydroquinone, and pyrocatechol were about five times higher in the oxygenated than in the deaerated solution.

Armentrout *et al.*[199] carried out trace determinations of phenolic compounds in water at the ppb level by reverse phase liquid chromatography using electrochemical detection with a carbon–polyethylene tubular anode.

Raman spectroscomy

Haverbeke and Herman[205] used laser excited resonance Raman spectroscopy to determine phenolic compounds at the 20–100 ppb level in distilled water, tap water, and pond water. The phenols were first converted to drivatives by reaction with the diazonium salt of 4-nitroaniline.

Haverbeke and Herman[205] used a closed-loop flow-through setup in their experiments. In this setup, a large volume of the sample solution, usually 650 ml, was continuously flowed through the laser beam. Since the individual dye molecules are only momentarily exposed to the laser beam, any decomposition occurring will be reduced to a minimum, thus assuring more accurate measurements.

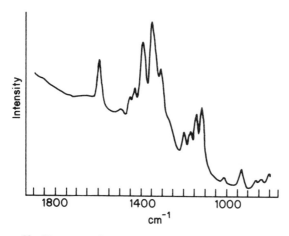

Figure 58 Resonance Raman spectrum of a 5 ppm solution of phenol after applying the 4-nitroaniline derivatization method. Reprinted with permission from Haverbeke and Herman.[205] Copyright (1979) American Chemical Society

The spectrum of a 5 ppm solution of phenol in distilled water and treated by the 4-nitroaniline derivatization method is given in Figure 58. As can be seen, there is a definite contribution from a fluorescence band of dye, on which the resonance Raman spectrum is superimposed. The individual resonance Raman bands are, however, clearly observed. The broad, weak band around $1640\,cm^{-1}$ is due to deformation vibration of the water solvent molecules.

When decreasing the concentration of phenol, some changes in the spectrum are noticed. Obviously, the $1640\,cm^{-1}$ water band becomes relatively more intense. Secondly, the bands around 1120, 1180, 1340, and $1590\,cm^{-1}$ decrease much less in relative intensity than the remaining bands. This is clearly seen in the spectrum of a 50 ppb solution of phenol as shown in Figure 59A. This effect can be explained by a contribution to this spectrum by the Raman spectra of the added reagents. Indeed the components of the buffer solution that is added, sodium carbonate and EDTA, are present in non-neglectable

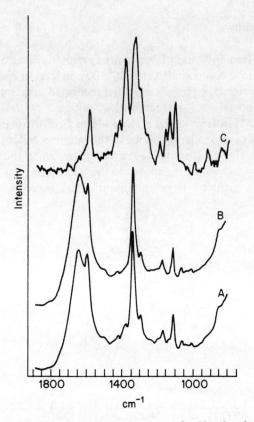

Figure 59 Resonance Raman spectra of a 50 ppb solution of phenol after applying the 4-nitroaniline derivatization method. (A) Original spectrum, (B) background spectrum, (C) spectrum obtained from the original spectrum after background subtraction. Reprinted with permission from Haverbeke and Herman.[205] Copyright (1979) American Chemical Society

concentrations. Furthermore, if the concentration of phenol is relatively low, a large excess of the diazonium salt is present. At ambient temperature and in basic conditions this diazonium salt is converted into the 4-nitrophenolate anion. Having a yellow colour, this compound may also be apt to exhibit preresonance Raman spectra as has been found for other members of this family.[206] Since the Raman spectra of the buffer components and/or the 4-nitrophenolate anion may be relatively important, they may contribute to the total Raman spectrum, especially if the concentration of the dye in solution is very low. The combination of the former spectra may be considered as a 'background spectrum' on which the spectrum of the dye is superimposed. The background spectrum is displayed in Figure 59A. It may be noticed that the most intense bands of this background spectrum coincide with the most intense bands of the spectrum of the 50 ppb phenol sample solution in Figure 59A.

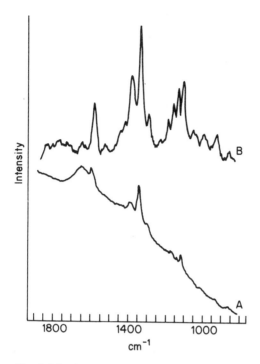

Figure 60 Original (A) and background subtracted (B) spectra of a 200 ppb solution of phenol in pond water. Reprinted with permission from Haverbeke and Herman.[205] Copyright (1979) American Chemical Society

To obtain the spectrum of the dye, the spectra of the sample solution and of the background have to be subtracted from one another. With a single beam instrument this can be done only by additional data manipulation. The data of the two spectra are digitized and the background spectrum is subtracted from the sample spectrum, based on equal intensity of the 1640 cm^{-1} water band in both spectra. The result of this subtraction, after the necessary baseline correction and a soft spectral smoothing (three times a triangular smoothing over three points), is displayed in Figure 59C. Comparison of this spectrum with the one in Figure 58 shows a remarkable agreement, except for the signal-to-noise which is obviously less in the spectrum of the lower concentration. Using this spectra subtraction procedure, concentrations of phenols down to 20 ppb could be detected and identified. Figure 60 shows the Raman Spectrum obtained with a 200 ppb solution of phenol in water.

Haverbeke and Herman[205] also carried out a preliminary investigation on the quantitative aspects of the method. For this purpose, they made up a series of standard solutions ranging from 100 ppb to 7 ppb. They determined the height ratio of the 1590 cm^{-1} band over the 1640 cm^{-1} water band. The heights were estimated as the distances between the maxima of the bands and a straight baseline, constructed through the minima around 1550 and 1720 cm^{-1}. A plot of this height ratio versus the concentration reveals a significant behaviour over the whole region. This straight line, however, does not go through the origin. This is due to the fact that the 1590 cm^{-1} band contains a contribution from the background spectrum. Furthermore, the error on the individual measurements becomes quite large when the higher concentrations are taken into consideration. This may be caused by the relatively small intensity of the water band for higher concentrations, and also by the rudimentary way in which the height ratio determinations are carried out. Nevertheless, this plot may be used as a calibration curve and concentrations can be determined within 10%.

Resins

The absorption of phenols from acidic solution on porous polymer resins such as Amberlite XAD-2 has been shown to form the basis for a simple and accurate method for determining organics, including phenols in water.[207] XAD-2 resin effectively sorbs phenols and other organic compounds from water, although low recoveries are obtained for phenol itself.

Lee et al.[208] separated phenolic compounds in industrial waste water on a column of Dowex 1-X8 resin (CI-form) using a 0.05 M aqueous cupric chloride solution of ethanol, methanol, or propanol as eluant. Compounds with low dissociation constants (e.g. ϵ-aminophenol) were only weakly adsorbed on the column, whereas strongly ionized compounds (such as salicylic and 4-aminosalicylic acids) were strongly adsorbed. Some compounds (e.g. catechol) form complexes with cupric copper ions in the eluant so that they are eluted more quickly than the others.

Carpenter et al.[209] separated phenolic materials from aqueous solutions on cross-linked aqueous insoluble N-vinyl-2-pyrrolidone polymer. The pH for maximal binding of the phenolic compound to the resin was found to be dependent on the acidity of the phenolic compound. Binding to resin was particularly favourable for polyhydroxyl and extended aromatic compounds. Columns packed with this resin removed more than 95% of simple phenolic compounds from aqueous solution, and quantitative recovery of the bound phenolic compound was possible by elution with 4 M urea solution.

Kawabata and Ohira[210] have studied the removal and recovery of phenols from aquatic samples using vinyl-pyridine-divinylbenzene copolymer as an adsorbent. Although the analytical implications of this work were not discussed these clearly exist. Elution of the concentrated phenols from the resin column was accomplished by a treatment with acetone or methanol.

Miscellaneous methods

Thielemann[211,212] has discussed the German standard bromination method for the determination of steam volatile phenols in water samples and industrial waste waters.

Stroehl[213] showed that in the distillation and steam distillation methods for the recovery of volatile phenols from surface water effluents, approximately 10% and 20% recovery respectively, of phenols occurred. He discusses a modified procedure which gives a 98% recovery of volatile phenols.

Fountaine et al.[214] have described an ultraviolet ratio spectrophotometric method for the determination of trace amounts of phenols in waters and effluents. The method is based on monitoring the u.v. bathochromic shift occurring when the sample is made alkaline with sodium hydroxide. Two ultraviolet lamps are employed, one to operate in the pH-sensitive region for phenols and the other in the pH-insensitive region, this giving a differential reading when the solution is made basic.

Goldberg and Weine[215] have described methods for the extraction and concentration of phenolic compounds from water and sediment.

Kjellen and Neujahr[216] have discussed an enzyme electrode for the determination of total phenols in sewage works samples. The most sensitive electrodes were obtained by immobilization of the enzyme covalently bound to AH-Sepharose 4B or to nylon nets, as well as by enzyme adsorption on ion-exchangers. Optimal conditions include pH 6.5–9.5, 40 °C and incubation in a buffer containing NADPH for a few minutes before addition of the sample in order to make the electrode response independent of the diffusion rate of the substrate. Readout is achieved within 30 seconds of sample addition and the maximum rate of oxygen consumption is linearly dependent on phenol concentration in the 0.5–50 μM range.

Thielemann and Wiss[217] have discussed a bromometric method for the determination of phenols in waste water and have discussed improvements in the standard German bromometric method for the determination of phenols in sewage.[218]

Dallakyan et al.[219] have described a method for the determination of low concentrations of phenols and substances containing sulphhydryl groups in microalgae secretions. The method is based on the electrochemiluminescent oxidation of luminol (3-aminophthalic hydrazide) at 14–16 °C, pH 6.5, with a potassium iodide electrolyte and a platinum electrode to determine phenols and thiols. The inhibition of the chemiluminescence, specific amongst substances studied to phenols and thiols, was used as a means of measurement. The inhibition varied as to redox potential, type and extent of substitution, position of hydroxyl and sulphhydryl groups, and the concentration of inhibitors. These workers studied the electrochemiluminescence of luminol as a function of the voltage on the electrodes, the concentration of luminol and potassium iodide, the pH of the medium, the temperature, the oxygen content, and the shape and dimensions of the electrodes. The optimum conditions for the detection of phenols and thiols proved to be aqueous solution of 10^{-5} moles of luminol in 10^{-1} moles of potassium iodide; a voltage on the electrodes of 0.56 V; a current density of 2.5×10^{-5} A cm^{-2}; a pH of the solution of 6.5; and a temperature of the electrolytic cell of 14–16 °C. When samples capable of changing the pH of the medium are studied, a solution of the same concentrations of luminol in potassium iodide in a 0.015 M phosphate buffer with pH 6.5 can be used.

Of the substances examined, phenolic compounds possessed the strongest inhibitory properties; among them monophenols—phenol and tyrosine—exhibited the lowest inhibiting effect. The introduction of a second hydroxyl group on the benzene ring (hydroquinone, pyrocatechol) increased the ability of the phenols of inhibit luminescence. The inhibitory activity varied depending on the position of the hydroxyl groups. Phenols with an *ortho-*and *para-*arrangement of the hydroxyl groups (hydroquinone, pyrocatechol, chlorogenic acid) inhibited luminescence more strongly than *meta-*phenols (resorcinol). With some phenols such as phenol, hydroquinone, pyrocatechol, and resorcinol, a direct connection was discovered between the redox potential and the ability to inhibit chemiluminescence of luminol. Phenols which contain three hydroxyl groups (propylgallate, gallic acid, pyrogallol) inhibit luminescence more weakly than diphenols. Benzoic acid, which lacks the hydroxyl group on the ring, did not influence luminescence. Of the complex phenols of plant origin which were studied, an inhibitory effect was exhibited by tannin with a mean molecular weight of 1700. Gossypol inhibited luminescence relatively weakly. This method is of high sensitivity with a relative error varying from 1.5 to 6%. Some phenols such as hydroquinone, pyrocatechol, chlorogenic acid, and tannin can be detected in water in concentrations of 10^{-9} mol l^{-1}.

Preservation of samples for phenol analysis

Carter and Huston[220] have compared preservation of phenolic compounds in waste waters using (a) copper sulphate and phosphoric acid, storage at 4 °C, with (b) the use of strong acids or bases and storage of samples at 25 and 4 °C. The addition of 2 ml concentrated sulphuric acid with sample storage at 4 °C was shown to be effective for 3-4 weeks, while other preservatives were effective for only 8 days. Loss of phenolic compounds occurred rapidly unless the preservative was added immediately after sampling. A correlation found between loss of phenolic compounds and microbial activity suggests that the latter is dominant in determining sample stability.

The stability of phenolics in three different waste waters preserved with copper sulphate-phosphoric acid and stored at 4 °C was studied. The results are shown in Figure 61. The raw sewage was fairly weak with a biochemical oxygen demand of only 95 mg l^{-1}, and the treated sewage sample was collected after secondary biological treatment but before chlorination. The most important result of this study was the rapid loss of phenolics from the samples at 4 °C with no addition of any chemical preservative. The percentage loss of phenolics within 24 h for the industrial waste, raw and treated sewage samples was 85, 80, and 40% respectively.

In a further study (Figure 62) Carter and Huston[220] examined the effectiveness of the combined copper sulphate-phosphoric acid preservative *vs* sample type. Activated sludge was added to raw sewage to create a sample

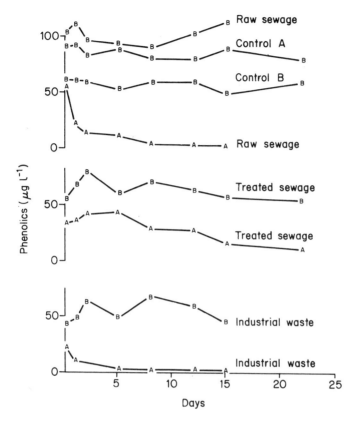

Figure 61 Plot of stability of phenolic compounds in several waste waters with time. All samples with points plotted as 'B' preserved with 1.0 g $CuSO_4 \cdot 5H_2O$ l^{-1}, pH brought to 4.0 with phosphoric acid, and then stored at 4 °C. Samples plotted as 'A' stored at 4 °C with no chemical preservatives. Both industrial waste, raw and treated sewage samples spiked with phenol to bring their initial concentrations to 50, 100, and 60 μg l^{-1} respectively. Reprinted with permission from Carter and Huston.[220] Copyright (1978) American Chemical Society

organically rich and biologically active. This sample was stable for 12 days but degraded to 85% of the original phenol concentration after 33 days. The other samples were stable for the duration of the study. The effectiveness of strong base or acid in preserving phenolic compounds was compared with copper sulphate in a further study (Figure 63). The concentration of phenolic compounds was stable in the raw sewage sample studied when stored at 4 °C regardless of the preservative used. However, the sulphuric acid and copper sulphate preserved samples deteriorated rapidly after 8 and 2 days, respectively, when stored at 25 °C.

Doetsch and Cook[221] reported that a common feature of acidophilic bacteria was a resistance to copper ions. Growth of acidophilic bacteria occurs at pH 2–5,

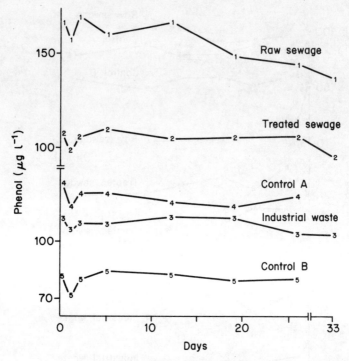

Figure 62 Plot of stability of phenolic compounds in several waste waters with time. All samples preserved with 1.0 g $CuSO_4 \cdot 5H_2O$ l_{-1}, pH adjusted to 4.0 with phosphoric acid, and stored at 4 °C. Industrial waste, raw and treated sewage samples spiked with phenol to bring their initial concentrations to 110, 165, and 110 $\mu g\ l^{-1}$, respectively. Reprinted with permission from Carter and Huston.[220] Copyright (1978) American Chemical Study

the pH range for the copper sulphate preservative. These facts make the use of copper sulphate at pH 4 suspect as a good preservative, especially if the samples are not stored at 4 °C. The same sample with 2 ml concentrated sulphuric acid per litre, which produces a pH of about 1.5, at 25 °C was stable for 8 days. Kushner[222] has reported that far fewer microorganisms can tolerate pH 1.5 than 4. Even at pH 1.5 and 25 °C the phenolic concentration decreased substantially. This observation indicates that while neither acidification nor cold storage stabilizes phenolic compounds in a waste water, the combination does.

To evaluate the biological-induced degradation of phenolic compounds Carter and Huston[222] measured microbiological activity on a raw and secondary treated sewage. Samples were preserved as indicated in Table 53, and total plate counts taken after 1 h (day 0), 8, and 20 days. The only secondary sewage aliquot that showed any significant activity was the chemically unpreserved sample stored at 4 °C. The microbiological activity noted corresponds very closely with the chemical stability of phenolics in treated sewage found by Carter and Huston.[222]

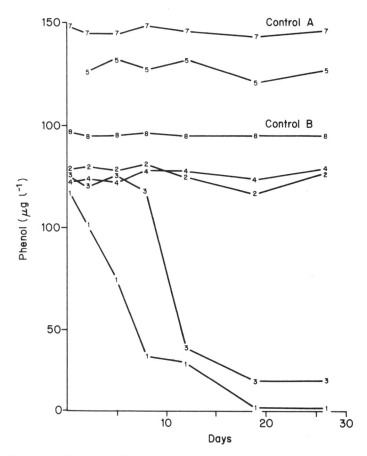

Figure 63 Plot of stability of phenolic compounds in raw sewage sample preserved with several chemicals; Aliquots 1 and 2 preserved with copper sulphate and phosphoric acid and stored at 25 and 4 °C respectively. Aliquots 3 and 4 preserved with 2 ml conc. H_2SO_4 l^{-1} and stored at 25 and 4 °C respectively. Aliquot 5 preserved with 2 ml 10 M NaOH l^{-1} and stored at 4 °C. Aliquots 1-4 spiked with phenol to bring their initial concentrations to 125 μg l^{-1}. Aliquot 5 spiked with phenol to bring its initial concentration to 130 μg l^{-1}. Reprinted with permission from Carter and Huston.[222] Copyright (1978) American Chemical Society

The unpreserved raw sewage sample stored at 4 °C showed very great microbiological activity that corresponds to the phenolic instability noted earlier (Figure 61). The addition of 2 ml of concentrated sulphuric acid per litre initially reduced the microbiological activity significantly. However, by day 8, the activity increased five-fold and then decreased slightly again by day 20. This trend corresponds closely with the rapid loss of phenolics after day 8 and then a moderate but continued loss thereafter. The same sample stored at 4 °C with 2 ml concentrated sulphuric acid per litre showed at least a ten-fold lower

Table 53 Effectiveness of preservatives in sterilizing sewage as indicated by total plate counts

Preservation method,* raw sewage	Total plate count colonies/ml		
	Day 0[†]	Day 8	Day 20
4 °C	≫ >30,000	≫ >30,000	≫ >30,000
2 ml conc. H_2SO_4, 25 °C	730	3500	2200
2 ml conc. H_2SO_4, 4 °C	...	70	200
4 ml conc. H_2SO_4, 25 °C	560	40	‡
$CuSO_4$, H_3PO_4, 4 °C	6300	800	600
2 ml 10 M NaOH, 4 °C	28,000	110	270
10 ml 10 M NaOH, 4 °C	230	90	100
Secondary treated sewage before chlorination			
4 °C	23,000	20,000	5400
2 ml conc. H_2SO_4, 25 °C	<30	<30	<30
2 ml conc. H_2SO_4, 4 °C	...	<30	<30
4 ml conc. H_2SO_4, 25 °C	<30	<30	<30
$CuSO_4$, H_3PO_4, 4 °C	<30	<30	<30
2 ml 10 M NaOH, 4 °C	40	<30	<30
10 ml 10 M NaOH, 4 °C	<30	<30	<30

*Volume of acid or base added per litre of sample. Temperatures refer to storage conditions.
[†] Plated within 1 h of preservation.
[‡] Confluent colonies.
Reprinted with permission from Carter and Huston.[220] Copyright (1978) American Chemical Society.

microbiological activity and a corresponding increase in chemical stability as shown in Figure 63.

The raw sewage sample with copper sulphate–phosphoric acid and stored at 4 °C exhibited greater microbiological activity than the aliquot with sulphuric acid. This observation corresponds to the moderate effectiveness of this preservative (Figures 61 and 63) and instability of the raw sewage (Figure 62). Addition of strong sodium hydroxide solution also lowered the microbiological activity of the raw sewage sample. However, the higher concentration of sodium hydroxide solution (10 ml 10 M sodium hydroxide l^{-1}) was required for a quick initial kill. The high initial microbiological activity of the 2 ml 10 M sodium hydroxide aliquot did not affect the chemical stability (Figure 62). However, to overcome the buffering capacity of many samples, the higher concentration of sodium hydroxide should be used.

Increasing the concentration of sulphuric acid two-fold with storage at 25 °C reduced the microbiological activity to the same level as the aliquot stored at 4 °C with 2 ml concentrated sulphuric acid. A further study was conducted to determine if a greater acid concentration could preserve phenolic stability without cold storage. The results in Figure 64 show good stability for the aliquots preserved with 2 ml sulphuric acid l^{-1} at 4 °C and 4 ml sulphuric acid l^{-1} at 25 °C. The aliquot with 2 ml sulphuric acid l^{-1} at 25 °C showed a substantial loss of phenolic compounds after the eighth day.

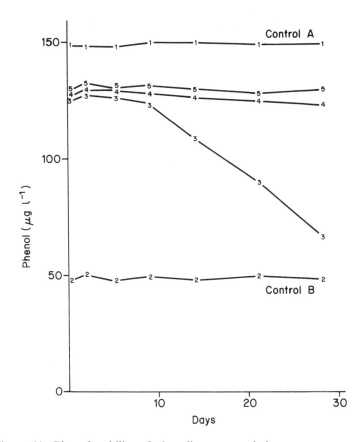

Figure 64 Plot of stability of phenolic compounds in raw sewage sample preserved with several concentrations of sulphuric acid. Aliquots 3 and 5 preserved with 2 ml conc. H_2SO_4 l^{-1} and stored at 25 and 4 °C respectively. Aliquot 4 preserved with 4 ml conc. H_2SO_4 l^{-1} and stored at 25 °C. All samples spiked with phenol to bring their initial concentration to 130 μg l^{-1}. Reprinted with permission from Carter and Huston.[220] Copyright (1978) American Chemical Society

The enhanced stability of the samples preserved with the higher acid concentration is excellent evidence that the greatest cause of sample instability is microbiological activity, not chemical activity.

REFERENCES

1. Cranwell, P. A. *Chem. Geol.*, **14**, 1 (1974).
2. Cranwell, P. A. *Geochim. Cosmochim. Acta*, **42**, 1523 (1978).
3. Matsuda, H. and Koyama, T. *Geochim. Cosmochim. Acta*, **41**, 1825 (1977).
4. Brooks, P. W., Eglington, G., Gaskell, S. J., McHugh, D. J., Maxwell, J. R. and Philip, R. P. *Chem. Geol.*, **18**, 21 (1976).
5. Brooks, P. W., Eglington, G., Gaskell, S. J., McHugh, D. J., Maxwell, J. R. and Philip, R. P. *Chem. Geol.*, **20**, 189 (1977).

6. White, D. C., Bobble, H. J., Morrison, S. J., Oosterhof, D. K., Taylor, C. W. and Meeter, D. A. *Limnol. Oceanogr.*, **22**, 1089 (1976).
7. Schnitzer, M. and Khan, S. U. *Humic Substances in the Environment*, Marcel Dekker, New York, 1972.
8. Esposito, G. G. and Schaeffer, K. K. H. *J. Am. Ind. Hyg. Assoc.*, **37**, 268 (1976).
9. Van Huyssteen, J. J. *Water Research*, **4**, 645 (1970).
10. Bethge, P. O. and Lindstroem, K. *Analyst (London)*, **99**, 137 (1974).
11. Ishwatari, R. and Hanya, T. *Gas Chromatography-Mass Spectrometric Identification of Organic Compounds in a River Water*. Proc. Int. Meeting Ed. Technig., 6th, 1051 (1974).
12. Baedecker, M. J., Nissenbaum, A. and Kaplan, I. R. *Geochim. Cosmochim. Acta*, **38**, 1185 (1972).
13. Carrol, K. K. in *Lipid Chromatographic Analysis*, G. V. Marinetti, (Ed.), Academic Press, New York, Vol. 1, pp. 174-212 (1976).
14. Johnson, R. W. and Calder, J. A. *Geochim. Cosmochim. Acta*, **37**, 264 (1973).
15. Thompson, S. and Eglinton, G. *Geochim. Cosmochim. Acta*, **42**, 199 (1978).
16. Van Hoevan, W., Maxwell, J. R. and Calvin, M. *Geochim. Cosmochim. Acta*, **33**, 877 (1969).
17. Nishimura, M. *Geochim. Cosmochim. Acta*, **41**, 1817 (1977).
18. Khomenkô, A. N., Goncharova, I. A., Stradomskaya, A. N. *Gidrokhim Mater.* 96 **(50)** 1969. Ref. *Zhur. Khim.* 19GD **(24)** (1969) Abstract No. 24G 262.
19. Khomenkô, A. N. *Anal. Abstr.*, **14**, 7927 (1967).
20. Richard, J. J. and Fritz, J. S. *J. Chromato. Sci.* **18**, 35 (1980).
21. Borch, R. F. *Anal. Chem.*, **47**, 2437 (1975).
22. Hoffman, N. W. and Liao, J. C. *Anal. Chem.*, **48**, 1104 (1976).
23. Hullett, D. A. and Eisenreich, S. J. *Anal. Chem.*, **51**, 1953 (1979).
24. Harrington, J. W. and Quinn, J. G. *Geochim. Cosmochim. Acta*, **37**, 735 (1971).
25. Metcalf, L. D. and Schmitz, A. A. *Anal. Chem.*, **33**, 363 (1961).
26. Hullett, D. A. *The Characterization of Biologically Important Organic Acids in the Upper Mississippi River*, MSc Thesis, University of Minnesota (1979).
27. Goncharova, I. A., Khomenkô, A. N. *Gidrokhim. Mater.* **53**, 36 (1970). Ref. *Zh. Khim.* 199D Abstract No. 59339 (1971).
28. Treguer, P., Le Corre, P. and Courtot, P. *J. Mar. Biol. Ass.* **52**, 1045 (1972).
29. Quinn, J. G. and Meyers, P. A. *Limnol. Oceanogr.*, **16**, 129 (1971).
30. Andrews, J. F. and Pearson, E. A. *Int. J. Air Wat. Pollut.*, **9**, 439 (1965).
31. McCarty, P. L., Jens, J. S. and Murdoch, W. *The Significance of Individual Volatile Acids in Anaerobic Treatment Preceedings*. 17th Purdue Industrial Wast Conference, (1962).
32. Mueller, H. F., Buswell, A. M. and Larsen, T. E. *Sew. Ind. Wastes*, **28**, 255 (1956).
33. Hunter, J. V. *The Organic Composition of Various Domestic Sewage Fractions*. Ph.D. Thesis, Rutgers Univ. (1962).
34. Hunter, J. W. and Heukelekian, H. *J. Wat. Pollut. Control Fed.*, **37**, 1142 (1965).
35. Painter, H. A. and Viney, M. *J. Biochem. microbiol. Technol. Engng.*, **1**, 143 (1959).
36. Hindin, E. *Wat. Sewage Wks.* **92**, 94 (1964).
37. Mueller, H. F., Larson, T. E. and Lennarz, W. J. *Anal. Chem.*, **30**, 41 (1958).
38. American Public Health Association, WPCF and AWWA. *Standard Methods for the Determination of Water and Wastewater*, 14th edn, New York (1976).
39. Buswell, A. M., Gilcreas, F. M. and Morgan, G. B. *J. Wat. Pollut. Control Fed.* **34**, 307 (1962).
40. Manganelli, R. M. and Brofazi, F. R. *Anal. Chem.* **29**, 1441 (1957).
41. Van Huyssteen, J. J. *Water Research*, **4**, 645 (1970).
42. Hunter, I. R., Orgeren, V. H. and Pence, J. W. *Anal. Chem.*, **32**, 682 (1960).

43. Murtaugh, J. J. and Bunch, R. L. *J. Wat. Pollut. Control Fed.*, **37**, 410 (1965).
44. Gehrke, G. W. and Larkin, W. M. *J. Agric. Food Chem.*, **9**, 85 (1961).
45. Harivank, J. *Voeni Hsparastvi Chem. Abstr.*, **14**, 394 (1964).
46. Smith, S. and Dila, R. *J. Pharm. Belg.*, **20**, 225 (1965).
47. Emery, E. M. and Koenrner, W. E. *Anal. Chem.*, **33**, 146 (1961).
48. Erwin, E. S., Marco, G. C. and Emery, E. M. *J. Dairy Sci.*, **44**, 1768 (1961).
49. Cochrane, G. C. *J. Chromatog. Sci.*, **13**, 440 (1975).
50. Baker, R. A. *J. Gas Chromat.* 418 (1960).
51. Narkis, W. and Henfield-Furie, S. *Water Research*, **12**, 437 (1978).
52. McKaveney, J. P. and Byrnes, C. J. *Anal. Chem.*, **42**, 1023 (1970).
53. Genkin, N. D. and Zel'manova, I. Y. A. *Zav. Lab.*, **38**, 1199 (1972).
54. Hueni, K. and Uebersax, P. *Landw. Forsch*, **26**, 125 (1973).
55. Farrington, J. W. and Quinn, J. G. *Geochim. Cosmochim. Acta*, **35**, 735 (1971).
56. Kawahara, F. K. *Anal. Chem.* **40**, 2073 (1968).
57. Kawahara, F. K. *Anal. Abstr.*, **17**, 1827 (1969).
58. Khomenkô, A. N. and Goncharova, I. A. *Gidrochim. Mater.*, **48**, 77 (1968). Ref. *Zh. Khim.* 199D. Abstract No. 13G 267 **(13)** (1969).
59. Bustin, R. M. and West, P. W. *Anal. Chim. Acta*, **68**, 317 (1974).
60. Smotrakov, V. G. and Stradomskaya, A. G. Goncharova Gidrochim. Mater., **49**, 202 (1969).
61. Smotrakov, V. G. and Stradomskaya, A. G. Zh. Khim. 19GD. Abst. No. 17G 239 **(17)** (1969).
62. Stradomskaya, A. G. and Goncharova, I. A. *Godrokhim. Mater.*, **48**, 72 (1968). Ref. *Zh. Khim.* 19GD (1969) Abstract No. 12G249 **(12)** (1969).
63. Hatsue, O. and Masayuki, I. *Br. J. Ind. Med.*, **27**, 150 (1970).
64. Hignite, C. and Azarnoff, D. L. *Life Sciences*, **20**, 337 (1977).
65. Goncharova, I. A. and Khomenkô, A. N. *Gidrokhim. Mater.*, **47**, 161 (1968). Ref. *Zh. Khim.* 19GD Abstr. No. 9G275 **(9)** (1969).
66. McDonald, K. L. *Tappi*, **61**, 77 (1978).
67. Brown, L. *Analyst (London)*, **104**, 1165 (1979).
68. Krotova, Z. A. *Zavod. Lab.*, **40**, 263 (1974).
69. Jordan, D. W. *Anal. Chim. Acta*, **113**, 189 (1980).
70. Igyarashi, S. *Japan Analyst*, **22**, 444 (1973).
71. Fox, M. W. *Environ. Sci. Technol.* **7**, 838 (1973).
72. Musselwhite, C. C. *Water Pollution Control*, **74**, 110 (1975).
73. Romantsova, G. I. *Zavod. Lab.*, **36**, 280 (1970).
74. Di Corcia, A. and Samperi, R. *Anal. Chem.*, **51**, 776 (1979).
75. Evans, W. H. and Dennis, A. *Analyst (London)*, **98**, 782 (1973).
76. Musselwhite, C. C. and Petts, K. W. *Water Pollution Control*, **73**, 443 (1974).
77. Karmil'chik, A. Ya., Stonkuz, V. and Korchargova, E. Kh. *Zhur. Analit. Khim.*, **26**, 1231 (1971).
78. Howe, L. H. *Anal. Chem.*, **48**, 2167 (1976).
79. Harrington, K. J. and Hamilton, J. *Can. J. Chem.*, **48**, 2607 (1970).
80. Vajta, L., Paimai, G., Vermes, E. and Szebenyi, I. *Periodica polytech. Chem. Engng.*, **15**, 263 (1974).
81. Vajta, L., Paimai, G., Vermes, E. and Szebenyi, I. *Anal. Abstr.*, **16**, 1948 (1969).
82. Karyakin, A. V. and Chirkova, T. S. *Zh. prikl. Spertrock*, **13**, 468 (1970). Ref. *Zh. Khim.* 19GD. Abstract No. 5G338 **(5)** (1971).
83. Hites, R. A. *J. Chromatog. Sci.*, **11**, 570 (1973).
84. Hites, R. A. and Biemann, O. *Science N. Y.*, **158**, 178 (1972).
85. Hites, R. A. *Environ. Hlth Perspect.*, **3**, 17 (1973).
86. Mori, S. *J. Chromat.*, **129**, 53 (1976).
87. Schouten, M. J., Copious Peereboom, J. W. and Brinkman, U. A. T. *Int. J. Environ. Anal. Chem.*, **7**, 13 (1979).

88. Schouten, M. J., Copious Peereboom, J. W., Brinkman, U. A. T., Schwart, H. E., Anzion, C. J. M. and Van Vleit, H. P. M. *Int. J. Environ. Anal. Chem.*, **6**, 133 (1979).
89. Fishbein, L. *Chromatography of Environmental Hazards*, Elsevier, Amsterdam, 579 (1973).
90. Sherman, J. *Adv. Chromat.*, **12**, 141 (1975).
91. Schwartz, H. W., Anzion, G. J. M., Van Vleit, H. P. M., Peerebooms, J. W. C. and Brinkman, U. A. T., *Int. J. Environ. Anal. Chem.*, **6**, 133 (1979).
92. Eglinton, G., Simoneit, B. R. T. and Zoro, J. A. *Proc. R. Soc. (London), B*, **189**, 145 (1975).
93. Cifrulak, S. D. *Soil Science*, **107**, 63 (1969).
94. Nammingas, H. and Wilhm, J. *J. Wat. Poll. Control Fed.*, **45**, 1725 (1977).
95. de Groot, A. J., de Goeif, J. J. M. and Zegers, C. *Geol. Mijnbouw*, **50**, 393, (1971).
96. Giam, C., Chan, H. and Nett, G. *Anal. Chem.*, **47**, 2225 (1975).
97. Steinle, G. *Zucker*, **26**, 589 (1973).
98. Shaova, L. G. and Kaplin, V. T. *Gidrokhim. Mater.*, **53**, 42 (1970).
99. Cavari, B. Z. and Phelps, G. *Appl. Environ. Microbiol.*, **33**, 1237 (1977).
100. Josefsson, B. O. *Anal. Chim. Acta*, **52**, 65 (1970).
101. Stabel, V. H. H. *Arch. Hydrobiol.*, **80**, 216 (1977).
102. Ochiai, M. *J. Chromat.*, **194**, 224 (1980).
103. Le B Williams, P. J. *Limnol. Oceanog.*, **9**, 138 (1964).
104. Josefsson, B. O. *Anal. Chim. Acta*, **52**, 65 (1970).
105. Eklund, G., Josefsson, B. and Roos, C. *J. Chromat.*, **142**, 575 (1977).
106. Mopper, K. and Regeus, E. T. *Anal. Biochem.*, **45**, 147 (1972).
107. Bark, L. S., Cooper, R. L. and Wheattone, K. C. *Water Research*, **5**, 1161 (1971).
108. Collins, J. W. and Webb, A. A. *TAPPI*, **55**, 1335 (1972).
109. Benefield, L. D. and Randall, C. N. *Water and Sewage Works*, **123**, 55 (1976).
110. Eremin, Yu. G. and Kopylova, G. A. *Zavod. Lab.*, **89**, 1065 (1973).
111. Kovalenko, P. N., Gavrilko, Yu M., Evstifeev, M. M. and Goborova, A. V. *Uhr. Khim. Zh.*, **35**, 965 (1969).
112. Thielemann, H. *Z. Anal. Chem.*, **253**, 39 (1971).
113. Thielemann, H. *Z. für Wasser and Ahwayer Forshung*, **7**, 91 (1974).
114. Suslov, S. N. and Stom, D. I. *Zh. Anal. Khim.*, **33**, 1423 (1978).
115. Environmental Protection Agency, *Quality Criteria for Water*, Superintendent of Documents, US Government Printing Office, Order No. 005-001-01049-4, Washington, DC (1976).
116. Sax, N. I. *Dangerous Properties of Industrial Materials*, 2nd edn, Reinhold, New York (1963).
117. Stecher, P. G. (Ed.) *The Merck Index*, 8th edn, Merck, Rahway, New Jersey (1968).
118. Baker, R. A. *J. Am. Water Works Ass.*, **55**, 913 (1963).
119. National Academy of Science and National Academy of Engineering, *Water Quality Criteria 1972: A report of the Committee on Water Quality Criteria*, Environmental Studies Board, Superintendent of Documents, US Government Printing Office, Order No. 5501-00520, Washington, DC (1972).
120. Environmental Protection Agency, *Manual of Methods for Chemical Analysis of Water and Wastes*, Office of Technology Transfer, Washington, DC (1974).
121. Cheeseman, R. V. and Wilson, A. L. The Water Research Association, Marlow, Buckinghamshire, SL7 2HD, UK, *The Absorbtometric Determination of Phenols in Water*, Report TP 84 August (1972).
122. *Standard Methods for the Examination of Water and Wastewater including Bottom Sediments and Sludges*, 12th edn, New York, American Public Health Association (P517-520) (1965).

123. Am. Public Health Ass., New York *Standard Methods for the Examination of Water and Wastewater*, 13th edn, Method 222 through 222E, American Water Works Ass., and Water Pollution Control Federation (1971).
124. American Society for Testing Materials, Philadelphia, Pa., *Manual on Industrial Water and Industrial Wastewater*, Part 23, Method D-1783-70 (1973).
125. Faust, S. D. and Mikulewicz, E. W. *Water Research*, 1, 405 (1967).
126. Mohler, E. F. Jr. and Jacob, L. N. *Anal. Chem.*, 29, 1369 (1957).
127. Fountaine, J. E., Joshipura, P. B., Keliher, P. N. and Johnson, J. D. *Anal. Chem.*, 46, 62 (1974).
128. Whitlock, R. C., Siggia, S. and Suida, J. E. *Anal. Chem.*, 44, 532 (1972).
129. Stroehl, G. W. *Staedteh Hygiene*, 19, 142 (1968).
130. Vinson, J. A. *Environ. Letts*, 5, 199 (1973).
131. Malz, A. *Foeder. Europ. Gemasserschutz*, 11, 19 (1964).
132. Gales, M. E. and Booth, R. L. *J. Am. Water Works Ass.*, 68, 540 (1976).
133. King, G. H. *Process Biochem.*, 7, 16, (1972).
134. Mohler, E. F. Jnr. and Jacob, L. N. *Anal. Chem.*, 29, 1369 (1957).
135. Afghan, B. K., Belliveau, P. E., Larose, R. H. and Ryan, J. F. *Anal. Chim. Acta*, 71, 355 (1974).
136. Ettinger, M. B., Ruchhoft, C. C. and Lishka, R. J. *Anal. Chem.*, 23, 1783 (1951).
137. Carter, M. J. and Huston, M. T. *Environ. Sci. Technol.* 12, 309 (1978).
138. Pays, M. and Bourden, R. *Annis Pharm. fr.*, 26, 681 (1968).
139. Freistad, H. O., Ott, D. E. and Gunther, F. A. *Anal. Chem.*, 41, 1756 (1969).
140. Freistad, H. O. *Anal. Chem.*, 41, 1753 (1969).
141. Goulden, P. D., Bronksbank, P. and Day, M. P. *Anal. Chem.*, 45, 2430 (1973).
142. Lei, C. F. and Pang, S. W. *Huaxue Tongbae (Chem. Bull.)* 19 (1) (1974).
143. Gales, M. E. *Analyst*, 100, 841 (1975).
144. Fountaine, J. E., Joshipura, P. B., Kelnher, P. N. and Johnson, J. D. *Anal. Chem.*, 46, 62 (1974).
145. Afghan, B. K., Belliveau, P. E., Larose, R. H. and Ryan, J. A. *Anal. Chim. Acta*, 71, 355 (1974).
146. Briedenbach, H. W., Lichtenberg, J. T., Heink, C. F., Smith, D. J. and Elchelberger, J. W. US Dept. of Interior Pub. WP-22 November (1966).
147. Kaplin, V. T., Semenchenko, L. V. and Fesenko, N. G. *Gidrokhim. Mater.*, 41, 42 (1966).
148. Semenchenko, L. V. and Kaplin, V. T. *Gidrokhim. Mater.*, 43, 74 (1967).
149. Kaplin, V. T. and Semenchenko, L. V. *Gidrokhim. Mater.*, 46, 182 (1968).
150. Szewczyk, J. and Desal, R. *Koks Smola Gaz.*, 13, 355 (1968).
151. Panova, V. A. *Ochistka. Proizvod Stochnykh Vod.*, 4, 184 (1969).
152. Semenchenko, L. L. and Kaplin, V. T. *Zhur. Anat. Khim.*, 23, 1257 (1968).
153. Eichelberger, J. W., Dressman, R. C. and Longbottom, J. *Environ. Sci. Technol.*, 4, 576 (1970).
154. Kawahara, F. K. *Environ. Sci. Technol.*, 5, 235 (1971).
155. Cooper, R. and Wheatstone, K. C. *Water Research*, 7, 1375 (1973).
156. Paklomova, A. D. and Berendeeva, V. L. *Ukr Zhur*, 40, 1211 (1974).
157. Baird, R. B., Kuo, C. L. and Shapiro, J. S. *Yankow Arch. Environ. Contam. Toxicol.*, 2, 165 (1974).
158. Plechova, O. A., Filippovy, Yu S. and Artemova, I. M. *Tekhnol. Ochiski Prirod Stochnvod* 200 (1) (1977).
159. Baker, R. A. *J. Am. Water Works Ass.*, 58, 751 (1966).
160. Baker, R. A. *J. Am. Water Works Ass.*, 1, 977 (1967).
161. Chriswell, D. C., Chang, R. C. and Fritz, J. S. *Anal. Chem.*, 47, 1325 (1975).
162. DiCorcia, A. *J. Chromat.*, 80, 69 (1973).
163. Goren-Stul, S., Kleijn, H. F. W. and Mostaert, A. E. *Anal. Chim. Acta*, 34, 322 (1966).

164. Grant, D. W. and Vaughn, G. A. in M. van Swaay (Ed.), *Gas Chromatography*, Butterworths, London, p. 305 (1962).
165. Kusy, V. *J. Chromat.* **57**, 132 (1971).
166. Dietz, F., Traud, J. and Koppe, P. *Chromatographia*, **9**, 380 (1976).
167. Smith, D. and Lichtenberg, J. J. *ASTM Spec. Techn. Publ.*, **448**, 78 (1968).
168. Baker, R. A. *Air Water Pollut.*, **10**, 591 (1966).
169. Derek, J. M. *J. Fish. Res. Bd Can.*, **32**, 292 (1975).
170. Nagasawa, K., Uchiyama, H., Ogamo, A. and Shinozuka, T. *J. Chromat.* **144**, 77 (1977).
171. Zerbe, J. *Chem. Anal.*, **22**, 575 (1977).
172. Matsumuto, G., Ishiwatari, R. and Hanya, T. *Water Research*, **11**, 693 (1977).
173. Meijers, A. P. and van der Leer, R. Ch. *Water Research*, **10**, 597 (1976).
174. Rump, O. *Water Research*, **8**, 889 (1974).
175. Murray, D. A. J. *J. Fish. Res. Bd Can.*, **32**, 292 (1975).
176. Yorkshire Water Authority, UK. *Determination of Phenols in Potable Waters by Gas Chromatography Report Method*, 655-01 (1977).
177. Voznakova, Z. and Popl, M. *J. Chromat. Sci.*, **17**, 682 (1979).
178. Goldberg, M. C. and Weiner, E. R. *Anal. Chim. Acta*, **115**, 373 (1980).
179. Pierce Chemical Co. Method 4. p. 11. In *Handbook in Silylation*. Pierce Chemical Co. Rockford Ill, pp. 48 (1972).
180. Voznakova, Z. and Popl, M. *J. Chromatog. Sci.*, **17**, 682 (1979).
181. Voznakova, Z., Popl, M. and Berker, M. *J. Chromatog. Sci.*, **16**, 123 (1978).
182. Brizova, E., Popl, M. and Coupek, J. *J. Chromat.*, **139**, 15 (1977).
183. Coupek, J., Unger, P. and Popl, M. *J. Chromat.*, **133**, 91 (1977).
184. Goldberg, M. C., DeLong, L. and Sinclair, M. *Anal. Chem.*, **45**, 89 (1973).
185. Gunther, G. A., Blinn, R. C., Kilbezen, M. J. and Barkeley, J. H. *Anal. Chem.*, **23**, 1835 (1951).
186. Kolesova, A. E., Nanmova, A. E. and Danilkova, U. A. *Khim. Tekhnol. Vysokomol. Soedin.*, **69**, 78 (1971).
187. Malz, F. and Gorlas, J. *Z Anal. Chem.*, **242**, 81 (1968).
188. Cooper, R. L. and Wheatstone, K. C. *Water Research*, **7**, 1375 (1973).
189. British Coke Research Association, Chesterfield, Derby. UK Research Report No. 55, pp. 21 (1969).
190. *Determination of Phenols in Coke Oven Effluents with Special Reference to Consent Conditions*, British Coke Research Association Coke Research Report No. 79 June (1973).
191. Thielemann, H. *Z. Chemie. Lpz*, **9**, 189 (1969).
192. Thielemann, H. *Pharmazie*, **25**, 365 (1970).
193. Thielemann, H. *Z Anal. Chemie.*, **253**, 38 (1972).
194. Edeline, F., Deswaef, R. and Lambert, G. *Tribune du Cebedeau*, **31**, 137 (1978).
195. Ragazzi, E. and Giovanni, V. *J. Chromat.*, **77**, 369 (1973).
196. Rump, O. *Water Research*, **8**, 889 (1974).
197. Thielemann, H. *Mikrochim. Acta*, **5**, 669 (1972).
198. Kishan, B. *Anal. Chem.*, **45**, 1344 (1975).
199. Armentrout, D. N., McLean, J. D. and Long, M. W. *Anal. Chem.*, **51**, 1039 (1979).
200. Hashimoto, S., Miyata, T., Washino, M. and Kawakami, W. *Environ. Sci. Technol.*, **13**, 71 (1979).
201. Bhatia, K. *Anal. Chem.*, **45**, 1344 (1979).
202. Carpenter, A. S., Siggia, S. and Carper, S. *Anal. Chem.*, **48**, 225 (1976).
203. Olsson, L., Nicolas, R. and Samuelson, O. *J. Chromat.*, **123**, 355 (1976).
204. Pietrizyk, D. J. and Chu, C. H. *Anal. Chem.*, **49**, 860 (1977).
205. Haverbeke, L. V. and Herman, M. A. *Anal. Chem.*, **51**, 932 (1979).
206. Thibeau, R., Van Haverbeke, L. and Brown, C. W. *Appl. Spectrosc.*, **32**, 98 (1976).

207. Junk, G. A., Richard, J. J., Grieser, M. D., Willink, D., Witlak, M. D., Argnello, M. D., Vick, R., Svec, H. J., Fitz, J. S. and Calder, G. V. T. *J. Chromat.*, **99**, 745 (1974).
208. Lee, K. S., Lee, D. W. and Chung, Y. S. *Anal. Chem.*, **45**, 396 (1973).
209. Carpenter, A., Siggia, S. and Carter, S. *Anal. Chem.*, **48**, 225 (1976).
210. Kawabata, N. and Ohira, K. *Environ. Sci. Technol.* **13**, 1396 (1979).
211. Thielemann, H. *Z. Chemie.*, **9**, 390 (1969).
212. Thielemann, H. *Pharmazie*, **25**, 202 (1970).
213. Storehl, G. W. *Mikrochim. Acta*, **1**, 130 (1969).
214. Fountaine, J. E., Joshipura, P. B., Keliher, P. N. and Johnson, J. D. *Anal. Chem.* **46**, 62 (1974).
215. Goldberg, M. C. and Weiner, E. R. *Anal. Chim. Acta*, **112**, 373 (1980).
216. Kjellen, K. G. and Neujahr, N. Y. *Biotechnology and Bioengineering*, **22**, 299 (1980).
217. Thielemann, H. and Wiss, Z. *Martin-Luther-University, Halle-Wittenb*, **19**, 111 (1970).
218. Thielemann, H. *Z Anal. Chem.*, **251**, 371 (1970).
219. Dallakyan, G. A., Veselovski, V. A., Tarusov, B. N. and Peogosyan, S. I. *Hydrobiol. J.*, **14**, 90 (1978).
220. Carter, M. J. and Huston, M. T. *Environ. Sci. Technol.*, **12**, 309 (1978).
221. Doetsch, R. N. and Cook, T. M. *Introduction to Bacteria and their Ecobiology*, University Park Press, Baltimore Md (1973).
222. Kushner, D. J. in *Inhibition and Destruction of the Microbial Cell*, W. B. Hugo (Ed.), Academic Press, London, UK (1971).
223. Kelus, J. and Waviernia, K. *Roczn. Panst. Zakl. Hig.*, **23**, 1 (1972).
224. Goldberg, M. C. and Weiner, E. R. *Anal. Chim. Acta*, **115**, 373 (1980).

Chapter 3

Nitrogen, Phosphorus, and Sulphur Compounds

NITROGEN COMPOUNDS

Aliphatic amines

Gas flame ionization chromatography has been used to determine dimethylamine,[1,3] dimethylformamide,[1] propylamine,[2] and dispropylamine[2] in river water and industrial effluents. To separate C_1-C_4 mono-, di-, and trialkylamines, Onuska[3] adjusted the pH of the sample to between 5 and 8. A 1 μl aliquot of the filtrate was injected on to a stainless steel column (185 cm × 2 mm i.d.) packed with 28% of Pennwalt 223 and 4% of potassium hydroxide on Gas-Chrom R (80–100 mesh) and maintained at 134 °C. A dual flame ionization detector was used and the carrier gas was helium (flow rate 52.2 ml min^{-1}). The detector response was rectilinear between 10 ng and at least 100 μg of dimethylamine, and the reproducibility was good. The column could be regenerated by increasing the column temperature to greater than 180 °C.

Kimoto et al.[4] used gas chromatography to determine simple aliphatic amines in dichloromethane extracts of water samples. They used a gas chromatograph equipped with a nitrogen-phosphorus detector and a 1.8 m × 2 mm i.d. glass column packed with Carbopack B—Carbowax 20—potassium hydroxide. The flow rates were helium 20 ml min^{-1}, hydrogen 3.5 ml min^{-1}, and air 80 ml min^{-1}. The injector and detector temperatures used were 200 and 250 °C respectively. The column was programmed from 70 to 150 °C at 4° min^{-1} and held at the upper temperature for 4 additional min. Then 2 μl samples were injected.

Hermanson et al.[2] used an aluminium column (276 cm × 4 mm) packed with 80–100 mesh Chromosorb W supporting 8.9% of amine 220 at 95 °C with nitrogen as carrier gas and flame ionization detection. A rectilinear response was obtained between peak area and amount of propylamine, dipropylamine, and propanol between 0.2 and 2.0 μg.

Florence and Farner[5] determined parts per billion (10^9) amines in waters, sea water, and raffinates from uranium processing using a spectrophotometric procedure.

Hexane 1,6-diamine

This substance has been determined in amounts down to 5 μg in water and industrial effluents by thin-layer chromatography and paper chromatography.[6]

Hexamine

Fishman and Pevzne[7] determined down to 0.15 g l^{-1} of this substance in effluents from polyformaldehyde manufacture by potentiometric titration with 0.2 M hydrochloric acid.

Hydrazine

Basson and Van Staden[8] determined low levels of hydrazine in water by reaction with 4-dimethyl-aminobenzaldehyde followed by spectrophotometric measurement of the absorbance at 460 nm.

Aromatic amines

Diphenlamine and other aromatic bases have been determined respectively in industrial effluents[9] and coal carbonization,[10] utilizing gas chromatography. Aniline derivatives at the microgram level have been determined in natural waters by a spectrophotometric procedure involving diazotization with sulphuric acid and naphthol.[11]

Benzidine

Jenkins[12] determined benzidine at the 10^{-9} concentration range in waste waters by spectrophotometric, gas chromatographic, and thin-layer chromatographic procedures. The American Public and Health Association[13] has also described a method for the determination of benzidine and its salts at the 0.3 μg l^{-1} level in water and waste water. In this method the water sample is made basic and the benzidine is extracted with ethyl acetate. Clean-up is accomplished by extracting the benzidine from the ethyl acetate with hydrochloric acid. Chloramine-T is added to the acid solution to oxidize the benzidine. The yellow oxidation product is extracted with ethyl acetate and measured with a scanning spectrophotometer. The spectrum from 510 nm to 370 nm is used for qualitative identification.

Heterocyclic nitrogen compounds

Gas chromatography has been used to determine down to 40 μg l^{-1} of pyridine and its homologues in waste water.[14] 2-Picoline and 2,5-lutidine have been determined spectrophotometrically in the ultraviolet region of the spectrum.[15]

N-*Methyl pyrolidone*

This substance which occurs as a biodegradation product in industrial waste water can be determined at the 1–10 ppm level by a method which involves concentration from an aqueous solution of sodium chloride by adsorption on XAD-2 resin, elution with methanol, and examination of the eluate by flame ionization gas-liquid chromatography.[16]

Method

500–1000 ml of cold water sample are saturated with sodium chloride and the saturated solution transferred to the column reservoir. The solution is percolated through the XAD-2 resin column at a flow rate of 5–10 ml min^{-1}. After all the solution has passed into the resin bed, the reservoir wall is rinsed with a jet of water, 10–15 ml, and this is allowed to just enter the resin bed. The adsorbed N-methyl pyrrolidone is eluted with methanol and the initial 50 ml of eluate collected. The gas chromatograph was operated with the injection port and detector maintained at 275 °C. A 10 foot × 1/8 inch stainless steel column was washed with acetone, dried and packed with Chromosorb 103, 80–100 mesh. The column was preconditioned for at least 24 hours at 15 °C higher than its final operating temperature of 250 °C. Helium carrier gas was used at 70 psig. A stable baseline is established with the chromatograph and 1–10 µl of sample are injected depending on the concentration level of N-methyl pyrrolidone expected.

Amino acids

Gardner and Lee[17] described an early gas chromatographic method for the estimation of dissolved free and dissolved total free and combined amino acids in lake water. The amino acids were first concentrated by ion- and ligand-exchange chromatography. Volatile N-trifluoroacetyl methyl esters were prepared and determined on glass columns (4 m × 3 mm o.d.) packed with 0.7% of XE-60, 0.5% of OV-101, and 0.2% of QF-1 on Diatoport S, with flame ionization detection. Internal standards are added before derivatization. Recoveries were from 36 to 97% for the free acids and 55 to 93% for the hydrolysed samples. Recoveries were poor for phenylalanine and lysine.

Various other analytical techniques have been used for the determination of amino acids in sea water (Table 54). Separation on an automatic amino acid analyser working on the ion-exchange principle coupled with a fluorimetric detection has emerged as the method of preference.

Table 54 Amino acids in sea water

Compounds investigated	Concentration range	Preconcentration method	Detection and quantification	Reference
18 Amino acids inclusive hydrolysed	3–130 µg l^{-6} (oceanic)	Coprecipitation with Fe(OH)$_3$, proteins hydrolysed, Fe and cations removed by ion-exchange	Paper and ion-exchange chromatography	18
Glycine, threonine, valine, phenylalanine	— —	Solvent extraction of 2.4 DNP AA	Circular thin-layer chromatography	19
16 Free and combined amino acids	FAA 16.3–123 µg l^{-1} CAA 6–20 µg l^{-1}	Extraction of dry acidified salts with 80% ethanol evaporated. Residue dissolved in water, desalted on ion-exchange	Paper chromatography	20
19 Amino acids FAA	10.5–87.5 µg l^{-1} (oceanic)	Desalted by strong acid ion-exchange at pH 1 or ethanol extraction as above	Paper chromatography	21
16 Amino acids FAA and CAA	FAA—38–77 µg l^{-1} CAA—185–290 µl l^{-1}	Ligand exchange on copper chelex	Ion-exchange chromatography	22
17 FAA	21–77 µg l^{-1}	Ligand exchange on Cu-Chelex pH raised to 9.5	Ion-exchange chromatography	23
17 FAA	38 µg l^{-1}	Same as Webb and Wood[23]	Ion-exchange chromatography	24
18 CAA	129 µg l^{-1}	Either sea water dried (acetone azeotropically) or hydrolysed sample desalted on Cu-Chelex	Gas chromatography; TAB amino acids	25
16 Amino acids FAA and CAA	FAA—2.3–4.6 µg l^{-1} CAA–5.8–105 µg l^{-1}	Extraction of dried salts with acid, 80% ethanol desalted on ion-exchange	Paper chromatography	26

Table 54 Amino acids in sea water *(continued)*

Compounds investigated	Concentration range	Preconcentration method	Detection and quantification	Reference
14 FAA	6.0–70 μg l^{-1} North Sea	Same as above	Ion-exchange chromatography	27
15 FAA	FAA—4.5–32.3 μg l^{-1}	Similar to above	Thin-layer chromatography	28
15 CAA Irish Sea	CAA—2.1–120 μg l^{-1}			
12 FAA	9.8–26.1 μg l^{-1} Total α-amino N$_2$ 20–80 μg l^{-1} — 124 μg result omitted	Same as Webb and Wood[23] lyophilization of extract	Ion-exchange chromatography	29
15 FAA	6–47 μg l^{-1}	Freeze drying of sea water extraction of salts with ether and ethanol (acidified). Ion-exchange clean-up	Gas chromatography	30
18 FAA	Surface 66–148 μg l^{-1} Deep 88–466 μg l^{-1}	Same as Webb and Wood[23]	Thin-layer chromatography	31
15 FAA and CAA	FFA—20–600 μg l^{-1} CFA—10–100 μg l^{-1}	Same as Webb and Wood[23] and extraction of liquids with chloroform	Ion-exchange chromatography	32
Total FAA	<50–200 μg l^{-1}	None	Colorimetric with ninhydrin	33
Primary amines, free, coastal	48–131 μg l^{-1}	None	Fluorimetric with fluorescamine	34
18 FAA FAA tank experiment with North Sea water	CAA—20–85 μg l^{-1} 20–310 μg l^{-1}	Same as Webb and Wood[23] Pre-extraction with ethyl acetate, Cu-Chelex method	Ion-exchange chromatography	35

Table 54 Amino acids in sea water *(continued)*

Compounds investigated	Concentration range	Preconcentration method	Detection and quantification	Reference
15 FAA	1.8–8.5 µg l⁻¹	Ligand-exchange Cu–Chelex and traces recovery technique	Ion-exchange chromatography	37
Up to 30 FAA and CAA identified, North Sea	FAA—20–180 µg l⁻¹ CAA—35–1350 µg l⁻¹	Desalting with Dowex 50WX8 at pH 3–4; elution with NH₄; lyophilization with added glycerine	Ion-exchange chromatography, ninhydrin detection 3.5 h analyses, 100 picomoles sensitivity	38
Up to 29 FAA and CAA identified, Baltic Sea water	FAA—4.5–84.5 µg l⁻¹ CAA—ca.500 µg l⁻¹	Desalting with Dowex as above; sample size 40 ml sea water; glycerine added before evaporation	Ion-exchange chromatography; OPA-fluorimetric detection—3 h analyses, sensitivity 20 picomoles	39,43
11 Amino acids free and combined oceanic	Mean *ca.* 1 µg l⁻¹; free 50 µg l⁻¹ CAA (totals)	Ligand-exchange on Cu–Chelex (no wash water)	Ion-exchange chromatography	40
11 FAA and CAA, oceanic CAA, oceanic	0–5 µg l⁻¹ FAA; 10–120 µg l⁻¹ CAA (totals)	Ligand-exchange Cu–Chelex (no wash water)	Ion-exchange chromatography	41
18 FAA and CAA	5–92 µg l⁻¹ FAA; 28–200 µg l⁻¹ CAA (Mediterranean)	Semi-automated Cu–Chelex method	Ion-exchange chromatography	42

FAA, free amino acids; CAA, combined amino acids.

Fluorimetry has been used to determine individual amino acids in marine waters.[18,19] Gardner[14] isolated free amino acids at the 20 nmole l^{-1} level in from as little as 5 ml of sample, by cation-exchange, and measured concentrations on a sensitive amino acid analyser equipped with a fluorimetric detector.

Dawson and Pritchard[45] determined α-amino acids in sea water using a standard amino acid analyser modified to incorporate a fluroimetric detection system. Sea water samples are desalinated on cation-exchange resins and concentrated prior to analysis. The output of the fluorimeter is fed through a potential divider and low-pass filter to a compensation recorder.

In this method the final separation and determination of individual amino acids is carried out using an automatic amino acid analyser (Locarte Amino Acid Analyser) modified for use with a fluorimetric reagent. The components are eluted with a stepwise buffer system using three buffers of increasing pH and molarity together with a mid-run temperature changeover and associated automatic regeneration and equilibration of the ion-exchange resin.

A full analysis of an extended hydrolysate requires about 165 min when using a 9 mm i.d. column and 100 min when using a 6 mm column. Using the 9 mm column it is possible to inject 2.5 ml of sample directly on to the column without significant loss of resolution. This amount is reduced to around 1.5 ml for the 6 mm column. The use of 6 ± 1 micron particle size resin allows shorter columns to be used without loss of separation and excessive increase in back pressures (normally around 25 atmospheres). The Locarte analyser used in this work was redesigned to incorporate a fluorimetric detection system by making the following modifications.

(1) The exit of the column was connected to a low-volume mixing block.
(2) The reagent pump was fitted with a pulse damper and pumped the reagent at one-half of the speed of the buffer pump into the mixing block.
(3) The output of the fluorimeter was fed through a potential divider and low-pass filter to the terminal of a compensation recorder.

The amino acids separated are displayed on the recorder as peaks and are identified by comparison of the retention times with those of known standards. Quantification is carried out using the digital output of integrated peak area which is linear with amino acid concentration.

Method

Reagent

The reagent consists of a mixture of 0.8 g *o*-phthalaldehyde predissolved in ethanol (8 g dl^{-1}) and 2 ml 2-mercaptoethanol dissolved in 1 l M sodium borate buffer adjusted to pH 9.5 with sodium hydroxide. To the solution is added 1% Brij 35 solution. As a precaution against the absorption of ammonia from the air, the reagent was kept in a sealed reagent bottle under a blanket of nitrogen delivered by the ballast system of the analyser. Following these precautions, the reagent was proved to be stable for up to a month. This reagent reacts with all α-amino acids without the necessity of heating, thus producing sharp peaks because of the reduction in dead volume. Proline and hydroxyproline are not detected unless they are oxidized with chloramine-T or sodium hypochlorite in strongly alkaline media.

The eluting buffers employed in the method are Durrum Pico Buffers System II (Durrum Chemical Corp., Palo Alto, Calif.). The buffer concentrates are diluted before use with freshly drawn deionized water and protected from the atmosphere with sulphuric acid traps.

Run Parameters

	6 mm column (minutes)	9 mm column (minutes)
Buffer A pH 3.19	1	20
Buffer B pH 3.86	40	50
Buffer C pH 4.60	45	55
Reagent 0.3 M NaOH	15	15
Temp. c/o 48–63 °C	20	35
Equilibrate buffer A	60	60

Buffer flow rates lie around 70 ml h^{-1} for the 9 mm column and 35 ml h^{-1} for the 6 mm column. Reagent flow rates are 35 ml h^{-1} and 17 ml h^{-1} respectively. All samples are injected on to the column in a loading buffer with a pH of 1.8.

Procedure

Samples of sea water were collected with a sterile Niskin 21 sampler and immediately filtered through 0.2 μm filters in a sterile, precleaned filtration apparatus. 120 ml of sample is transferred to a glass bottle with Teflon lined cap and disinfected and adjusted to pH 3 by the addition of pentachlorophenol in acidified aqueous ethanol. The samples are then stored at -20 °C whilst awaiting analysis. All glassware used for the sampling and work-up of the samples was cleaned by rinsing with 20% hydrofluoric acid solution followed by rinses with bidistilled (Quartz glass) water. The glassware is then dried at 450 °C.

Desalting of sea water samples

The samples of sea water were desalted on cation-exchange resin (Dowex 50W-X8) in the H$^+$ form after acidification to pH 3. A maximum of 100 ml needs to be desalted since the detection system is sensitive to less than 20 picomoles

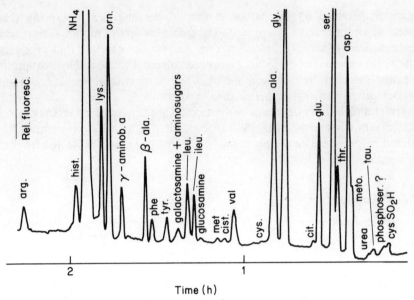

Figure 65 Chromatogram of a seawater extract (20 ml sample) for amino acids collected at 6 m in the Kiel Fjord. The concentrations of the individual acids were quantified as follows: in nmoles/l: meto, II; asp, 34.4; thr, 23.2; ser, 88; glu, 36; gly, 100; ala, 56; val, 16; ileu, 9.6; leu, 12; galactosamine and aminosugars, 4; tyr, 6.8; phe, 7.2; β-ala, 20.8; γ-aminoba, 14.4; orn, 44; lys, 12; hist, 7.2; arg, 9.6; cysSO$_2$H, 4; cit, trace; tan, cys, trace; glucoseamine, trace; met, trace; urea, trace; phosphoser., trace; OH-lys, trace. The total concentration of amino acid in the sample lies around 51 μg l^{-1}, assuming a mean molecular weight of 100. Reprinted with permission from Dawson and Pritchard.[45] Copyright (1978) Elsevier Science Publishers

of an amino acid. For a sample size of 100 ml, approximately 20 ml bed volume of resin was employed. The cation-exchange resin should preferably be well aged or cleaned and regenerated over a long period of time and packed into glass columns 20 × 2 cm with glass frits and Teflon stopcocks. A dropping funnel with a ground-glass joint serves to contain the sample during dropwise passage through the column. The sea water sample was passed through the column (regenerated and converted to H$^+$ and rinsed with bidistilled water) at a flow rate of about 5 ml min^{-1}. The column was then rinsed with two or three bed volumes of distilled water and then eluted with 80 ml (four bed volumes) of 3 M ammonia into a rotary evaporator flask acting as a receiver. 0.1 ml of a solution containing 50% glycerine in ethanol was added and the sample evaporated under reduced pressure at 40 °C. The sample was transferred with about 5 ml of distilled water to a 25 ml flask where it is concentrated twice to dryness with intermediate washings with distilled water to remove the ammonia. The sample is taken up to 0.5 ml pH 1.8 buffer and is ready for injection on to the column of the analyser. Usually only 0.2 ml of sample needs to be injected to give reasonably sized peaks.

Mean recoveries obtained by this procedure ranged from as low as 30% (urea), 50–75% (Cy SO$_2$H, taurine, tyrosine, phenylalanine, β-alanine, arginine, histidine) to 80–113% (aspartic acid, theonine, serine, glutamic acid, glycine, alanine, cystine, valine, methionine, alloisoleucine, leucine, orthinine, and lysine). The basic amino acids are only poorly recovered by the procedure as are the more acidic components: cysteic acid, taurine, and urea. Urea reacts erratically with the reagent and has a very low response (20 times less than other acids).

An example of a chromatogram obtained from a sea water sample and the mole percentage of each amino acid in the sample is depicted in Figure 65.

Dawson and Pritchard[45] point out that all procedures used for concentrating organic components from sea water, however mild and uncontaminating, are open to criticism, simply because of the ignorance as to the nature of these components in sea water. It is for instance feasible that during the process of desalting on ion-exchange resins under weakly acidic conditions, labile peptide linkages are disrupted or metal chelates dissociated, and thereby larger quantities of 'free' components are released and analysed.

Bajor and Bohling[46] have investigated the use of piperidine for desalting amino acid extracts of sediments and sea water.

Tusek et al.[47] isolated two high molecular weight polypeptides from the biomass of activated-sludge microorganisms.

Activated-sludge microorganisms were cultured in a laboratory system modelling the operation of the activated-sludge process, using a mixture of glucose and glutamic acid as the substrate. The high molecular weight products were isolated from the biomass and subjected to a fractionation procedure using Sephadex gel columns, and further purification by dialysis and gel filtration. Two components were isolated, with molecular weights of ca. 70,000 and 10,000 respectively. Both compounds were almost entirely composed of amino acids (principally lycine and glycine) with no evidence of any saccharide residues in either fraction. The presence of polypeptides in the biomass is considered in relation to reports of physiological effects such as growth stimulating properties towards certain algal species.

Nitrosamines

Many N-nitrosamines are toxic and carcinogenic, and furthermore the carcinogenic action exhibits a high degree of organ specificity. Nitrosamines are formed by interaction between nitrite and an amine with varying ease, depending on the nature of the amine and the prevailing conditions. The reaction is not restricted to secondary amines, but also occurs with primary and tertiary amines and even quaternary ammonium salts. Thus, the precursors are widespread, both as naturally occurring compounds and in many commercial and industrial processes nitrosamines are generated and it is therefore conceivable that trace amounts may be present in air and water in the vicinity of industrial sites. Nitrosamines in minute amounts have been found in deionized water, generated from the resins.

Mills and Alexander[48] have discussed the factors affecting the formation of dimethylnitrosamine in samples of water and soil. Dimethylnitrosamine was formed as readily in sterilized samples as in non-sterile samples, indicating that, although microorganisms can carry out an enzymatic nitrosation in some soils and waters, dimethylnitrosamine can be formed by a non-enzymatic reaction, even at near neutral conditions. The presence of organic matter appears to be important in promoting nitrosation in the presence of the requisite precursors.

Fine et al.[49] have described a gas chromatographic method for the determination of parts per trillion of N-nitroso compounds in drinking water. Nikaido et al.[50] give details of procedures for the recovery of low levels of dialkylnitrosamines from natural waters, including lake water and sewage effluents before subsequent detection by gas chromatography. The recovery technique involves the addition of potassium carbonate to the sample and concentration of the nitroso compounds on Amberlite XAD-2 resin. Greater than 90% recoveries were obtained for dimethylnitrosamine and diethylnitrosamine.

Method

To 2 l of sample was added 800 g potassium carbonate. This liquid was maintained in an ice-water bath until required and stirred. Two XAD-2 columns were used. To clean up the columns they were thoroughly washed with 300 ml of methylene chloride, 600 ml of 95% ethanol, and then water until the effluent was free of ethanol. Then the resin was left under reduced pressure for 2 h. This procedure eliminated those impurities which have retention times similar to those of dimethyl- and diethylnitrosamine. The sample solution containing potassium carbonate was then passed through the columns at 3 ml min^{-1}. The solutions containing higher nitrosamine levels were passed through two 46 × 2.5 cm Amberlite XAD-2 columns and eluted with 300 ml of methylene chloride, while the solution containing lower nitrosamine levels was passed through two 25 × 2.5 cm Amberlite XAD-2 columns and eluted with 150 ml of methylene chloride. These extracts were gas chromatographed as described below.

A gas chromatograph equipped with a flame ionization detector was used. A 3.6 m × 3 mm stainless steel column packed with 10% diethyleneglycol succinate on 80–100 mesh Chromosorb W (H.O.) was used, the nitrogen flow rate was 20 or 300 ml min^{-1} and the column was maintained at either 100 or 125 °C. The injector and detector temperatures were 230 °C and 240 °C, respectively.

As shown in Table 55 the recoveries obtained by this procedure were 77% or higher and were particularly good for nitrosamine levels of about 10 parts per 10^9. Richardson et al.[51] have applied gas chromatography with a chemiluminescent detection system to the determination of microgram levels of nitrosamines (N-nitrosodimethylamine, N-nitrosodiethylamine, N-nitrosomorpholine and N-nitrosodiethanolamine, N-nitrosopyrrolidine, N-nitrosopiperidine, and N-nitroso-5-methyl-1,3-oxazolidine) in cutting fluid recovery plant effluents, inputs to sewage plants, and potable water supplies.

Method

Nitrosamines may be removed from aqueous media by solvent extraction and subsequently concentrated by evaporation of the solvent, in order to detect levels as low as 0.01 µg l^{-1}.

Table 55 Recovery of nitrosamines using a salting out procedure with Amberlite XAD-2 resin but no alumina

Test no.	Nitrosamine	Amount added (parts per 10^9)	Recovery (%)
1	Dimethyl	102	96
	Diethyl	87.5	90
	Di-n-propyl	102	83
2	Dimethyl	50.9	100
	Diethyl	43.8	97
	Di-n-propyl	50.4	77
3	Dimethyl	9.8	110
	Diethyl	8.0	91
	Di-n-propyl	8.8	98

Reprinted with permission from Fine et al.[53] Copyright (1975) Springer Verlag, N.Y.

For the estimation of volatile dialkyl nitrosamines and N-nitrosopiperidine, N-nitrosopyrrolidine, and N-nitrosomorpholine, 400 ml of each sample was taken. To this was added 80 g of sodium chloride and 4 ml of 10 N sulphuric acid. This was extracted with 4×40 ml of redistilled dichloromethane, and 70 ml of 1.5 M sodium hydroxide was added to the combined extract. After separation, the organic layer was dried over sodium sulphate and evaporated to 2.5 ml at 46 °C on a water bath, using a Kuderna-Danish flask. Hexane (800 μl) was added, and evaporation continued to about 250 μl. This volume was accurately measured using a 500 μl calibrated syringe and transferred to a 300 μl capacity vial, and 5 μl aliquots were analysed for volatile nitrosamines using gas chromatography.

N-Nitrosodiethanolamine can be estimated by gas chromatography after conversion to the volatile silyl derivative. Sample (200 ml) was extracted with 3×30 ml of ethyl acetate in the presence of 10 g of sodium chloride. The organic extract was evaporated to about 3 ml at 80 °C and transferred to a 5 ml capacity vial, and evaporation was continued to dryness. The residue was stored in a desiccator overnight, after which 1 ml of Trisil reagent was added. The extract was vigorously shaken and allowed to stand overnight, and 5 μl aliquots were analysed by gas chromatography.

In the chemiluminescent procedure, which detects all nitrosamines amenable to gas chromatography, effluent from a chromatograph passes into a catalytic chamber whereupon the nitrosamine is fragmented to give rise to nitric oxide. This interacts with ozone and results in a chemiluminescent emission in the near infrared, which is detected with a photomultiplier tube. Interferences are minimized by placing a cold trap between the catalyst and ozone chamber, and by incorporating an optical filter in front of the photomultiplier. All extracts were analysed using a Pye 104 chromatograph connected to a chemiluminescent detector, Model TEA 502 (Thermo Electron Corp., Waltham, Mass.). The volatile nitrosamines were separated on a 4 m × 1.8 mm i.d. stainless steel column containing 5% Carbowax 20 M on Diatomite C-AW-DMCS at 150 °C. For the N-nitrosodiethanolamine derivative the column was 1.5 m × 1.8 mm i.d.

containing 5% SE-30 on the same support at 170 °C. A carrier gas flow rate of 11 ml min^{-1} of argon was used for both analyses.

The presence and amounts of nitrosamines were confirmed in some of the samples using the gas chromatograph coupled to an AEI MS902 mass spectrometer operating at a resolution of 7000. The nitrosamines were detected by parent ion monitoring using peak matching, in the manner described by Gough and Webb.[52] The detection limit was 0.1 µg l^{-1} for each of the nitrosamines. While measurement of the nitric oxide fragment by mass spectrometry is applicable to all nitrosamines, it results in a significantly poorer detection limit. Further, the mass spectrometer will respond to any compound giving rise to NO$^+$, including C- and N-nitroso compounds, nitro compounds, and nitramines.

The upper limits of N-nitrosodimethylamine, N-nitrosodiethylamine, N-nitrosomorpholine, and N-nitrosodiethanolamine detected were 0.2, 2.0, 100, and 60 µg l^{-1} respectively. It was considered that even with factory waste waters containing relatively high levels of nitrosamines, dilution results in much lower levels reaching the sewage treatment plant. No nitrosamines were detected in water before entering into the potable water supply.

Fine et al.[53] used gas chromatography to analyse volatile N-nitroso compounds at the parts per trillion level in American potable water supplies. Two different concentration and extraction procedures were used by these workers: one based on liquid-liquid extraction, and the other based on the adsorption of the organic fraction on carbon and its subsequent extraction with chloroform and alcohol. In both cases, final quantitative analysis and identification were carried out on a single column gas chromatograph equipped with a N-nitroso compound specific thermal energy analyser.[54,55] The chromatographic column was prepared from 6. mm × 2 mm i.d. stainless steel tube packed with 15% FFAP (15 g FFAP on 100 g Chromasorb W, acid washed, DMCS treated, 80–100 mesh) and conditioned overnight at 220 °C with carrier gas flowing prior to use. The column was operated isothermally at 200 °C, with carrier gas flow rate in the range 10–30 ml min^{-1}. Argon was used as the carrier gas.

For the liquid-liquid extraction, 500 ml of water was extracted three times with 25 ml of dichloromethane. A second 500 ml of water was extracted in a similar manner, and the combined extracts dried over 100 g of sodium sulphate. The sodium sulphate was filtered on a vacuum filter, the vacuum line being trapped at −151 °C to condense out the more volatile nitrosamines. The vaccum trappings were combined with the filtrate. The extract was then concentrated on a Kuderna-Danish apparatus at 58 °C to a final volume of 0.8 ml. A 10 µl sample of the concentrate was injected on to the gas chromatograph.

For the carbon adsorption technique, water was drawn through several cyclindrical columns packed with activated charcoal. The charcoal from the traps was dried and then extracted with chloroform, and then alcohol.[56]

Figure 66(a) is a chromatograph of 10 µl of the mixture containing seven N-nitroso compounds that were added to the water. Figure 66(b) is the chromatogram of 10 µl of the final dichloromethane extract following concentration on the Kuderna-Danish evaporator. From a comparison of the

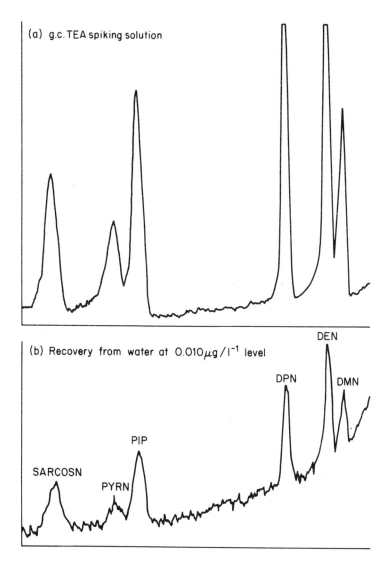

Figure 66 Gas chromatography of N-nitroso compounds in water, dimethylnitrosamine (DMN), diethylnitrosamine (DEN), dipropylnitrosamine (DPN), N-nitrosopiperdine (PIP), N-nitrosopyrrolidine (PYRN) and N-nitrososarcosinate (SARCOSN). Reprinted with permission from Fine et al.[53] Copyright (1975) Springer Verlag, NY

two chromatograms it is possible to calculate the efficiency of the extraction and analytical processes. Recovery efficiencies for the various N-nitroso compounds are tabulated in Table 56. At approximately 0.200 μg l^{-1} (200 ppt) concentration level the recovery efficiency is 30% for dimethylnitrosamine, increasing to about 90% for the less volatile species. At a concentration of approximately 0.020 μg l^{-1} (20 ppt) similar recoveries were observed.

Table 56 Recovery efficiency of N-nitroso compounds from water

Compound	Concentration µg l⁻¹ (ppb)	Recovery efficiency (%)	Concentration µg l⁻¹ (ppb)	Recovery efficiency (%)
Dimethylnitrosamine	0.128	30	0.013	30
Diethylnitrosamine	0.126	60	0.016	84
Dipropylnitrosamine	0.190	70	0.020	84
Dibutylnitrosamine	0.200	90	0.020	84
N-Nitrosopiperidine	0.206	88	0.021	84
N-Nitrosopyrrolidine	0.174	50	0.017	—
N-Nitrososarsosinate	0.263	89	0.026	130

Reprinted with permission from Fine et al.[53] Copyright (1975) Springer Verlag, N.Y.

In every case, volatile g.c.-amenable N-nitroso compounds were not detected in American water samples. The chromatograph was also operated in the temperature programme mode, with an initial oven temperature of 100 °C, increasing at the rate of 5 °C min⁻¹ to 240 °C, and then left for several hours; again, no trace of any N-nitroso compound was detected. For the liquid-liquid extraction procedure, this would indicate that volatile N-nitrosamines are definitely not present down to the 0.001 µg l⁻¹ level. For the carbon extracts, assuming an efficiency of 100%, volatile N-nitrosamines were not detected at concentration levels down to 10 µg l⁻¹.

Nitrosamines have been reported in deionized water by Fiddler et al.,[57] Cohen and Backman,[58] and Gough et al.[59] N-Nitrosodimethylamine and N-nitrosodiethylamine have been confirmed in water exposed to deionizing resins by Fiddler et al.[57] Levels of N-nitrosodimethylamine detected were 0.03–0.34 ppb and N-nitrosodiethylamine 0.33 and 0.83 ppb. The highest levels of nitrosamines were found after resin regeneration. N-Nitrosodimethylamine levels of 0.25 ppb and lower concentrations were also detected in deionized water by Cohen and Backman.[58] The origin of the nitrosamines in these reports was not identified. However, Gough et al.[59] suggested that N-nitrosodimethylamine present in the ion-exchange resins (up to 125 µg kg⁻¹) as manufacturing contaminants, not the reaction of the components in the water with the resins, was responsible for the occurrence of trace levels of N-nitrosodimethylamine (0.01 ppb) detected in deionized water. N-Nitrosodimethylamine was detected in the effluent when a 1 M sodium nitrite solution was passed through a mixed strong anion and cation resin column by Angeles et al.[60] Formation of N-nitrosodimethylamine was postulated by the cation acid-catalysed nitrosation of the amine/ammonium functional group on the strong anion resin during the deionization process.

Kimoto et al.[61] attempted to ascertain whether one of the explanations offered by these earlier workers[59,60] for the occurrence of nitrosamines in water deionized with strong cation and anion resins is correct or whether other factors are involved. They showed that N-nitrosodimethylamine was formed when tap water was passed through a column containing only the anion exchange resin, thereby indicating that nitrosamine formation by the conventional acid-catalysed nitrosation reaction, as would be expected in the case of a mixed strong anion and cation resin system, was not the primary mechanism. The quarternary ammonium ion of the strong anion resin may be the amine precursor for nitrosamine formation. Strong anion and strong cation resins (Amberlite and Dowex brands) were ineffective in concentrating low levels of N-nitrosodimethylamine that were added to the influent. Accumulation of extremely low levels of nitrosamine already present in the water, therefore, also did not contribute importantly to the concentration of nitrosamine found in deionized water. In addition, the normal levels of cations and anions in water, and heavy metal ions were also not primarily responsible for this occurrence. There was, however, an unknown substance(s) in the tap water that promoted nitrosamine formation. This unknown soluble substance(s) can be removed by activated carbon treatment or degassing techniques.

Kimoto et al.[61] extracted nitrosamines from 1 litre water samples with 3×150 ml portions of dichloromethane with shaking for 5 min per extraction. The extracts were combined, dried over anhydrous sodium sulphate and concentrated in a Kuderna-Danish evaporator to 1.0 ml. These workers determined nitrosamines in dichloromethane extracts by a gas chromatographic procedure employing a Thermal Energy Analyser as detector. Samples (7-8 μl) from 1.0 ml concentrates of dichloromethane extracts were injected into a gas-liquid chromatograph interfaced with a Thermal Energy Analyzer. A $2.7 \text{ m} \times 0.5 \text{ mm}$ nickel column tubing was packed with 15% Carbowax 20 M-TEA on 60-80 Gas Chrom P. The operating conditions were: injector temperature 200 °C; helium flow rate, 40 ml min^{-1}; column temperature programmed from 130 to 220 °C or 110 to 210 °C at 4 °C min^{-1}. The Thermal Energy Analyser conditions were similar to those that have been reported by Fine and Rounbehler.[62] Chromatographic peaks obtained from the samples were compared with those from standard nitrosamines for their precise retention times and quantitation.

Extracts containing apparent nitrosamines were sealed in melting point capillary tubes and subjected to u.v. light at 365 nm for 1-2 h. These samples were again subjected to gas chromatography to determine if the photolabile nitrosamines disappeared. Kimoto et al.[61] used a Varian Aerograph Model 2700 gas chromatograph equipped with a $1.8 \text{ m} \times 2 \text{ mm}$ i.d. glass column packed with 15% Carbowax 20 M-TPA on 60-80 Gas Chrom P connected to a Varian MAT 311A mass spectrometer to confirm the identity of nitrosamines. The helium flow rate was 15 ml min^{-1}. The temperatures used were: detector, 200 °C; injector port, 200 °C; gas chromatograph-mass spectrometer interface system, 180 °C; and column programmed from 90 to 130 °C at 4 °C min^{-1} for N-nitrosodimethylamine and N-nitrosodiethylamine. The mass spectrometer was operated in the peak matching mode adjusted to a resolution of 1 in 10,000 or 12,000. The mass spectra were obtained at an ionizing voltage of 70 eV and an ion source temperature of 150 °C. The mass-to-charge ratios (m/e of 74.04799) for nitrosodimethylamine and (m/e of 102.07930) for N-nitrosodiethylamine were determined with the m/e 69.99857 and m/e 99.99361 perfluorokerosene reference peaks, respectively, by measuring the difference in m/e. The signal was recorded on an oscilloscope and a recording oscillograph.

Acrylamide

The earliest reference found to the determination of acrylamide is that of Croll[63,64] who described an extraction–gas chromatographic method for the determination of down to 0.01% acrylamide in acrylamide polymer polyelectrolytes. Thin-layer chromatography and infrared spectroscopy both confirmed conclusions reached by gas chromatography.

Method

Extraction procedure

One gram of polymer powder is weighed into a 1 oz bottle and 10 ml of 80:20 methanol–water mixture are added. The bottle is capped horizontally in a flask shaker and shaken vigorously for 24 hours, then left 15 minutes to settle. A 5 µl aliquot is injected into a gas chromatograph which has previously been calibrated for acrylamide. The area of the acrylamide peak is measured and the equivalent quantity of acrylamide read off from the calibration graph.

Gas chromatographic conditions

A glass column 1 m long and 3 mm i.d. packed with 60–80 mesh acid-washed dichlorodimethylsilane treated Chromosorb W, supporting 20% by weight of

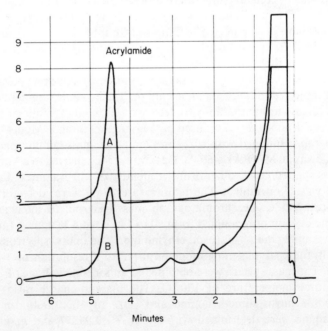

Figure 67 Gas chromatography of acrylamide. Reprinted with permission from Croll.[63] Copyright (1971) Royal Society of Chemistry

Carbowax 20 M, was used. When operated at 170 °C with a nitrogen flow rate of 32 ml min^{-1} the retention time of acrylamide was 4 min 20 s. A chromatogram of 1 µg of acrylamide in 5 µl of 80:20 methanol–water mixture is shown in Figure 67A. Polymer extracts chromatographed without any difficulties, a typical trace being shown in Figure 67B. This trace (Figure 67B) was of a 5 µl aliquot of an extract of a polymer containing 0.032% acrylamide. At the lowest levels of acrylamide analysed (0.02% in polymer) some small peaks were apparent which sometimes interfered with the estimation of the acrylamide peak areas.

This method was applied to water samples as follows. The method involved a preliminary bromination of acrylamide monomer in the aqueous sample to α,β-dibromoproprionamide which is then extracted with diethylether prior to gas chromatographic analysis.

Method

Apparatus and reagents

Gas Chromatograph—with an electron-capture detector g.l.c. Column— 1 m × 3 mm i.d., packed with 60–80 mesh AW DMCS Chromosorb W, supporting 10% by weight FFAP operated at 180 °C and a nitrogen flow of 40 ml min^{-1}.

Ultraviolet fluorescence lamp, giving emissions from 235 nm to 1140 nm.

Bromine Water at pH 1—prepared by mixing together bromine and distilled water at pH 1 until a saturated solution is formed.

Sodium Sulphite—approximately 0.1 M solution.

Sodium hydroxide—2 M solution.

Sulphuric acid—1:3 concentrated sulphuric acid:water solution.

Buffer Solution—pH 4 range.

Diethylether, redistilled using fractionating column.

α,β-Dibromopropionamide.

Acrylamide—reagent grade.

Potassium bromide—reagent grade.

Potassium bromate—reagent grade.

Procedure

A 100 ml volume of sample is measured into a 250 ml beaker. The pH of the solution is adjusted to 1 with 1:3 sulphuric acid. The beaker and its contents are placed under the u.v. lamp at the predetermined height with the beam directed on to the surface of the solution and 4 ml of the saturated bromine water are added, stirring vigorously. The solution is allowed to react for the predetermined time, then the excess bromine is destroyed with 4 ml of 0.1 solution of sodium sulphite. This aqueous solution is poured into a 250 ml glass-stoppered separating funnel (solution at pH 1) and 20 ml of redistilled diethylether are added. The solution is shaken vigorously for approximately 2 minutes, then allowed to stand for approximately 10 minutes to ensure a good phase separation. The ether phase is discarded. The aqueous phase is transferred to a 250 ml beaker and the pH of the aqueous phase adjusted to pH 3 using 2 M sodium hyroxide. The solution is poured into a 250 ml glass-stoppered separating funnel, 100 ml of diethylether added, and the funnel shaken vigorously for approximately 2 minutes. The phases are allowed to separate and the ether phase is run into a glass-stoppered conical flask. The aqueous phase is re-extracted with two further 50 ml quantities of diethylether. The ether extracts are combined and the volume obtained is measured. If the acrylamide concentration in the sample is above 50 μg l^{-1} this extract may be analysed directly by injecting 5 μl aliquots into the gas chromatograph. For lower concentrations the extract may be concentrated by a factor of up to 100 before the analysis by directing a jet of dry air (200 ml min^{-1}) on to the surface of the liquid. The flask is immersed in a bath of water at room temperature during the evaporation to prevent undue cooling. The exact degree of concentration depends on the sample and the sensitivity of the chromatograph but is usually a factor of 10 in the range 1–50 μg l^{-1} and 100 in the range 0.1–1 μg l^{-1} acrylamide.

5 μl quantities of standards are injected into the chromatograph. The areas of the acrylamide responses are calculated by multiplying the peak height by the peak width at half its height. Quantity of acrylamide versus peak area is plotted, using log–log ordinates.

5 μl samples of the organic extract are injected into the gas chromatograph and the quantity of α,β-dibromopropionamide present estimated by measuring the area of the peak obtained. As some day-to-day variation may occur in the response of the detector to α,β-dibromopropionamide it is necessary to recalibrate the chromatograph each day.

Calculation of results

The quantity of α,β-dibromopropionamide (a) present in the water sample may be calculated as

$$a = 20 \times w \ b \times 10^{-3} \ \mu g \ l^{-1} \ water$$

assuming a 100 ml water sample, 100% extraction efficiency, and where w is the quantity of α,β-dibromopropionamide (pg) in a 5 μl injection of solvent extract and b is the volume in ml of the solvent extract after concentration.

The approximate quantity of acrylamide present in solution before bromination (c) may then be calculated as

$$c = a \frac{71}{230} \times \frac{100}{y} \ \mu g \ l^{-1}$$

where $71/230$ is the ratio of the molecular weights of acrylamide and α,β-dibromopropionamide, and y is the approximate % yield of α,β-dibromopropionamide in the solvent extract after the complete analysis calculated from standards. This would be:

40% from 0.1 to 1.0 µg l^{-1},
50% from 1.0 to 10 µg l^{-1},
60% from 10 to 1000 µg l^{-1}.

In the range 1.0–1000 µg l^{-1}, c would probably give a sufficiently accurate figure for most purposes. However, for levels below 1.0 µg l^{-1} the yields drop sharply and are somewhat variable with conditions. It is desirable, therefore, to use acrylamide as an internal standard in order to obtain a more accurate estimate of yield of α,β-dibromopropionamide. The procedure is as follows.

An analysis is performed and c calculated as above. To another 100 ml aliquot of water sample is added sufficient acrylamide to double the quantity calculated from the analysis, i.e. $\frac{c}{10}$ µg dl^{-1} sample is added. The 'spiked' sample which will contain approximately $2c$ µg acrylamide l^{-1} of water is analysed. Calculate a_2, the quantity of α,β-dibromopropionamide per litre of water. If the quantity of dibromopropionamide from the original analysis was a_1 µg l^{-1}, then the quantity Z, from c µg l^{-1} acrylamide added, is given by

$$Z = a_2 - a_1$$

the % yield of α,β-dibromopropionamide (y^1) is given by

$$y^1 = \frac{Z \times 71 \times 100}{c \times 230}$$

and the correct quantity of acrylamide in the sample (q) is

$$q = \frac{a_1, c}{z} \mu g\ l^{-1}$$

If peaks which interfere with the acrylamide are present in the water before the addition of acrylamide or polymer, the blank from this source must be subtracted as µg l^{-1} of α,β-dibromopropionamide from the value of a.

Polarography has been used to determine acrylamide in aqueous solution.[67] Arkell and Croll[65] and Croll and Simpkins[66] determined acrylamide monomer in amounts down to 0.1 µg l^{-1} by a gas chromatographic procedure.

Hashimoto[68] used gas chromatography to determine acrylamide monomer in water. This method utilizes the response of the brominated form of acrylamide (α,β-dibromopropionamide) to electron capture detection. The acrylamide is brominated by an ionic reaction using bromine in the presence of potassium bromide and hydrobromic acid. The α,β-dibromopropionamide thus formed is extracted from water using ethyl acetate and the concentrate is then analysed by electron capture gas-liquid chromatography. Electron capture detectors are, however, prone to contamination. Difficulties in maintaining optimum electron capture sensitivity over extended periods may lead to some decrease in sensitivity during analysis.

Earlier high performance liquid chromatographic methods[69,70] for the determination of acrylamide monomer had detection limits of approximately 0.1 mg l^{-1}. Brown and Rhead[71] improved the sensitivity of high performance liquid chromatography to 0.2 µg l^{-1}. The procedure consists of bromination, extraction of the α,β-dibromopropionamide with ethyl acetate and quantification using high performance liquid chromatography with ultraviolet detection. Samples tested included river, sea, and estuarine waters, sewage and china clay works effluents, and potable waters.

Method

Apparatus

A Perkin Elmer Series 2 Liquid chromatograph was used fitted with a Spherisorb 10 µm octadecylsilane (ODS) column (250 × 4 mm); a variable-wavelength Perkin Elmer LC 55 spectrophotometer was used. A Rheodyne Model 7105, injection valve with a 175 µl sample loop was used to inject the sample on to the analytical column. An ultrasonic bath was employed to deaerate the solvents used as the eluants.

Reagents

All standards and solvents were of reagent or nanograde quality unless otherwise stated.

Solvents—ethyl acetate, hexane, and diethylether.

Saturated bromine water—distilled water was shaken with bromine and allowed to stand overnight in a refrigerator. The aqueous phase was removed and used.

Anhydrous sodium sulphate—Anhyrous sodium sulphate was heated at 600 °C for 24 h, dried, and stored in a grease-free desiccator.

Sodium thiosulphate solution, 1 M.

Potassium bromide—potassium bromide was heated at 400 °C for 24 h, dried, and stored in a grease-free desiccator.

Concentrated hydrobromic acid, sp. gr. 1.48.

Acrylamide monomer—electrophoresis-grade, purity greater than 99%. A solution of acrylamide (100 mg l^{-1}) in distilled water was prepared for each set of analyses. A diluted acrylamide solution (1 mg l^{-1}) was prepared from the former and aliquots of the latter solution were used to prepare spiked samples.

Procedure

All glassware was soaked in a chromic acid bath, washed with distilled water, and oven dried at 250 °C.

Acrylamide (5 g) and potassium bromide (7.5 g) were dissolved in distilled water (50 ml) in a glass-stoppered flask (300 ml) and three drops of concentrated hydrobromic acid were added. Saturated bromine water was added, dropwise, to the mixture with stirring until the yellowish colour of bromine persisted. A further volume of saturated bromine water (5 ml) was added with stirring, and the flask set aside overnight, in the dark, at 0 °C to complete the reaction. The excess of bromine was decomposed by adding 1 M sodium thiosulphate solution dropwise. The resulting white solid was filtered with suction, dried *in vacuo* and recrystallized from benzene until a constant melting point was obtained (132–134 °C).

Samples were collected in glass-stoppered bottles (250 ml) and stored in a refrigerator. Boiling prior to storage, to prevent bacteriological degradation, was carried out if the lapse time between collection and analysis was longer than 16 h. Each sample was filtered through Whatman GF/F filters. An aliquot of sample (100 ml) was pipetted into a glass-stoppered conical flask (250 ml) and extracted with ethyl acetate (20 ml) and hexane (20 ml). No grease was used on any glass joints. The organic phase was removed after each extraction using a Pasteur pipette. Samples containing high organic loadings (sewage, etc.) required a further extraction using hexane (20 ml). Hydrobromic acid (0.5 ml) and potassium bromide (15 g) were dissolved with stirring. Saturated bromine water (5.0 ml) was added with stirring and the flask set aside overnight, in the dark, at 0 °C to complete the reaction. The excess of bromine was decomposed by adding 1 M sodium thiosulphate solution dropwise. Anhydrous sodium sulphate (30 g) was added while using a magnetic stirrer to effect vigorous stirring.

The brominated sample was extracted with three portions of ethyl acetate (15 ml) and the organic phases were transferred into a glass beaker (100 ml) by means of a Pasteur pipette and dried by addition of sodium sulphate (2 g). The dried organic phase was transferred into a glass beaker (100 ml). The remaining sodium sulphate was rinsed three times with ethyl acetate (1 ml) and the rinsings were transferred into a glass beaker. The organic phase was concentrated to approximately 3 ml by use of a hot air blower, which was positioned so that the maximum temperature above the solvent was 70 °C. The concentrate was transferred into a glass vial (15 ml). The beaker was rinsed twice with ethyl acetate (1 ml) and the rinsings were transferred into the vial. The concentrate was evaporated to dryness using a nitrogen flow (1000 ml min^{-1}).

The residue was dissolved in distilled water (120 μl). An aliquot (12 μl) of the concentrate was injected into the liquid chromatograph's rheodyne valve and chromatographed under the following conditions: column, Spherisorb 10 μm ODS (250 × 4 mm): mobile phase, distilled water; flow rate, 4 ml min^{-1}; pressure, 1000 lb in^{-2}; detector wavelength, 196 nm; and chart speed 2 cm min^{-1}.

Quantification was achieved by peak-height measurement. From the test injection, the approximate concentration of α,β-dibromopropionamide was obtained. Depending on the sensitivity required, either duplicate aliquots of the concentrate (1–50 μl) or one aliquot (100 μl) was injected. If the latter volume was used the detector response was manipulated to bring it on to the scale shortly before the α,β-dibromopropionamide peak by use of the recorder zero and/or the spectrophotometer zero. The concentration of acrylamide in the original sample was calculated by using a calibration graph of average absorbance (peak height) of α,β-dibromopropionamide as calculated for a 120 μl injection versus the original concentration of acrylamide. To check the u.v. detection calibration, 10 µl of freshly prepared α,β-dibromopropionamide standard (100 mg l^{-1}) were injected prior to analysis.

The concentration of acrylamide in the water sample was calculated from the total absorbance of the concentrate, i.e.

$$\text{Total absorbance} = \frac{\text{mean absorbance of the concentrate}}{\text{volume of concentrate injected}} \times 120$$

The concentration of acrylamide was found by comparison with a calibration graph of total α,β-dibromopropionamide absorbance versus original acrylamide concentration.

The recovery of α,β-dibromopropionamide from spiked samples, R, was calculated from the equation

$$R = 100 \times \frac{\text{Conc}_s}{\text{Abs}_s} \times \frac{(\text{Abs}_{t2} - \text{Abs}_{t1})}{\text{Conc}_t}$$

where Conc_s (µg) is the concentration of α,β-dibromopropionamide in the standard injection (10 µl of freshly prepared 100 mg l^{-1} α,β-dibromopropionamide), Abs_s is the average absorbance of the standard injection (as above), Conc_t (µg) is the equivalent amount of α,β-dibromopropionamide added to the spiked portion of the sample and Abs_{t2} and Abs_{t1} are the measured total absorbances for the spiked and unspiked sample, respectively.

Acrylamide stock solutions should be prepared freshly for each batch of determinations, as loss of acrylamide from 1.0 mg l^{-1} acrylamide standards in distilled water may occur (Figure 68).

The levels of inorganic ultraviolet absorbing impurities found in water samples did not interfere in this procedure.

The solvent extraction procedure described above lowered interferences in all samples tested without removal of acrylamide or excessive use of solvents. A second extraction with hexane is recommended for highly coloured peat bog runoff and sewage effluents (Figure 69).

The experimental yields of α,β-dibromopropionamide encountered gave a mean of 70.13 ± 8.52% (95% confidence level) for acrylamide-spiked estuarine waters, sea water, sewage effluent, china clay process waters, and potable waters over the concentration range 0.2–8.0 µg l^{-1} of acrylamide monomer.

Bezazyan et al.[72] have described a phototurbidimetric method for the determination of polyacrylamide fluocculants in effluents.

Nitriles

Stefanescu and Ursu[73] determined acrylonitrile and acetonitrile in residual waters by spectrophotometric and titrimetric procedures after separation from the sample by azeotropic distillation. Acrylonitrile and acetonitrile are determined together (as ammonia) after alkaline hydrolysis and acrylonitrile is

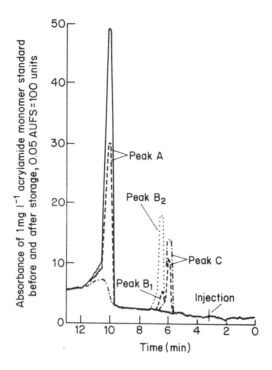

Figure 68 Degradation of 1 ml l^{-1} acrylamide standard in single distilled water stored at room temperature in the dark. Chromatographic conditions: column, Spherisorb 10 μm ODS (250 × 4 mm); eluant, distilled water; flow rate, 1 ml min^{-1}; chart speed, 1 cm min^{-1}; wavelength, 196 nm; absorbance scale, 0.05 aufs (100 units); response, normal; injection volume, 100 μl. ———, Original trace of standard 11.5.78; - - - - - -, trace 2.6.78; . . . . ., trace 2.6.78 spiked with acrylic acid; and ············, trace 3.7.78. Peaks: A, acrylamide; B$_1$, unknown B$_2$, acrylic acid; and C, unknown. Acrylamide (peak A) was significantly reduced when stored at room temperature in the dark. Unknown peaks (B$_1$ and C) appeared as the acrylamide was degraded. Peak B$_1$ occurred at the same position as that of acrylic acid (peak B$_2$). Peak C coincided with the front owing to species not retained by the Spherisorb 10 μm ODS column and thus could not be assigned. Reprinted with permission from Brown and Rhead.[71] Copyright (1979) Royal Society of Chemistry

determined separately with mercaptoacetic acid. Down to 2 mg of each substance per litre can be determined.

Ghersin *et al.*[74] compared colorimetric and titrimetric procedures for the determination of acrylonitrile in effluent waters. A titration method based on addition of sodium sulphite to the acrylonitrile followed by titration of liberated sodium hydroxide gave a sensitivity of 20 mg l^{-1} acrylonitrile.

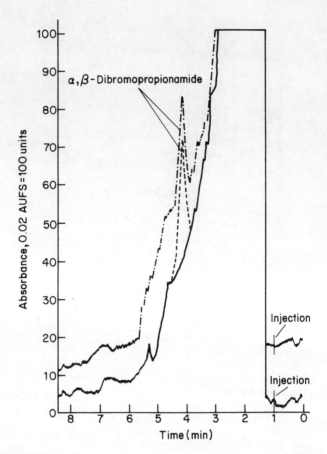

Figure 69 Resolution of α,β-dibromopropionamide from interferences present in environmental samples using a Spherisorb 10 μm ODS column and eluting with distilled water. Chromatographic conditions; column, Spherisorb 10 μm ODS (250 × 4 mm); eluant, distilled water; flow rate, 4 ml min^{-1}; chart speed, 2 cm min^{-1}; wavelength, 196 nm; absorbance scale, 0.02 aufs (100 units); and response, 1.0 s. Samples analysed on July 14th, 1978. ———, 78 μl of brominated Ivybridge tap water concentrate; --------, 78 μl of brominated Ivybridge tap water (spiked with acrylamide, 3 μg l^{-1}) concentrate; and —.—.—.—. 12 μl of brominated Camel's Head sewage effluent concentrate. This sewage effluent was cleaned prior to analysis using one ethyl acetate and two hexane extractions. Reprinted with permission from Brown and Rhead.[71] Copyright (1979) Royal Society of Chemistry

Provided that the sample is not yellow or brown in colour, a titrimetric method involving the use of mercaptoacetic acid has a sensitivity of 2 mg acrylonitrile per litre, or down to 0.4 mg per litre if the acrylonitrile is separated by a preliminary distillation from methanol–sulphuric acid medium; such a distillation also separates acrylonitrile from interfering substances.

Urea and substituted ureas

Nehring[75] has described an enzyme method, based on the use of urease for the determination of traces of urea in potable water. McCarthy[76] has described a urease method for the determination of the microgram amounts of urea in sea water. Urea and ammonia in natural waters have been determined[77] in natural waters in amounts down to 0.2 μg N l^{-1} by a method based on reaction with phenol and sodium hypochlorite followed by spectrophotometric determination at 454 μm (urea) and 630 μm (ammonia). Tyrosine is the only amino acid which interferes in this procedure.

Urea has been determined[78] in the presence of ammonia at the μg l^{-1} level in raw sewage using an ammonia-gas-selective electrode. Determinations of ammonia with and without the addition of urease enable the urea content to be obtained by difference.

Dodecylguanidines

These substances have been determined by sewage effluents by thin-layer chromatography.[79] Thin-layer chromatography was carried out on 0.25 mm layers of Adsorbasil-1 that had been activated at 100° for 1 hour. The best results were obtained with butanol–acetic acid–water (8:1:1) as solvent (R_f 0.6); iodine vapour was used as locating reagent. The sensitivity was equivalent to 0.5 mg of dodecylguanidine per litre.

Nitro compounds

Nitrophenols

Micro amounts down to 4 μg l^{-1} of dinitro-o-cresol have been determined in water by oscillographic polargraphy.[80] Chambou and Chambon[81] have described a thin-layer chromatographic method for the determination of down to 5 μg l^{-1} of 4,6-dinitro-o-cresol in diethylether extracts of water samples. The extract is applied to a 0.25 mm layer of Cellulose MN and the chromatogram is developed with ammonia-saturated butanol (in a chamber equilibrated with the solvent) for 14 cm ($=2$ h). The nitrophenol spots are located by the bright yellow colour formed with ammonia, removed from the plate and extracted with 50% acetic acid (0.05 ml)–methanol (5 ml); the extinction of the extract is measured at 370 nm.

Cranmer[82] has described a gas chromatographic method for the determination of down to 0.05 ppm 4-nitrophenol in human urine and this procedure would, no doubt, be applicable to water samples. The sample is adjusted to pH II and twice extracted with benzene–ethyl ether (4:1); the extracts are discarded, the aqueous phase is acidified to pH 2 and the extraction is repeated. This extract is dried over sodium sulphate and hexamethyldisilazane is added. To effect on-column silanization, the solution is injected into a gas chromatograph equipped with an inlet operated at 200 °C, an electron capture detector and a column of either 5% of DC-200 on Gas-Chrom Q or 1.5% of OV-17 plus 1.95% of QF-1 on Chrom W, operated at 140 °C.

Ethylene glycol dinitrate

This substance has been isolated and identified in American drinking water supplies using high performance liquid chromatography with a thermal energy analyser.[83]

Trinitrotoluene

This substance has been determined in effluents from munitions plants by spectrophotometric[84,85] and fluorimetric procedures.[86] The fluorimetric procedure described by Heller et al.[86] is capable of determining down to 0.07 ppm trinitrotoluene in munition plant wastes. The basis of the method is the absorption of trinitrotoluene on a quaternary ammonium ion-exchange resin which darkens on contact with trinitrotoluene. The resin is irradiated with ultraviolet radiation and the fluorescent output is monitored by a photomultiplier.

Ethylene diamine tetraacetic acid

EDTA salts are present in low concentrations in detergent preparations and some food products. They are not biodegradable and might have an effect in mobilizing trace metals in river waters, i.e. in reducing their tendency to be removed from solution by adsorption and precipitation reactions and possibly causing desorption from contaminated river sediments. Hence, there is an interest in determining EDTA and its salts in river water, sewage, and sewage effluents.

Kunkel and Manahan[87] have described an atomic absorption method for determining strong heavy metal chelating agents, such as EDTA and nitriloacetic acid in natural and waste waters. The method involved solubilization of cupric ions (added as 0.05 M $CuSO_4$) by the chelating agents at pH 10 in the boiling solution, filtration of the cool mixture, and then determination of cupric ions in the filtrate. The concentration of total strong heavy metal chelating agents is proportional to the amount (in mg) of copper chelated in a standard volume of sample.

Rudling[88] and Chau and Fox[89] have described a method in which the methyl ester was determined by gas-liquid chromatography using 1,2-cyclohexanediaminetetraacetic acid as internal standard. This presented problems as the retention times of the methyl esters of EDTA and those of certain C-18 fatty acids (stearic and oleic), which were also usually present in the samples being investigated, were very similar and the peaks were not resolved under the conditions employed. It was considered unlikely that any stationary phase would completely resolve EDTA from the range of C-18 fatty acids that might be found in samples of sewage and sewage effluent. It was also found that humic acid in the sample, most of which failed to react with the esterifying agent, produced a greater than theoretical recovery of EDTA, apparently because of interaction between EDTA and the non-reactive fraction of humic acid.

Korsunovskii[90] has developed a procedure for determining very small amounts of EDTA, which can be used to analyse turbid and coloured solutions. The method is based on the proportionality of the photoelectric current generated in the photogalvanic cell to free EDTA concentration. Differential pulse polarography has also been used to determine EDTA and nitriloacetic acid in synthetic sea water and phytoplankton media.[91] Cadmium is used to convert a large fraction of either ligand to the reducible cadmium complex. The presence of competing metal cations, including copper, is not detrimental if the method of standard additions is used. The method was used in correlating the concentration of complexed and uncomplexed species of copper with phytoplankton productivity and the production of extracellular metal-binding organic compounds.

Gardiner[92,93] described a gas chromatographic method for the determination of ethylenediaminetetraacetic acid in aqueous environmental samples. The separation of the major peaks is increased by preparing the ethyl derivatives of the sample compounds, 1,6-hexanediaminetetraacetic acid being used as internal standard. The lower limit of detection of the method is approximately 15 μg l^{-1} with 25 ml samples. This limit can be improved, if necessary, by using a larger sample volume. In this method the ethyl derivatives of the sample components were prepared so that the major peaks would be well separated. The ethyl esters of fatty acids up to and including the C-18 fatty acids eluted well before the EDTA derivative and did not interfere. Fatty acids of greater chain length are invariably present at insignificantly low concentrations in environmental samples. It was also found that the linear analogue of EDTA, 1,6-hexanediaminetetraacetic acid (HDTA), when used as internal standard, did not interact with the humic acid present in the sample. A series of small peaks occurring in the region of the peak for HDTA did not interfere; these peaks may have originated from detergent components.

Method

Reagents

EDTA, disodium salt — laboratory-reagent grade.

HDTA — laboratory-reagent grade.

Esterifying solution — slowly, with stirring and cooling, 2.5 ml of sulphuric acid (Aristar grade) are added to approximately 40 ml of absolute ethanol; 25 μl of acetic acid (Aristar grade) are added and the solution made up to 50 ml with ethanol.

Phosphate buffer solution — 14.1 g of anhydrous disodium hydrogen orthophosphate (AnalaR grade) are dissolved in approximately 75 ml of distilled water, hydrochloric acid (AnalaR grade) is added until pH 7.0 is reached and the solution made up to 100 ml with distilled water.

Chloroform containing 2% of ethanol—laboratory-reagent grade, redistilled before use.

All other reagents should, if possible, be of AnalaR grade.

Apparatus and conditions

All glassware with which the sample came into contact was rinsed with concentrated nitric acid followed by distilled water in order to reduce potential interferences to a minimum.

Gas-liquid chromatography

Instrument—a gas chromatograph with dual flame ionization detectors or equivalent. Use of a reference column to compensate for column bleed at high temperature is optional.

Columns—one or two 1.5 m glass columns of 4 mm i.d. containing 3% OV-1 on 80–100 mesh Chromosorb W(HP).

Gas flow rates—nitrogen (carrier) 65 ml min^{-1}; hydrogen 60 ml min^{-1}; and air 730 ml min^{-1}.

Temperature programme—the temperature of the column oven was maintained at 150 °C for 4 min after injection, then increased at a rate of 10 °C min^{-1} to 285 °C. This temperature was maintained for 2 min. The injection-port temperature was approximately 200 °C at the time of injection.

Ionization amplifier—an attenuation factor of between 500 and 5000, depending on the concentrations of EDTA and HDTA, was used.

Procedure

The sample is filtered through a 0.45 μm Millipore filter with a Whatman GF/A prefilter and the filtrate extracted once with two-thirds of its volume of chloroform. With a few drops of 90% formic acid, a 25–200 ml aliquot of the aqueous layer (containing 10^{-8}–10^{-7} mol of EDTA) was acidified and the internal standard (10^{-8}–10^{-7} mol of HDTA) added. After evaporation on a water bath to approximately 2 ml, and before precipitation of solids has occurred, the concentrate is rinsed into a Pyrex test-tube (150 × 10 mm). Evaporation is continued in the tube at 100 °C under a stream of nitrogen and finally in an oven at 110 °C. The middle of the tube is drawn out in a gas-oxygen flame, between 1.0 and 1.5 ml of freshly prepared esterifying solution are added, the tube is sealed and placed with the bottom in 2–3 cm in boiling water for 5 h. After cooling, the tube is opened and the contents rinsed, using one ml of chloroform, into a centrifuge tube containing 3.0–4.5 ml of phosphate buffer solution. The tube and contents are shaken for a minute. The layers are separated by centrifuging, the chloroform layer is transferred into a small sample tube and the solvent evaporated at 50 °C under a stream of nitrogen. The residue is dissolved in 50 µl of acetone and approximately 7 µl of the solution are injected on to the gas-liquid chromatographic column.

Table 57 shows recoveries of EDTA from aqueous samples obtained by the above procedure.

Table 57 Recovery of EDTA from aqueous samples to which known concentrations of EDTA had been used

Sample number	Sample type	Concentration of EDTA (µg l^{-1})		Apparent original concentration of EDTA (µg l^{-1})	Recovery (%)
		Added	Found		
1	Distilled water	46	40	0	87
		92	86	0	94
		221	202	0	91
		368	370	0	101
2	Tap water	46	54	0	117
		92	83	0	90
		221	194	0	88
		368	325	0	88
3	River water	0	860	860	112
		368	1180	810	112
		736	1350	610	79
		1105	1710	600	85
4	Sewage effluent	0	187	187	104
		96	263	167	89
		184	360	176	99
		293	473	180	100
5	Sewage effluent	0	251	251	115
		294	554	260	114
		589	777	188	96
		883	1007	124	91
		1178	1361	183	97
6	Settled sewage	0	109	109	136
		96	166	70	91
		184	226	42	79
		293	356	63	95
7	Settled sewage	0	176	176	88
		184	402	218	109
		293	539	246	115

Each of the results given in Table 57 is from a single determination. The apparent original EDTA concentration in samples 1 and 2 has been assumed, from a knowledge of the origin of the samples, to be zero. Where the concentration found exceeds the concentration added, this is probably due to experimental error.

From the n estimates ($n = 3$, 4, or 5) of the apparent original EDTA concentration in each of samples 3–7, the weighted average was calculated by giving the first figure, in the order given in the table, a weighting of n, the second of $n-1$, and so on. This procedure, which was designed to compensate for the increase in experimental error in the difference between the 'added' and 'found' concentrations of EDTA as both increased, gave, for samples 3–7, values of 769, 178, 219, 80, and 202 µg l^{-1} respectively. These values were used to calculate the recoveries given in the last column.

Reprinted with permission from Gardiner.[92] Copyright (1977) Royal Society of Chemistry

All except Samples 1 and 2 contained EDTA on collection and the recovery was calculated from the weighted average of the apparent original concentration of EDTA. The mean recovery was 98.6% with a standard deviation of 13.1%. Approximately half of the experimental error arises from the measurement of the chromatographic peak areas. The sensitivity obtained was approximately 5×10^{-8} mol l^{-1} (15 µg l^{-1}) with 25 ml samples. Increasing the sample volume up to a probable convenient maximum of 200 ml gave a corresponding increase in sensitivity.

A typical chromatogram is shown in Figure 70. Peaks due to compounds other than EDTA and the internal standard were usually relatively small. Primary sewage that was processed immediately after collection, however, showed prominent peaks closely following the solvent peak and in positions corresponding to long chain fatty acids (Table 58).

The method is sensitive to nitriloacetic acid (NTA), but none was detected in the environmental samples examined by Gardiner.[92,93] A series of small peaks was obtained in the region of the peak for EDTA in positions corresponding to the alkanes C_{24}–C_{30}. These peaks may have derived from alkyl sulphonate detergents; for example, under the esterification conditions

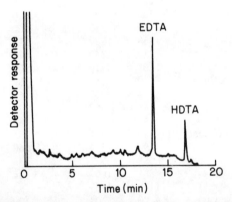

Figure 70 Typical chromatogram showing EDTA in 25 ml of primary (settled) sewage 24 h after collection. (EDTA) = 4.5×10^{-8} mol per 25 ml (520 µg l^{-1}); (HDTA) = 1.25×10^{-8} mol per 25 ml (internal standard). Reprinted with permission from Gardiner.[92] Copyright (1977) Royal Society of Chemistry

Table 58 Relative retention times of the ethyl esters of EDTA, HDTA and nitriloacetic acid (NTA)

Chelating agents			Long chain fatty acids			
HDTA	EDTA	NTA	C18	C16	C14	C12
1.00*	0.80*	0.32	0.73, 0.71, 0.69	0.58	0.47	0.35

*Kovats retention indices. HDTA, 2737.86; EDTA, 2320.42.
Reprinted with permission from Gardiner.[92] Copyright (1977) Royal Society of Chemistry

used, a C13 alkyl sulphonate gave a large peak in the position corresponding to the alkane C26. The detergents appeared to originate at least partly from laboratory glassware, despite rinsing with concentrated nitric acid before use.

Nitriloacetic acid (NTA)

Methods for the determination of this substance in water samples are mainly based on polarography and gas chromatography.

Polarography

Polarographic methods for the determination of nitriloacetic acids have been described by several workers.[94,100] Wernet and Wahl[94] removed interfering cations and heavy metals that form complexes with nitriloacetic acid from the surface water and effluent samples by equilibrating at pH 3 with Dowex 50W−X8 resin (sodium form) prior to polarography in the presence of ammoniacal cadmium buffer solution at pH 8. Afghan and Goulden[95] used linear potential sweep chronoamperometry of the nitriloacetic acid lead complex to determine down to 10 µg l^{-1} of nitriloacetic and in water.

To determine total nitriloacetic acid, the sample is acidified to pH 1 to release the nitriloacetic acid from heavy metal complexes, and these metals are masked by adding EDTA before bringing the pH back to 8 to form the lead complex. Large amounts of *bis*-(2-aminoethoxyl)-ethane-N,N,N',N'-tetraacetic acid interfere with the lead–nitriloacetic acid method.

Haberman[96] determined down to 0.02 ppm of nitriloacetic acid in water and sewage by first passing the sampled, adjusted to pH 3, through a cation exchanger. After adjustment to pH 7.0, the percolate is then passed through an anion exchanger to absorb the nitrilotriacetate, which is subsequently eluted with sodium chloride in 0.1 M sodium acetate 0.1 M acetic acid buffer of pH 4.7. Nitriloacetic acid is then determined by adding a known amount (in excess) of trivalent indium to the solution and measuring the height of the wave due to reduction of the In^{3+}-nitriloacetic acid complex at −0.79 V *vs* the SCE. An isotope dilution technique usually ^{14}C-labelled nitriloacetic acid is used to correct for incomplete recovery.

Low concentrations (1–10 ppm) of nitriloacetic acid have been determined in lake water by a method[97] which involved conversion to its 1:1 cadmium complex. The resulting solution (adjusted to pH 9) is subjected to polarography between −0.25 and 1.25 V (*vs* the SCE) in 0.1 M potassium nitrate, as supporting electrolyte; Cd^{2+} gives a wave at −0.60 V and the Cd–nitriloacetic acid complex gives a wave at 0.97 V. To determine nitriloacetic acid in concentrations between µM and 10 µM it is advantageous to use a pH of 8.5 to avoid precipitation of cadmic hydroxide.

Afghan *et al.*[98] developed an automated method for the determination of nitriloacetic acid in natural water and sewage samples. The method is based on the formation of the bismuth nitrilotriacetic acid complex at pH 2 followed by determination by twin-cell oscillographic d.c. polarography. As little as 10 μg l^{-1} of nitrilotriacetic acid can be determined, with no preconcentration of the sample being required. The coefficient of variation for 100 mg l^{-1} was 1.3%.

Dietz[99] used polarography to determine nitriloacetic acid, EDTA and other complexing agents in surface and ground waters using bismuth complexes at pH 2. Concentrations of nitriloacetic acid and EDTA in the range 0.1-3 mg l^{-1} could be determined selectively using cathode ray, impulse, or modified alternating voltage polarography.

Haring and Van Delft[100] studied the application of derivative pulse polarography at a hanging mercury drop electrode to the determination of nitriloacetic acid in water.

Gas chromatography

Murray and Povoledo[101] converted nitriloacetic acid to its trimethyl ester prior to gas chromatographic determination on a column packed with 2% of poly(ethanediol adipate) on Chromosorb W. Down to 25 μg l^{-1} nitriloacetic acid could be determined by this procedure.

Chau and Fox[102] concentrated nitriloacetic acid in lake water samples by passing them down a Dowex 1 column (formate form) and elution with 2.5 M to 8 M formic acid. The nitriloacetic acid is then esterified, with heptadecanoic acid added as internal standard, by heating for 1 h at 100 °C in a sealed ampoule with propanol saturated with hydrogen chloride. The propyl esters are analysed on a stainless steel column (6 ft $\times$ 0.25 in) packed with 3% of OV-1 on Chromosorb WHP (80-100 mesh), temperature programmed from 180 to 225 °C min^{-1}, and operated with nitrogen as carrier gas (65 ml min^{-1}) and flame ionization detection. The calibration graph is rectilinear for up to 20 μg of nitriloacetic acid as ester. The limit of detection is 0.01 μg and at the level of 20 μg l^{-1} the standard deviation was 1.3 μg ml^{-1} and the coefficient of variation was $\pm 6.3\%$.

Warren and Malec[103] determined nitriloacetic acid and related aminopolycarboxylic acids (iminodiacetic acid, glycine, and sarcosine in inland waters and sewage effluents by converting to the butyl or the *N*-trifluoroacetyl esters followed by chromatography on dual glass U-shaped columns (1.9 m $\times$ 2 mm) packed with 0.65% of ethanediol adipate on acid-washed Chromosorb W (80-100 mesh), temperature-programmed from 80 to 220 °C and operated in the differential mode with flame ionization detectors. The signal was fed to a digital integrator and then to both channels of a dual-pen recorder operated at high and low sensitivities.

Aue *et al.*[104] determined parts by billion (US) of nitriloacetic acid and citric acid in tap water and sewage effluents. Following a preliminary clean-up and concentration procedure the acids are converted to their butyl esters and chromatographed on a Carbowax on Celite 545 column at 183 °C.

By converting nitriloacetic acid to its trimethylsilyl ester, Stolzberg and Hume[105] were able to extend the limit of detection of this substance down to 1 μg l^{-1} without a concentration step.

Williams et al.[106] applied a nitrogen specific detector to a survey of the levels of nitriloacetic acid as its tri-n-butyl ester in US tap water supplies at concentrations approaching 1 ppb (US). These workers based their method on that described by Aue,[104] described earlier. Williams et al. used a Perkin Elmer Model 910 gas chromatograph, equipped with a single column, a two-way effluent splitter, a flame ionization detector and a nitrogen–phosphorus detector operating in the nitrogen mode. The column was 6 ft × ¼ in. o.d. glass, packed with either 5% OV-101 or 3% OV-210 on 80–100 mesh Chromosorb WHP. The carrier gas was helium at a flow rate of 60 ml min^{-1} and the effluent splitter diverted 60% to the flame ionization detector and 40% to the nitrogen detector. Hydrogen and air flows were optimized for each detector. The injector and detector temperatures were 240 to 280 °C respectively and the column and interface temperatures 200 and 250 °C.

Aue et al.[104] claimed a limit of detection of 1 ppb nitriloacetic acid for a 50 ml water sample, but preliminary investigations of Williams et al.[106] with standard solutions of the tri-n-butyl ester showed that quantitation at this level was difficult due to interference from the solvent peak (Figure 71(a)) when using acetone as the injection solvent as specified by Aue et al.[104] The use of alternate injection solvents gave some improvement but quantitation was still difficult. Analysis of standard solutions of the tri-n-butyl ester, equivalent to 1 ppb nitriloacetic acid in a 50 ml water sample, showed that the sensitivity of nitrogen specific detector was adequate, quantitation was straightforward, and there was minimal interference from the injection solvent, acetone (Figure 71(b)). The nitrogen-selective detector gave a linear response over the range 1–1000 ng injected of the tri-n-butyl ester of nitriloacetic acid.

The isolation procedure of Aue et al.[104] gave a satisfactory chromatogram (Figure 72(a)) for a control blank water sample provided that all solvents were redistilled in glass and the ion-exchange resin and glassware were thoroughly washed before use. The lower detection limit is four times the level of the blank which would give a detection limit of ca 0.2 ppb nitriloacetic acid for a 50 ml water sample. A typical chromatogram obtained from a 50 ml raw water sample analysed as containing 0.4 ppb nitriloacetic acid is shown in Figure (72(b)). Recoveries of nitriloacetic acid from water samples spiked with 1–1000 ppb nitriloacetic acid were greater than 90%.

Reichert and Linckens[107] have reviewed gas chromatographic methods for the determination of nitriloacetic acid in drinking waters. They point out that esterification of the nitriloacetic acid is required to enable it to be volatilized in the gas-liquid chromatographic column, and they compared a range of esterification reagents and conditions for simplicity and speed of operation. The method chosen involved treatment of a concentrated sample with a mixture of n-propanol–acetyl chloride (10:1) and the resulting nitriloacetic acid–propyl ester injected into the column, which was fitted with a nitrogen-sensitive detector. The detection limit for nitriloacetic acid in drinking water is about 1 μg l^{-1}.

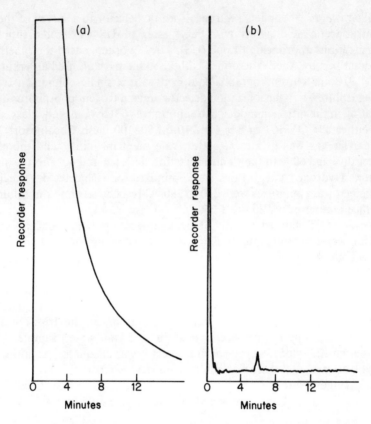

Figure 71 Gas chromatograms of tri-n-butyl ester of nitriloacetic acid, retention time 6 min. Column, 5% OV-101 at 220 °C; 2.6 ng injected: (a) flame ionization detector, 60% of effluent, attentuation 10×4; (b) nitrogen-selective detector, 40% of effluent, attentuation 10×1. Reprinted with permission from Williams et al.[106] Copyright (1977) Elsevier Science Publishers

Miscellaneous methods

Robinson and Lott[108] used a fluorimetric method for the determination of down 0.2 ppb nitrilotriacetic acid in tap water. The method is based on the displacement of 8-hydroxyquinoline from its fluorescent complex with Ga^{3+} by nitrilotriacetic acid, the gallium complex of which is not fluorescent. The reaction is carried out in a medium buffered by acetate at pH 6 and containing hydroxylammonium chloride (to reduce Fe^{3+}) and $KI-NaH_2PO_4-NaCN$ masking reagent. Standard additions of nitriloacetic acids are made, and the organogallium complexes are extracted into chloroform for fluorescence measurement (and thereby separated from the 8-hydroxyquinoline complexes of magnesium and calcium.

Longbottom[109] used high speed ion-exchange chromatography to determine nitriloacetic acid with a sensitivity of $1 \mu g \, l^{-1}$. Nitrilotriacetic acid was

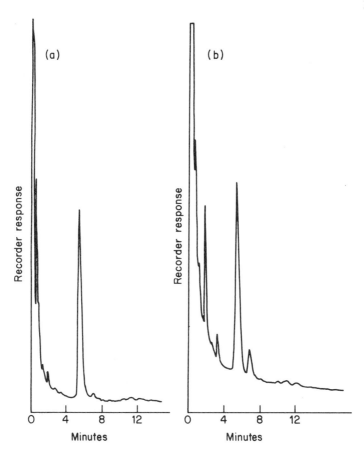

Figure 72 Gas chromatograms on 5% OV-101 column at 235 °C, nitrogen-selective detector, attentuation 10 × 1. (a) Control water blank, butylated residue dissolved in 100 μl acetone and 4.8 μl injected; (b) raw water sample containing 0.4 ppb nitriloacetic acid butylated residue dissolved in 100 μl acetone, 4.9 μl injected, retention time 3.4 min. Reprinted with permission from Williams et al.[106] Copyright (1977) Elsevier Science Publishers

separated on a stainless steel column packed with SAX, a strong anion-exchange resin, coated on Zipax. The mobile phase was 0.02 M $Na_2B_4O_7$ (pH 9). Possible interference from metal ions was overcome by converting all metal–nitriloacetic acid complexes into the Fe^{3+}-nitriloacetic acid complex. The nitriloacetic acid was monitored at 254 nm with an ultraviolet spectrophotometric detector; maintenance of the pH at 9 was essential because the extinction of nitriloacetic acid at this wavelength varies widely with pH. The method was applied to the analysis of sewage samples.

Coombs et al.[110] analysed mixtures of aminopolycarboxylic acids at the ppb level (US) by chemical kinetics. The method is based on the reaction of the nickel complexes of these acids with cyanide ion and the large differences in reaction rates for the formation of $Ni(CN)_4^{2-}$. The reaction is monitored spectrophotometrically (at 267, 285, or 310 nm) with use of a stopped-flow system, and the results are calculated from a computer program that provides on-line data acquisition and performs a regressive differential kinetic analysis for two components. The acids can be determined singly or up to three in admixture. The error is within ±5 to 10% for mixtures of ligands at the μM level and the sensitivity is 0.04 μM (8 parts per 10^9) in water.

Nitriloacetic acid in sewage effluents

Rudling[111] determined nitriloacetic acid gas chromatographically at the 0.1 μg l^{-1} level in sewage. Following separation of the nitriloacetic acid on an anion-exchange column, it is derivatized with boron trifluoride in 2-chloroethanol to produce 2-chloroethylnitrilotriacetate and gas chromatographed on a column packed with 2% QF-1 on Varaport 30. Detection is achieved by electron capture. Chloride, sulphate, calcium, and magnesium did not interfere in this procedure.

Rudling[112] simultaneously determined 0.01–0.02 μg of nitriloacetic acid, EDTA, and diethylenetriaminepentaacetic acid in sewage samples as their methyl esters by gas chromatography. The sample, adjusted to pH 7, is extracted with chloroform, 1,2-diaminocyclohexanetetraacetic acid, and internal standard added and the mixture treated with methanolic boron trifluoride-chloroform extracts are injected into a column packed with 5% V-17 on Aeropak, temperature programmed from 150 to 285 °C at 10 °C min^{-1} using helium as carrier gas. Cadmium, copper, zinc, nickel, and iron did not interfere.

Longman et al.[113] compared five procedures for removing interferences in the determination of nitriloacetic acid in sewage and sewage effluents. None of these methods completely overcame interference by metals. These workers then developed an alternate procedure based on passing the filtered sample through a column of Chelex-1.00 chelating resin (N_a+ form). To the eluate was added zincon reagent and the extinction was measured at 620 nm. The method was applicable in the range 0–10 μg l^{-1} nitriloacetic acid in sewage. Ferric iron did not interfere.

Sekerka et al.[114] described a potentiometric titration procedure for the determination of down to 0.1 μg l^{-1} of nitriloacetic acid in waste water and sewage. To the sample (100 ml) was added 0.01 M thallium nitrate (0.1 ml) and 10 M sodium hydroxide, the mixture was filtered, the filtrate adjusted to pH 7 by addition of 10 M nitric acid and hexamine (2 g), and 0.01 M ammonium pyrrolidine-l-carbodithioate added (this preliminary treatment removes interfering cations and reducing substances). The solution was titrated with 0.1 M thallium nitrate (Tl^{3+} forms a 1:2 complex with nitrilotriacetate and 1:1 complexes with EDTA) with the use of a combination platinum redox electrode (Orion 96-78). A second 100 ml sample is treated similarly, but in the first step 0.1 ml of 0.01 M cupric nitrate is added in place of the thallium nitrate and titrated with 0.1 mM or mM cupric nitrate (Cu_{2+} forms 1:1 complexes with all the complexing agents tested). The titration curves exhibit two end points corresponding to

(1) Tl^{3+} or (Cu^{2+}) consumed by ammonium pypollidine-l-carbodithioate and (2) nitriloacetic acid (or other ligands) present in the sample.

The content of nitriloacetic acid is calculated from the differences in titres. The recovery of nitriloacetic acid in admixture with EDTA in synthetic sewage samples ranged from 90 to 102%.

Melamine and cyanuric acid

These substances have been determined[115] polarographically in waste water.

Chlorinated isocyanurates

These substances have been determined in amounts down to 10 ppm in swimming pool water by a spectrophotometric procedure.[116] The method is based on the production of isocyanuric acid and reaction with cobalt acetate, in the presence of isopropylamine. The water sample is evaporated and the residue dissolved in methanol; 5 ml is mixed with 5 ml chloroform. Cobalt acetate solution is added followed by 0.5 ml isopropylamine. After 30 min, the extinction is read at 565 nm against a blank of methanol–chloroform–isopropylamine (10:10:1).

Nucleic acids

Seibert and Zahn[117] used precipitation titration with the cationic detergent N-cetyl-N,N,N-trimethylammonium bromide to quantify the nucleic acid content in activated sludge. With the introduction of toxins into the influent of a sewage treatment plant the ribonucleic acid content of the activated sludge fell sharply, and it was possible to correlate the deterioration in effective purification with a reduction in the ribonucleic acid fraction precipitable by N-cetyl-N,N,N-trimethylammonium bromide. Advantages of the method are simplicity and speed; DNA and RNA can be determined in one operation and the process is not affected by the presence of dyes or compounds absorbing ultraviolet light.

Hicks and Riley[118] have presented a method for determining the natural levels of nucleic acids in lake and sea waters, which involves preconcentration by adsorption on to hydroxyapatite, elution of the nucleic acids, and then photometric determination of the ribose obtained from them by hydrolysis.

Determination of organic nitrogen

Trifonova[119] has described a diffusion–isothermal method of distillation of ammonia for use in the determination of organic nitrogen in natural waters. After Kjeldahl digestion, the digest is placed in a Conway diffusion cell containing sodium hydroxide solution in one compartment to trap the ammonia. The analysis is completed photometrically or by titration of ammonia with 0.01 M hydrochloric acid.

Mancy et al.[120] described a method involving ultraviolet combustion of organic nitrogen compounds in lake waters prior to determination of the resulting nitrate, nitrite, and ammonia. The samples were irradiated, by means of a 450 W mercury-arc lamp enclosed in a water-cooled silica well immersed directly in the sample, for 1.5–3 hours, with continuous oxygenation, at pH 4.0 (buffered with boric acid). The 'combustion' products were determined spectrophotometrically; nitrate plus nitrite by the method of Wood,[121] except that the hydrochloride of N-1-naphthylethylenediamine was used instead of the dihydrochloride, and ammonia by the indophenol blue method. The sensitivities were 2–5 µg l^{-1} for nitrate- plus nitrite-N and for ammonia-N. The rate of release and the relative proportions of the combustion products were found to depend on the molecular structure of the dissolved organic nitrogen compounds, and the concentration of total dissolved organic carbon and inorganic salts.

Methods have been described for the determination of organic nitrogen in water, waste water, and sludge,[122] waters and waste waters,[123] and sewage plant effluents.[124] Jenkins[123] has reviewed the present state of the art in the analysis of organic nitrogen, ammonia, nitrite, and nitrate in natural waters, over various concentration ranges, with particular reference to the precision of the different methods.

Helfott and Mazurek[124] give details of techniques for rapid determination of nitrogen compounds at various points in a sewage treatment plant, using a nitrite probe, an ammonia probe, and a pyrolysis-microcoulometric analyser for total nitrogen. Total nitrogen can be determined within 15 min, and nitrate and ammonia within 1 min; organic nitrogen is estimated by subtracting ammonia and nitrate from total nitrogen.

Adamski[125] has reported a simplified Kjeldahl nitrogen determination of organic nitrogen in sea water using a semiautomated persulphate digestion procedure. The method gave an accuracy of ±8.1% and a precision of ±8.2% for sea water samples spiked with 3.35 µg l^{-1} of organic nitrogen.

Method

Apparatus

A modified version of the indophenol procedure as described by O'Connor and Miloski[126] was employed using a Technicon-AutoAnalyzer II system with the appropriate accessories as shown in Figure 73. Sample digestion was accomplished on a hot plate in standard 125 ml Erlenmeyer flasks.

Reagents

Phenate reagent was prepared by dissolving 35 g of phenol and 0.4 g of sodium nitroprusside in approximately 250 ml of distilled water and diluting to one

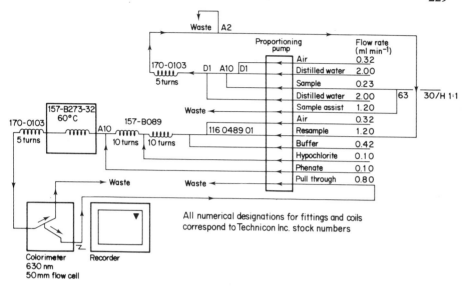

Figure 73 AutoAnalyzer II analysis scheme for the determination of Kjeldahl nitrogen in sea water via indophenol colorimetry. Reprinted with permission from Adamski.[125] Copyright (1976) American Chemical Society

litre. The phenate reagent was stored in an amber glass bottle to inhibit decomposition and refrigerated when not in use. Preparation of the phenate reagent with fresh reagent grade phenol was necessary for proper colour development. Once a bottle of phenol has been opened, reagent quality can be maintained by storing the remainder of the phenol under nitrogen gas.

Hypochlorite reagent was prepared by dissolving 20 g of sodium hydroxide and 2 g of sodium dichloro-S-triazine-2,4,6-($1H,3H,5H$)-trione in approximately 250 ml of distilled water and diluting to one litre. This reagent is generally stable for about 3 days if refrigerated.

Buffer reagent is prepared by dissolving 9 g of boric acid, 2 g of sodium hydroxide, and 120 g of sodium citrate in 500 ml of distilled water and diluting to one litre. A wash solution approximating the sodium sulphate content and final colour of a digested blank was prepared by adding 20 ml of concentrated sulphuric acid and six drops of 0.1% methyl red in ethanol to about 500 ml of distilled water and neutralizing to the methyl red end point with 4 M sodium hydroxide ($\approx$ 190 ml). Back titration with 4 M sulphuric acid was then required to duplicate the pink colour of digested samples. The final volume was adjusted to one litre with distilled water. This solution was maintained in the wash reservoir of the automatic sampler throughout routine analytical procedures.

Procedure

A 25 ml aliquot of sea water sample and 1.0 ml of concentrated sulphuric acid were pipetted into a 125 ml Erlenmeyer flask and boiled down on a hot plate in the presence of glass beads until the white fumes of sulphur trioxide first appeared. While the sample was sufficiently warm to maintain the salt residue in solution, a measured amount of doubly recrystallized potassium persulphate (approximately 1 g) was added to the digestion mixture. The sample digest was then swirled gently to ensure complete contact between the persulphate and the sample residue. If salt precipitation occurred prior to persulphate addition, gentle heat was again applied until all of the precipitate redissolved. After persulphate addition was completed, the digestion mixture was strongly heated at fuming for 10 min. The data in Table 59 for a series of replicate distilled water standards digested over various time intervals indicate the undesirable effects of prolonged digestion.

After allowing sufficient time for cooling, 15–20 ml of distilled water were added to the residue with gentle application of heat until solution was complete. The digest was then neutralized with 4 M sodium hydroxide to the yellow methyl red end point (one drop 0.1% methyl red in ethanol) and back-titrated dropwise to the first pink colour (final pH approximately 5.5). The sample was then quantitatively transferred to a 50 ml volumetric flask and brought to volume with distilled water. The final step accomplished a two-fold dilution on the initial sample aliquot to ensure that all of the salt residue remained in solution. Blanks must be analysed with each set of digested samples in order to correct for any nitrogen contamination in the sulphuric acid and potassium persulphate reagents. The nitrogen concentration in the blank was independently determined and this result subtracted from all sample analyses before any final results were reported.

The sea water digest now containing nitrogen as ammonium ions was finally analysed colorimetrically using the Technicon AutoAnalyzer II system schematically illustrated in Figure 73. Calibration of the automated system was accomplished by employing a dilution loop to maintain a routine operating range from 0 to 5.6 mg l^{-1} as N. The nitrogen concentration in digested distilled

Table 59 Effects due to extended digestion time on ammonium chloride standards* prepared in distilled water

Sample	N concentration (mg l^{-1})		
	10 minute digestion	20 minute digestion	50 minute digestion
1	1.51	1.44	1.06
2	1.48	1.58	1.75
3	1.50	1.47	1.46
4	1.51	1.47	1.91
5	1.48	1.55	0.90
6	...	...	1.61
Std dev (mg l^{-1})	±0.016	±0.06	±0.37
Rel std dev (%)	±1.1	±4.0	±29

*Known concentration of standards, 1.40 mg l^{-1}.
Reprinted with permission from Adamski.[125] Copyright (1976) American Chemical Society

water blanks varied between 0.21 and 0.31 mg l^{-1} as N. In the operating range indicated, a digested blank is approximately equal to a 5% chart recorder deflection.

Various other workers have described automated procedures for the determination of low levels of organic nitrogen in sea water,[127-129] samples,[130] activated sludge,[131] and water and waste water[132,133] and Elkei[132] has described a detailed procedure using the AutoAnalyzer. The samples are digested continuously at 300 °C with sulphuric acid and hydrogen peroxide to convert the organic nitrogen to ammonia, which is determined colorimetrically with sodium salicylate and sodium dichloroisocyanurate (as chlorine source).

Lowry and Mancy[133] give details of an automated system for the determination of dissolved organic nitrogen in natural waters, based on ultraviolet irradiation of the sample followed by a heterogeneous reduction of the nitrogen-containing irradiation products to ammonia which is detected by an ammonia probe. Quantitative recoveries of different organic compounds were obtained after irradiation permits separate determination of total and inorganic nitrogen. This procedure utilizes the Devadas alloy reduction technique described by Mertens et al.[134] for the heterogeneous reduction of the irradiation products to ammonia. Using the combination of ultraviolet irradiation and reduction with Devadas alloy considerably reduces the duration of irradiation required to effect quantitative recoveries of organic nitrogenous compounds.

The autoanalysis system for total nitrogen is shown diagrammatically in Figure 74. The system consisted of an automatic sampler, a proportionating pump, a water bath, a series of quartz coils irradiated by an ultraviolet lamp, a reduction column, an electrode assembly, a high impedance pH/mV meter and a strip chart recorder. The system was driven by a Technicon II proportionating pump. A Gilson automatic sampler utilizing a sample time of 25 min and a rinse time of 3.5 min facilitated sampling. The ultraviolet light assembly consisted of quartz delay coils centred around a 0.75×12 in. 1200 W high pressure quartz mercury vapour arc lamp housed in a sheet metal reflector. The mercury lines as a function of radiated energy are given in Table 60. The quartz delay coils were in two sections, each 2.4 mm i.d. with a coil diameter of 5.5 in. and each section approximately 15 coils. A high speed fan placed at one end of the housing maintained the temperature at the coils of about 32 °C. An Orion 95-10 ammonia selective electrode filled with an Orion 94-00-25 flowthrough cap was the sensing assembly. The potentiometric output of the electrode assembly was continuously recorded using a digital pH/mV meter and a strip chart recorder.

Column

The reduction column, usually 5.5 mm i.d. $\times$ 30 mm formed from disposable Pasteur pipettes, was packed with Devarda's alloy that had been sized to 100-140 mesh and washed 10-15 times with distilled water to eliminate fines. To prevent channelling, quartz wool was packed every 10 mm. Conventional slurry packing

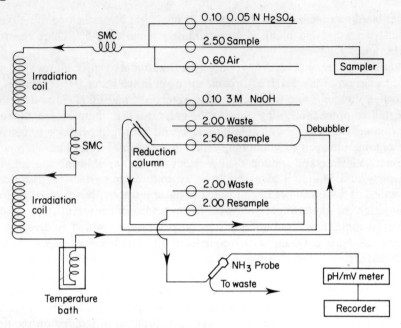

Figure 74 Flow diagram of the automated system. Reprinted with permission from Lowry and Mancy.[133] Copyright (1978) Pergamon Press

Table 60 Spectral energy distribution of the mercury vapour lamp

Mercury lines (A)		Radiated energy (W)
13,673–10,140	(infrared)	48.68
5780–4045	(invisible)	187.07
3660–3341	(near u.v.)	104.03
3130–2804	(med. u.v.)	117.01
2753–2224	(far u.v.)	116.15

Reprinted with permission from Lowry and Mancy.[133] Copyright (1978) Pergamon Press

techniques utilizing a vibrator allowed maximum packing of the alloy. The degassing system, necessary due to the large hydrogen gas production, consisted simply of having the column open to the air and drawing off the sample below a set volume (0.1 ml).

Reagents

All the reagents were reagent grade. Stock solutions (1000 ppm N) were prepared and stored at 4 °C in amber glass bottles. Dilutions were made with ammonium free water made by double acid distillation of dionized water in a silica apparatus.

Procedure

Aliquots from a discrete automatic sampler were aspirated into a stream of air-segmented solutions flowing at a constant rate. The air was bubbled through a 1 M sulphuric acid solution to prevent ammonia uptake. Sample irradiation was done in acidic medium (pH 2.0) in one of the coils of the irradiation assembly. After leaving the first coil, the solution's pH was adjusted to above 13.0 with alkali and passed through the second coil. After irradiation, the solution was debubbled and passed through the reduction column. Finally, the sample stream was resampled and pumped to the electrode cell.

The inorganic nitrogen analysis was accomplished by bypassing the irradiation assembly. The ammonium content or the irradiation products was determined by simply bypassing the reduction column. A 300 ppb N as NO_3^- blank usually was utilized to facilitate a better sample to wash ratio over the concentration range 300–3000 ppb N. A sampling rate of $10\,h^{-1}$ was employed. Organic nitrogen in a mixed sample was obtained by the difference in total nitrogen analysis and the total inorganic analysis.

Lowry and Mancy[133] confirmed the decomposition of different organic compounds by irradiation of these compounds in distilled water and by spiking the same amounts of these compounds in actual natural waters. Table 61 reveals that the carbon–nitrogen bonds were quantitatively cleaved, even in a ring such as pyridine. The more stable nitrogen–nitrogen bonds seemed to resist the conditions employed.

Spiking experiments on natural waters spiked with 1 ppm nitrogen as urea and glycine, gave recoveries, respectively of 93 and 98%. The precision of the entire organic nitrogen system, based on replicate measurements of the concentrations of urea and glycine expressed as percentage relative standard deviations was between 3.1 and 4.5%. The accuracy of the entire organic

Table 61 The recovery of nitrogen from various organic compounds following u.v. irradiation

Name of compound	Known concentration of nitrogen (ppb)	Actual concentration found (ppb)
Glycine	500	485
Urea	500	505
di-Alanine	500	500
Leucine	500	500
EDTA	1000	950
NTA	500	420
Pyridine	1000	900
S'-Diphenylcarbazone	1000	140
Semicarbazone	1000	300
Dinitrophenylhydrazine	1000	200
3-Methyl-1-phenyl-2-pyrazolin-5-one	1000	310

Reprinted with permission from Lowry and Mancy.[133] Copyright (1978) Pergamon Press.

nitrogen system based on the average recovery of nitrogen from urea (Table 61) was 97% for carbon–nitrogen bound compounds and 24% for the nitrogen–nitrogen compounds.

A principal shortcoming of this method is the relatively long time of analysis required at low concentration levels, e.g. less than 50 ppb N. This is due to the slow diffusion of ammonia gas across the membrane of the electrode system. A dichotomy exists in that a main advantage of the system is the high selectivity of the sensor electrode which renders the system applicable to a wide variety of aqueous solutions. This reduces interferences more effectively than in colorimetric techniques. Furthermore, this system has the advantage of a short irradiation time, facilitated by a unique reduction system, which allows this system to be used for both total inorganic and organic nitrogen dissolved in natural waters.

Nydah[135] has proposed a method for the determination of total nitrogen in natural waters, involving oxidation of nitrogen to nitrate at high temperatures, using peroxidisulphate in alkaline medium; and nitrate is then reduced to nitrite in alkaline medium; the nitrate is then reduced to nitrite by cadmium, and the nitrite is determined as an azo dye.

Middelboe[136-138] has investigated the determination of nitrogen-15 by photospectrometry and applied this technique to the determination of low concentrations of organic nitrogen compounds in natural waters. The sample nitrogen must be in the ammonium form. If this is not already the case, the transformation is brought about by Kjeldahl digestion (which simultaneously yields a total N value). Subsequently, sample nitrogen as a gas is evolved at room temperature by mixing the ammonium solution with a solution of hypobromite in an evacuated flask. Water vapour and other condensable contaminants are frozen out by means of liquid air, and finally the sample nitrogen in the gaseous state is expanded into a previously degassed discharge tube to give a pressure of 3–5 mmHg. The dried sample is heated in the presence of copper oxide, whereby the sample nitrogen is liberated together with other gases (e.g. H_2O and CO_2) which are absorbed by calcium oxide. The liberation process may take place inside a previously evacuated glass or quartz container, in which case an appropriate amount of sample nitrogen is subsequently transferred to a pretreated discharge tube via a vacuum line. Alternatively, a sample containing an appropriate amount of total N (preferably as simple nitrogen compounds) may be processed by the modified Dumas method inside the degassed and sealed-off discharge tube itself.

In certain experiments the amount of total N obtainable from the individual sample is restricted to the order to 1 μg or less. In such cases it may still be possible, as demonstrated by Cook et al.,[139] to carry out a ^{15}N analysis by utilizing discharge tubes of quartz and carrier-gas mixtures. However, during the preparation of samples containing submicro quantities of total N, the problem of contamination by traces of natural nitrogen becomes serious.

The optical measurement of ^{15}N abundance by emission spectrometry is based on the rather large isotopic shifts found in the band spectrum of molecular nitrogen.[140-143] Due to the high cost of highly enriched ^{15}N-labelled compounds and the tendency of the labelled nitrogen to become diluted with natural nitrogen during the course of the tracer experiment, the ^{15}N abundance in the final samples are as a rule low. For various reasons an electronic-vibrational band in the near ultraviolet region (at about 300 nm) is normally selected for analysis of ^{15}N abundances in the range 0.3–30 atom %. This band shifts approximately 1.2 nm towards longer wavelengths when the light is emitted by $^{15}N^{15}N$ molecules instead of $^{14}N^{14}N$ molecules, and for the intermediate molecule, $^{14}N^{15}N$, about half this shift is observed. Thus, a chromatic resolving power of 10^3–10^4 is adequate for separating the isotopically shifted band-heads.

The electrodeless discharge tube containing the gaseous nitrogen is brought to emit light by means of a high frequency transmitter operating in the range 100–2500 Mhz, and a prism—or grating—monochromator is used to disperse the light. The desired section of the optical spectrum is scanned across the exit slit of the monochromator by mechanical rotation of the prism or grating. Alternatively, a swinging exit slit is used to scan the desired spectral region. The light emerging from the exit slit is transformed to an electronic signal by the use of a photomultiplier tube. Finally, the signal from the photomultiplier tube is amplified and either registered by a pen recorder or transferred to a computer.

Assuming the N_2 molecules in the discharge tube to represent random combinations of ^{14}N and ^{15}N atoms, the ^{15}N abundance in the sample can easily be calculated, for instance, from the ratio of the number of $^{14}N^{14}N$ molecules to the number of $^{14}N^{15}N$ molecules. A value for this ratio is obtained by observing the relative intensity of the light emitted by the two types of N_2 molecule. For example, if the $^{14}N^{14}N$–$^{14}N^{15}N$ ratio is 1:1, then the ^{15}N abundance is 33.3 atom %. A statistical derivation of the general relationship, based on the laws inherent in binomial distributions, yields the following formula:

$$A = \frac{100}{2 \cdot R + I} \text{ atom \%}$$

where A is (number of ^{15}N atoms/total number of N atoms) $\times$ 100 and R is number of $^{14}N^{14}N$ molecules/number of $^{14}N^{15}N$ molecules.

The absolute accuracy of an optical ^{15}N analysis is 10–100 times inferior to that of a corresponding analysis by mass spectrometry. However, the reproducibility as expressed by the standard deviation of a single measurement, of an optically determined ^{15}N value is ± 0.01 atom % in the difficult region to analyse close to natural abundance (0.37 atom %). Hence, by careful calibration with samples of known ^{15}N abundances, corrected values can be determined (even for samples with low abundances) to a relative accuracy of 2–3%.[144]

One of the main accuracy limiting factors in optical ^{15}N analysis is the uncertainty involved in determining the exact contribution of the spectral background to each of the observed band-heads. This uncertainty is caused by the presence in the discharge tube of foreign gases, in particular carbon monoxide, since the spectra of such contaminating gases overlap and interfere with the nitrogen spectrum. The amount and composition of the unwanted gases is a function of both sample type and sample preparation procedure. Further uncertainty is caused by the fact that the tail of the $^{14}N^{15}N$ band overlaps the head of the $^{14}N^{14}N$ band; however, a fairly accurate correction can be made for this effect.[145]

The lower limit of sample size is a point of particular interest in connection with the pollution of natural waters by relatively low concentrations of nitrogenous material. Suppose, as an example, that the nitrogen fraction under consideration represents a concentration of 0.1 ppm. In this case, a minimum sample of 100 ml water would be needed for a routine manual optical ^{15}N analysis, whereas the corresponding amount for a mass spectrometrical analysis would be about 1 litre.

Johansen and Middelboe[138] have described a method for the determination of trace amounts of total nitrogen in aquatic samples using the isotope dilution method (in which ammonium chloride labelled with nitrogen-15 is added to the water sample) followed by liberation of gaseous nitrogen in a discharge tube and analysis of the isotopic composition of the gas by optical emission spectroscopy. An aliquot of the labelled ammonium chloride solution is analysed similarly and the content of total nitrogen in the water sample can then be calculated. Satisfactory results have been obtained, with a detectivity in the region of 100 ng of total nitrogen. These workers make suggestions for improving the detectivity to approach or surpass the value of 5 ng of nitrogen achieved by the micro-Kjeldahl method.

Only about 1 μmole nitrogen is needed for the analysis of isotopic composition by optical emission spectroscopy. Thus, so-called optical analysis is particularly suited to determination of trace amounts of total nitrogen by the isotope dilution method.

The procedures used for liberating gaseous nitrogen in the discharge tube and optically analysing the nitrogen for isotopic composition are the same as described elsewhere by Leicknam et al.,[146] Karlsson and Middelboe,[147] and Fiedler and Proksch,[148] except

(1) a copper shaving is added to the discharge tube to facilitate the reduction of nitrates, etc. and
(2) the 0 to +2 vibrational transition in the second positive nitrogen system is utilized in order to obtain the relatively small $^{14}N^{15}N$ band-head on the tailless side of the prominent $^{15}N^{15}N$ band-head.

In the procedure 4.0 μmole ammonium chloride standard (labelled with about 96 atom % ^{15}N) is added to a 10.0 ml aliquot of water sample. This mixture is acidified with hydrochloric acid and pre-evaporated at a temperature well below boiling point to about 0.2 ml, whereafter about 0.05 ml is taken up automatically from the concentrate by a preheated capillary section of appropriate dimensions. The contents of the capillary section are neutralized by the addition of a chip of pure calcium carbonate in stoichiometric excess and then evaporated to dryness at a temperature below 50 °C.

The capillary section containing the dry residue is transferred to a degassed discharge tube, and the nitrogen liberated by the usual procedure is optically analysed for ^{14}N abundance. An aliquot of pure standard (labelled ammonium chloride solution) is treated and analysed in the same manner as described for the mixture of water and standard. Finally, the content of total N in the water aliquot is calculated by use of the following equation.

$$N_{nat} = (N_{std}) \frac{A^{14}_{mix} - A^{14}_{std}}{A^{14}_{nat} - A^{14}_{mix}}$$

where

N_{nat} = unknown moles of total N in water aliquot taken;
N_{std} = moles of standard (labelled) N added to water aliquot;
A^{14}_{mix} = abundance of ^{14}N in mixed N sample;
A^{14}_{std} = abundance of ^{14}N in standard N sample;
A^{14}_{nat} = abundance of ^{14}N in nature = 99.6 atom %.

In a series of nine replicates an average value of 0.42 ± 0.02 mg total N per litre of water (ppm) was obtained, which is satisfactory agreement with the Kjeldahl value of 0.38 ± 0.03 ppm. The detectivity of the isotope dilution method is in the vicinity of 100 ng total N.

Strict observance of nitrogen-14 (i.e. natural nitrogen) hygiene during sample preparation appears to be of paramount importance. Assuming that ^{14}N contamination and ^{15}NH$_3$ loss can be kept at an insignificant level, it should be possible to improve the detectivity by one or two orders of magnitude through the introduction of silica capillaries as discharge tubes, a higher resolution spectrometer; and/or carrier gas(es) to sustain the discharge. The time-consuming pre-evaporation step could then be avoided, and the detectivity of the isotopic method would approach or surpass that of the classical Kjeldahl method.

In a more recent development Middelboe[137] improved the optical emission spectrometry and isotope dilution methods by the addition of perchlorate for the enhancement of organic matter oxidation during combustion, and the introduction of a 'frozen sample' technique which makes it possible to perform the analysis without complete drying of the water sample. The improved methodology is described, and the results of determinations on ground water samples are compared with those obtained using ion-selective electrodes and the reductive semimicro-Kjeldahl method.

Middelboe[137] found that far from all the organic nitrogen in ground water samples is released by the standard Dumas–Faust procedure. However, this problem can be overcome by the addition of 0.5 mg sodium perchlorate to the discharge tube in accordance with a suggestion put forward by Fiedler and Proksch.[149] Furthermore the usual 10 mg of copper oxide can, to some

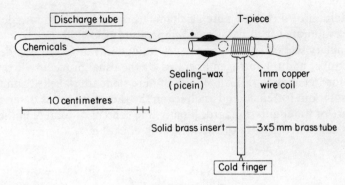

Figure 75 Part of the sample preparation apparatus that has been modified for use of the 'frozen sample' technique (Carl Zeiss). Reprinted with permission from Middelboe.[137] Copyright (1978) Pergamon Press

advantage, be substituted by 30 mg 'Cuprox' (Coleman reagent containing platinum catalyst). A few fine copper shavings, weighing about 30 mg should be added for the dual purpose of reducing nitrates and absorbing surplus dioxygen. Finally, the discharge tube is supplied with 30 mg of activated calcium oxide granules (*ca.* 1 mm diameter) this amount being sufficient to absorb completely at least 1 mg of water.

Investigation showed that ground water samples containing both nitrate and ammonium ions should not be dried down completely, the reason being that towards the end of the drying process, dependent on pH and buffers, a significant quantity of nitric acid fumes or ammonia gas is likely to be lost. Accordingly, the sample preparation procedure has been modified as follows. First, the pH of the water sample is adjusted to 4.5 ± 0.2 units using $1-0.1$ M hydrochloric acid or potassium hydroxide (low in nitrogen content). To a 1000 µl aliquot (containing 1–100 µg N) of water sample is then added (a) 5.0 µmol ammonium chloride labelled with ^{15}N at close to 100 atom % and (b) 5 mg potassium or sodium perchlorate (low in N content). This mixture is preconcentrated to a volume of 100 µl at 50 °C *in vacuo* in the presence of silica gel at room temperature. Next, a sub-aliquot of 10 µl is taken from the 100 µl preconcentrate by a 10 mm section of thin-walled glass tubing calibrated to hold 1 µl mm^{-1}. The sub-aliquot in the capillary is concentrated in the same way as before to a final volume of 1 µl and then frozen solid at about -100 °C. Subsequently, the capillary containing the frozen sample is transferred to the usual place in the T-piece of the standard sample preparation apparatus (Figure 75). The end-cap of the T-piece is replaced, and the sample immediately cooled down again, to about -100 °C with the aid of a copper-wire coil and cold finger attached to the T-piece, as shown in Figure 75. The T-piece is thereafter evacuated continuously, and the discharge tube containing the chemicals is heated for 10–15 min at 600–650 °C by the use of a cylindrical furnace. Finally, after cooling the discharge tube to room temperature and testing the system for air-tightness, the T-piece is turned to the vertical position thereby allowing

Table 62 Results of 21 ground water samples analysed for nitrogen content by ion-selective electrodes, reductive Kjeldahl and isotopic dilution

Sample symbol	NO_3^- (ppm N)	NH_4^+ (ppm N)	$NO_3^- + NH_4^+$ (ppm N) (A)	Reductive semimicro-Kjeldahl* (ppm total N) (B)	Isotope dilution analysis (ppm total N) (C)	Standard deviation on B-C (ppm total N) (D)
a-1	3.9	0.0	3.9	1.9	2.1	±0.3
d-2	3.2	0.0	3.2	1.8	1.7	±0.3
c-3	16.0	0.0	16.0	15.9	15.8	±0.5
d-4	13.0	0.3	13.3	14.2	13.4	±0.5
e-5	3.7	0.0	3.7	3.5	3.9	±0.3
a-6	2.4	0.0	2.4	2.3	2.2	±0.3
b-7	3.1	1.0	4.1	5.8	5.1	±0.4
e-8	5.2	0.9	6.1	6.9	6.7	±0.4
b-9	5.4	0.0	5.4	5.0	5.0	±0.4
e-10	5.7	0.0	5.7	5.8	5.7	±0.4
1+1	3.6	50	54	43	44	±1.5
1+2	0.2	0.2	0.4	1.5	1.0	±0.3
K+1	4.5	0.0	4.5	4.6	4.6	±0.3
K+2	5.3	0.2	5.5	6.0	5.8	±0.4
M+½	4.7	38	43	44	46	±1.5
M+1	0.1	77	77	78	79	±2
N+1	4.6	0.1	4.7	5.3	5.1	±0.4
N+2	2.0	0.5	2.5	2.3	2.5	±0.3
O+2	0.1	42	42	46	46	±1.5
P+1	0.2	0.5	0.7	3.5	4.1	±0.3
P+2	13.0	75	88	88	91	±3

*Carried out using the method of Nelson and Sommers.[150]
Reprinted with permission from Middelboe.[137] Copyright (1978) Pergamon Press.

the sample to fall to the bottom of the discharge tube. The distal half of the discharge tube is then immediately cooled to about −100 °C, whereupon there is ample time to swing the T-piece back to the horizontal position and seal the discharge tube before the sample begins to thaw out. During the sealing process the copper-wire coil is kept at about −100 °C, but after pulling off the discharge tube, the coil is allowed to warm up to room temperature, under vacuum pumping, in order to facilitate the removal of condensates inside the T-piece.

The sample in the standard discharge tube (made of Rasotherm glass) is heated in a muffle furnace for 3–4 h. The baking period is terminated with at least 1 h at 600–650 °C, which is feasible because the pressure of the contained water vapour at that temperature approximately balances the pressure of the atmosphere. Finally, the discharge tube is allowed to cool slowly overnight, whereafter measurement of the ^{14}N abundance and calculation of the total N content in the sample is carried out as described by Johansen and Middelboe.[138]

The total nitrogen values obtained by duplicate isotopic dilution analysis of 21 ground water samples are given in column C of Table 62. Column A indicates the concentration of nitrate plus ammonium in the various samples obtained using ion-selective electrodes. Column B shows the total nitrogen values using the reductive semimicro-Kjeldahl method developed by Nelson and Sommers.[150] Finally, column D gives for each sample the calculated standard deviation on the difference between the values in columns B and C, assuming these differences to be entirely due to statistical deviations (i.e. no systematic deviation is considered to be significant). Nelson and Sommers[150] indicate that the relative standard deviation on the mean of a duplicate semimicro-Kjeldahl analysis is 2-3%; hence, column D shows that the precision of the isotope dilution method is equal to or better than that of the semimicro-Kjeldahl method, except perhaps for samples containing less than *ca.* 10 ppm total N.

PHOSPHORUS COMPOUNDS

Adenosine triphosphate (ATP)

An estimate of the quantity of living microorganisms (biomass) in an aquatic environment can be a useful tool to assess water quality. Both phytoplankton and microzooplankton biomass can be estimated by direct microscopic determination of volume and number. Other microorganisms (for example, bacteria) have been estimated by various plating techniques (colony counting) and extinction–dilution techniques. The techniques are often time-consuming, expensive, and subject to inherent sources of error. Some of the major problems include the following. Direct microscopic counting of bacteria may yield higher estimates because of the difficulty in distinguishing bacteria from bacteria-sized inert particles. There is also the inability to differentiate between viable and non-viable cells, and the problems of cell aggregation, for bacteria and algae. Low estimates may be produced by plating and extinction techniques which are selective because of the chemical composition of the media and inherent physical parameters such as temperature and pressure. The extinction–dilution technique may also be biologically selective in that only types capable of growth in specialized media will grow to measurable size.

For a total biomass determination, it is desirable to measure some biochemical that is present in all living cells but is not associated with non-living particulate material. This cellular constituent must have a short survival time after death, so that it would be specific for viable biomass. It must likewise be present proportional to some measure of the total biomass for all microorganisms— algae, bacteria, fungi, and protozoans.

A biochemical which seems to meet these requirements is adenosine triphosphate. Adenosine triphosphate is the primary energy donor in cellular life processes. Its central role and biological and chemical stability make it an excellent indicator of the presence of living material. The level of endogenous adenosine triphosphate which is the amount of adenosine triphosphate per unit biomass in bacteria, algae, and zooplankton is relatively constant when compared to the cellular organic carbon content in several species and throughout all phases of the growth cycle. In studies where cell viability was determined the concentration of adenosine triphosphate per viable cell remains relatively constant during periods of starvation. The quantity of adenosine triphosphate, therefore, can be used to estimate total living biomass.

Shoaf and Lium[151] compared methods for the extraction and determination of adenosine triphosphate in pure algal cultures and natural aquatic samples. They found that three methods for the extraction of adenosine triphosphate, namely neutral dimethyl sulphoxide, boiling TRIS buffer, and butanol–octanol extraction, were equally effective on the alga *Chlorella vulgaris*, and measurement of the activity by either peak height or integration of the area under the peak was equally sensitive and reproducible. Determination of adenosine triphosphate was inhibited by mercuric chloride, cadmium chloride, calcium chloride, potassium or sodium phosphates, and high concentrations of the extracted dimethyl sulphoxide. Of the methods tried for preserving samples for analysis, field extraction of the adenosine triphosphate followed by quick freezing in an acetone–dry ice bath is recommended.

Method

Reagents

Buffer preparation Morpholinopropane sulphonic (MOPS) acid buffer (sodium salt), 0.01 M, at pH 7.4, was prepared with freshly distilled water, autoclaved, and capped until used.

ATP standards Adenosine triphosphate was prepared at a concentration of 0.1 μg ml^{-1} in 0.5 mmol l^{-1} (ethylenedinitrilo)tetraacetic acid disodium salt (Na$_2$EDTA), 5 mmol l^{-1} MgSO$_4$, and 5 mmol l^{-1} TRIS-Cl at pH 7.7 using freshly distilled water. One-half millilitre was placed in 6 × 49 mm glass vials, capped, and quick frozen in an acetone–dry ice bath. The standards were then maintained at -20 °C until used. A vial was thawed once and then discarded after use.

Neutral dimethyl sulphoxide extraction procedure

This procedure utilizes self-supporting membrane filters (13 mm diameter, 0.45 μm pore size). The filter assembly is placed in the stand and rinsed with 1 ml of 0.01 M MOPS, pH 7.4. A volume of sample is placed on the filter and filtered. The cells are then immediately rinsed with 1 ml of 0.01 M MOPS, pH 7.4. Adhering droplets are shaken from the bottom of the filter assembly. A clean graduated tube is placed in the filter stand to collect the extract. A 0.2 ml mixture of 90% dimethyl sulphoxide and 10% 0.01 M MOPS, pH 7.4, is added to the filter assembly so that all cells are covered. After 20 s, the vacuum is applied and the sample is filtered 'dry'. One millilitre of 0.01 M MOPS, pH 7.4 is added to the filter assembly. After 10 s the vacuum is applied and the sample is filtered dry. One millilitre of 0.1 M MOPS, pH 7.4 is again added and the sample is filtered dry. The final volume, approximately 2.2 ml, is recorded and used in the calculations for initial concentration. The sample is mixed and assayed.

Boiling TRIS buffer method

The boiling TRIS method consists of pipetting a volume of sample (usually 100 μl) into 5.9 ml of boiling 0.05 M TRIS-Cl, pH 7.7, in a 15 ml graduated tube. After 2 min the sample is cooled, volume adjusted to 6 ml, and mixed. The sample is then assayed.

The adenosine triphosphate assay method reported by Shoaf and Lium[151] is based on the observation that luminescence in fireflies has an absolute requirement for adenosine triphosphate. Adenosine triphosphate is determined by measuring the amount of light produced when adenosine triphosphate reacts with reduced lucifern (LH_2) and oxygen in the presence of firefly luciferase and magnesium, producing adenosine monophosphate (AMP), inorganic pyrophosphate (PPi), oxidized luciferin (L), water and carbon dioxide and light (hv) by the following reaction.

$$ATP + O_2 + LH_2 \xrightarrow[Mg^{2+}]{\text{luciferase}} AMP + PPi + L + H_2O + CO_2 + hv$$

The bioluminescent reaction is specific for adenosine triphosphate and the reaction rate is proportional to the adenosine triphosphate concentration with one photon of light emitted for each molecule of adenosine triphosphate hydrolysed. When adenosine triphosphate is introduced to suitably buffered enzyme and substrates, a light flash follows which decays in an exponential fashion (Figure 76). Either the peak height of the light flash or integration of the area under the decay curve can be used to form standard curves.

The enzyme–substrate mixture used in this work contained 100 Dupont units of purified luciferase, 0.71 mmol l^{-1} luciferin, 10 mmol l^{-1} magnesium sulphate, and 10 mmol l^{-1} morpholinopropane sulphonic acid, at pH 7.4, in a volume of 100 μl. This solution was stable at room temperature (22–24 °C) for at least 4 h. The enzyme–substrate solution was not used for adenosine triphosphate analysis until at least 15 min after it was dissolved, as the background luminescence decreased rapidly during this period. A 10 μl aliquot of sample containing the adenosine triphosphate was injected into the standard 100 μl volume of enzyme substrate in a cuvette (50 mm length by 5 mm I.D.).

Various spectrophotometers were able to detect adenosine triphosphate in extracts in amounts as low as 10–15 mole (5×10^{-3}) with good reproducibility. The peak height measurements were more dependent on a uniform rate of delivery and mixing of adenosine triphosphate in the luciferin–luciferase

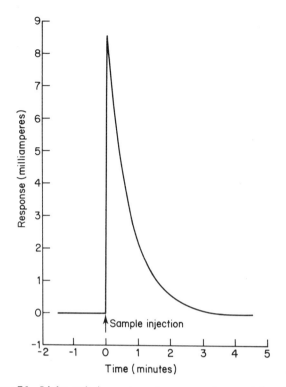

Figure 76 Light-emission curve when a sample containing ATP is injected into the luciferin–luciferase mixture. Reprinted from Shoaf and Lium.[151]

solution. The instruments produced linear responses from approximately 10^{-12} to 10^{-15} mol of adenosine triphosphate.

The neutral dimethyl sulphoxide extraction procedure was found to be the most rapid and convenient method. This eliminates the possible problem of heat-gradient formation (and thus lower recovery of adenosine triphosphate) when the amount of filtered biomass placed in the boiling TRIS buffer is large. This extractant inhibits the luciferase enzyme by about 25%. This inhibition may be eliminated by a 10-fold dilution of the extractant with low-response distilled water before analysing for adenosine triphosphate content.

Several metal salts, also inhibit the enzymatic determination of adenosine triphosphate (Table 63). This inhibition can usually be eliminated, however, by washing the filtrate with 0.01 M MOPS buffer immediately after completing filtration. Although high concentrations of most salts will inhibit the reaction, a few were very potent inhibitors. Mercuric ion was the most potent, reducing the amount of ATP measured to 29% of the correct amount (Table 63) at a concentration of 8.3 μmol l^{-1} mercuric chloride. Of those tested, cadmium was the next most potent, followed by calcium, and sodium and potassium phosphate.

Table 63 Inhibition of the ATP assay

Compound	Final concentration in assay cuvette	Activity
Distilled water	—	100
Mercuric chloride (μmol l^{-1};7	8.3	28
Cadmium chloride (mmol l^{-1})	0.42	38
Calcium chloride (mmol l^{-1})	4.2	63
Potassium phosphate (pH 7.4) (mmol l^{-1})	21	33
Sodium phosphate (pH 7.4) (mmol l^{-1})	21	41

Note: The standard volume and concentration of luciferin–luciferase was used. Then 10 μl of low-response water or the appropriate inhibitor was added and mixed. Samples were then immediately assayed by addition of 10 μl of 0.1 μg ATP per millilitre.
Reprinted from Shoaf and Lium.[151]

Siegrist[152] has described a method for the determination of adenosine triphosphate in lake water. The procedure can be applied in routine limnological work by freezing the filter residue samples in liquid nitrogen for transport to the laboratory for analysis.

Tobin *et al.*[153] give details of two extraction procedures for the determination of adenosine triphosphate in environmental samples by luciferin–luciferase assay. One method is suitable for natural waters, sediments, and sludges while the other can be used for water or sludge, but is unsuitable for sediments with a high content of humic acid or metals.

Patterson *et al.*[154] carried out determinations of adenosine triphosphate in activated sludge. The method, again, involved the use of firefly lantern extract.

Kucnerowicz and Verstraete[155] carried out direct measurements of microbial adenosine triphosphate in activated sludge samples. The method uses an activated-sludge apparatus designed for determining the biodegradability of anionic detergents. Mixed liquors are diluted with triethanolamine buffer, homogenized, mixed with adenosine triphosphate releasing agent and the luminescence of the mixture is measured after addition of luciferase. Results obtained were in agreement with literature data for adenosine triphosphate in activated sludge. Relationships established between adenosine triphosphate content of sludges and other sludge parameters indicate that adenosine triphosphate determination could be used as a method of monitoring activated-sludge treatment processes.

André *et al.*[156] discuss the determination of adenosine triphosphate by luciferin–luciferase assay. This method was applied to the determination of adenosine triphosphate in bacterial colonies filtered from samples of polluted water after incubation for different periods. The adenosine triphosphate was extracted from the residue in the filter and the amount compared with the BOD of the filtered water. The oxygen uptake rate and the rate of formation of adenosine triphosphate were then plotted against time, the two curves being similar up to 3–4 days' incubation, after which adenosine triphosphate production declined markedly, although oxygen uptake continued to increase.

Hysert *et al.*[157-159] have applied a bioluminescence adenosine triphosphate assay method to brewery waste water studies. They described an assay using partially purified luciferase and synthetic firefly luciferin for the bioluminescence reaction, a liquid scintillation counter in the out-of-coincidence phase and a sludge adenosine triphosphate extraction technique involving dimethyl sulphoxide at room temperature. Experiments with several pure bacteria cultures showed good linear correlation between adenosine triphosphate and mixed-liquor suspended solids, return sludge suspended solids and effluent suspended solids.

In this method[158,159] 0.1 ml of waste water sample was added to 0.9 ml dimethyl sulphoxide at room temperature. The mixture was swirled for several seconds, then stored at $-25\,°C$ until assayed. The extracts were thawed, swirled, and diluted 10-fold with pH 7.4 TRIS buffer prior to assay.

When adenosine-5′-diphosphate and adenosine-5′-monophosphate as well as adenosine triphosphate assays were to be done, the extractions were done with boiling acetone.[159-161] Typically, 0.1 ml sample was added to 1 ml boiling acetone in a 7 ml vial immersed in a $90\,°C$ water bath and the acetone was removed by blowing a stream of air over the liquid,[159] 10 μl of extracts were added to 1.0 ml, pH 7.40. 10 mM tris(hydroxymethyl)aminomethane (TRIS)-3.5 mM $MgSO_4$ buffer and 10 μl of a purified luciferase–synthetic D-luciferin premix was added to initiate the bioluminescent reaction. The light output of the reaction was measured with a scintillation counter.

Hysert et al.[157] state that the firefly bioluminescence adenosine triphosphate assay has several attractive features including high sensitivity, selectivity, and freedom from sample interferences are a consequence of the use of purified luciferase and synthetic D-luciferin at optimum concentrations.[158] The high assay sensitivity permitted very high sample dilutions thus further reducing the possibility of interference and/or inhibition of the bioluminescent reaction by sample components.[158,162,163] Typically, samples were diluted 10-fold in extraction, a further 10-fold prior to assay, and 100-fold in the assay itself—a total of 10,000. Furthermore, the assay is fast (approximately 1 min per assay) and reproducible (relative standard deviations for standard adenosine triphosphate solutions vary from 2 to 4%). Its wide dynamic range (over five decades of adenosine triphosphate concentration) permits direct adenosine triphosphate determination over a broad range.

These workers[157] demonstrated that interference and inhibition of the bioluminescence assay by extract components was negligible by a standard adenosine triphosphate method.[163] This circumstance undoubtedly resulted from the aforementioned high extract dilution as well as from the use of purified luciferase. The reproducibility of the overall method for determining the adenosine triphosphate content of activated sludge, which includes sampling, dimethyl sulphoxide extraction, and adenosine triphosphate assay, was considerably poorer than that observed for the adenosine triphosphate assay alone or for adenosine triphosphate assays of pure cultures. The relative standard deviations for the latter assays were 2–4%, whereas those for the activated-sludge determinations were in the 7–11% range. This greater method variance no doubt results from difficulties of reproducibly sampling the heterogenous, clumped biological flocs that comprise activated sludge.

Glyphosphate residues

Brøstad and Hakon[164] have described a polarographic method for the determination of glyphosphate residues in natural waters as their *N*-nitroso derivatives.

Inositol phosphate esters

These have been determined in lake sediments.[165]

Triaryl phosphate esters

Murray[166] has described a gas chromatographic method for the determination in water and fish tissues of triarylphosphate esters (1 mol S-140, tricresyl phosphate, cresol diphenyl phosphate). These substances are used commercially as lubricant oil and plastic additives, hydraulic fluids, and plasticizers. The method involves extraction from the samples, hydrolysis, and measurement of the individual phenols by gas chromatography as the trimethylsilyl derivatives. The lower detection limit was about 3 ppm.

Method

Equipment

A gas chromatograph with flame ionization detectors was used for the analysis and the column used was 8 ft × ⅛ in stainless steel packed with 5% Imol on Chromsorb W, AW, and DCMS treated 80–100 mesh. A temperature programme from 80–120 °C at 4 °C min^{-1} was used.

Operating conditions

Injector, manifold and detector temperature	150 °C
Carrier gas (nitrogen)	25 ml min^{-1}
Hydrogen	35 ml min^{-1}
Air	400 ml min^{-1}

Procedure

(1) Triaryl phosphate esters—A weighed (3–4 drop) sample was placed in a 20 ml ampoule and 10 ml of 5% potassium hydroxide in 95% methanol added. The ampoule was sealed and autoclaved at 25 psi for 90 min. When cool, the ampoule was opened and the contents washed into 950 ml of distilled water, acidified with about 5 ml 6 M hydrochloric acid to pH 1–2 and made up to 1000 ml. This was transferred to a separatory funnel, 5 µl of o-xylene were added by syringe as an internal standard, and the mixture was extracted once with 50 ml of chloroform. The solvent layer was evaporated to 1–2 ml and treated

with Tri-Sil concentrate to form the trimethylsilyl derivatives. This was allowed to react overnight and analysed by gas chromatography the following day. The concentrations of the individual phenols were calculated from calibration graphs and the composition of the ester was determined.

(2) Imol in fish tissue — A weighed (5–10 g) sample of fish tissue was placed in a Waring blender with 100 ml of 95% methanol and mixed at high speed for 30 s. The slurry was filtered through a No. 1 Whatman filter paper into a 240 ml Quickfit flask, the filtrate reduced to 3–4 ml on a rotary evaporator and 0.5–0.7 g potassium hydroxide added. The flask was shaken gently to dissolve the pellets and the solution quantitatively transferred to a 20 ml ampoule with 10 ml of 95% methanol. After sealing and autoclaving, the previously described procedure was followed. A second extract was performed on the residue collected on the filter paper and the filtrate analysed in the same way. The areas of the two main peaks, m- and p-cresol, which comprise 57% by weight of Imol, were measured, combined and calculated to give the concentration of Imol. Tissues from untreated control fish were spiked with known weights of Imol and extracted using the above method.

(3) Imol in water — one litre samples of water containing dispersed Imol were extracted with 50 ml of chloroform and the solvent layer was evaporated to dryness in a small flask on a rotary evaporator. The contents of the flask were quantitatively washed into an ampoule with 5% potassium hydroxide in 95% methanol. The ampoule was sealed, autoclaved, and the above procedure followed.

The chromatograms (Figure 77) show the separation of phenols in hydrolysed commercial samples, tricresolyl phosphate (TCP), cresol diphenylphosphate (CDP), and Imol S-140. The peaks on the Imol chromatogram were identified by retention time data relative to the standard compounds.

Determination of organic phosphorus

Olsen[167] has reviewed the determination of inorganic, organic and total phosphorus in water, soil, and sediments. Determination of total phosphorus in aqueous samples commonly involves a hot acid–oxidation type digestion procedure, although various other dry-ashing, fusion, and u.v. irradiation methods have been reported and evaluated, e.g. Harwood et al.,[168] Grasshoff,[169] Osburn et al.[170] Use of an autoclave to facilitate the digestion has been reported by Harvey,[171] Lee et al.,[172] and Menzel and Corwin[173] while analytical methods manuals available from both the US Environmental Protection Agency and Environment Canada give procedures employing an autoclave promoted acidic persulphate digestion for the analysis of 'acid-hydrolysable phosphorus'. Although both agencies note the interferences and shortcomings of the molybdate colorimetric test when arsenate, heavy metals, etc. are present, neither fully discusses the recovery limitations, or the expected precisions obtained under routine test conditions. Goulden and Brookbank[174] reported a relative standard deviation of 3.4% (at the 0.010 mg l^{-1} concentration level) for a procedure employing autoclave digestion followed

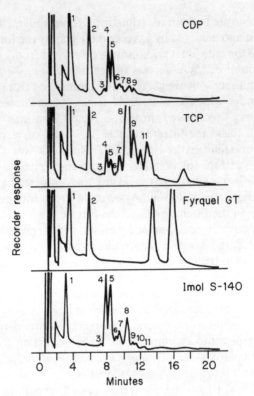

Figure 77 Temperature programme chromatograms (80–120 °C at 4 °C min^{-1}) of trimethylsilyl phenol derivatives prepared from commercial samples. 1, o-Xylene; 2, phenol; 3, o-cresol; 4, m-cresol; 5, p-cresol; 6, o-ethylphenol; 7, 2,5-dimethylphenol; 8, 2,4-dimethylphenol, 3,5-dimethylphenol; 9, 2,6-dimethylphenol; 10, 2,3-dimethylphenol; 11, 3,4-dimethylphenol. Reprinted with permission from Murray.[166] Copyright (1975) Fisheries Research Board, Canada

by automated colour development, solvent extraction, and absorbance measurement.

The total phosphorus test used by the Water Quality Branch Inlands Water Directorate, Canada,[175] is a modification of the 'Standard Methods' sulphuric acid/persulphate procedure employing a manual digestion on a hot-plate and neutralization prior to automated colour development and absorbance measurement. Quality control method evaluation has shown that a between-run precision of 0.008–0.015 mg l^{-1} (95% confidence) in the 0.000–0.200 mg l^{-1} range may be obtained when analysing river and lake samples.[176] For many water systems where total phosphorus concentrations are commonly less than 0.020 mg l^{-1}, such precision values are unacceptable.

In an attempt to improve this precision, Jeffries et al.[177] developed a procedure which minimizes sample handling by eliminating sample transfers and uses an autoclave-promoted acidic persulphate digestion. The procedure was subjected to extensive testing for precision and phosphorus recovery, not only for natural surface waters, but also for waste waters and synthetic solutions — a precision of about 0.001 mg l^{-1} P was achieved.

Method

Sample collection including field filtration if desired, digestion, and analysis were carried out using a single container — a precalibrated (35 ml) 25×150 mm Pyrex screw top culture tube fitted with a Teflon lined cap. Sample digestion was achieved by vacuum aspiration of excess sample, followed by the addition of 2.5 ml of digestion reagent (55 ml concentrated sulphuric acid, plus 60 g potassium persulphate in 1 l of solution) and autoclaving the capped tubes for 1 h. An autoanalyser AAII system employing autoclave digestion was used for total phosphorus analysis (Figure 78).

Solutions of a number of pure phosphorus-containing compounds and surface water and waste water samples were analysed using both the autoclave and hot plate digestion techniques (Table 64). Assuming 100% compound purity the

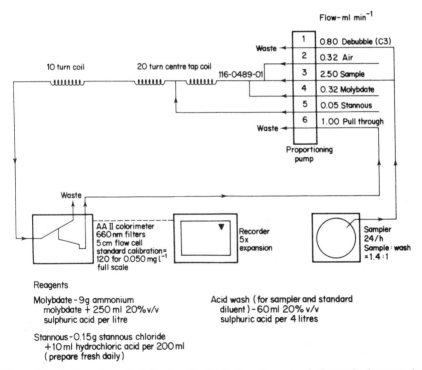

Figure 78 Autoanalyser AAII system for total phosphorus analysis employing autoclave digestion

Table 64 Phosphorus recovery—synthetic samples

Compound	No. of replicates	Hot plate digestion mean mg l^{-1} concentration	Standard deviation	No. of replicates	Autoclave digestion Standard concentration	Standard deviation
Glucose-1-phosphoric acid (dipotassium salt)	7	0.041	0.0023	9	0.039	0.0009
Glucose-6-phosphoric acid (dipotassium salt)	7	0.043	0.0024	9	0.042	0.0005
DNA (sodium salt)	6	0.047	0.0041	9	0.046	0.0004
Adenosine-5'-monophosphoric acid	8	0.041	0.0036	7	0.038	0.0004
Adenosine-5'-diphosphate (sodium salt)	8	0.041	0.0035	9	0.041	0.0004
Adenosine-5'-triphosphate (sodium salt)	8	0.043	0.0026	9	0.043	0.0005
Phosphoserine	8	0.041	0.0038	8	0.040	0.0005
Sodium-β-glycerophosphate	8	0.046	0.0033	9	0.043	0.0007
Tetrasodium pyrophosphate	8	0.026	0.0031	9	0.025	0.0006
Sodium tripolyphosphate	8	0.043	0.0038	9	0.044	0.0003
Sodium metaphosphate	8	0.040	0.0034	9	0.040	0.0003
Disodium hydrogen orthophosphate	8	0.040	0.0028	9	0.039	0.001

calculated phosphorus concentration was 0.040 mg l^{-1} P throughout except tetrasodium pyrophosphate which was 0.025 mg l^{-1} P.

From Table 64, it is apparent that comparable recoveries were obtained in all cases. Application of a t-test to this data showed that no significant difference in mean concentrations of total phosphorus could be found for any solution at the 95% confidence level. An overall F value of -42 confirms that the precisions of the two procedures is different, the autoclave technique yielding a more precisely known mean concentration.

The autoclave technique was further compared to the hot plate procedure by conducting a parallel analysis study on a variety of natural samples. The methods' intercomparison was evaluated by using regression analyses which allow separation of the variance of each technique (see Figures 79 and 80). The comparisons show the following

(1) The phosphorus concentrations determined by the two procedures are nearly equal, but given the range of sample concentrations and lack of homogeneity,

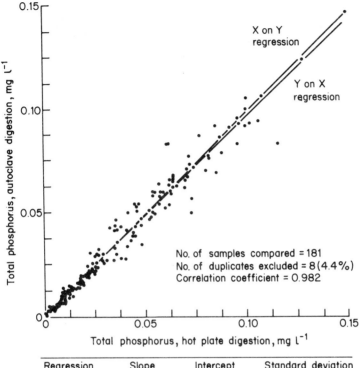

Figure 79 Phosphorus recovery: comparison of hot plate digestion recovery vs autoclave digestion recovery for river- and lake-type samples

the superior precision of the autoclave promoted persulphate procedure was suppressed since $S_{xy} = S_{yx}$.

(2) The precision of the analyses performed on low concentration (0.000–0.025 mg l^{-1}) samples (i.e. a subset of those given in Figure 79) is threefold better than the precision determined for the whole concentration range (0.000–0.150 mg l^{-1}), as expected.

(3) From Figure 80 it is evident that the autoclave digestion procedure is suitable for effluent sample types, but may be less acceptable for raw sewages, lagoon samples, digestor samples, etc., i.e. those having a higher proportion of suspended particulates. In general, the presence of large amounts of particulate material should suggest the use of a more chemically rigorous digestion procedure.

Using samples containing 0.0000–0.2000 μg l^{-1} P a precision of approximately 0.0008 μg l^{-1} P was obtained. Precision deteriorated considerably if the conditions under which samples were collected were not controlled and only unfiltered samples were analysed (precision 0.141 μg l^{-1} P).

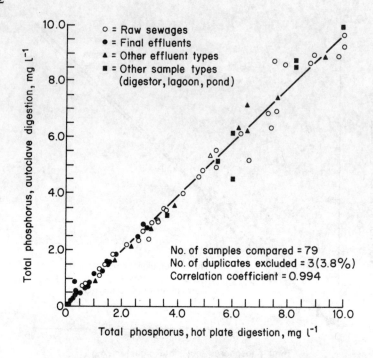

Figure 80 Phosphorus recovery: comparison of hot plate digestion recovery *vs* autoclave digestion recovery for processed water-type samples

Goossen and Kloosterboer[178] have described a method for the determination of organic phosphate and condensed inorganic phosphate in natural and waste waters. The combined action of ultraviolet radiation and heat decomposes organic and condensed phosphate to orthophosphate, enabling the determination of total phosphate without rigorous chemical pretreatment.

Phosphorus in sediments

Comparative studies of different methods for determination of total phosphorus in sediment have been made by Sommers *et al.*,[179] Nordforsk,[180] and Mehta.[181] Digestion with perchloric acid,[182] Standard Methods,[183] is the most common and generally accepted procedure. A more rapid method for determination of total phosphorus in water samples by digestion with persulphate was introduced by Koroleff,[184] but this method has not been widely used for sediment samples. Preliminary measurements of phosphorus in lake sediments using the persulphate digestion method gave considerably lower values than the perchloric acid method.

Determination of total phosphorus in lake sediments by ignition of samples in a muffle furnace at 550 °C, boiling of the residue from ignition in 1 M hydrochloric acid and subsequent determination of orthophosphate gave approximately the same values as the perchloric acid digestion.

The ignition principle is recommended by Stainton et al.[185] for determination of particulate phosphorus in water after the particulate matter has been retained on a glass fibre filter, and by Saunders and Williams[186] for determination of organic phosphorus in soils. Anderson[187] used this method for sediment samples but he did not compare his results with those obtained by other methods.

Anderson[188] further investigated ignition methods for the determination of total phosphorus in lake sediments and biogenic materials and compared results with those obtained by perchloric acid digestion. The organic matter is destructed by ignition. Material remaining after ignition is boiled in hydrochloric acid and orthophosphate determined after dilution. The method generally gave lower results than the perchloric acid digestion method (97.7, 98.7, 94.4, 100.5, and 97.3%) for five sediment samples. The reproducibility of the ignition method was slightly less than that observed with the perchloric acid method.

Perchloric acid method[189]

Ten millilitres of demineralized water and 2 ml of concentrated nitric acid were added to 0.15–0.2 g of dry sediment (predried at 103 °C) or plant material in a 100 ml Erlenmeyer flask. After a preliminary oxidation by evaporation of water and nitric acid on a hot plate 2 ml of concentrated perchloric acid were added, and the sample was boiled until clear. After cooling the sample was diluted to 100 ml and an aliquot was withdrawn for orthophosphate determination by the ascorbic acid reduction method of Murphy and Riley.[189] Blanks and standards were treated as samples.

Ignition method

Dry sediment or plant material (0.15–0.2 g) was ignited in a muffle furnace in a porcelain crucible (550 °C for 1 h). After cooling the residue was washed into a 100 ml Erlenmeyer flask with 25 ml 1 M hydrochloric acid and boiled for 15 min on a hot plate. The sample was diluted to 100 ml and orthophosphate was determined as in the perchloric acid method. Standards and blanks were not ignited.

The 95% confidence limits vary from ± 0.5 to $\pm 1.4\%$ of the average for the perchloric acid method and from ± 1.0 to ± 2.1 for the ignition method (Table 65). Thus, the reproducibility of the two methods is similar, but a little better for the perchloric acid method.

Digestion with perchloric acid has been investigated by other workers as has fusion with sodium carbonate.[190–192]

Aspila et al.[193] have described a simple, rapid, and semi-automated method for the determination of inorganic, organic, and total phosphorus in lake and

Table 65 Total phosphorus contents of dried sediment and plant material determined by the perchloric acid digestion and by ignition. Ten determinations on each sample (mg pg^{-1} dry matter)

	Esrom So		Fureso		Gribso		Kvind so		Dried leaves of *Glyceria*	
	Perchloric acid	Ignition	Perchloric acid	Ignition	Perchloric acid	Ignition	Perchloric acid	Ignition	Perchloric acid	Ignition
Average	2.109	2.061	1.387	1.369	1.356	1.280	2.495	2.507	3.188	3.101
% of the perchloric acid method		97.7		98.7		94.4		100.5		97.3
Standard deviation	0.031	0.034	0.017	0.041	0.027	0.019	0.026	0.048	0.023	0.055
Standard error	0.010	0.011	0.005	0.013	0.009	0.006	0.008	0.015	0.007	0.017
95% confidence limit	0.022	0.023	0.012	0.029	0.019	0.013	0.018	0.033	0.016	0.038
Dissolved oxygen, % of average	1.0	1.1	0.9	2.1	1.4	1.0	0.7	1.3	0.5	1.2

Reprinted with permission from Anderson.[188] Copyright (1975) Pergamon Press.

river sediments. Total phosphorus is extracted from sediments with 1 M hydrochloric acid after ignition at a high temperature (550 °C) or by digestion with sulphuric acid–potassium persulphate at 135 °C in a sealed PTFE-lined Parr bomb. Organic phosphorus is determined by the difference in phosphorus content of the 1 M hydrochloric acid extract measured before and after ignition of the dry sediments at 550 °C. Orthophosphate is determined by using standard Technicon AutoAnalyzer II techniques. The interferences caused by silica and variable acid concentrations on the determination of phosphorus were studied. Freedom from interferences under the chosen experimental conditions as well as the good results obtained for recovery and precision indicate that the methods are suitable for monitoring inorganic, organic, and total phosphorus in sediments.

Method

Reagents: method 1—detection system

Ascorbic acid 17.6 g of L-ascorbic acid (reagent grade) are dissolved in a small volume of distilled water. To this solution is added 0.50 ml of Levor IV (Technicon No. T 21-0332) wetting agent and then the solution is diluted to 1 litre and mixed thoroughly; the solution is stable for about 1 week.

Ammonium molybdate The ammonium molybdate solution was prepared by mixing 10 g of the ammonium salt $(NH_4)_6Mo_7O_{24} \cdot 4H_2O$ with about 500 ml of distilled water and 62 ml of concentrated sulphuric acid. The mixture was diluted to 1 litre with distilled water. This molybdate solution is 2.2 N in sulphuric acid. For the preparation of molybdate solutions that were 1.1 N in sulphuric acid, 31 ml of the concentrated acid were used.

Method 2—detection system

Mixed reagent the following reagents are mixed in the order given: 50 ml of 4.9 N sulphuric acid (136 ml l^{-1} of concentrated sulphuric acid), 15 ml of ammonium molybdate solution (40 g l^{-1}) and 5.00 ml of an antimony potassium tartrate solution (3.0 g l^{-1} of $K(SbO) \cdot C_4H_4O_6 \cdot \frac{1}{2}H_2O$). Prior to mixing, each of the above reagents was diluted to 1 litre with distilled water. The mixed reagent prepared as above is stable for not more than 6–8 h.

Extraction of sediment samples with hydrochloric acid

Aliquots (0.3–0.5 g) of dry sediment (passing 100 mesh) were weighed and then transferred into 10 cm^3 Coors alumina crucibles. The uncovered crucibles, contained in a suitable tray, were placed into a warm muffle furnace and ignited at 550 °C. The samples were maintained at 550 °C for 1.5 h, then removed, allowed to cool, and transferred into 100 ml calibrated flasks; 50 ml of 1.0 N hydrochloric acid were then added to the flasks. The mixtures were next shaken for 14–18 h at about 22 °C.

For determinations of 1 N hydrochloric acid extractable inorganic phosphate an identical aliquot of sediment was used except that no ignition was performed. After extraction, aliquots of the ignited and non-ignited mixtures were transferred into 15 ml test-tubes and centrifuged at 2000 rev min^{-1} for about 5 min. The clarified extracts were finally diluted ten times and analysed by the two automated Technicon procedures described below.

An alternate extraction procedure used by Aspila et al.[193] involves ignition of the sediment with potassium persulphate in a PTFE-lined Parr bomb.

Extraction by the bomb method

For the extraction of total phosphate a weighed aliquot of sediment (0.3–0.5 g), together with 3 ± 0.1 g of potassium persulphate (Analar) and 5.00 ml of concentrated sulphuric acid, was added to the bomb, which was then heated in an oven at 135 ± 5 °C for 2 h. The contents of the bomb were then transferred quantitatively into a 500 ml calibrated flask. After dilution to volume with distilled water, the extract (containing 1% v/v of sulphuric acid) was analysed for total phosphate by the automated procedure outlined below.

Automated procedures

Detection systems

The determination of the orthophosphate was carried out by using the automated systems described by the Technicon Instruments Corporation. The manifolds used are shown in Figure 81. The procedures referred to below as methods I and II are Technicon industrial methods Nos. 94-70W and 155-71W, respectively. Method I includes ascorbic acid alone for the reduction of the molybdophosphoric acid whereas in method II the mixed reagents[194-196] ascorbic acid, sulphuric acid, ammonium molybdate, and antimony potassium tartrate are used. Method I is intended for use for high levels of phosphorus (up to 10 μg ml^{-1}) and method II for low levels (less than 0.5 μg ml^{-1}). The wetting agent (Levor IV) used in order to obtain a smooth bubble pattern, is present in the ascorbic acid reagent line for method I whereas it is added externally (see Figure 81) in the water line (0.5 μg ml^{-1} of Levor) in method II.

Results obtained by method I were found to be linear over the range 0.5–5 μg ml^{-1} of phosphorus and by method II from 0.05 to 1 μg ml^{-1}. As solution extracts contain 1 mg ml^{-1} of sediment, the above concentration ranges allow direct analyses of sediments containing from 100 to 5000 ppm (μg g^{-1}) of phosphorus to be made, which encompasses the entire range of sediment phosphate levels expected.

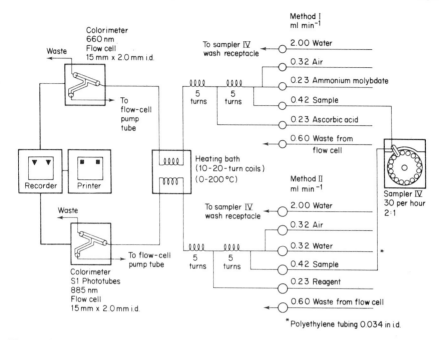

Figure 81 Manifold for phosphorus determination (Technicon). Reprinted with permission from Aspila et al.[193] Copyright (1976) Royal Society of Chemistry

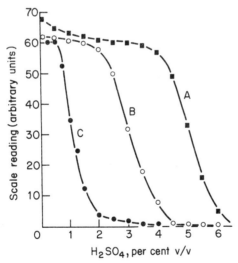

Figure 82 Effect of concentration of sulphuric acid (% v/v) in sample solution on the response for 1 mg ml^{-1} of phosphorus. A, Method I (1.1 N H_2SO_4*); B, method I (2.2 N H_2SO_4*); and C, Method II.
*Concentration of sulphuric acid in the ammonium molybdate reagent solution used. Reprinted with permission from Aspila et al.[193] Copyright (1976) Royal Society of Chemistry

Table 66 Comparison of results for phosphorus (ppm) obtained by detection with methods I and II

Sediment sample	Bomb method of extraction		Extraction with 1 M HCl after ignition		Extraction with 1 M HCl before ignition, method I
	Method I	Method II	Method I	Method II	
1	1800	—	1850	1800	1400
	1870	1770	1820*	—	
2	1080	1030	1090	1100	810
	1120	1070	1090*	1180	
3	870	800	860	860	640
	920	860	860*	950	
4	650	(570)	605	620	495
	680	610	755*	815	
5	1060	1000	1060	1030	805
	1120	1070	1080*	1150	
6	1390	1360	1230	1220	820
	1370	1300	1310*	1390	
7	1040	990	1130	1080	840
	1060	1000	1020*	1100	
8	1130	1080	1200	1130	880
	1160	1080	1150*	1070	
9	—	—	855	810	750
	850	780	820*	890	
10	1020	990	980	970	750
	1060	970	930*	1010	
11	970	930	910	900	690
	970	900	930*	1010	
12	690	640	675	660	525
	730	660	675*	730	

*Sample ignited for 16 h at 550 °C (normal time is 2 h at 550 °C). Reprinted with permission from Aspila et al.[193] Copyright (1976) Royal Society of Chemistry

Aspila et al.[193] found that the concentration of sulphuric acid in the sample solution had an appreciable effect on the response for 1 μg l^{-1} of phosphorus (Figure 82): hence, the need to carefully control acidity levels during the analysis. Additionally, acidity levels which are too low allow serious interference in the method by silica.

Figure 83 illustrates the complex interactions induced by the presence of silica between reaction temperature, sulphuric acid acidity, and the apparent level of phosphorus found. By careful control of acid concentration and reaction temperature, interference by silica can be minimized. Arsenic, germanium, and bismuth would interfere in the method but not at the low levels normally encountered in sediment in water samples.[197,198]

Table 66 compares results for phosphorus determinations in sediments obtained by methods I and II.

The coefficient of variation obtained for the determination of total phosphorus in sediment at the 1400 ppm level was 2.5%. Some 98–100% recovery of inorganic phosphate was obtained in spiking experiments carried out on sediments.

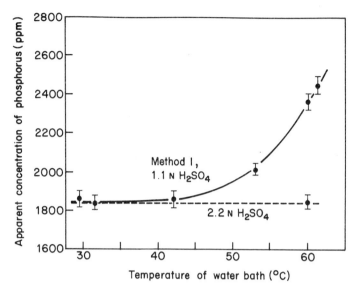

Figure 83 Effect of temperature of heating water bath on the apparent concentration of phosphorus. The 1.1 N H_2SO_4 refers to the concentration of sulphuric acid in the ammonium molybdate reagent solution. Reprinted with permission from Aspila et al.[193] Copyright (1976). Royal Society of Chemistry

SULPHUR COMPOUNDS

Organic sulphides and disulphides

Vitenberg et al.[199] have described a gas chromatographic method for the determination of traces (down to 10^{-6}–10^{-7} %) of sulphur compounds, such as hydrogen sulphide, mercaptans, sulphides, and disulphides, in industrial waste waters (kraft paper mill effluents) by a combination of head-space analysis and microcoulometry. This method, described below, increases the analytical sensitivity 10^2–10^3 times without any preliminary concentration of the sample.

A sample of the solution to be analysed of volume V_L without the gas phase is drawn into the variable-volume device.[200] A 1:1 KCl–HCl buffer solution of pH 2 containing 14% of sodium sulphate is added through the elastic rubber plug with the help of a hypodermic syringe to the sample. After this, air of volume V_G (an inert gas is preferable for prolonged equilibration) is drawn into the vessel with the solution to be analysed and the system is maintained for 30 min at a constant temperature under periodic shaking. After equilibration, the concentration of the sulphur compounds in the gas phase above the solution is determined. To do this, the equilibrium gas is displaced by the plunger from the vessel, filling the same loop of the gas sample valve which is used to inject the analyte into the chromatographic column. On measuring the areas or heights of the corresponding peaks on the chromatogram I_s, one determines the concentration of each compound of interest in the gas phase from the calibration data $I_s = f(C_G)$ available for a given sample loop. The conditions for the gas chromatographic analysis are specified below. The content of each sulphur compound in the solution is calculated by equation 1 taking into account dilution by the buffer solution. The values of K for different temperatures for the buffer solution (pH2) containing 7% of sodium sulphate which are needed in the calculations are presented in Figure 84.

$$C_L^\circ = C_G \left(K + \frac{V_G}{V_L}\right) \tag{1}$$

$$K = \frac{C_L}{C_G}$$

where K is partition coefficient and
C_L° is concentrations of microimpurity in the original solution.

Equation 1 implies that the gain in sensitivity α of the analysis of an impurity in solution obtained by determining C_G gas chromatographically as compared with direct injection of the liquid, depends on the magnitude of the partition coefficient and is

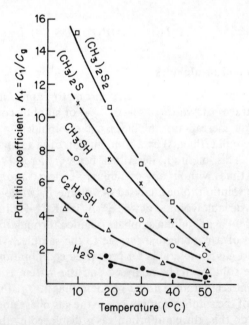

Figure 84 Partition coefficient of sulphur compounds vs temperature for a pH 2 buffer solution with 7% of sodium sulphate. Reprinted with permission from Vitenberg et al.[199] Copyright (1977) American Chemical Society

$$\alpha = \frac{10^3}{K + V_G/V_L} \text{ times} \qquad (2)$$

According to Equation 1, and assuming the error in the liquid and gas determination to be much smaller than that of the concentration measurement, the error in the analysis of a microimpurity in the solution will be

$$\frac{\Delta C_L^\circ}{C_L^\circ} = \frac{\Delta C_G}{C_G} + \frac{\Delta K}{K} \cdot \frac{K}{K + V_G/V_L} \qquad (3)$$

whence it follows that the error in determining the content of an impurity in the liquid can be practically reduced to that of measuring the concentration of this impurity in the equilibrium gas phase if the analysis is performed under the condition $V_G > V_L$. However, this involves a substantial decrease in the sensitivity of the analysis (equation 2). At the same time, if one takes comparable volumes of the liquid and of the gas, the contribution of the error in the determination of K to the total error of analysis becomes essential, increasing with increasing numerical value of K. Therefore, in cases where the detector of the chromatograph has a margin of sensitivity in determining an impurity, one should take as small liquid samples as possible.

As seen from Table 67 which lists the results of an analysis of test solutions close in composition to kraft mill effluents and containing sulphur compounds at the concentration level of 1–0.01 ppm, at contents of about 1 ppm, the analytical error does not exceed 8% for the flame ionization detector and 12% for the microcoulometric detector, of the given amount of the compound. When analysing solutions with concentrations of not more than 0.1 ppm, the analytical error in the region of the highest sensitivity reaches 15% and 20% for the flame ionization and the microcoulometric detector, respectively. This is apparently associated with the large error of the detecting device which operates near its sensitivity threshold at such concentrations.

Gas chromatography

This work was carried out on a gas chromatograph employing a flame ionization detector. For selective determination of sulphur-containing compounds, a microcoulometric detector was used. When both detectors were used, the gas flow emerging from the chromatographic column was divided by means of a T-shaped stream splitter into two parts in a ratio close to 1:1. The contents of the component of the mixture under analysis were calculated taking into account the accurate value of the ratio of the gas flows delivered to the flame ionization and the microcoulometric detectors.

A mixture of the simplest mercaptans, sulphides and disulphides was separated in a glass column 2 m long with an inner diameter of 3 mm. The packing consisted of 15% polyethylene glycol adipate on C-22 celite, 80–100 mesh grain

Table 67 Analytical data on test solutions of sulphur compounds approaching the composition of concomitant impurities to kraft mill effluents

Compound	Flame ionization detector				Microcoulometric detector			
	Introduced (%)	Found (%)	Analytical error Absolute	Rel. %	Introduced (%)	Found (%)	Analytical error Absolute	Rel. %
Methyl mercaptan	3.8×10^{-5}	3.5×10^{-5}	-0.3×10^{-5}	7.9	3.8×10^{-5}	3.3×10^{-5}	-0.5×10^{-5}	13.2
	1.6×10^{-5}	1.7×10^{-5}	$+0.1 \times 10^{-5}$	6.3	1.6×10^{-5}	1.4×10^{-5}	-0.2×10^{-5}	12.5
	7.4×10^{-6}	8.2×10^{-6}	$+0.8 \times 10^{-6}$	10.8	7.4×10^{-6}	8.8×10^{-6}	$+1.4 \times 10^{-6}$	18.9
Ethyl mercaptan	1.5×10^{-4}	1.6×10^{-4}	$+0.1 \times 10^{-4}$	6.7	1.5×10^{-4}	1.4×10^{-4}	-0.1×10^{-4}	6.7
	4.0×10^{-5}	3.8×10^{-5}	-0.2×10^{-5}	5.0	8.6×10^{-5}	1.0×10^{-4}	$+1.4 \times 10^{-5}$	16.3
	4.0×10^{-6}	4.4×10^{-6}	$+0.4 \times 10^{-6}$	10.0	4.0×10^{-5}	3.2×10^{-5}	-0.8×10^{-5}	20.0
Dimethyl sulphide	5.5×10^{-4}	5.4×10^{-4}	-0.1×10^{-4}	1.8	5.5×10^{-4}	4.7×10^{-4}	-0.8×10^{-4}	14.5
	9.9×10^{-4}	1.0×10^{-5}	$+0.1 \times 10^{-4}$	1.0	7.4×10^{-5}	8.6×10^{-5}	$+1.2 \times 10^{-5}$	16.2
	7.1×10^{-6}	6.3×10^{-6}	-0.8×10^{-6}	11.3	6.9×10^{-6}	7.4×10^{-6}	$+0.5 \times 10^{-6}$	7.2
Dimethyl disulphide	5.9×10^{-4}	6.2×10^{-4}	$+0.3 \times 10^{-4}$	5.1	5.9×10^{-4}	5.6×10^{-4}	-0.3×10^{-4}	5.1
	7.1×10^{-6}	6.3×10^{-4}	-0.8×10^{-6}	11.3	3.9×10^{-4}	4.3×10^{-4}	$+0.4 \times 10^{-4}$	10.3
	6.4×10^{-6}	5.8×10^{-6}	-0.6×10^{-6}	9.4	7.1×10^{-6}	5.7×10^{-6}	-1.4×10^{-6}	19.7
Diethyl disulphate	6.5×10^{-4}	6.5×10^{-4}	—	—	6.5×10^{-4}	7.3×10^{-4}	$+0.8 \times 10^{-4}$	12.3
	4.9×10^{-5}	4.8×10^{-5}	-0.1×10^{-5}	2.0	4.9×10^{-5}	5.7×10^{-5}	$+0.8 \times 10^{-5}$	16.3
	9.9×10^{-6}	1.1×10^{-5}	$+0.2 \times 10^{-6}$	2.0	9.9×10^{-4}	8.3×10^{-6}	-1.7×10^{-6}	17.2
Hydrogen sulphide					4.2×10^{-4}	4.5×10^{-4}	$+0.3 \times 10^{-5}$	7.1
					3.5×10^{-5}	3.9×10^{-5}	$+0.4 \times 10^{-5}$	11.4
					5.7×10^{-6}	4.9×10^{-6}	-0.8×10^{-6}	14.0

Reprinted with permission from Vitenberg et al.[199] Copyright (1977) American Chemical Society.

size. The column temperature was 90 °C, the flow rate of the carrier gas (helium) was 50 ml min^{-1}.

Quantitative analysis was carried out by the absolute calibration method. For the flame ionization detector, the measured parameter on the chromatograms was the peak height, while for the microcoulometric detector, it was the peak area calculated by an analogue integrator.

For calibration of the instrument, Vitenberg et al.[199] used standard ethanol or water solutions of the sulphur compounds with concentrations ranging from 100 to 0.0 ppm which were prepared by dilution of concentrated (0.1%) solutions obtained by introducing an accurately weighed amount of the compound in question into the solvent of known volume. Calibration solutions of volatile compounds (e.g. methyl mercaptan) were prepared by crushing sealed thin-walled ampoules containing a precisely known amount of the compound in a solution in a closed vessel.

Of considerable importance for the achievement of the maximum sensitivity and a reasonably high accuracy in the analysis of trace impurities is the correct choice of conditions for the gas chromatographic separation of the mixture (column efficiency, temperature, carrier gas flow rate), characterized by the spread of the peak in the eluate which increases with increasing retention time of the compound of interest under isothermal conditions. To reach the highest possible sensitivity, Vitenberg et al.[199] chose the conditions of separation under which the duration of the analysis did not exceed 20 min. Figure 85 presents an example of separating an air–vapour mixture of the simplest mercaptans, sulphides, and disulphides in concentrations of 10–100 ppm. The compounds most difficult to separate are dimethyl sulphide and ethyl mercaptan which are not separated under the conditions specified above. Complete separation of these compounds can be effected after a fairly long time using a 6 m long column with a non-polar stationary phase (Apiezon L). Vitenberg[199] suggests that dimethyl sulphide and ethyl mercaptan might be separated in a few minutes using programmed temperature control and a flame ionization detector. Vitenberg et al.[199] did manage to resolve these two substances involving analysis of the gas phase above solution at various pH values. At pH 2, one determines the total content of these compounds, after which pH is increased up to 10 where mercaptan is mostly converted into mercaptide, the remainder of ethyl mercaptan undergoing oxidation to diethyl sulphide at room temperature (Figure 86).

The key condition for a successful use of the head-space analysis of the sulphur impurities in solutions is the exclusion of adsorption losses of the compounds of interest, as well as of losses involved in sampling, injection in the chromatograph, and in the chromatographic process itself. The existing devices for establishing equilibrium distribution of a compound between a liquid and gas and the techniques of injecting the equilibrium gas in the chromatograph do not preclude such losses since they have rubber membranes and the injection is accompanied by a change in the analyte concentration in the gas phase.[201,202]

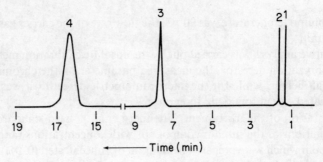

Figure 85 Chromatogram of vapours of the simplest sulphur compounds — in air contents in the range 10–100 ppm: (1) mercaptan; (2) dimethyl sulphide + ethyl mercaptan; (3) dimethyl disulphide; (4) diethyl disulphide. Detector: flame ionization. Conditions of the gas chromatographic analysis are specified in the text. Reprinted with permission from Vitenberg et al.[199] Copyright (1977) American Chemical Society

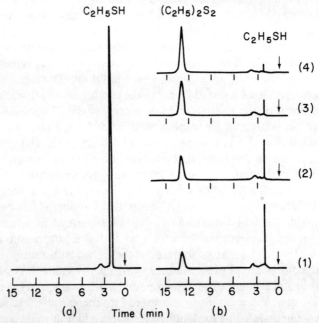

Figure 86 Chromatogram illustrating the change of ethyl mercaptan content in an alkaline solution in time: (A) pH 2; (B) pH 10; (1) after 20 min; (2) after 45 min; (3) after 70 min; (4) after 90 min. Conditions of the chromatographic analysis are specified in the text. Reprinted with permission from Vitenberg et al.[199] Copyright (1977) American Chemical Society

Losses of small amounts of sulphur compounds in the course of the gas chromatographic analysis can be prevented only by using chromatographs in which the vapours of the compounds of interest do not come in contact with metal. Some workers[203] believe that the use of metal in columns and the evaporator may result in losses of hydrogen sulphide and/or mercaptans.

Therefore Vitenberg et al.[199] employed a gas chromatograph whose parts coming in contact with the vapours are made of glass or Teflon.

To avoid losses of the analyte in the course of sampling, preparation for analysis, and injection in the chromatograph, Vitenberg et al.[199] used glass vessels of variable volume. For this purpose, they chose large volume hypodermic syringes (50–100 ml) in which a sealing device is attached to the tip.[204]

It is well known that adsorption losses of analytes in trace analysis may result in systematic error amounting to tens and even hundreds of per cent.[205] In the device described by Vitenberg et al.,[199] adsorption during equilibration may occur on the surface of the elastic rubber plug and (from the solution) on rough walls of the syringes. To reduce the adsorption of the compounds of interest on the surface of the elastic plug made of silicone rubber, it is separated from the cylinder by means of a capillary Teflon tube which practically excludes the loss of the analyte from the gas phase onto the rubber plug.

Various workers have studied the occurrence of volatile odorous sulphur compounds in waste waters.[206,207]

The method described by Jenkins et al.[206] involves gas chromatography coupled with the sulphur-specific flame photometric detector. The method has been used to trace sources of odour due to mercaptans in the air surrounding a sewage works and in waste waters, sewer gases. They used these techniques to study the fate of various organosulphur compounds in raw sewage and activated sludge.

Jenkins et al.[206] used a gas chromatograph equipped with a Melpar flame photometric detector and a Perkin Elmer 3920B gas chromatograph equipped with a linearized Perkin Elmerflame photometric detector. Each instrument also had a flame ionization detector. A FEP Teflon column (18 in × 0.0625 in i.d.) containing acetone-treated Porapak QS,[208] was used for all analyses. Instrumental conditions were as follows: detector 200 °C; injector, 140 °C; column, 80–200 °C at 16 °C min^{-1}, 10 min hold at 200 °C; column flow, 30 ml helium min^{-1}; detector flows, 68 ml hydrogen min^{-1} (3920B), 115 ml hydrogen min^{-1} (550), 110 ml air min^{-1} (3920B), 200 ml air min^{-1} (550). The PE 3920B used a glass-lined flash vaporizer injector and the interface lines were in stainless steel. For the Tracor 550, the Porspak QS column was connected to glass capillary tubing which was inserted into the injector and outlet barrels. Teflon-backed septa were used in the injectors of both instruments.

Jenkins et al.[206] describe a vacuum line system for the storage and preparation of calibration mixtures of mercaptans.

Figure 87 shows the design of the gas sampling device which was used in the field. All glass parts which came in contact with sample gases were coated with Siliclad solution (Clay Adams, Parsippany, New Jersey) to minimize adsorptive losses. Prior to field use, the bulb was evacuated to <1 Torr and 10 ft of FEP Teflon tubing (0.188 in i.d.) was attached to the stopcock. The volume of this tubing was negligible compared to the 1 litre bulb volume. In the field, the tubing was inserted through manhole covers and the stopcock opened, allowing the

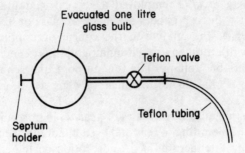

Figure 87 Apparatus for sampling gases in the field. Reprinted with permission from Jenkins *et al.*[206] Copyright (1980) Pergamon Press

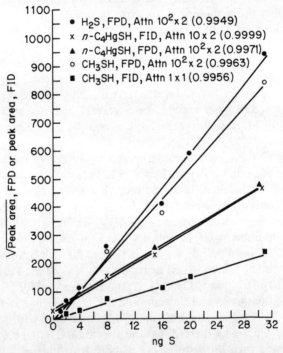

Figure 88 Flame photometric and flame ionization detector plots square root of peak area and peak area respectively *vs* ng sulphur. Numbers in parentheses are linear correlation coefficients. Reprinted with permission from Jenkins *et al.*[206] Copyright (1980) Pergamon Press

sample gas to be drawn into the glass bulb. A syringe was then used to withdraw gas samples through the septum entry for chromatography.

To assess the performance of the calibration system, Jenkins *et al.*[206] plotted the square root of the peak area and peak area for the flame photometric and

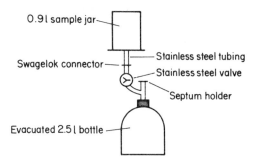

Figure 89 Apparatus for sampling waste water. Reprinted with permission from Jenkins *et al.*[206] Copyright (1980) Pergamon Press

flame ionization detectors *vs* ng sulphur for methyl mercaptan and ethyl mercaptan using a column effluent splitter. Hydrogen sulphide was plotted using the flame photometric detector only (Figure 88). Linearity of the detector plots was good. All curves gave positive ordinate intercepts except for hydrogen sulphide with an abscissa intercept of 0.33 ng sulphur, indicating that adsorptive and chemical losses were not significant for the lower sulphur quantities of the plot. On a practical basis, assuming that a signal to noise ratio of 2 defines the detection limit, 0.25 ng of hydrogen sulphide sulphur and 0.50 ng methyl mercaptan sulphur were detectable by the flame photometric detector with approximately a 1:1 effluent split. Using a signal to noise ratio of 4 to define flame photometric detector sensitivity hydrogen sulphide and methyl mercaptan were quantifiable at 1 and 2 ng sulphur respectively.

The flame photometric detector sensitivity was better than that of the flame ionization detector for hydrogen sulphide and methyl mercaptan, but sensitivity to *n*-butyl mercaptan was about two times better with the flame ionization detector relative to the flame photometric detector.

Liquid samples were taken in a 0.9 litre glass jar which was completely filled with sample and equipped with a Swagelok fitting through the cap. The jar was connected to an evacuated 2.5 litre glass bottle by means of Swagelok connectors, stainless steel tubing, and a Whitey stainless steel valve (Figure 89). The valve was opened and the liquid from the jar was aspirated into the bottle. After closing the valve, the sample jar was removed and aliquots of the head-space gas in the 2.5 litre bottle were withdrawn through the septum using gas syringes.

Jenkins *et al.*[206] used the waste water technique to examine waste water from a treatment plant containing petroleum refinery mercaptans. They used the apparatus shown in Figure 89. From the respective chromatograms of the plant and oil refinery waste waters (Figures 90(a) and (b)) the primary odorant in both samples was identified by retention time as ethyl mercaptan. Although the other organosulphur compounds were not identified, the chromatographic profiles were similar for both samples which rather decisively identified the oil refinery as the odour source. The same treatment plant experienced a moderate

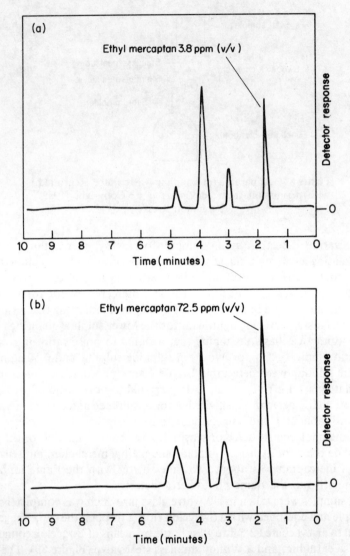

Figure 90 (a) FPD chromatogram of head-space gas of oil refinery suspected of discharging odorous waste water. Attentuation 128×10. (b) FPD chromatogram of odorous head-space gas at treatment plant. Attentuation 128×10. Reprinted with permission from Jenkins et al.[206] Copyright (1980) Pergamon Press

odour problem which appeared to be concentrated near the head-end of the aeration tanks. Analysis of the mixed liquor gave the chromatogram in Figure 91, showing hydrogen sulphide and methyl mercaptan. Figures 92 and 93 respectively, show chromatograms for the head-space gas above activated sludge dosed to 10 mg l^{-1} with hexyl mercaptan immediately and 3 h after dosing. The

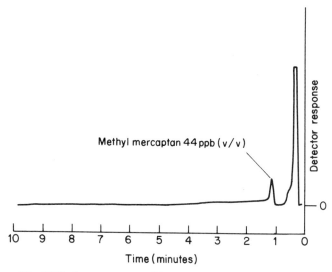

Figure 91 FPD chromatogram of head-space gas from mixed liquor sample. Attenuation 128 × 10. Reprinted with permission from Jenkins et al.[206] Copyright (1980) Pergamon Press

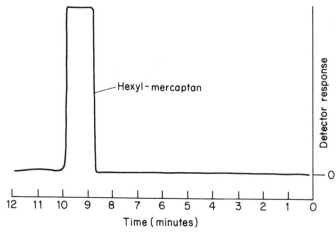

Figure 92 FPD chromatogram of head-space above activated sludge immediately after dosing with n-hexyl mercaptan. Attentuation 128 × 10. Reprinted with permission from Jenkins et al.[206] Copyright (1980) Pergamon Press

chromatograms in Figures 92 and 93 demonstrated that methyl mercaptan was produced from aerobic decomposition of hexyl mercaptan.

Bailey and Viney[207] have also applied gas chromatography to the investigation of odours produced at sewage treatment plants. Samples of ambient air at a sewage treatment works were taken using Tedlar bags and also by using traps containing Tenax GC on site. Analyses were performed using gas

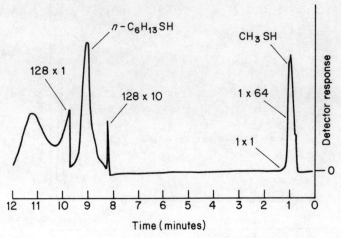

Figure 93 FPD chromatogram head-space above activated sludge after 3 h reaction time with n-hexyl mercaptan. Attenuation 128×10. Reprinted with permission from Jenkins et al.[206] Copyright (1980) Pergamon Press

chromatography coupled to each of the following: an odour port with flame ionization detector, a microwave plasma detection system, and a mass spectrometer. The results indicated that hydrogen sulphide was a major component in samples taken close to sludge pumping operations, and that other odorous compounds could be detected by odour port and the more sophisticated detectors. The presence of hydrocarbons, however, rendered the results from ambient trapping on site with Tenax GC followed by analysis by flame ionization detector useless.

Beehard and Rayburn[209] determined volatile organic sulphides in fresh water algae.

Benzthiazole and 2-mercaptobenzothiazole

Jungclaus et al.[210] used gas chromatography–mass spectrometry to identify benzthiazole and 2-mercaptobenzothiazole in tyre manufacturing plant waste waters. The gas chromatograph was equipped with a flame ionization detector. Separations were achieved on a 180 cm × 0.32 cm of stainless steel column packed with 3% SP 2100 (methyl silicone fluid), of 80–100 mesh supelcoport and was temperature programmed for 70–300 °C at 16 °C min^{-1}. Approximately 44 compounds were identified in the waste water samples including benzothiazole and 2-mercaptobenzothiazole, at concentrations, respectively of 0.06 and 0.03 µg l^{-1}.

Cox[211] determined 2-mercaptobenzothiazole in waste water dump effluents using high performance liquid chromatography. The detector was a variable wavelength u.v. monitor operated at a wavelength of 325 nm. A stopped flow injection system similar to that reported by Cassidy and Frei[212] was used. The column was constructed from stainless steel tubing (15 cm × 4.0 mm i.d.) and was packed with Merckosorb SI 60 (5 μm) silica gel at 3500 psi pressure from a slurry in 2,2,4-trimethylpentane. Ethanol–2,2,4-trimethylpentane (1:9) was used as the mobile phase with a flow rate of 1 ml min^{-1}. The aqueous sample (2 ml) was acidified with two drops of concentrated hydrochloric acid. This mixture was shaken with chloroform (2 ml) for 1 min using a flask shaker. Aliquots (2 μl) of the chloroform layer were used for the chromatographic analysis.

A column packed with silica gel bonded with octadecyl groups gave the best results. 2-Mercaptobenzothiazole was eluted from the column in 4 min with a capacity factor (k') of 1.4 and a height equivalent to a theoretical plate (HETP) of 20 μm. A graph plotting peak height against sample size for this system showed a rectilinear relationship between the parameters up to at least 1.5 μg 2-mercaptobenzothiazole injected; the minimum quantity detectable (defined as a peak with height equal to three times the noise level) was 0.6 ng.

Two procedures were investigated for the extraction of 2-mercaptobenzothiazole from aqueous samples. The first involved acidification of the sample followed by several extractions with chloroform; this resulted in a quantitative recovery. Similar recoveries were also obtained by the more convenient and rapid procedure of shaking equal volumes of acidified sample and chloroform for 1 min using a flask shaker.

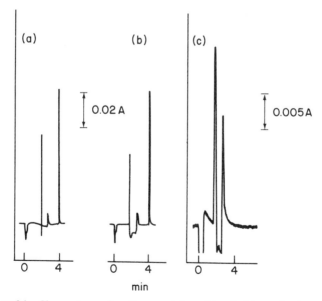

Figure 94 Chromatography of mercaptobenzthiazole (a) standard, 80 mg l^{-1}, (b) spiked effluent, 80 mg l^{-1}, (c) unspiked effluent. For conditions, see text. Reprinted with permission from Cox.[211] Copyright (1976) Elsevier Science Publishers

Figure 94 shows chromatograms obtained for a 2-mercaptobenzothiazole standard and those obtained for the effluent and for the effluent spiked with 2-mercaptobenzothiazole.

Hydroxymethanesulphinite salts

These substances and their decomposition products (formaldehyde, zinc, and sulphites) have been determined by polargraphic procedure in waste waters[213] originating from textile industries. In an alkaline medium it is possible to have well defined and separated waves relative to the oxidation of hydroxymethyane sulphinite ion and to the reduction of formaldehyde and zincate ion. The quantitative determination of these compounds is carried out with standard addition methods since the concentration–current relationship in linear. The sulphite is determined in 01 M perchloric acid and in the same sample it is possible to determine even hydromethane sulphite salts, formaldehyde, and zinc after addition of sodium hydroxide up to 0.1 M. An indirect method for the determination of hydroxymethanesulphinite salts is described. This method is based on the determination of formaldehyde released by means of a weak oxidation of these salts which is stoichiometrically proportional to total hydroxylmethylsulphinite content.

Dimethyl sulphoxide

Andreae[214] has described a gas chromatographic method for the determination of nanogram quantities of dimethyl sulphoxide in natural waters, sea water, and phytoplankton culture waters. The method involves a chemical reduction to dimethyl sulphide, which is then determined gas chromatographically using a flame photometric detector.

Andreae[214] investigated two different apparatus configurations. One consisted of a reaction/trapping apparatus connected by a six-way valve to a gas chromatograph equipped with a flame ionization detector, the other apparatus combined the trapping and separation functions in one column, which was attached to a flame photometric detector. The gas chromatographic flame ionization detector system was identical to that described by Andreae[215] for the analysis of methylarsenicals, with the exception that a reaction vessel which allowed the injection of solid sodium borohydride pellets was used. The flame photometric system (Figure 95) is modified after a design by Braman et al.[216]

The first stage in both systems is a reaction vessel, which contains the sample and buffer solutions. Helium is bubbled through the solution by a glass diffuser. A side port allows the injection of the reducing solutions by a hypodermic syringe through a Teflon-coated silicone septum attached by a Teflon Swagelok fitting. In a modified design, a short, bent piece of glass tubing which can hold a borohydride pellet is attached to the upper part of the reaction vessel by a ground glass joint. By turning this tubing, the pellet can be dropped into the solution without opening the system to the atmosphere.

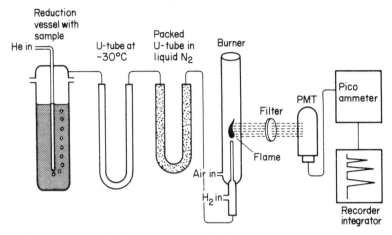

Figure 95 Apparatus for the reduction of DMSO to DMS and the flame photometric detection of DMS (FPD system). Reprinted with permission from Andreae.[215] Copyright (1977) American Chemical Society

From the reaction vessel, the gas stream passes through a 25 cm long, 13 mm o.d. Pyrex U-tube, which is immersed in an isopropyl alcohol bath at $-35\,°C$. This trap removes most of the water vapour from the gas stream without condensing any dimethyl sulphide.

In the gas chromatographic system, the gas stream now passes through a 15 cm long, 6 mm o.d. Pyrex U-tube which is filled with silanized glass wool and immersed in liquid nitrogen. This trap is interfaced by a six-way valve to a gas chromatograph, equipped with a flame ionization detector. The separation was performed on a 4.8 mm o.d. 6 m long stainless steel column, packed with 16.5% silicone oil DC-550 on 80–100 mesh Chromosorb W AW DCMS. The helium carrier gas flow is 80 ml min^{-1} air and 30 ml min^{-1} auxiliary helium.

In the flame photometric system, the sample gas stream is passed through a treated 6 mm o.d. 30 cm long glass U-tube filled with 15% OV3 on Chromosorb W AW DCMS 60–80 mesh. This trap is connected to a quartz tube burner with an interior tubing drawn out to a flame tip of *ca*. 1 mm i.d. The flow rates for this burner are 157 ml min^{-1} helium carrier, 115 ml min^{-1} hydrogen, and 200 ml min^{-1} air. This burner was mounted in an aluminium housing, which is flanged to a photomultiplier system containing a high sensitivity multialkali photomultiplier tube with u.v.-glass window operated at 700 V. A Varian coloured glass filter[217] was used to monitor the emission from the S_2 band system centred at 390 nm.

For the borohydride reduction, 0.1 ml concentrated hydrochloric acid was added per 25 ml sample; for the reduction with chromium (II) chloride, 2.5 ml of acid were added. The reaction vessel was then attached to the system and purged for 5 min with the helium stream to remove air and volatile sulphur compounds. Scrubbing is then continued until a zero blank is obtained.

After the purging period, the cold trap is immersed in liquid nitrogen, and the reluctant (chromium(II) chloride, sodium borohydride pellet or solution) is injected. When sodium borohydride solution is used, it has to be slowly injected into the reactor (*ca.* 1 min per 2 ml) to avoid an excessively fast reaction. The reluctant amounts used are: sodium borohydride, one 0.25 g pellet or 2 ml of 4% solution; chromium(II) chloride, 5 ml of a 2 M solution. The reaction times are 6 and 30 min for the borohydride and chromium(II) reductions, respectively. The helium stream is continuously purging the solution during this period in order to remove the volatile dimethyl sulphide formed by reduction of dimethyl sulphoxide from the solution.

The gas stream was dried by passing through a U-tube at $-35\,°C$ which removes all of the water and none of the dimethyl sulphide. The dimethyl sulphide and other reaction products in the gas stream were collected in a cold trap immersed in liquid nitrogen. For the gas chromatographic–flame ionization system, this trap was filled with silanized glass wool. After the collection time, the trap was switched into the carrier gas stream of the gas chromatograph by the six-way valve, then heated rapidly by immersing into hot water. The dimethyl sulphide was then separated on the column from other reaction products and detected by the flame ionization detector. In the flame photometric system, the trap serves both to collect the reaction products and to achieve the gas chromatographic separation. For this purpose, a trap filled with a gas chromatographic packing is used. After the reaction time, the liquid nitrogen is removed, and the variable transformer is switched on to provide 7 V to the heating coil. The dimethylsulphide elutes in a sharp peak after 1.1 min and enters the flame photometric detector.

Using 100 ml samples, a detection limit of 0.01 $\mu g\ l^{-1}$ (as 5) of dimethyl sulphoxide was achieved by these procedures. The flame ionization detector shows a completely linear response in the lower range with a decrease in slope starting at the 100 ng level. The flame photometric detector shows a logarithmic slope of 1.6 at the lower analyte levels, which also shows a significant decrease above 100 ng dimethyl sulphoxide. These variations in the response characteristics of the flame photometric detector are dependent on burner design and gas flow rates.[217] Fluorescence self-absorption effects become important at high analyte concentrations in the flame, and limit the detector response. Andreae[215] emphasizes that samples should be analysed immediately to prevent any changes in dimethyl sulphoxide concentrations, especially if the samples contain live microorganisms. If immediate analysis is not possible, steps have to be taken to prevent the biological interconversion of dimethyl sulphoxide and dimethyl sulphide, and the oxidation of dimethyl sulphide to dimethyl sulphone by bacteria.

If samples contain large amounts of dimethyl sulphide, it should be removed by purging with inert gas to prevent oxidation to dimethyl sulphoxide during storage and consequently erroneously high dimethyl sulphoxide concentrations.

Two types of interference were observed in this method. A negative interference due to loss of dimethyl sulphide to the glass container and tubing surfaces, and the positive interference due to reactions of compounds other than dimethyl sulphoxide to produce dimethyl sulphide. At the nanogram level, losses of over 50% are commonly encountered. These losses can be completely prevented by thorough deactivation of all surfaces that contact the dimethyl sulphide vapour, including the chromatographic support, by application of silylating reagents, such as SILYL-8 (Supelco Inc.).

Andreae[215] tested a large number of sulphur compounds in order to investigate potential positive interferences due to the formation of dimethylsulphide from the reaction of sulphur compounds other than dimethyl sulphoxide to form dimethyl sulphide. The results are summarized in Table 68. No compound other than dimethyl sulphoxide showed a volatile reaction product with chromium(II). The only compound other than dimethyl sulphoxide which gave a positive reaction with sodium borohydride is dimethylpropiothetin $(CH_3)_2S^+CH_2CH_2COO^-$) an organosulphur compound occurring in some algae. These results show that the method is essentially free from positive interferences, except in the rare cases when dimethylsulphonium compounds are present. In these situations the chromium(II) reaction has to be used to determine dimethyl sulphoxide.

Table 68 Reaction of sodium borohydride and chromium(II) chloride with various organic sulphur compounds*

Compound	Sodium borohydride	Chromium(II) chloride
Dimethyl sulphoxide	DMS	DMS
Dimethyl sulphone	NR	NR
Dimethylpropiothetin	DMS	NR
Cystine	NR	NR
Cysteine	NR	NR
Homocystine	NR	NR
Homocysteine	—	NR
S-Methylcysteine	NR	NR
Glutathione	NR	NR
Methionine	NR	NR
Methionine sulphoxide	NR	NR
Methylmethionine	—	NR

*NR = No detectable reaction product
Reprinted with permission from Andreae.[215] Copyright (1977) American Chemical Society.

Andreae[215] showed that dimethyl sulphoxide is a common constitutent in natural waters. Its occurrence in sea water is restricted to the zone of light penetration. This fact and the abundance of dimethyl sulphoxide in the medium after the growth of phytoplankton suggest that it occurs as an end product of algal metabolism. The occurrence of dimethyl sulphoxide in rain may be due either to the release of this substance from the biosphere into the atmosphere or to the atmospheric oxidation of dimethyl sulphide to dimethyl sulphoxide.

Tetramethylthiuram disulphide (Thiram)

This substance can be determined in waters spectrophotometrically.[218] A chloroform extract of the sample is tested with a methanolic suspension of cupric iodate and the extraction of the cream-yellow colour evaluated spectrophotometrically at 440 nm.

Calcium Lignosulphonate

This substance has been determined by ultraviolet spectroscopy in sulphite containing effluents.[219]

Organosulphur compounds in marine sediments

Dichloroethane extraction of the sediment, followed by elimination of elemental sulphur on a copper column is followed by a flame photometric estimation of the organosulphur compounds. The detection limit is 1 ng as sulphur with a precision of ±10%.[220]

REFERENCES

1. Metera, J., Rabhl, V. and Mostecky, J. *Water Research*, **10**, 137 (1976).
2. Hermanson, H. P., Helrich, K. and Carey, W. F. *Anal. Lett.*, **1**, 941 (1968).
3. Onuska, F. I. *Water Research*, **7**, 835 (1973).
4. Kimoto, W. I., Dooley, C. J., Canne, J. and Fiddler, W. *Water Research*, **14**, 869 (1980).
5. Florence, T. M. and Farner, Y. J. *Anal. Chim. Acta*, **63**, 255 (1973).
6. Dyatlovitskaya, F. G. and Botvinova, L. E. *Khim. Volokna*, **1**, 64 (1970) Ref. *Z. Khim.* 19GD (13) Abstr. No. 13G308 (1970).
7. Fishman, G. I. and Pevzner, I. D. *Zavod. Lab.*, **36**, 926 (1970).
8. Basson, W. D. and Van Staden, J. F. *Analyst*, **103**, 998 (1978).
9. Zatkovetskii, V. M., Safonova, Z. B., Koren'kov, V. N., Nevskii, A. B. and Sosin, S. L. *Zavod. Lab.*, **37**, 1434 (1971).
10. Bark, L. S., Cooper, R. L. and Wheatstone, K. C. *Water Research*, **6**, 117 (1972).
11. El-Dib, M. A. *J. Ass. Off. Anal. Chem.*, **54**, 1383 (1971).
12. Jenkins, R. L. *Bull. Environ. Contam. Toxicol.*, **13**, 436 (1975).
13. American Public Health Association. Supplement to the 15th Edition of *Standard Methods for the Examination of Water and Wastewater Method III*, p.548.
14. Jaroslav, N. *Chemicky Prum.*, **20**, 575 (1970).
15. Kupor, V. G. *Nauch. Trudy Omsk. Med. Dist.* **160**(88) (1968). Ref. *Zh. Khim.* 19GD Abstr. No. 24G203 (1969).
16. Scoggins, M. W. and Skurcenski, L. *J. Chromatg. Sci.*, **15**, 573 (1977).
17. Gardner, W. S. and Lee, G. F. *Environ. Sci. Technol.*, **7**, 719 (1973).
18. Tatsumoto, M., Williams, W. T., Prescott, J. M. and Hood, D. W. *J. Mar. Res.*, **19**, 89 (1961).
19. Palmork, K. H. *Acta Chem. Scand.*, **17**, 1456 (1963).
20. Degens, E. T., Reuter, J. H. and Shaw, N. F. *Geochim. Cosma. Acta*, **28**, 45 (1964).
21. Rittenberg, S. C., Emery, K. O., Hulsemann, J., Degens, E. T., Fay, R. C, Reuter, J. H., Grady, J. R., Richardson, S. H. and Bray, E. E. *J. Sediment. Petrol.*, **33**, 140 (1963).

22. Siegel, A. and Degens, E. G. *Science*, **151**, 1098 (1966).
23. Webb, K. L. and Wood, L. In: Scova, N. B. (ed.), *Automation in Analytical Chemistry* (Technicon Symposium 1966), Mediad Incorporated, New York, N.Y., 1, 440-444 (1967).
24. Hobbie, J. E., Crawford, C. C. and Webb, K. L. *Science*, **159**, 1463 (1968).
25. Palmork, K. H. *Int. Counc. Explor. Sea. Hydrography Committee*, C16 CM (1969).
26. Starikova, N. D. and Korzhikova, R. I. *Oceanol.*, **9**, 509 (1966).
27. Bohling, H. *Mar. Biol.*, **6**, 213 (1970).
28. Riley, J. P. and Seagar, D. A. *J. Mar. Biol. Assoc. U.K.*, **50**, 713 (1970).
29. Andrews, P. and Williams, P. J. LeB., *J. Mar. Biol. Assc. U.K.*, **51**, 11 (1971).
30. Pocklington, R. *Nature*, **230**, 374 (1971).
31. Clark, M. E., Jackson, G. A. and North, W. J. *Limnol Oceanogr.*, **17**, 749 (1972).
32. Bohling, H. *Mar. Biol.*, **16**, 281 (1970).
33. Coughenower, D. D. and Curl, H. C. Jr. *Limnol. Oceanogr.*, **20**, 128 (1975).
34. North, B. B. *Limnol. Oceanogr.*, **20**, 20 (1975).
35. Crawford, C. C., Hobbie, T. E. and Webb, K. L. *Ecology*, **55**, 551 (1974).
36. Brockmann, U. H., Eberlein, K., Junge, H. D., Maier-Reimer, E., Siebers, D. and Trageser, H., 1974. Entwicklung naturlicher Planktonpopulationen in einem outdoor-Tank mit nahrstoffarmem Meerwasser. II. Konzentrationsveranderungen von gelosten neutralen Kohlenhydraten und freien gelosten Aminosauren. *Berichte aus dem Sonderforschungsbereich Meeresforschung*. SFB 94. Universitat Hamburg., 6, 166-184 (1974).
37. Williams, P. J. LeB., Berman, T. and Holm-Hansen, O. *Mar. Biol.*, **35**, 41 (1976).
38. Garrasi, C. and Degens, E. T. Analytische Methoden zur saulenchromatographischen Bestimmung von Arminosauren und Zuckern im Meerwasser und Sediment. Berichte aus dem Projekt DFG-DE 74/3: *Litoralforschung - Abwasser in Kustennahe*, DFG-Abschlusskolloquium, Bremerhaven (1976).
39. Dawson, R. and Mopper, K. *Anal. Biochem.*, **83**, 100 (1977).
40. Lee, C. and Bada, J. L. *Earth Planet Sci. Lett.*, **26**, 61 (1975).
41. Lee, C. and Bada, J. L. *Limnol. Oceanog.*, **22**, 502 (1977).
42. Daumas, R. A. *Mar. Chem.*, **4**, 225 (1976).
43. Dawson, R. and Gocke, K. *Oceanol. Octa*, **1**, 80 (1977).
44. Gardner, W. S. *Mar. Chem.*, **6**, 15 (1978).
45. Dawson, R. and Pritchard, R. G. *Mar. Chem.*, **6**, 27 (1978).
46. Bajor, M. and Bohling, H. *Z. Anal. Chem.*, **249**, 190 (1970).
47. Tusek, F., Lische, P. and Chudova, S. *Zeit für Wasser in Abwasser Forschung*, **12**, 242 (1979).
48. Mills, A. I. and Alexander, M. *J. Environ. Qual.*, **5**, 437 (1976).
49. Fine, D. E., Rounbehler, D. P., Hurffman, F., Garrison, A. W., Wolfe, N. L. and Epstein, S. S. *Bull. Environ. Contam. Toxicol.*, **14**, 404 (1975).
50. Nikaido, M. M., Raymond, D. D., Francis, A. J. and Alexander, M. *Water Research*, **11**, 1085 (1977).
51. Richardson, M. L., Webb, K. S. and Gough, T. A. *Ecotoxicol. Environ. Safety*, **4**, 207 (1980).
52. Gough, T. A. and Webb, K. S. *J. Chromat.*, **79**, 57 (1973).
53. Fine, D. H., Rounbehler, D. P., Huffman, F. and Epstein, S. S. *Bull. Environ. Contam. Toxicol.*, **14**, 404 (1975).
54. Fine, D. H., Lieb, D. and Rounbehler, D. P. unpublished work.
55. Fine, D. H. and Rounbehler, D. P. unpublished work.
56. *Draft Report New Orleans Area Water Supply Study*, US Environmental Protection Agency, Dallas, Texas (November 1974).
57. Fiddler, W., Pensabene, J. W., Doerr, R. C. and Dooley, C. J. *Food Cosmet. Toxicol.*, **15**, 441 (1977).

58. Cohen, J. B. and Backman, J. D. *Environmental Aspects of N-Nitroso Compounds* (edited by Walker, E. A. *et al.*), pp. 257-372, International Agency for Research on Cancer (1978).
59. Gough, T. A., Webb, K. S. and McPhail, M. F. *Food Cosmet. Toxicol.*, **15**, 437 (1977).
60. Angeles, R. M., Keefer, L. K., Roller, P. P. and Uhm, S. J. *Environmental Aspects of N-Nitroso Compounds* (edited by Walker, E. A. *et al.*), pp. 357-372. International Agency for Research on Cancer (1978).
61. Kimoto, W. K., Dooley, C. J., Carre, J. and Fiddler, W. *Water Research*, **14**, 869 (1980).
62. Fine, D. H. and Rounbehler, D. P. *J. Chromat.*, **109**, 271 (1975).
63. Croll, B. T. *Analyst (London)*, **96**, 67 (1971).
64. Croll, B. T. Water Research Association, Medmenham, Harlow, U.K. Report TP70 *The Determination of Acrylamide in Polyelectrolytes by Extraction and Gas Chromatographic Analysis,* December (1969).
65. Arkell, G. H. and Croll, B. T. The Water Research Association, Medmenham, Harlow, U.K., Report TP78 *The Determination of Acrylamide in Water*, December (1970).
66. Croll, B. T. and Simpkins, G. M. *Analyst (London)*, **97**, 281, (1972).
67. Croll, B. T., Arkell, G. M. and Hodge, R. P. *Water Research*, **8**, 989 (1974).
68. Hashimoto, A. *Analyst (London)*, **101**, 932 (1976).
69. Husser, E. R., Stehl, R. H., Price, D. R. and DeLap, R. A. *Anal. Chem.*, **49**, 154 (1977).
70. Ludwig, F. J. and Besand, M. F. *Anal. Chem.*, **50**, 185 (1978).
71. Brown, L. and Rhead, M. *Analyst (London)*, **104**, 391 (1979).
72. Bezazyan, I. I., Kulikova, O. I. and Wicheva, I. A. *Zavod. Lab.*, **38**, 415 (1972).
73. Stefanescu, T. and Ursu, G. *Materiale Plate*, **10**, 330 (1973).
74. Ghersin, Z., Stitzl, H. and Wanea, R. *Revta. Chim.*, **20**, 689 (1969).
75. Nehring, H. *Pharmazie*, **27**, 741 (1972).
76. McCarthy, J. J. *Limnol Oceanog.*, **15**, 309 (1970).
77. Emmet, R. T. *Anal. Chem.*, **41**, 1648 (1969).
78. Rogers, D. S. and Pool, K. *Anal. Lett.*, **6**, 801 (1973).
79. Chu, J. P., Kirsh, E. J. and Born, G. S. *Bull. Environ. Contam. Toxicol.*, **6**, 343 (1971).
80. Supin, G. S., Vaintraub, F. P. and Makarova, C. U. *Cig. Sanit.*, **5**, 61 (1971). Ref. *Zh. Khim.* 19GD (**20**) Abstr. No. 20G191 (1971).
81. Chambou, P. and Chambon, R. *J. Chromat.* **87**, 287 (1973).
82. Cranmer, M. *Bull. Environ. Contam. Toxicol.*, **5**, 329 (1970).
83. Fan, T. Y., Ross, R., Fine, D. H., Keith, L. H. and Garrison, A. W. *Environ. Sci. Technol.*, **12**, 692 (1978).
84. Hess, T. L., Gurdry, L. J. and Sibley, S. D. *Bull. Environ. Contam. Toxicol.*, **13**, 579 (1975).
85. Watsh, J. T., Chalk, R. C. and Merritt, C. *Anal. Chem.*, **45**, 1215 (1973).
86. Heller, C. A., McBride, R. R. and Ronning, M. A. *Anal. Chem.*, **49**, 2251 (1977).
87. Kunkel, R. and Manahan, S. E. *Anal. Chem.*, **45**, 1465 (1973).
88. Rudling, L. *Water Research*, **6**, 871 (1972).
89. Chau, A. Y. and Fox, M. E. *J. Chromatog. Sci.*, **9**, 271 (1971).
90. Korsunovskii, G. A. *Z. Anal. Khim.*, **29**, 1244 (1974).
91. Stolzberg, R. J. *Anal. Chim. Acta*, **92**, 139 (1977).
92. Gardiner, J. *Analyst (London)* **102**, 120 (1977).
93. Gardiner, J. *Water Research Centre*, Stevenage Laboratory, Herts. U.K. Technical Memorandem TM101 (1975).
94. Wernet, J. and Wahl, K. *Z. Anal. Chem.*, **251**, 373 (1970).
95. Afghan, B. K. and Goulden, P. D. *Environ. Sci. Technol.*, **15**, 60 (1971).

96. Haberman, J. P. *Anal. Chem.*, **43**, 63 (1971).
97. Asplund, J. and Wanninen, E. *Anal. Lett.*, **4**, 267 (1971).
98. Afghan, B. K., Goulden, P. D. and Ryan, J. F. *Anal. Chem.*, **44**, 354 (1972).
99. Dietz, F. *Zeitschrift für Wasser und Aberwasser Forschung*, **7**, 74 (1974).
100. Haring, B. J. A. and Van Delft, W. *Anal. Chem.*, **94**, 201 (1977).
101. Murray, D., Povoledo, D. and Fish, J. *Research Board, Canada*, **28**, 1043 (1971).
102. Chau, Y. K. and Fox, M. E. *J. Chromatog. Sci.*, **9**, 271 (1971).
103. Warren, C. B. and Malec, E. J. *J. Chromat.*, **64**, 219 (1972).
104. Aue, W. A., Hastings, C. R., Gerhardt, K. O., Pierce, J. D., Hill, H. M. and Moseman, R. F. *J. Chromat.*, **72**, 259 (1972).
105. Stolzberg, R. J. and Hume, D. N. *Anal. Chem.*, **49**, 374 (1977).
106. Williams, P. J., Benoit, F., Muzcka, K. and O'Grady, R. J. *Chromatography*, **136**, 423 (1977).
107. Reichert, J. K. and Linckens, A. H. M. *Environ. Technol. Lett.*, **1**, 42 (1980).
108. Robinson, J. L. and Lott, P. F. *Microchem. J.*, **18**, 128 (1973).
109. Longbottom, J. E. *Anal. Chem.*, **44**, 418 (1972).
110. Coombs, L. C., Vasiliades, J. and Margerum, D. W. *Anal. Chem.*, **44**, 2235 (1972).
111. Rudling, L. *Water Research*, **5**, 831 (1971).
112. Rudling, L. *Water Research*, **6**, 871 (1972).
113. Longman, G. F., Stiff, M. J. and Gardiner, D. K. *Water Research*, **5**, 1171 (1971).
114. Sekerka, I. L., Lechner, J. and Afghan, B. K. *Anal. Lett.*, **6**, 977 (1973).
115. Zhantalai, B. P. and Slisarenko, V. P. *Trudy naucho-issled, proekt. Inst. Azot. Prom. Prod. org. Sint.*, **16**, 126 (1972). Ref *Zh. Khim.*, 19GD. Abstr. No. 9GI94 (1973).
116. Van de Haar, G., Pijper-Noordhoff, F. M. and Strikwerda, K. H_2O, **12**, 420 (1972).
117. Seibert, G. and Zahn, R. K. *Gas-in-Wasserfach (Wasser, Abwasser)*, **117**, 184 (1976).
118. Hicks, E. and Riley, J. P. *Anal. Chim. Acta*, **116**, 137 (1980).
119. Trifonova, N. A. *Trudy Iust. Biol. Unutr. Vod. Akad. Nauk SSSR*, **18**, 247 (1968). Ref. *Zh. Khim.* 19GD (12) Abstr. No. 12G248 (1969).
120. Manny, B. A., Miller, M. C. and Welzel, R. G. *Limnol. Oceang.*, **16**, 71 (1971).
121. Wood, O. *Anal. Abstr.*, **15**, 7628 (1968).
122. Stephenson, R. L. *J. Water Pollution Control*, **49**, 2499 (1977).
123. Jenkins, D. *Progress in Water Technology*, **8**, 31 (1977).
124. Helfgott, T. and Mazurek, J. S. *Progress in Water Technology*, **8**, 433 (1977).
125. Adamski, J. H. *Anal. Chem.*, **48**, 1194 (1976).
126. O'Conner, B. and Miloski, O. unpublished work. Suffolk County Department of Environmental Control, Suffolk, UK, October (1974).
127. Armstrong, F. A. J., Williams, P. M. and Strictland, J. D. H. *Nature*, **211**, 481 (1966).
128. Stevens, R. J. *Water Research*, **10**, 171 (1975).
129. Tenny, A. M. *Autom. Anal. Chem.*, 580 (1966).
130. Kai, O. E. *Anal. Chim. Acta*, **86**, 63 (1976).
131. Harwood, J. E. and Huyser, D. J. *Water Research*, **4**, 539 (1970).
132. Elkei, O. *Anal. Chim. Acta*, **86**, 63 (1976).
133. Lowry, J. H. and Mancy, K. H. *Water Research*, **12**, 471 (1978).
134. Mertens, J., Van den Winkel, P. and Massert, P. L. *Anal. Chem.*, **47**, 522 (1975).
135. Nydahl, F. *Water Research*, **12**, 1123 (1978).
136. Middelboe, V. *Progress in Water Technology*, **8**, 447 (1977).
137. Middelboe, V. *Int. J. Appl. Radiat. Isotopes*, **29**, 752 (1978).
138. Johansen, H. S. and Middelboe, V. *Int. J. Appl. Radiat. Isotopes*, **27**, 591 (1976).
139. Cook, G. B., Goleb, J. A. and Middelboe, V. *Nature*, (London) **216**, 475 (1967).
140. Hoch, M. and Weisser, H. R. *N. Helv. Chim. Acta*, **33**, 2128 (1950).

141. Broida, H. P. and Chapman, M. W. *Anal. Chem.*, **30**, 2049 (1958).
142. Meier, G. and Müller, G. *Isotopen-praxis*, **1**, 53 (1965).
143. Leicknam, J. P., Middelboe, V. and Proksch, G. *Anal. Chim. Acta*, **40**, 487 (1968).
144. Middelboe, V. *Appl. Spectrosc.*, **28**, 274 (1974).
145. Ferraris, M. and Proksch, G. *Anal. Chim. Acta*, **59**, 177 (1972).
146. Leicknam, J. P., Figdor, H. C., Keroe, E. A. and Muehl, A. *Int. J. Appl. Radiat. Isotopes*, **19**, 235 (1968).
147. Karlsson, L. and Middelboe, V. *Isotopes and Radiation in Soil-Plant Relationships including Forestry*, p.211. IAEA, Vienna (1972).
148. Fiedler, R. and Proksch, G. *Anal. Chim. Acta*, **78**, 1 (1975).
149. Fiedler, R. and Proksch, G. *Anal. Chim. Acta*, **78**, 29 (1975).
150. Nelson, D. W. and Sommers, L. E. *J. Environ. Qual.* **4**, 465 (1975).
151. Shoaf, W. T. and Lium, B. W. *J. Res. US Geol. Survey*, **4**, 241 (1976).
152. Siegrist, L. H. *Schweizerische Beitschrift für Hydrologie*, **38**, 49 (1976).
153. Tobin, R. S., Ryan, J. F. and Afghan, B. K. *Water Research*, **12**, 783 (1978).
154. Patterson, J. W., Brezonik, P. L. and Putnam, H. D. *Environ. Sci. Technol.*, **4**, 569 (1970).
155. Kucnerowcz, F. and Verstraete, W. *J. Chem. Technol. Biotechnol.*, **29**, 707 (1979).
156. André, M., Van Beneden, P. and Bassleer, J. *Tribune du Cebedeau,* **31**, 251 (1978).
157. Hysert, D. W., Knudson, F. B., Morrison, M. N., Van Gheluwe, G. and Lom, T. *Biotechnol. Bioengng,* **21**, 1301 (1979).
158. Hysert, D. W., Kovecses, F. and Morrison, N. M. *J. Am. Soc. Brew. Chem.*, **34**, 145 (1976).
159. Hysert, D. W. and Morrison, N. M. *J. Am. Soc. Brew. Chem.*, **35**, 160 (1977).
160. Chappelle, E. W., Picciolo, G. L., Curtis, C. A., Knust, E. A., Nibley, D. A. and Vance, R. B. *Laboratory Procedures Manual for the Firefly Luciferase Assay for Adenosine Triphosphate (ATP)*, Goddard Space Flight Center, NASA, TM X-70926 (1975).
161. Knust, E. A., Chappelle, E. W. and Picciolo, G. L. in *Analytical Applications of Bioluminescence and Chemiluminescence.* Chappelle, E. W. and Picciolo, G. L. (eds) NASA Scientific and Technical Information Office, Washington, DC, p. 27 (1975).
162. Lundin, A. and Thore, A. *Appl. Microbiol.*, **30**, 713 (1975).
163. Lundin, A. and Thore, A. *Anal. Biochem.*, **66**, 47 (1975).
164. Brøstad, J. O. and Hâkon, O. *Analyst (London)*, **101**, 820 (1976).
165. Weimer, W. C. and Armstrong, D. E. *Anal. Chim. Acta*, **94**, 35 (1977).
166. Murray, D. A. J. *J. Fish. Res. Bd, Can.*, **32**, 457 (1975).
167. Olsen, O., Gotterman, H. L. and Clymo, R. S. (Eds) *Chemical Environment in the Aquatic Habitat.* Proc. of an IBP Symposium in Amsterdam 10–16 Oct. 1966. NV Noord-Hollandsche. Uitgevers Moat schappi Amsterdam (1967).
168. Harwood, J. E., van Steerderen, R. A. and Kuhn, A. L. *Water Research*, **3**, 425 (1969).
169. Grasshoff, K. Z. *Anal. Chem.*, **220**, 89 (1966).
170. Osburn, Q. W., Lemmel, D. E. and Downey, R. L. *Environ. Sci. Technol.*, **8**, 363 (1974).
171. Harvey, H. W. *J. Mar. Biol. Ass., UK*, **27**, 337 (1948).
172. Lee, G. F., Clesceri, N. L. and Fitzgerald, G. P. *J. Air Water Pollut.*, **9**, 715 (1965).
173. Menzel, D. W. and Corwin, N. *Limnol. Oceanog.*, **10**, 280 (1965).
174. Goulden, P. D. and Brookbank, P. *Anal. Chim. Acta*, **80**, 183 (1975).
175. *Environment Canada Analytical Methods Manual.* Inland Waters Directorate, Water Quality Branch, Ottawa, Ontario, Canada (1974).
176. King, D. E. and Fellin, P. Ontario Ministry of the Environment, Laboratory Services Branch, Data Quality Report Series, Section I. *The Water Quality Laboratories, Data Quality Summary* (1975).

177. Jeffries, D. S., Dieken, F. P. and Jones, D. E. *Water Research*, **13**, 275 (1979).
178. Goossen, J. T. H. and Kloosterboer, J. G. *Anal. Chem.*, **50**, 707 (1978).
179. Sommers, L. E., Harris, R. F., Williams, J. D. H., Armstrong, D. E. and Syers, J. K. *Limnol. Oceanog.*, **15**, 301 (1970).
180. Nordforsk, O. *Interkalibrering av sedimentkemiska analysmetoder*. Nordforsk. Miljovardssekretariatet (1974).
181. Mehta, O. *Proc. Soil Science. Soc. Am.*, **18**, 443 (1954).
182. Jackson, M. I. *Soil Chemical Analysis*, 498 pp. Englewood Cliffs, N.J. (1958).
183. Standard Methods — *Standard Methods for the Examination of Water and Waste Water*, 13th edn, American Public Health Association, pp. 1–874, New York (1971).
184. Koroleff, F. *Determination of Total Phosphorus in Natural Waters by Means of Persulfate Oxidation*. International Council for the Exploration of the Sea (ICES). Report No. 3 (1970).
185. Stainton, M. P., Capel, M. J. and Armstrong, F. A. J. *The Chemical Analysis of Fresh Water*. Department of the Environment (Canada). Fisheries and Marine Service. Miscellaneous special publication No. 25 (1974).
186. Saunders, W. M. H. and Williams, E. G. *J. Soil Sci.*, **6**, 254 (1955).
187. Anderson, J. M. *Arch. Hydrobiol.*, **7**, 528 (1974).
188. Anderson, J. M. *Water Research*, **10**, 329 (1975).
189. Murphy, J. and Riley, J. P. *Z. Anal. Chim. Acta*, **12**, 162 (1962).
190. Jackson, M. L. *Soil Chemical Analysis*, Prentice-Hall Inc. Englewood Cliffs, N.J. (1958).
191. Hesse, P. R. *A Textbook of Soil Chemical Analysis*, Chemical Publishing Co. Inc., New York (1971).
192. Black, C. A. *Methods of Soil Analysis*, Part 2, *Chemical and Microbiological Properties*, American Society of Agronomy Inc., Madison, Wisc. (1965).
193. Aspila, K. I., Agemian, H. and Chau, A. S. Y. *Analyst (London)*, **101**, 187 (1976).
194. Griffiths, E. J. (Ed.) *Environmental Phosphorus Handbook*, John Wiley & Sons, New York (1973).
195. Golterman, H. L. and Clymo, R. S. (Eds) *Chemical Environment in the Aquatic Habitat, Proceedings of an IBP Symposium in Amsterdam and Nieuwersluis, 10–16 October, 1966*, N. V. Noord, Hollandsche Uitgevers Maatschappij, Amsterdam (1967).
196. Degens, Egon T. *Geochemistry of Sediments, A Brief Survey*, Prentice-Hall Inc., Englewood Cliffs, N.J. (1965).
197. Johnson, D. L. *Environ. Sci. Technol.*, **5**, 411 (1971).
198. Shulda, S. S., Syers, J. K. and Armstrong, D. E. *J. Environ. Qual.*, **1**, 292 (1972).
199. Vitenberg, A. G., Kuznetsova, L. H., Butaeva, I. L. and Ishakov, M. D. *Anal. Chem.*, **49**, 128 (1977).
200. Vitenberg, A. G., Butaeva, I. L. and Dimitova, Z. St., *Chromatographia*, **8**, 693 (1975).
201. Vitenberg, A. G., Ioffe, B. V. and Borisov, V. N. *Zh. Anal. Khim.*, **29**, 1795 (1974).
202. Vitenberg, A. G., Ioffe, B. V. and Borisov, U. N. *Chromatographia*, **7**, 610 (1970).
203. Williams, H. and Murray, F. E. *Pulp, Pap. Mag. Can.*, 347 (1966).
204. Vitenberg, A. G., Butsaeva, I. L. and Dimitrova, Z. St., *Chromatographia*, **8**, 693 (1975).
205. Kaiser, R. *Int. Congr. Chromatography 1972 Montreuse, Switzerland, 9–13 October* (1972).
206. Jenkins, L. L., Gute, J. P., Krasner, S. W. and Baird, R. B. *Water Research*, **14**, 441 (1980).
207. Bailey, J. C. and Viney, N. J. *Gas Chromatographic Investigation of Odour and a Sewage Treatment Plant*, Water Research Centre, Menmenham Laboratory, Medmenham, UK Technical Report No. TR125, December (1979).
208. de Souza, T. L. C., Lane, D. C. and Bhatia, S. D. *Anal. Chem.*, **47**, 543 (1975).

209. Beehard, M. J. and Rayburn, W. R. *J. Phycol.*, **15**, 379 (1979).
210. Jungclaus, G. A., Game, L. H. and Hites, R. A. *Anal. Chem.*, **48**, 1894 (1976).
211. Cox, G. B. *J. Chromat.*, **116**, 244 (1976).
212. Cassidy, R. M. and Frei, R. W. *Anal. Chem.*, **44**, 2250 (1972).
213. Piccardi, E. B. and Cellini, P. L. *Water, Air and Soil Pollution*, **9**, 301 (1978).
214. Andreae, M. O. *Anal. Chem.*, **52**, 150 (1980).
215. Andreae, M. O. *Anal. Chem.*, **49**, 820 (1977).
216. Braman, R. S., Ammons, J. M. and Bricker, J. L. *Anal. Chem.*, **50**, 992 (1978).
217. Patterson, P. L., Howe, R. L. and Abu-Shumays, A. *Anal. Chem.*, **50**, 339 (1978).
218. Bilikova, A. *Vod Hospod, B.*, **21**, 16 (1971).
219. Ban, S., Glauser-Soljam, M. and Smailagic, M. *Biotechnol. Bioengng*, **21**, 1917 (1979).
220. Bates, T. S. and Carpenter, R. *Anal. Chem.*, **51**, 551 (1979).

Chapter 4
Halogen Compounds

ALIPHATIC HALOGEN COMPOUNDS

The application of gas chromatography to the determination of chlorinated hydrocarbons in water and effluents, with particular reference to the types of these compounds used in industry, has been reviewed by Hassler and Rippa.[1] Glaze et al.[2] used flame ionization, electron capture, and Coulson electrolytic detectors with gas chromatography to study the formation of chlorinated aliphatics during the chlorination of waste waters. Chlorinated normal paraffins up to C_{30} carbon number range are of low volatility and are thermally unstable, producing hydrogen chloride on decomposition; hence direct gas chromatography is not attractive. Zitko[3] has devised a method based on column chromatography followed by microcoulometric detection. The procedure is not specific. Zitko has also described[4] a confirmatory method in which the chloroparaffins are reduced to normal hydrocarbons which are then analysed by gas chromatography. Both methods lack sufficient senstivity for trace (sub-ppm) analysis and the confirmatory method may be difficult to apply. Friedman and Lombardo[5] have described a gas chromatographic method applicable to chloroparaffins that are slightly volatile; the method is based on microcoulometric detection and photochemical elimination of chlorinated aromatic compounds that otherwise interfere.

Hollies et al.[6] have carried out a very extensive study of the determination of chlorinated long chain normal paraffins (13–30 carbon atoms) in water, sediments, and biological samples. They considered liquid chromatography, gas chromatography with a Coulson conductivity detector, and thin-layer chromatography.

Studies were carried out of the liquid chromatographic approaches used on a Pye moving wire transport system with a ^{63}Ni electron capture detector. However, studies of this technique were discontinued because of the poor sensitivity to chloro-n-paraffins with low chlorine contents. The second technique involving gas chromatography with a Coulson conductivity detector led to decomposition of the chloro-n-paraffins when they were volatilized. The peaks from the decomposition products were broad and irreproducible.

Hollies et al.[6] found that chloro-n-paraffins could be chromatographed on a silica gel plate from which an image of the chromatogram could be 'printed' on an aluminium oxide plate by heating the two face to face so that the high sensitivity of detection on aluminium oxide could be utilized. This detection procedure preceded by suitable preliminary clean-up and separation steps is the basis of the methods described below.

In these methods the samples are cleaned up by liquid-solid adsorption chromatography and thin-layer chromatography but those rich in lipids require preliminary solvent extraction. The methods distinguish between chloro-n-paraffins based on long carbon chains (C_{20}–C_{30}) and those based on shorter chains (C_{13}–C_{17}). The methods cover the ranges 500 ng l^{-1} to 8 μg l^{-1} for water (i.e. from about the solubility limit upwards) and 50 μg kg^{-1} to 16 mg kg^{-1} for sediments and biota. The precision of the methods ranges from $\pm 50\%$ relative at the lowest concentrations to $\pm 12\%$ relative at the highest. Recoveries are about 90% for water, 80% for sediments, and between 80 and 90% for biota according to sample type.

For all samples liquid adsorption chromatography clean-up was essential and in the methods described below a non-polar solvent, 60/80 petroleum spirit, is described for separating mobile impurities from chloroparaffins (the adsorbate). The latter are then desorbed with a polar solvent, toluene or carbon tetrachloride. According to sample type, two packings are prescribed. For water and sediments, aluminium oxide is effective and the column can be prepared simply by dry packing. However, lipidous samples are cleaned up more effectively on silica gel which must be slurry-packed to achieve satisfactory efficiency. Furthermore lipidous samples require pre-extraction based on solvent partition with dimethylformamide and 60/80 petroleum spirit. To get rid of any remaining impurities, which can spoil the subsequent thin-layer chromatographic stage by altering the chromatographic properties of the plate, a preliminary thin-layer chromatographic clean-up procedure is valuable; the impurities are separated from the chloroparaffin and then discarded by cutting off that half of the plate which contains them. Omission of this procedure leads to spots that are distorted and therefore difficult to quantify.

The final stage is to separate the chloroparaffins according to carbon chain length, Cereclor S45 and Cereclor 42 being used to calibrate the resulting chromatograms. These grades contain carbon chains in the ranges C_{13}–C_{17} and C_{20}–C_{30} respectively, and have chlorine contents of 45% and 42% (w/w) respectively. Using a desitometer a detection limit (50 ng per spot) similar to the visible method is achievable, giving linear response up to 200 μg of chloroparaffin.

Two methods are described below, one for water and sediments, the second for biological and animal materials. R_f values obtained for two chlorinated paraffins (Cereclor S45 and Cereclor 42) and a range of other chlorinated compounds (Table 69) show that little or no interference is to be expected in the thin-layer chromatographic method for substances other than n-chloroparaffins.

Table 69 R_f values of halogenated compounds

Halogenated compounds	R_f value	Halogenated compounds	R_f value
Cereclor S45	0.74	Dichlorophen	ND
Cereclor 42	0.80	Dieldrin	0.49
Aldrin	0.88	DDD	0.69
Araclor 1254	ND	DDE	ND
α-BHC	0.67	DDT	0.65
β-BHC	0.71	Endrin	0.52
γ-BHC	0.69	Endosulfan	ND
Chlordane	0.64	Heptachlor	0.51
1-Chloroeicosane	0.54	Methoxychlor	0.47
1-Chlorohexadecane	ND	Mirex	ND
1-Chlorooctadecane	ND	Strobane	0.65
p-Dichlorobenzene	ND	Toxaphene	0.63

ND did not respond to the silver nitrate spray agent.
Reprinted with permission from Hollies et al.[6] Copyright (1979) Elsevier Science Publishers

Method

Solvents

n-Hexane is used for thin-layer chromatography plate development. Petroleum spirit (60/80) and toluene are used for extractions and column clean-up exclusively. Fisons' Distol grade is satisfactory for the three solvents but they must be checked as being free from interfering impurities. For this purpose, 200 ml of solvent are evaporated to dryness, 0.2 ml of the same solvent are added to dissolve any residue, the whole of this is applied to a silica gel t.l.c. plate using microapplicator and any interfering impurity is detected as recommended under 'Detection and measurement of t.l.c. spots'. Any batch of solvent that gives a spot is rejected.

Spray reagent

General-purpose laboratory reagents are used. Industrial methylated spirit (45 ml), 2.5 ml of aqueous silver nitrate (10% w/v) and 2.5 ml of aqueous ammonia ($d = 0.88$) are mixed and used within 2 days.

Calibration compounds

Imperial Chemical Industries Cereclor grade S45 and Cereclor grade 42 are used.

A standard solution (50 mg l^{-1}) in 60/80 petroleum spirit is prepared for each grade of Cereclor as follows: a stock solution is made by dissolving 1 g in 50 ml of 60/80 petroleum spirit in a 100 ml volumetric flask, made up to the mark with petroleum spirit and mixed; 0.5 ml of this solution is diluted to 100 ml with petroleum spirit. The stock solution is stable for several months, but the dilute standard must be prepared freshly each week.

Drying agent

Anhydrous sodium sulphate (analytical-reagent grade) is decontaminated by heating at 300–350 °C overnight.

Apparatus

T.l.c. plates

Silica gel F_{254} and aluminium oxide type E, F_{254}, both types 0.25 mm thick × 200 mm × 200 mm (Merck) are used without pretreatment.

Chromatographic columns

The glass column (10 mm bore × 300 mm long) is fitted with a glass sinter and tap (Gallenkamp, CR 12/30) and decontaminated by heating at 250 °C for 24 h. To pack a column, it is half-filled with petroleum spirit, 10 g of aluminium oxide (Laporte Limited, grade UG) are added and allowed to settle. One gram of anhydrous sodium sulphate is added to form a layer on top of the aluminium oxide. Surplus petroleum spirit is drained off until the meniscus just touches the sodium sulphate layer. The petroleum spirit is retained and the column checked to ensure that it is free from interfering impurities by evaporating the petroleum spirit and chromatographing the residue as prescribed (see 'Solvents').

Glassware

For collection of samples, ground-glass stoppered bottles (1.5 l) and jars (250 ml) are suitable for water and sediment, respectively.

Other glassware comprises Soxhlet apparatus with 100 ml boiler; Soxhlet thimbles (glass microfibre, decontaminated by baking at 300 °C for 1.5 h); vials, 2 ml; separating funnels, all glass, 250 ml, with graphite-lubricated taps; and conventional equipment.

Before use, all glassware is decontaminated by heating at 250 °C for 24 h.

Miscellaneous equipment

This consists of: vacuum oven; steam bath; t.l.c. tanks, 275 × 125 × 20 mm air blower; microapplicator fitted with a No. 17 needle bent through 90 °C; t.l.c. spot drier (see Figure 96); air oven settable to 240 °C; t.l.c. plate drier; spray gun; u.v. lamp switched to 350 nm and with filter removed.

Sample pretreatment

Water

Water from rivers, canals, and reservoirs often contain suspended solids or sediments. Sometimes it is impracticable to apply the procedure for water

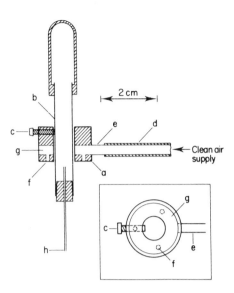

Figure 96 The t.l.c. spot drier (a) in position on the Shandon micropipette assembly (b), and a plan of the spot drier (inset). The drier is made from Perspex and has a locking screw (c) to adjust its position on the pipette holder; clean air is fed in from 3 mm o.d. silicone rubber tubing (d) to side arm (e) and enters chamber (g); the air exits from three passages (f) which focus on the micropipette tip (h). Reprinted with permission from Hollies et al.[6] Copyright (1979) Elsevier Science Publishers

directly to this suspension because the particles would interfere in the solvent extraction stage. Therefore, settling must be allowed, followed by separation of the sedimentary and supernatant liquid layers. The chloroparaffin contents of these two layers are determined separately and the results are combined, in relation to the amounts of the two layers, to give the overall concentration of chloroparaffin in the original sample. Such water samples are treated as follows: the bottle of sample is allowed to stand until any suspended solids or sediments have settled. The supernatant water is decanted or syphoned off, the volume noted and 1 litre retained in a glass-stoppered conical flask for subsequent analysis and for the following operation. The sedimentary layer is poured into a tared 250 ml beaker washing residual traces of sediment into the beaker with a little of the retained supernatant water. The beaker is reweighed to obtain the weight of the sedimentary layer.

Sediment

The beaker or sample jar of sediment is thoroughly stirred and mixed. The beaker of sediment or 20 g from the sample jar of sediment is taken in a tared 250 ml beaker, placed in a vacuum oven at 70 °C and dried to constant weight. The weight of dried sediment is calculated.

Extraction of chloroparaffin from water and sediments

Water

A 100 ml aliquot of the supernatant water sample is extracted with 20 ml of petroleum spirit in a 250 ml separating funnel by shaking for a minute. The lower, aqueous phase is run into a similar separating funnel. The organic phase is run into a 100 ml beaker via a filter funnel containing a 3 g bed of anhydrous sodium sulphate supported by a short plug of silica wool, freshly washed with 50 ml of petroleum spirit before use. The separating funnel and sodium sulphate bed are rinsed with a further 5 ml of petroleum spirit and run into the 100 ml beaker. The extraction is completed by repeating this sequence of operations on the aqueous phase in the separating funnel. The aqueous phase is discarded and the bulked extract evaporated to about 5 ml on a steam bath.

At this stage, a check for contamination is made by starting a blank determination in parallel with the sample analysis. Using duplicate apparatus, 50 ml of petroleum spirit are treated in the same way as a sample extract in this and the subsequent stages.

Sediment

A 10.0 ± 0.1 g sample of dried sediment is weighed into a tared Soxhlet thimble, enclosed with a pad of silica wool and transferred to the Soxhlet apparatus. The sample is extracted with about 60 ml of petroleum spirit for 24 h. The extract is transferred to a 100 ml beaker and evaporated to about 5 ml on a steam bath. At this stage, 60 ml of petroleum spirit are introduced as a blank and, starting with a dummy Soxhlet extraction, using duplicate apparatus, treated in the same way as a sample extract in this and subsequent stages.

Adsorption chromatography clean-up

The petroleum spirit concentrate from the evaporation step is transferred into a prepared aluminium oxide column using a 10 ml pipette. The excess of solvent is slowly run off until the meniscus just touches the surface of the sodium sulphate plug. The remains of the concentrate are washed from the beaker with two consecutive 5 ml portions of petroleum spirit, adding them to the adsorption column as above. Petroleum spirit (100 ml) is passed through the column to remove gross impurities from the chlorinated paraffin that remains adsorbed. This eluant is discarded. The chlorinated paraffin is desorbed by eluting with 50 ml of toluene and collecting in a 100 ml beaker. The toluene eluant is evaporated on the steam bath, assisting the evaporation with a jet of clean nitrogen and the concentrate is transferred quantitatively to a 2 ml vial, using toulene for washing. The concentrate is evaporated to dryness on a steam bath, cooled and 0.2 ml of petroleum spirit added from a calibrated pipette and swirled to dissolve the residue. This clean-up stage is performed on the 'blank'.

The t.l.c. separation

Two t.l.c. tanks are used; into the first tank is poured pour n-hexane to a depth of 10 mm and into the second toluene similarly. The ends of the tanks are lined with filter paper wetted with the solvent in the tank. The tank covers are replaced and allowed to equilibrate for 30 min.

A silica gel plate is prepared as follows (see Figure 97). The plate is marked at lines (a) and (b) with a soft pencil. The plate is rested horizontally, the air blower is clamped above it and a stream of cold air is allowed to play along the origin line.

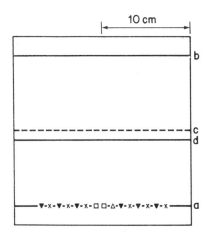

Figure 97 Diagram of silica gel F_{254} t.l.c. plate. (a) Origin line, 20 mm from plate edge; (b) first solvent limit line, 160 mm from origin; (c) 'cut line', 80 mm from origin; (d) second solvent limit line 70 mm from origin. (▼) Cereclor S56; (x) Cereclor, 42; (□) sample; (△) blank. Reprinted with permission from Hollies et al.[6] Copyright (1979) Elsevier Science Publishers

Water extracts

All the 200 µl of water extract and blank fractions from the adsorption chromatography clean-up stage are applied with the microapplicator as shown in Figure 97. On the same plate are spotted 4, 2, and 1 µl aliquots of the standard Cereclor S45 and S42 solutions progressively to the left of the sample spots (Figure 97) from a 1 µl pipette using the spot drier (Figure 96) and alternating the spots according to grade. Similarly, 8, 12, and 16 µl portions of the standard solutions are spotted to the right of the sample spots. With proper use of the spot drier, the spots should not exceed 3 mm diameter. The calibration spots cover the range 50–800 ng of Cereclor.

Sediment extracts

Spots of 1 µl, 10 µl, and 20 µl of sediment extracts and 20 µl of blank extract from the adsorption chromatography clean-up stage are applied to the t.l.c. plate using 1 µl micropipettes and the spot drier. Calibration spots are applied as described for water extracts.

For development, the plate is transferred to the *n*-hexane tank and developed to the first solvent limit line (Figure 97). The plate is removed and dried in the t.l.c. plate drier until no odour of solvent is apparent (*ca.* 10 min) The plate is cut in half the lower half is placed in the toluene tank, and eluted up to the second solvent limit line (Figure 97). The plate is removed and dried as above. The plate is reversed by putting it back in the hexane tank with the cut edge dipping in the hexane. The plate is eluted back to the origin line, removed and dried as above.

Detection and measurement of the t.l.c. spots

The positions of the origin and solvent limit lines of an aluminium oxide half-plate are marked in exactly the same way as for the silica gel half-plate. The two half-plates are clamped face to face with spring clips and the chloroparaffin spots 'printed' on to the aluminium oxide plate by heating the plates at 240 °C for 8 min. The half-plates are cooled to ambient temperature, unclamped, and the silica gel half-plate discarded. The aluminium oxide half-plate is sprayed evenly with the silver nitrate reagent and then placed under the u.v. lamp for 10 min to develop the chromatogram. The half-plate is removed and inspected under ordinary light. Any spots from chlorinated compounds will have a grey-to-black colour on a nearly white background. Any chlorinated paraffin in the sample is identified by reference to the R_f values which are approximately 0.74 and 0.80 for $C_{13}-C_{17}$ and $C_{20}-C_{30}$ chlorinated paraffin respectively (see Table 69).

Within 30 min (the plate turns grey in strong light) any chlorinated paraffin in the sample is estimated by visual comparison between the sample and standard spot intensities. However, if the blank has given a spot or if the plates show other serious interference, the results are rejected and contamination of the materials and equipment is looked for.

If the intensities of the sample spots are above the range of calibration spots, the t.l.c. procedure is repeated using a smaller aliquot of sample extract or a dilution of it as required. The half-plate is stored in the dark but for a permanent record, photographed within 30 min. Estimation of the chlorinated paraffin from the photograph should not be attempted.

Calculation of results

Water

The concentration (C_w) of chloroparaffin in the water sample is calculated from

$$C_w = 1000\ W_w/V_w\ (\text{ng l}^{-1})$$

where V_w (ml) is the volume of water taken (100 ml as prescribed) and W_w is the weight of chloroparaffin estimated in the t.l.c. spot (ng).

Sediment

The concentration (C_s) of chloroparaffin in the sample expressed on a dried basis is calculated from

$$C_s = 200 \times W_s/M_s \times v\ (\mu\text{g kg}^{-1})$$

where W_s is the weight of chloroparaffin estimated in the t.l.c. spot (ng); M_s is the weight of dried sediment taken (g); v is the volume of extract spotted onto the t.l.c. plate (μl).

Water containing suspended sediment

The concentration expressed in terms of the predetermined concentrations in water and dried sediment is calculated by the equation:

$$(V \times C_w + M_{ST} \times C_s)/(M + 1000 \times V)\ (\mu\text{g kg}^{-1})$$

where V is the volume of supernatant water in the sample (l); M is the weight of wet sediment in the sample (g); M_{ST} is the total weight of dried sediment (g); C_w and C_s have the meanings above.

Kaiser and Oliver[7] have determined volatile halogenated hydrocarbons at the 0.1–10 μg l^{-1} level in water by head-space and gas chromatography. Hrivňak et al.[8] determined chlorinated C1–C4 hydrocarbons in water using capillary gas chromatography. For the isolation of chlorinated hydrocarbons (*n*-butyl chloride, di-, tri-, and tetrachloromethane, 1,2-dichloroethane, 1,2-dichloropropane, and trichloethylene), a stripping technique was used. The hydrocarbons were analysed in a capillary stainless steel column at 80 °C. Using electron capture it is possible to determine down to 0.1 μg l^{-1} of these substances.

Dowty et al.[9] used gas chromatography–mass spectrometry to identify halogenated aliphatic hydrocarbons and aromatics in drinking water. About 70 compounds were identified.

Dawson et al.[10] have described samplers for large volume collection of sea water samples for chlorinated hydrocarbon analyses. The samplers use the macroreticular absorbent Amberlite XAD-2. Operation of the towed 'fish' type sampler causes minimal interruption to a ship's programme and allows a large area to be surveyed. The second type is a self-powered *in situ* pump which can be left unattended to extract large volumes of water at a fixed station.

Methods have been described for determining chlorinated aliphatic hydrocarbons in fish,[11] environmental samples,[12] and soil and chemical waste disposal site samples.[13] The latter method involves a simple hexane extraction and temperature programmed gas chromatographic analysis using electron capture detection and high resolution glass capillary columns. Combined gas chromatography–mass spectrometry was used to confirm the presence of the chlorocarbons in the samples.

SATURATED POLYCHLORO COMPOUNDS

Murray and Riley[14,15] described gas chromatographic methods for the determination of trichloroethylene, tetrachloroethylene, chloroform, and carbon tetrachloride in natural waters, sediments, marine organisms, and air. These substances were separated and determined on a glass column (4 m × 4 mm) packed with 3% of SE-52 on Chromosorb W (AW DMCS) (80–100 mesh) and operated at 35 °C, with argon (30 ml min^{-1}) as carrier gas. An electron capture detector was used, with argon–methane (9:1) as quench gas. Chlorinated hydrocarbons were stripped from water samples by passage of nitrogen and removed from solid samples by heating in a stream of nitrogen. In each case the compounds were transferred from the nitrogen to the carrier gas by trapping on a copper column (30 cm × 6 mm) packed with Chromosorb W (AW DMCS) (80–100 mesh) coated with 3% of SE-52 and cooled at −78 °C, and subsequently sweeping on to the gas chromatographic column with the stream of argon. A limitation of this procedure is that compounds which boil considerably above 100 °C could not be determined.[16]

A different approach is to pass the water through a bed of activated carbon which was subsequently extracted exhaustively in a Soxhlet unit, and the extract was evaporated and analysed; this measured perchloroethylene and hexachloroethane but the results are uncertain quantitatively.[17] A method has been published by which the water sample was codistilled with cyclohexane and the organic phase was then injected into an electron capture detector gas chromatograph.[18] Extraction with n-pentane followed by gas chromatography has also been used[19] but although the extraction was easy and effective, the chromatographic conditions described were time-consuming and unsuitable for compounds heavier than perchloroethylene.

Deetman et al.[20] have devised an electron capture gas chromatographic technique, applicable to water, mud, fish, and air samples, for the determination of down to 1 ng l^{-1} of 1,1,1-trichloroethane, trichloroethylene, perchloroethylene, 1,1,1,2-tetrachloroethane, 1,1,2,2-tetrachloroethane, pentachloroethane, hexachloroethane, pentachlorobutadiene, hexachlorobutadiene, chloroform, and carbon tetrachloride. These workers used extraction of the water samples with n-pentane as a means of isolating the chlorinated compounds from the sample. Recoveries of 95% were obtained in a single extraction. To dry the extract anhydrous sodium sulphate was found to be effective. Furthermore this drying agent could be freed from electron-capturing contaminants by heating[21] and did not absorb the chlorinated compounds. Under the specific conditions (i.e. using a temperature programmed Dexsil-300 column) all the compounds are separated with the exception of carbon tetrachloride and 1,1,1-trichloroethane which are resolved only on the Apiezon-L column. This column is an alternative for the analysis of water with the proviso that it is not suitable for samples containing the less volatile compounds. If the water sample contains chlorobromomethanes which can interfere with the determination of chloroform and trichloroethylene, it is advisable to augment the analyses by repeating the chromatography with a column containing oxydipropionitrile packing which will separate the bromine compounds from the chlorinated solvents. To avoid contamination use of a glove box is recommended for the preparation of water samples. In general, it is wise to exclude chlorinated solvents from the laboratory and if the ambient air is suspect, to blanket the inject port of the chromatograph with clean nitrogen.

Method

Scope and field of application

This method covers the determination in sea and fresh water of the compounds chloroform, carbon tetrachloride, trichloroethylene, perchloroethylene, 1,1,1,2-tetrachloroethane, pentachlorobutadiene, hexachlorobutadiene in the concentration range $0.01-10\,\mu\mathrm{g\ kg^{-1}}$.

The column specified in this method, Dexsil-300, does not separate carbon tetrachloride from 1,1,1-trichlorethane, but these compounds can be separated by the chromatographic system used by Hollies et al.[6] (see earlier in this chapter). However, if the sample is thought to contain high boiling components which require purging from the column at high temperature, then the thermal stability limitation of Apiezon-L makes it unsuitable and the sample must be chromatographed as described below.

Principle

A 200 ml sample of water is extracted with *n*-pentane and the extract is dried with anhydrous sodium sulphate. A portion of this solution is injected into the gas chromatograph fitted with a 1.5 m (5 mm i.d.) stainless steel or glass column packed with 15% Dexsil-300 on Diatomite-C (180–212 μm) and a ^{63}Ni electron capture detector. The carrier gas is purified nitrogen at 50 ml min^{-1}. The oven is maintained at 65 °C for 6 min and then temperature programmed at 10 °C min^{-1} up to 150 °C and held until hexachlorobutadiene has eluted. The column is then purged at 250 °C. The concentration of the chlorinated hydrocarbons are determined by comparison of peak areas in the sample chromatogram with those of an external standard mixture.

Materials

The materials are as specified by Hollies et al.[6] with the following exceptions.

Calibration compounds

Chloroform, carbon tetrachloride, 1,1,1-trichloroethane, trichloroethylene, perchloroethylene, 1,1,1,2-tetrachloroethane, 1,1,2,2-tetrachloroethane, pentachloroethane, hexachloroethane, pentachlorobutadiene and hexachlorobutadiene (purity ⩾99.0%) were used. Each gave only one peak when used as a standard in the method described.

Drying agent

Anhydrous sodium sulphate is heated at 300–350 °C overnight.

Column packing materials

Diatomite-C (180–212 μm equivalent to British Standard mesh 72–85: see BS 410 and LSO/TC 24) and Dexsil-300, both available commercially, are used.

Apparatus

The apparatus specified by Hollies et al.[6] is suitable with the following exceptions.

Injection device

Liquid syringes (10 μl, e.g. Hamilton) are suitable.

Columns

The stainless steel or glass columns (1.5 m long, 5 mm i.d., 6.4 mm o.d.) are pretreated as described by Hollies et al.[6] To pack one column, 10.2 g of Diatomite-C (180–212 μm) and 1.8 g of Dexsil-300 are weighed and the Dexsil-300 is dissolved in 10 ml of acetone (chloroform, the solvent recommended by the manufacturers, is not used because it is strongly electron capturing and would entail lengthy conditioning of the column). This solution is stirred into the Diatomite-C which has been slurried in 50 ml of acetone, heated gently under a ventilated hood and stirred to remove the solvent until the powder is dry and free-flowing. The powder is sieved and the fraction sized between 180 and 212 μm retained. The column is packed as described by Hollies.[6]

For conditioning, the column is connected directly to the injector, without a sample vaporizer, the detector disconnected, and the column heated at 350 °C for 48 h, with the nitrogen carrier gas. When not in use, the column is kept purged with carrier gas.

Bottles for collection of water samples

Brown glass, ground-glass stoppered bottles (1 l) are suitable.

Glove box

This must be of sufficient size (about 0.5 m^3) to accommodate a flask shaker with a timer and fitted with nitrogen lock. During use it is purged with 2000 l h^{-1} of purified nitrogen.

Other equipment needed includes separating funnels (250 ml, all-glass, with ungreased stopcocks), sample bottles (amber glass, 30 ml capacity with narrow necks) and 1 ml graduated pipettes. The serum caps to fit the sample bottles must be refluxed with *n*-pentane for 4 h to extract impurities.

Sampling

A fresh bottle must be used for each sample. Before use bottles are purged with purified nitrogen and stoppered. The water to be analysed is not filtered; the bottle is always fully filled with sample, leaving minimum ullage. The sample is stored in the dark at 0–5 °C and analysed within 2 days.

Extraction of the sample

All the following operations are performed inside the glove box: 200 ml of sample are transferred to a 250 ml separating funnel which has been stored unstoppered in the glove-box or purged with purified nitrogen. To this is added, accurately, 10 ml of *n*-pentane, stoppered and shaken mechanically for 5 min. The liquid phases are allowed to separate not centrifuged, as this would involve removal of the sample from the glove box and possible contamination, and would also cause warming, resulting in loss of volatiles. The (lower) aqueous layer is run off and then the *n*-pentane layer was run into a 30 ml sample bottle containing 1 g of anhydrous sodium sulphate. The bottle is closed with a clean serum cap and removed from the glove-box.

Setting the optimal conditions for the method

Follow the general procedure as described by Hollies *et al.*[6] but modify where necessary as follows.

The following temperature programme is set: initial isothermal period 6 min at 65 °C, then set at 10 °C min^{-1} up to 150 °C and held until hexachlorobutadiene has eluted (after about 5 min). The oven temperature is raised to 250 °C and held at that temperature for 10 min or a longer period as required to elute any heavy impurities.

For the programming test, the general instructions described by Hollies *et al.*[6] are followed but the programme prescribed above is used, and the stabilization temperature is maintained at not greater than 55 °C.

Calibration and preparation of standards

The external standard method is used. In the extraction stage, there is a 20-fold enrichment factor, thus, for measurement of detector sensitivities a suitable concentration for standards in n-pentane is $2\,\mu g\ l^{-1}$ for each compound. Standards of other concentrations are required for calibration within the range of concentration for each compound found in samples. In contrast to the earlier method[6] these should be individual solutions, not mixtures, because of the very variable composition of water samples.

A standard solution ($10\,\mu g\ l^{-1}$) in n-pentane is prepared for each compound as follows: by means of the $10\,\mu g$ liquid syringe a volume of the calibration compound equivalent to 10 mg is added or, for solids, 10 mg are dissolved in about 5 ml of n-pentane in a 10 ml graduated flask, made up to the mark with n-pentane and mixed. A $10\,\mu l$ aliquot of this solution is diluted similarly to 10 ml with n-pentane. A $100\,\mu l$ aliquot of this latter solution is diluted further to 10 ml with n-pentane. These standard solutions are prepared freshly each week.

To prepare calibration solutions from the standard solutions, a calculated volume of the standard solution is diluted with n-pentane. For example, to prepare the solutions ($2\,\mu g\ l^{-1}$) required for the sensitivity test (see below), 0.2 ml of the above standard solution is collected in a 1 ml pipette and diluted to 1 ml in a graduated flask.

Gas chromatography system tests

The test described previously by Hollies *et al.*[6] is performed with the modifications of column, temperature programming, and conditioning appropriate for the analyses.

Test for sensitivity to chlorinated hydrocarbons

Chromatograms on the $2\,\mu g\ l^{-1}$ standards of each compound of interest are run by the following procedure with the minimum prescribed attenuation, and the limits of detection are measured as described by Hollies *et al.*[6]

Recording the sample chromatograms

Using the above optional conditions, the amplifier attenuation is set to the prescribed minimum and the initial isothermal temperature at 65 °C, and 5 min allowed for stabilization. The syringe is flushed with sample solution and then filled, discharged to the $1\,\mu l$ mark and then the plunger is drawn back until the residual solution is in the syringe barrel. This is injected into the gas chromatograph. After withdrawing the syringe, it is inverted and the plunger drawn back. The residual volume of solution in the syringe is noted and the difference between the first reading and this is taken as the volume of solution injected. At the end of the chromatogram, the column oven temperature is raised

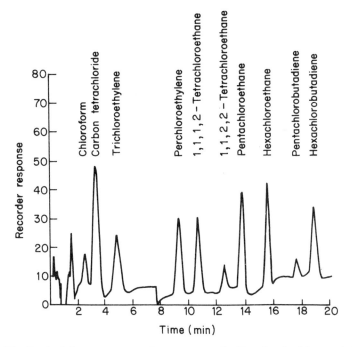

Figure 98 Typical chromatogram of a water sample. Reprinted with permission from Deetman et al.[20] Copyright (1976) Elsevier Science Publishers

to 250 °C and held for 10 min or a longer time if required to allow heavy impurities to elute. Then the oven is cooled and restabilized as described by Hollies,[6] ensuring that the stabilization temperature does not exceed 55 °C. If the chromatogram shows peaks that have gone off the scale, the amplifier attenuation is reset to a suitable higher value and another aliquot of sample injected. The attenuation is noted when an acceptable chromatogram has been obtained. Figure 98 shows a typical chromatogram.

Recording the standard chromatograms

The standards and their concentrations described earlier by Hollies et al.[6] are selected. A suitable set of calibration solutions is prepared and the chromatograms run as prescribed above. Peak identity is confirmed and peak areas measured as described by Hollies et al.[6]

Calculations of results

The concentration of any particular compound in the sample is given by the equation

$$C = \frac{10 \times C' \times A \times \mu'}{A' \times \mu \times W}$$

as the volume of sample extract is 10 ml and where C is the concentration of the compound in the sample (μg kg^{-1}); C' is the concentration of the compound in the standard (μg l^{-1}); A and A' are the peak area × attentuation of the compound in the sample and standard, respectively; μ and μ' are the volumes of the sample extract and the standard injected (μl), respectively; and W is the weight of the sample (g). This simple relationship holds over the concentration range 0.01–10 μg kg^{-1}. Duplicate results on any one sample would be expected to agree within 20%.

Other workers who have studied the determination of this class of compounds in water include Kummert et al.[22] who described a method for the trace determination (down to 0.06 μmol) of tetrachloroethylene in natural waters using direct aqueous injection–high pressure liquid chromatography and Stozek and Beumer[23] who determined chlorinated degreasing solvents in water.

Dilling et al.[24] have studied the evaporation rates in aqueous solution of various polychlorinated compounds such as methylene dichloride, chloroform, 1,1,1-trichloroethane, trichloroethylene, and tetrachloroethylene. The compounds were studied at concentrations of 1 ppm in water. All the compounds examined had evaporated by 50% in less than 30 min and by 90% in less than 90 min when stirred in an open container at 25 °C. The addition of salt, clay, limestone, sand, peat moss, and kerosine to the water has relatively little effect on the rates of disappearance. These workers conclude that low-molecular-weight chlorinated hydrocarbons would not persist in agitated natural water bodies owing to evaporation.

CHLOROLEFINS INCLUDING VINYL CHLORIDE

Simonov et al.[25] have described an ultraviolet spectrophotometric method for determining down to 1 ppm of tetrachloroethylene, hexachloropropene, hexachlorobutadiene, and hexachlorocyclopentadiene from their extinction at 202 nm, 240 nm, 220 and 255 nm, and 335 nm respectively.

Renberg[26] has described a method utilizing XAD-4 microreticulin resin for the determination of chloroethenes in water. Haloalkanes do not interfere in this procedure.

Burgasser and Calaruotolo[27] have described a gas chromatographic method for determining semi- or non-volatile chlorinated organics such as hexachlorobutadiene, hexachlorocyclopentadiene, octachlorocyclopentene, and hexachlorobenzene in amounts down to 0.1 ppb in water. These compounds fall into the category of those which are capable of being analysed by purge or trap techniques but which are preferentially soluble in non-aqueous solvents. These workers used a Brinkmann Polytron Homogenizer to perform the extraction and a Sorval refrigerated centrifuge to speed up the phase separation process. The extraction of chlorinated organic compounds from water can be carried out in a single vessel in one step, taking only 10 min to complete.

Method

Apparatus

The extractions were carried out in Pierce 125 ml hypovials using a Brinkmann Polytron Homogenizer, Model PT-1035, equipped with a model PT-10ST generator. Phase separation was carried out by centrifugation of the sample in the same vessel using a Sorval RC2-B centrifuge. Analysis of the extracts was carried out using a Hewlett-Packard 5840 gas chromatograph with an electron capture detector. The gas chromatographic conditions used are shown in Table 70.

Table 70 Gas chromatographic conditions

Column	6 ft × 2 mm i.d. glass
	3% Dexsil 300 on 80–100 Supelcoport
G.c.	Hewlett-Packard 5840 A
	Detector: electron capture (^{63}Ni) 300 °C
	Injector: 220 °C
	Column temp.: 150 °C
	Carrier gas: 10% methane argon
	Flow: 30 ml min^{-1}

Reprinted with permission from Burgasser and Calaruotolo.[27] Copyright (1977) American Chemical Society.

Reagents

Water used for preparation of standards was triple distilled from potassium permanganate solution and nitrogen purged to ensure that no organic residues were present. All organic solvents used (acetone, hexane, benzene) were pesticide distilled in glass. All glassware was cleaned by treatment with chromic acid and subsequent rinsing with water, acetone, and hexane. The glassware was then oven dried at 250 °C.

Procedure

One hundred (100) ml of the water to be analysed were placed into a Pierce 125 ml hypovial and the pH was adjusted to suit the nature of the compounds being analysed (for the chlorinated compounds, the pH was adjusted to 9–11). Then 10 ml of 15% benzene in hexane were added to the vial. The sample was extracted for 30 s with the homogenizer at 50% of full speed (approximately 11,000 rpm). Following each extraction, the PT-10ST generator was cleaned by successive washings in acetone, hexane, acetone, and hexane. The hypovial containing the emulsified solution was then centrifuged for 5 min at 1500 rpm and 4 °C. This was sufficient time to completely separate the organic and aqueous layers. The organic layer was removed with a Pasteur pipette and sealed in an autosampler vial for analysis on the gas chromatograph. The extracts can also be concentrated further by dry nitrogen purge for gas chromatographic-mass spectrometric examination.[27]

Extraction efficiencies are 90–100% following a single 30 second extraction. By using hexane, hexane–benzene, or hexane–toluene instead of hexane–methylene chloride, the concentration step can be eliminated except for gas chromatographic–mass spectrometric analysis where only concentration by nitrogen purge would be required. For compounds with less favourable distribution coefficients, pH adjustments and multiple extractions might be necessary and these factors should be examined for individual cases.

Because of the very short extraction time and the design of the Polytron Homogenizer, essentially no heat is transferred to the sample during the extraction process. This eliminates the potential for thermal degradation or evaporation loss of the compounds of interest, an effect occasionally observed when ultrasonic extraction techniques are used.

Table 71 is a summary of the statistical analysis of the data. The percentage recovery and standard deviation at each concentration level and the correlation coeffcients for the linearity of each compound over the range studied are given. This demonstrates that the procedure produces results which meet or

Table 71 Precision and accuracy data*

Compound	Recovery (%)	std.dev.(%)
	0.1 ppb	
Hexachlorobutadiene	90	0.03
Hexachlorocyclopentadiene	80	0.01
Octachlorocyclopentene	88	...
Hexachlorobenzene	109	0.02
	1.0 ppb	
Hexachlorobutadiene	85	0.15
Hexachlorocyclopentadiene	94	0.23
Octachlorocyclopentene	99	0.15
Hexachlorobenzene	96	0.24
	10 ppb	
Hexachlorobutadiene	125	...
Hexachlorocyclopentadiene	89	0.24
Octachlorocyclopentene	119	0.12
Hexachlorobenzene	86	0.13
	Correlation coefficient[†]	
Hexachlorobutadiene	0.997	
Hexachlorocyclopentadiene	0.995	
Octachlorocyclopentene	0.996	
Hexachlorobenzene	0.996	

*Data represent three replicates for each compound at each concentration.
†Correlation coefficient represents how closely the experimental fits with the expected values and is equal to $m\alpha_x/\alpha_y$ where m is slope of the line and α_x and α_y are the standard deviations of x and y array of the data points.
Reprinted with permission from Burgasser and Caruotolo.[27] Copyright (1977) American Chemical Society.

exceed the accuracy and precision requirements necessary for trace level environmental monitoring.

The technique has been used for the determination of the four compounds discussed above in both plant effluent and ground water samples. Additionally the technique has also been used to extract chlorinated organics from soils efficiently with the emulsions formed easily broken with centrifugation.

Vinyl chloride

Direct aqueous injection gas chromatography using flame ionization, microcoulometry, electrolyric conductivity, and mass spectrometry for detection has been used for the identification and measurement of vinyl chloride in industrial effluents.[28] The reported lower limits of detection vary, but 100 μg l^{-1} appears to be conservative for vinyl chloride using a flame ionization detector. Halogen-specific detectors, for example, the microcoulometric and electrolytic conductivity, are less sensitive (approximately 1000 μg l^{-1}). However, they do improve the qualitative accuracy of the determination.

A method for liquid-liquid extraction of vinyl chloride from aqueous solution has also been reported.[28] As much as 500 ml of water is extracted with 1 ml of carbon tetrachloride. One microlitre of the extract is analysed by gas chromatography. The reported lower limit of detection is approximately 0.1 μg l^{-1}. Extraction efficiencies for vinyl chloride are reported to be about 77% at 1-10 μg l^{-1} and near 100% at 0.2-3 mg l^{-1}.

Another method for determining ppb of vinyl chloride in surface waters is that of Alberti and Jonke.[29] These authors describe a gas chromatographic method for its determination using a flame ionization detector and a Porapak-Q or Chromosorb-101 column. The detection limit is 0.3 mg l^{-1} and samples of waste waters from vinyl chloride or PVC factories can be injected direct into the gas chromatograph, while water samples with lower concentrations require preliminary enrichment for which a gradient-tube method is described.

Workers at the National Environment Research Centre, US Environmental Protection Agency[30] have described a method for determining vinyl chloride at the μg l^{-1} level in water. An inert gas is bubbled through the sample to transfer vinyl chloride to the gas phase, and the vinyl chloride is then concentrated on silica gel or Carbosieve-B under non-cryogenic conditions, and determined by gas chromatography with a halogen-specific detector. Gas chromatography–mass spectrometric methods were used to provide confirmatory identification of vinyl chloride.

Method

Apparatus

A Perkin Elmer 900 gas chromatograph was equipped with a dual-flame ionization detector, a microcoulometric detector (halide mode) and a Hall

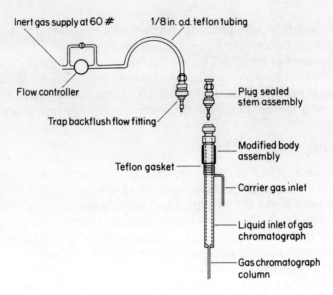

Figure 99 Desorber #1. Reprinted with permission from Bellar et al.[30] Copyright (1976) American Chemical Society

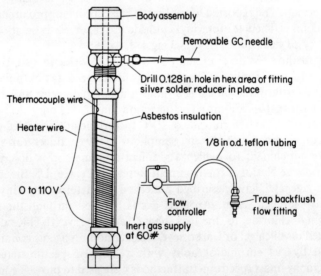

Figure 100 Desorber #2. Reprinted with permission from Bellar et al.[30] Copyright (1976) American Chemical Society

electrolytic conductivity detector (halide mode). Dual stainless steel columns, 180 cm (6 ft) long × 2.67 mm (0.105 in) i.d., were packed with Chromosorb-101 (60–80 mesh). The oven temperature was isothermal at 90 °C or programmed from 90 to 200 °C at 10 °C min^{-1}. Nitrogen, at 60 ml min^{-1}, was employed as the carrier gas. Desorber #1 (Figure 99) was used with this instrument.

A Varian Aerograph 1400 gas chromatograph with a Finnigan 1015C quadrupole mass spectrometer controlled by a System Industries 150 data acquisition system was employed. The glass column, 240 cm (8 ft) long × 2 mm (0.078 in) i.d. was packed with Chromosorb-101 (50–60 mesh). Helium, at 30 ml min^{-1} was employed as the carrier gas. The initial oven temperature of 90 °C was held for 3 min and then programmed to 220 °C at 4 °C min^{-1}. Desorber #2 (Figure 100) was used with this instrument.

Reagents

Water that is free of interfering organics was prepared by passing distilled water through a Millipore Super-Q water treatment system.

Standard solutions of vinyl chloride were prepared as follows. Approximately 8 ml of acetone were placed into a 10 ml volumetric flask. The flask and contents were carefully weighed. Vinyl chloride was slowly bubbled into the acetone from a finely drawn glass tube for about 2 min. The flask and contents were reweighed, diluted to volume, and stoppered. The weight gain, 50–200 mg, was used to calculate the concentration. A secondary dilution of 10 ng μl^{-1} of vinyl chloride in acetone was prepared from this standard. These standard solutions, when stoppered and stored at 4 °C, were stable for at least 1 week.

Procedure

Trap conditioning

Newly packed silica gel traps were conditioned at approximately 200 °C with a nitrogen backflush flow of 20 ml min^{-1} for 16–24 h with one of the desorbers vented to atmosphere. Each day before use, traps were placed into the desorber and conditioned at 150 °C for approximately 10 min while being backflushed with nitrogen at 20 ml min^{-1}.

Purging and trapping

Unless otherwise stated, samples were purged and trapped as follows. With nitrogen flowing through the purging device (Figure 101) at 20 ml min^{-1}, the trap inlet (Figure 102) was attached to the purging device exit end of the trap. Five millilitres of sample were injected into the purging device using a 5 ml syringe. After purging the sample for 10 min, the trap was removed from the purging device, and the trap vent was removed from the exit end of the trap. All samples were analysed within 10 min of trapping.

Desorption and Analysis

Desorber #1 (Figure 99) The gas chromatographic oven was cooled below 80 °C with the oven door open. After removing the plug from the desorber, the

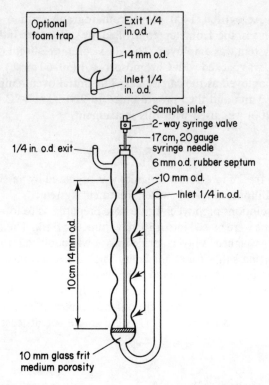

Figure 101 Purging device. Reprinted with permission from Bellar et al.[30] Copyright (1976) American Chemical Society

trap was inserted into the desorber. The trap backflush flow fitting was then locked into place on the trap exit. In this manner the trap was backflushed with nitrogen at 20 ml min^{-1} for 4 min at 150 °C. Then the trap backflush flow fitting was removed (trap still locked into place), the oven lid closed, and the oven rapidly heated to its normal or initial operating temperature. Gas chromatographic analyses were carried out under these conditions.

After analysis, the trap was removed by inserting the trap vent into the trap exit fitting (to vent inlet system), removing the trap, resealing the gas chromatographic (g.c.) inlet system with the 'plug', removing the trap vent, and resealing the trap inlet with a removable compression fitting cap.

Desorber #2 (Figure 100) The gas chromatographic oven was cooled to below 30 °C with the oven door open. The needle was inserted into the liquid inlet system on the gas chromatograph. The trap was then inserted into the desorber and locked into place. The trap backflush flow fitting was locked into the trap exit flow fitting. The trap was then backflushed with nitrogen at 20 ml min^{-1} for 3 min at 150 °C. After 3 min the needle was removed from the liquid inlet system, the oven lid closed, and the oven rapidly heated to the normal or initial operating temperature. Gas chromatographic analyses were performed under

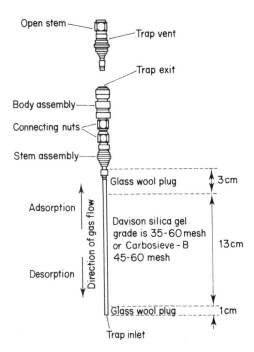

Figure 102 Trap. Reprinted with permission from Bellar et al.[30] Copyright (1976) American Chemical Society

these conditions. After sample transfer, the trap was removed from the desorber and sealed for future use.

Data obtained by Bellar et al.[30] showed that a quantitative recovery of vinyl chloride is obtained on silica gel and carbosorb B with purge volumes of 150–400 ml at 20 ml min^{-1}.

To determine the effect of sample collection and storage on the accuracy of the method, a 1 litre sample of river water contained in a 1 litre separatory funnel was dosed with vinyl chloride at 20 μg l^{-1}. This mixture was then used to fill several 50 ml glass-stoppered bottles. Care was taken so that no air passed through the sample as the bottles were filled. The bottles were over-filled, and part of the sample was displaced with the ground-glass stopper so that no headspace was trapped in the bottle. The bottles were then stored under ambient conditions. Seven of the samples, having no head-space, were randomly selected and analysed over a period of 93 h. The data show that the recoveries were constant over the period of study (Figure 103). The average recovery was 15.1 ± 0.4 μg l^{-1}. The initial 25% loss is attributed to the head-space above the dosed sample while it was contained in the separatory funnel. Losses due to head-space or exposure to the atmosphere are further illustrated below.

The time zero sample from the above experiment, now containing 5 ml of head-space, was reanalysed at 15 min and again at four additional times over a period of 300 min (Figure 104). Each time 5 ml of sample was withdrawn

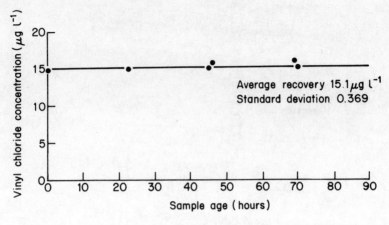

Figure 103 Recovery of vinyl chloride from dosed Ohio River water stored in glass-stoppered bottles with zero head-space at ambient temperature. Reprinted with permission from Bellar et al.[30] Copyright (1976) American Chemical Society

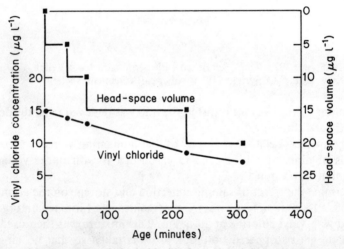

Figure 104 Recovery of vinyl chloride from dosed Ohio River water stored with variable head-space at ambient temperature. Reprinted with permission from Bellar et al.[30] Copyright (1976) American Chemical Society

leaving an additional 5 ml of head-space. Care was taken not to agitate the sample during the storage period. The results show that as the head-space increases, the recovery of vinyl chloride decreases. The total loss over the time period was about 50% or about 10% h^{-1}. These observations indicate the extreme care that is essential when dealing with the analysis of volatile organics in water samples.

The loss of vinyl chloride from water in an open narrow neck container at ambient temperature was observed by dosing 50 ml of tap water in a 50 ml

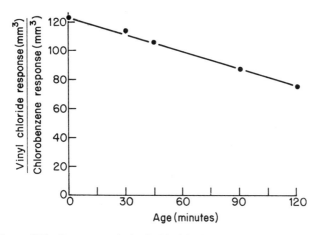

Figure 105 Recovery of vinyl chloride from dosed tap water stored unstoppered at ambient temperature. Reprinted with permission from Bellar et al.[30] Copyright (1976) American Chemical Society

volumetric flask with 10 mg l^{-1} of vinyl chloride and 20 mg l^{-1} chlorobenzene. Chlorobenzene is relatively non-volatile and was used as an internal standard. These analyses were done by direct aqueous injection gas chromatography, not by the purge and trap technique. The recovery of vinyl chloride relative to the chlorobenzene is shown in Figure 105. The loss of vinyl chloride was linear throughout the time period with a total loss of 35% or about 17% h^{-1}. The recovery of chlorobenzene was constant throughout the study. To test the procedure over a wide concentration range, a standard curve was prepared by injecting known amounts of a 10 ng μl^{-1} vinyl chloride in acetone solution into the purging device containing 5.0 ml of organic-free water. Each mixture was then purged and analysed. The response obtained by microcoulometric titration gas chromatography was linear over a concentration range of 4–40 μg l^{-1} (Figure 106). Based on data collected for similar halogenated hydrocarbons, the method may be useful up to 2500 μg l^{-1}.

Figure 107 represents a typical gas chromatogram obtained from chlorinated tap water which has been dosed with vinyl chloride. The chloroform, bromidichloromethane, and dibromochloromethane are common to chlorinated drinking waters and result from the chlorination process. Low levels of methylene chloride are often observed in samples analysed by this technique. These are attributed to method background. Figure 108 represents the chromatogram obtained from a sea water sample dosed with vinyl chloride and other organohalides. Using the Hall electrolytic conductivity detector, response was obtained for the acetone used to prepare the vinyl chloride standard solution.

Bellar et al.[30] used a computer to scan the data and construct a selected ion current profile consisting of peaks that produce a m/e 62 ion. Other compounds likely to be present in the water sample which produce m/e 62 ions are easily resolved using the gas chromatographic conditions recommended

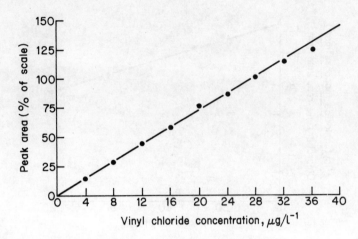

Figure 106 Response curve for vinyl chloride using microcoulometric detector. Reprinted with permission from Bellar et al.[30] Copyright (1976) American Chemical Society

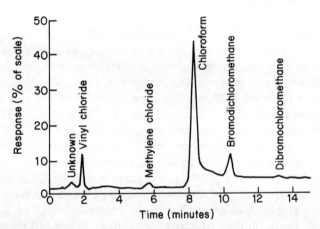

Figure 107 Microcoulometric gas chromatogram of organohalides recovered from tap water dosed with vinyl chloride (sensitivity 150 ohms). Reprinted with permission from Bellar et al.[30] Copyright (1976) American Chemical Society

by these workers, so, in this sense, the method is specific for vinyl chloride.

Fujii[31] has also combined mass spectrometry with gas chromatography for the direct determination of sub-ppb amounts of vinyl chloride in tap and river waters. The method is based on mass fragmentography followed by chromatography–mass spectrometry by simultaneously recording m/e 62 and 64.

Apparatus

Analyses were performed on a Finnigan 2300F gas chromatography–quadrupole mass spectrometer equipped with a multiple ion detector, by which mass

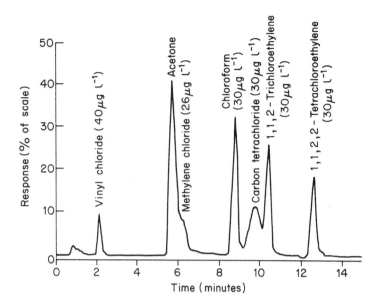

Figure 108 Electrolytic conductivity gas chromatogram of organohalides recovered from dosed sea water (full-scale response, 160 μmhos). Reprinted with permission from Bellar et al.[30] Copyright (1976) American Chemical Society

fragmentography can be carried out. The interface between the gas chromatograph and the mass spectrometer was an all-glass jet-type enrichment device. The mass spectrometer was set to unit resolution (10% valley between adjacent nominal masses). The resulting ion currents were recorded on a multichannel strip chart recorder. The instrument was operated in the electron impact mode. Other conditions held constant throughout the analysis were: helium carrier gas at a flow rate of 34 nl min^{-1}; temperature of the gas chromatograph injection port at 200 °C; pressure in the mass spectrometer of 1×10^{-5} torr; ionization voltage of 70 eV; emission current of 490 μA.

Column

A 600 cm × 2 mm i.d. metal coiled main column (5% SE-30 on 60–80 mesh Chromosorb W AW DMCS) in simple conjunction with a 40 cm × 6 mm i.d. metal straight precolumn (10% diglycerol on 60–80 mesh Chromosorb G NAW) was used. The diglycerol precolumn was used to strip water from the samples. This situation is made possible by the very long elution time of water through the diglycerol precolumn in comparison to the elution time of vinyl chloride. The diglycerol has the further property of repetitive use whereas a calcium sulphate precolumn used for the stripping of water should be replaced each time. The large volume injection made the big precolumn necessary. The long main column was chosen to meet the required separation. The column temperature was maintained isothermally at 55 °C.

A vacuum diverter[32] for venting was installed to prevent high volume effluent water in the sample (eluting after vinyl chloride) from entering the mass spectrometer. Venting of the water allows continuous mass spectrometer operation without the possibility of damage to the filament or electron multiplier.

Reagents and standardization

A 125 ml hypovial (Pierce) was filled with ethyl alcohol and sealed with a silicone septum by means of an aluminium crimp seal. Using a 1 ml Pressure Look syringe, 500 μl (the vapour density of vinyl chloride at 21 °C is 2.56 mg ml^{-1}) at atmospheric pressure was injected into this hypovial to give the vinyl standard.

Working solution over the range of 0.1–10 ppb (w/w) was prepared by adding the appropriate quantity of this standard to the organic-free water in a volumetric flask and diluting. This solution was immediately transferred to fill several 20 ml, glass-stoppered bottles. The bottles were over-filled, and part of the solution was displaced with the glass stopper so that no head-space was in the bottle. The working solution bottles, when stored at 5 °C, were maintained stably for at least 1 week. Milli-Q water was used throughout.

Procedure

Vinyl chloride analysis was performed as follows. The water sample was injected directly with a 100 μl or a 1000 μl Hamilton syringe. Positive identification of vinyl chloride in the water samples is supported not only from known retention times of the standards but also from the selectivity afforded by selected ion monitoring. The ions chosen to monitor vinyl chloride are $CHCl=CH_2+$. Thus, from the chlorine isotope clusters, a ratio of 3:1 would be expected at masses 62 and 64.

Figure 109 illustrates mass fragmentograms of specific ions of vinyl chloride spiked tap water samples, indicating the vinyl chloride peak shapes of a 1000 μl sample of water containing 0.1 ppb vinyl chloride and a 100 μl sample of water containing 1 ppb were found to be virtually identical. Some components in water other than vinyl chloride responded to m/e 62 and 64. As may be seen from Figure 109 the initial peak becomes higher as the injection volume is increased. It was identified as nitrogen and oxygen by interpretation of mass spectra.

The lower detection limit of the method is considered to be 0.1 ppb vinyl chloride. The extension of the detection limit was possible, because of the ability of the large diglycerol precolumn to handle water samples as large as 1000 μl with no detrimental effects to the separation performance and because of the property of the column system to elute vinyl chloride before the overload water peak (trace components are not easily determined when appearing on the tail of an overload water peak). Detection response (peak area) was linear over the chosen range of 0.1–10 ppb vinyl chloride with a 1000 μl injection. The

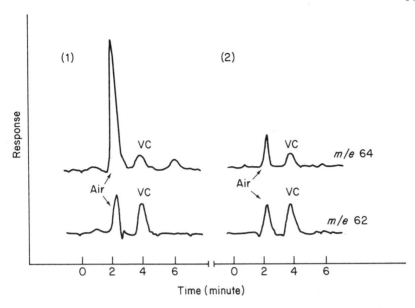

Figure 109 Mass fragmentograms of vinyl chloride VC spiked tap water samples. (1) a 1000 μl water sample containing 0.1 ppb monomer, (2) a 100 μl water sample containing 1 ppb monomer. Reprinted with permission from Fujii.[31] Copyright (1977) American Chemical Society

Table 72 Reproducibility of vinyl chloride analysis

Concentration (ppb)	Amount injected (μl)	Mean peak area* (mm^2)	Std. dev. (mm^2)	Rel. std. dev. (%)
0.1	1000	188	23	12
1	100	196	12	6
10	10	208	12	6

*Represents the mean of five consecutive injections of each standard.
Reprinted with permission from Fujii.[31] Copyright (1977) American Chemical Society.

reproducibility of the method is presented in Table 72. Although a little decrease in precision was observed at the large volume injections, the method has a good reproducibility.

Rivera et al.[33] have also described a direct mass spectrometric method for determining volatile chlorinated hydrocarbons, including vinyl chloride, in water. They give details of a highly sensitive technique for the determination of aliphatic chlorinated hydrocarbons, based on concentration by adsorption by stripping on a charcoal filter and quantitation by a mass spectrometric integrated ion-current procedure, with desorption from the charcoal inside a temperature programmed inlet probe. The apparatus used for the concentration step, an all-glass system, is shown in Figure 110.

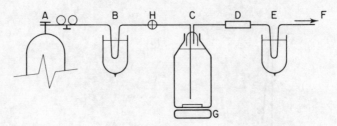

Figure 110 Stripping system. Reprinted with permission from Rivera et al.[33] Copyright (1977) Springer Verlag, NY

Glassware was cleaned, before use, with sulphuric acid–dichromate mixture, then thoroughly rinsed with distilled water and methanol, and dried.

Helium from a gas-pressure cylinder (A) (Figure 110) was purified on a liquid nitrogen cooled trap fitted with activated charcoal (B), and was then bubbled through the water contained in a 1 l Pyrex glass jar (C). Bubbling was maintained for one hour at a gas flow rate of 150 ml min^{-1}, monitored by means of a flow meter at the end of the system (F). Stirring of the water was accomplished by a magnetic stirrer (G).

The gas from the stripper, after drying over 300 mg of magnesium perchlorate (D) was conducted into a dry ice cooled adsorption tube (2 mm i.d.) (E), containing 5–10 mg of commercial activated charcoal (80–100 mesh), kept in position by a small plug of stainless steel grid. Water condensation in the cooled adsorption trap was prevented by a magnesium perchlorate drying filter. Besides water, this desiccating agent will retain only some oxygenated compounds (ketones, esters) and accordingly it would not interfere with the analysis.

The system was provided with a three-way valve (H) to run several samples simultaneously, or standard solutions for quantitative analyses. Similar gas flows were ensured through all traps by restricting the diameter of the tubing at the end of each system. When the bubbling was finished, adsorption tubes were disassembled, sealed with glass stoppers, and stored in a refrigerator until the analysis was to be performed.

The charcoal filter was quantitatively transferred to the previously cooled direct inlet probe of a MS-902S AEI high resolution mass spectrometer, that could be temperature programmed from -150 to $+350\,°C$. Vinyl chloride desorption took place in the range of -30 to $+100\,°C$. In this temperature interval and with the use of the peak matching technique (resolution 1000), the signals at m/e 62 and 64, corresponding to the molecular ions $C_2H_3{}^{35}Cl$ and $C_2H_3{}^{37}Cl$, were recorded.

Figure 111 shows the recording obtained from a sample of water (0.2 ppb of vinyl chloride), together with the signal displayed in a memory oscilloscope during desorption. Quantitative measurements of vinyl chloride in water were made by interpolating measured curve areas on a linear plot obtained by running standard water samples in the 0.05–10.0 ppb range. Standard deviation was 9%. Quantitation became difficult below 0.05 ppt since the method could not be easily

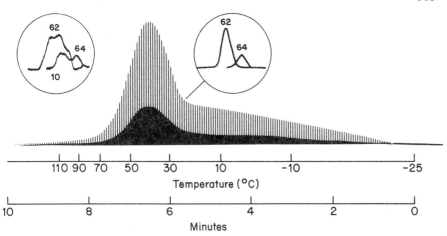

Figure 111 Desorption profile of the vinyl chloride in the mass spectrometer inlet probe. Reprinted with permission from Rivera et al.[33] Copyright (1977) Springer Verlag, NY

standardized at such a low level. The selectivity of the method can be observed in Figure 111. Complex signals, due to desorption of hydrocarbon (high resolution measurements showed the fragments to be C_5H_4 and C_5H_2) from the water or from the adsorbent, arose only at high temperatures, and did not interfere with the analyses.

This method was also applied to the determination of vinylidine chloride in water.

MISCELLANEOUS AROMATIC CHLORO COMPOUNDS

Polychlorinated styrenes

Kuehl et al.[34] have identified this substance in fish.

Hexachlorobenzene

Kapila and Aue[35] have studied the determination of hexachlorobenzene using electron capture gas chromatography. The technique involves routing of gas chromatographic peaks of chlorinated hydrocarbons to a built-in flow through reactor where they are partially dechlorinated in carrier gas doped with hydrogen, over a nickel catalyst, and the reaction products flow on to a second column for separation.

Residues of hexachlorobenzene in fish have been determined at the $\mu g\ kg^{-1}$ level using gas chromatography combined with mass spectrometry.[36]

Chlorinated alkylnaphthalenes

Bjorseth et al.[37] have determined these substances in bleach works effluents by a combination of glass capillary gas chromatography, gas-liquid chromatography–mass spectrometry, and neutron activation analysis.

Chlorophenols

Gas chromatography

Chau and Coburn[38] have described an electron capture gas chromatographic method for the determination of pentachlorophenol in natural waste waters. The phenol is extracted from the sample (1 litre) into benzene and subsequently from the latter into 0.1 M potassium carbonate. Addition of acetic anhydride to the aqueous solution gives the acetate derivative, which is extracted into hexane and analysed by gas chromatography with the use of a conventional polar column. The limit of detection is 10 pg l^{-1} of sample.

Rudling[40] has described a gas chromatographic procedure capable of determining one part of pentachlorophenol in 10^{10} parts of water.

Chriswell and Cheng[39] have showed that chlorophenols and alkylphenols in the ppb to ppm range in natural waters and treated drinking water can be determined by sorption on macroporous anion-exchange resin, elution with acetone, and measurement by gas chromatography. These workers describe techniques preventing phenol losses caused by chlorination, oxidation, and other reactions during their determination. Common inorganic ions and many organic substances cause no interference; neutral organics that are retained by the resin can be removed by a methanol wash. Phenols are effectively retained from water when samples of water containing low concentration of phenols are made basic and passed through a column containing an anion-exchange resin. The basis of the method, described below, is as follows. Phenols are taken up as phenolate ions by passing an alkaline water sample through a column of A-26 anion-exchange resin in the hydroxyl form. Any neutral organic compounds retained by the resin are removed by washing with alkaline methanol. Phenolate ions continue to be held by the resin during this washing step and are then converted to the molecular form by washing the column with aqueous hydrochloric acid. The phenols are subsequently eluted from the column with acetone–water. The hydrochloric acid and acetone–water effluents are each extracted with methylene chloride. The organic phases are concentrated by evaporation and the phenols are separated by gas chromatography.

Method

Reagents and apparatus The A-26 resin obtained from the Rohm and Haas Chemical Company was first screened to remove any resin beads smaller than

60 mesh, which restrict flow through adsorption columns, and then subjected to a thorough cleaning procedure to remove organic impurities left by the manufacturing process. To do this the resin is placed in a sintered glass filter attached to a suction flask, and with the vacuum adjusted so that solvents will flow slowly through the resin, the resin is washed in sequence with 2 M sodium hydroxide, purified distilled water, 4 M hydrochloric acid, purified distilled water, and acetone. This sequence of washings is repeated until no colour is apparent in the final acetone wash. The resin is then extracted with acetone for 24 hours in a Soxhlet extractor. Following this cleaning, the resin requires only routine generations to maintain its effectiveness for at least a hundred phenol determinations.

A ½ in × 6 in glass chromatographic column with a 1000 ml reservoir is prepared by first placing a glass wool plug in the bottom to retain the resin and then pouring a water slurry of the cleaned resin into the column until the resin bed is within about one-half inch of the top of the column. The column is placed in the hydroxy form by passing approximately 20 ml of 0.1 M sodium hydroxide solution through the resin. Excess sodium hydroxide is washed from the resin bed with 50 ml of purified distilled water.

Distilled water is further purified to remove trace organic contaminants by passing it through a column packed with XAD-2 resin and activated charcoal. Acetone is distilled to remove any high boiling impurities before use. All other chemicals are of reagent quality.

Solutions of phenols in acetone–methylene chloride are concentrated by evaporation in specially designed flasks with an attached Snyder column.[41]

A Hewlett-Packard Model 5711A gas chromatograph equipped with dual flame ionization detectors and a 6 foot stainless steel column, ⅛ in o.d. (5% OV-17 on Chromosorb W AW DSCS 80–100 mesh), and a Hewlett-Packard Model 5750B equipped with dual flame ionization detectors and an 18 in × ¼ in o.d. stainless steel column packed with Tenax GC 60–80 mesh, were used for the determination of the phenols. A DuPont Model 21-490 l combination gas chromatograph–mass spectrometer was used to identify certain chloro-substituted phenols.

Procedure

A 500 ml sample of water is used. If phenol standards are added to water containing chlorine, 15–25 mg of hydroxylamine hydrochloride are added and allowed to stir for at least 5 min before proceeding. To the sample is added 15–25 mg of sodium hydrosulphite and the pH adjusted to between 12.0 and 12.5 with 2 M sodium hydroxide. If a precipitate forms, it is coagulated by allowing the sample to sit for about 15 min. The supernatant liquid is decanted through the medium porosity 150 ml sintered glass filter attached to a suction flask. The precipitate is washed into the filter with a minimum amount of water, then washed thoroughly with approximately 50 ml of distilled water. The filtered water and washings are poured into the reservoir of the adsorption column and allowed to flow through the resin columns at a rate of 10–15 ml min^{-1}. When the liquid level reaches the top of the resin bed, the column is washed with 25 ml of basic methanol (2 ml of 2 M sodium hydroxide in 23 ml of methanol) and 25 ml of distilled water. A 125 ml separatory funnel is placed under the column and the column eluted with 25 ml of 4 M hydrochloric acid, then with 25 ml of distilled water. The solution is extracted in the separatory funnel with 25 ml of methylene chloride. The phases are allowed to separate and become clear, then the lower methylene chloride layer is drained into a second 125 ml separatory funnel containing the methylene chloride. The funnel is shaken, the phases allowed to separate and clear, then the lower organic layer is drained separatory funnel containing the methylene chloride. The funnel shaken, the phases allowed to separate and clear, then the lower organic layer is drained into an evaporating flask. A small boiling chip is added to the flask, a Snyder column is attached, and the solvent evaporated over a steam bath until the volume is reduced to approximately 0.5 ml. The evaporating flask is removed from the steam bath and the outside immediately sprayed with acetone to condense the vapours inside. The volume is adjusted to exactly 1.0 ml with acetone, 2 μl of the acetone solution are injected into the gas chromtograph, with the temperature programmed from 115 to 230 °C at a rate of 16 °C min^{-1} and the chromatograph is held at 230 °C for 4 min (OV-17 column). The phenols are identified by comparing their retention times with standards. Their concentration is determined by comparing either their peak heights or areas with a previously prepared calibration curve.

If the acetone solvent peak interferes with some of the individual phenol peaks, the acetone may be removed by adding approximately 7 ml of pentane to the evaporated (1.0 ml) acetone extract. By re-evaporating and diluting to exactly 1.0 ml with pentane, the acetone is volatilized by azeotropic distillation. Injection of a pentane solution permits operation of the gas chromatograph at significantly lower attenuation settings without interference from the tailing edge of the solvent peak.

Some very hard water samples contain sufficient bicarbonate to form a copious precipitate of calcium or magnesium bicarbonate when the sample is made basic with sodium hydroxide, which restricts flow through the column. The carbonate precipitate can be effectively removed by filtration through sintered glass. Copper sulphate, commonly used as a preservative for phenol containing samples did not interfere at the 1000 ppm level.

Table 73 gives recoveries obtained by this procedure for potable waters spiked with various phenols (see Figure 112).

A number of neutral organic compounds (aliphatic, alcohols, 2-phenoxy-ethanol, methyl cellosolve, naphthalene, C_4–C_8 alphatic acids) are retained by A-26 resin. However, none of these compounds affected recovery of phenols. They are eluted from a resin by basic methanol prior to elution of phenols and thus do not interfere with the gas chromatographic determination of phenols. Carboxylic acids also cause no interference with the recovery of phenols using the procedure but can interfere with the gas chromatographic determination

Table 73 Recovery of phenols added to tap water

Compound	Concentration (ppb)	Recovery (%)*	Concentration (ppb)	Recovery (%)†
Phenol	500	93	25	95
o-Cresol	300	94	15	90
p-Cresol	800	96	40	80
p-Chlorophenol	900	100	45	95
4-Chloro-3-methyl-phenol	800	100	40	95
2,4,6-Trichlorophenol	1100	102	55	95
Pentachlorophenol	1700	89	85	80
3,5-Dimethylphenol	700	95	35	90
2-Naphthol	500	95	...	...

*Using calibration curve: average of 9 analyses except for 2-naphthol.
†Using a single standard, a single analysis is reported to the closest 5%. The results in column five of the table show that the method is applicable for analysis of phenols at concentration levels as low as 15–85 ppb.
Reprinted with permission from Chriswell and Cheng.[39] Copyright (1975) American Chemical Society

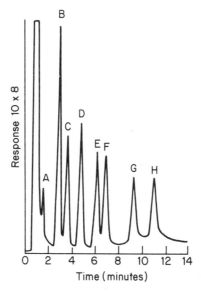

Figure 112 Separation of a standard mixture of phenols on a Tenax GC column. Peak order: (A) solvent impurity; (B) phenol; (C) o-Cresol; (D) 3,5-dimethylphenol; (E) 4-chloro-3-methylphenol; (F) 2,4,6-trichlorophenol; (G) 2-naphthol; and (H) pentachlorophenol. Separations were obtained on a ¼ in × 18 in S.S. Tenax GC column. Temperature held at 190 °C for 1 min, then programmed at 10 °C min^{-1} to 270 °C and held at 270 °C for 4 min. Reprinted with permission from Chriswell and Cheng.[39] Copyright (1975) American Chemical Society

of phenols on OV-17 columns by overlapping the phenol and cresol peaks. When a Tenax-GC column is used, acids elute well before phenols.

Chriswell and Cheng[39] observed that low recoveries of phenols in chlorinated drinking water coincided with enhanced recoveries of chlorinated phenols. Addition of low concentrations of chloramine T to very dilute aqueous solutions of phenols caused similar results. Letting 3,5-dimethylphenol solutions prepared in chlorinated tap water stand for a few minutes resulted in 40% loss. New gas chromatographic peaks from extracts from this solution were positively identified by mass spectrometry as being the di- and trichlorodimethylphenols. Chlorination reactions during the determination of phenols can be prevented by addition of hydroxylamine hydrochloride. On real samples, however, it must be recognized that chlorination reactions may have occurred before the sample was taken. The recovery of pentachlorophenol was affected by the amount of reductants used to prevent oxidation and chlorination. The more reductant added, the lower its recovery. If the amounts of reductants specified in the recommended procedure are used, pentachlorophenol losses will be negligible as will losses due to chlorination and oxidation of other phenols.

Farrington and Mundy[42] used gas chromatography to determine trace amounts of chlorophenols in water.

Morgade et al.[43] have described an electron capture gas chromatographic method for the determination of polyhalogenated phenols in chlorinated and unchlorinated drinking water. The particular phenols studied were: 2,4-dichlorophenol (2,4-DCP); 2,3,5-trichlorophenol (2,3,5-TCP); 2,4,5-trichlorophenol (2,4,5-TCP); 2,4,6-trichlorophenol (2,4,6-TCP); 2,5-dichloro-4-bromophenol (2,5-DC-4-BP); 2,3,4,5-tetrachlorophenol (2,3,4,5-TTCP); 2,3,4,6-tetrachlorophenol (2,3,4,6-TTCP); and pentachlorophenol (PCP).

Method

Apparatus A gas chromatograph equipped with a tritium electron capture detector (ECD) and a 1.8 m × 4 mm i.d. glass column packed with 4% SE-30 + 6% QF-1 on 80–100 mesh Supelcoport® was used. A nitrogen carrier gas flow rate of 30 ml min^{-1} and inlet, column, and detector temperatures of 215, 165, and 210 °C, respectively were used.

Solvents and reagents All solvents were Nanograde® quality. The deionized water was extracted twice with benzene. N-Ethyl-N'-nitro-N-nitrosoguanidine was used to prepare the ethylating reagent.[44] The silica gel (Woelm, activity grade I) was prepared according to Shafik et al.[45]

Procedure

The pH of 100 ml of water contained in a separatory funnel is adjusted to 1.5 by adding 1 ml of concentrated hydrochloric acid. After the addition of 10 ml of methylene dichloride the mixture is shaken vigorously and allowed to separate.

Meanwhile, a small pad of glass wool is placed at the bottom of a 2.2 cm × 30 cm Pyrex glass filtering column, and sodium sulphate is added to a depth of 2 cm. The tip of the column is positioned over a 25 ml concentrator tube, and the lower (organic) layer is drained through the sodium sulphate. The aqueous layer is re-extracted with 10 ml of methylene dichloride. The second organic layer is also passed through the sodium sulphate and into the same concentrator tube. The filtering column is rinsed with 5 ml of methylene dichloride, which is similarly collected in the tube. Ten drops of 'keeper' solution (1% USP paraffin oil in hexane), which suppresses sample loss during the concentrating steps, is added to the tube and the sample is concentrated to 0.2 ml by using a nitrogen stream evaporator. Two ml of hexane are then added, and the sample is reconcentrated to 0.5 ml. Another 2 ml of hexane are added, and sufficient freshly prepared diazoethane solution is added to give a persistent yellow coloration to the mixture. After allowing the solution to remain at room temperature for 20 min nitrogen is bubbled through the solution and it is concentrated to 0.5 ml. Finally the sample is passed through a silica-gel column using the method of Shafik et al.[45] The first fraction is concentrated to 1 ml and analysed by gas chromatography using an electron capture detector.

Recovery studies are performed by adding known amounts of reference standards to 100 ml of deionized water devoid of the halogenated phenols for which the analyses are being made, then running the fortified sample through the entire procedure.

Using this method Morgade et al.[43] found levels of pentachlorophenol between less than 0.14 ppm and 340 ppt in Florida drinking water and 0.14 ppm and 110 ppt in well water. No other chlorophenols were found. Table 74 lists recoveries, detector sensitivity, and limits of detection for various polyhalophenols obtained using this method.

Ashiya et al.[46] combine the phenols in the water with bromine, and measure the concentration of phenols and chlorophenols by electron capture detection gas chromatography, obtaining their concentrations separately to the extent of 1 ppb.

Renberg[47] has used an ion-exchange technique for the determination of chlorophenols and phenoxyacetic acid herbicides in water, soil, and fish tissues.

Table 74 Percentage recovery, detector sensitivity, and limits of detectability of halogenated phenols in water

Compound	Recovery (%)	Detector sensitivity (pg)	Limit of detectability (ppb)
2,4-DCP	52	400	8.1
2,3,5-TCP	79	22	0.4
2,4,5-TCP	87	29	0.6
2,4,6-TCP	73	8	0.2
2,5,-DC-4-BP	94	14	0.3
2,3,4,5-TTCP	98	9	0.9
PCP	64	12	0.3

Reprinted with permission from Morgade et al.[43] Copyright (1980) Springer Verlag, NY.

The water samples or soil extracts are mixed with Sephadex QAE A-25 anion exchanger and the adsorbed materials are then eluted with a suitable solvent. The chlorinated phenols are converted into their methyl ethers and the chlorinated phenoxy acids into their methyl or 2-chloroethyl esters for gas chromatography.

Method

Reagents and equipment Hexane, acetone, diethylether (anhydrous), methanol, potassium chloride (0.2 M), sodium hydroxide (0.2 M and 0.1 M), hydrochloric acid (1.0 M and 0.2 M), sodium sulphate (anhydrous), 2-chloroethanol, diazomethane in diethylether solution, Sephadex QAE, A-25 anion exchanger. All reagents should be tested in a blank procedure.

An acidic buffer was prepared by mixing equal volumes of the hydrochloric acid (0.2 M) and potassium chloride solutions. The ion exchanger is swollen in distilled water at least 2 hours before use.

A 10 cm × 1 cm (i.d.) column was used when analysing water samples. For the batch procedures, 15 ml test tubes with screw caps and Teflon packings were used.

A Varian 1400 gas chromatograph, equipped with a tritium electron capture detector was used. The 160 cm × 0.18 cm (i.d.) glass columns were filled with either OV-17 (1%), SF 96 (1%), or a mechanically prepared mixture of 67 parts QF 1 (8%) and 33 parts of SF 96 (4%) on acid-washed, silanized Chromosorb W 100–120 mesh. The column temperature and the corresponding relative retention times are shown in Table 75. Injector and detector temperatures were held about 10 °C above the column temperature.

Table 75 Levels of substances in fortified samples and corresponding recoveries

	Fish tissue, 5 grams		Water, 1000 ml		Soil 1 gram		
	Level (ppm)	Rec (%)	Level (ppb)	Rec (%)	Level (ppm)	Humus rec (%)	Clay rec (%)
Fungicides							
2,4,6-Trichlorophenol	0.10	74	0.50	>97	0.50	>97	>97
2,3,4,6-Tetra-chlorophenol	0.10	90	0.50	>97	0.50	>97	>97
Pentachlorophenol	0.30	92	1.5	>97	1.5	>97	>97
Bactericides							
2,Hydroxy-2',4,4'-trichlorodiphenyl ether	1.0	79	5	94	...	...	...
Hexachlorophen	1.0	83	5	92	...	...	...
Herbicides							
2,4,-D-acid	3.2	70	16	>97	16	70	74
2,4,5-T-acid	1.6	82	8.0	>97	8.0	86	84

Reprinted with permission from Renberg.[47] Copyright (1974) American Chemical Society.

Analysis of water samples A suspension of the ion exchanger is transferred into the column (see equipment above). The ion exchanger is allowed to settle and the upper end of the column is connected to a separatory funnel containing the water sample. The bed volume should be about 3–4 ml. If an increase of the outflow is desired, the lower end of the column is connected to a water suction pump (without a pump the outflow usually varies between 0.6 and 0.8 ml min^{-1}). Two different methods can now be used for eluting the substances from the ion exchanger.

Alternative 1: 3–4 ml of distilled water are added to the column, shaken carefully and the suspension decanted into a 15 ml test tube. The column is washed with another portion of water to ensure that all ion exchanger is transferred to the test tube. After centrifugation, the water is removed and the procedure continued as above.

Alternative 2: The column is eluted with 10 ml of acidified methanol (1 gram sulphuric acid/50 ml methanol). To one part of the eluate (in a test tube) is added an equal volume of benzene and four parts of the hydrochloric acid solution (1.0 M). The test tube is shaken and, after centrifugation, the benzene phase is transferred into a graduated test tube and the substances converted into suitable derivatives as described under the preparation of derivatives below.

Analysis of soil samples The sample is shaken with 0.2 M sodium hydroxide (4 ml g^{-1} soil) in a test tube, for 30 min. After centrifugation the liquid is removed and re-extracted with a new portion of sodium hydroxide solution. The volume of the combined alkaline extracts is estimated. The extract (2 ml) and 8 ml of water are shaken for 10 min with the ion exchanger (3 ml bed volume). After centrifugation, the liquid is discarded and the ion exchanger rinsed with 5 ml of distilled water. The water is discarded and the procedure continued as described above.

Analysis of fish tissues The sample (5 g) is homogenized in a mixture of hexane and acetone (5 + 10 ml) by means of an insertable homogenizer in a dropping funnel with a glass filter disk. The liquid is dropped into a separatory funnel containing 1.0 M hydrochloric acid (5 ml); nitrogen pressure is used if necessary. The sample is homogenized once more with a mixture of hexane and diethylether (10 + 5 ml) and the mixture collected in the separatory funnel. The funnel is shaken and the upper phase transferred into a centrifuge tube, then the water phase is re-extracted twice with a mixture of diethylether and hexane (2 + 2 ml) and the extracts transferred to the centrifuge tube. Sodium sulphate (100–300 mg) is added to bind any water present. After centrifugation, the extract is transferred into a weighed flask, the sodium sulphate is rinsed with diethylether (2 ml) and the solvents gently evaporated on a water bath in a nitrogen stream. The flask is reweighed and the fat content calculated; the fat is then dissolved in benzene (about 1 ml per 25 mg fat is used).

The suspension of the ion exchanger is transferred into a 15 ml test tube and after centrifugation the water is discarded. The bed volume should be about

3 ml. To the test tube is added 3 ml of the benzene solution and 3 ml of sodium hydroxide solution (0.1 M). The test tube is shaken carefully for 5 min, and the liquid phases removed after centrifugation (the benzene phase can be used for analysis of non-acidic pesticides). To the ion exchanger is added 3 ml of distilled water, the test tube is shaken for about 30 seconds, and, after centrifugation, the water is discarded.

To the test tube is added 3 ml of benzene (containing a suitable internal standard) and 3 ml of the acidic buffer. This is shaken carefully for about 5 min and the benzene phase transferred into a graduated test tube. The substances are converted into suitable derivatives as described under the preparation of derivatives below.

Preparation of Derivatives Methyl ethers of the phenols and methyl esters of the phenoxyacetic acids Diazomethane in ether solution is prepared from, e.g., N-methyl-N- nitroso-p-tolouene-sulphonamide. The diazomethane solution is added to the benzene extract until the extract becomes pale yellow. After about 1 hour, the solution is evaporated to the original volume. The extract is injected into the gas chromatograph and the result compared with a standard treated the same way.

2-Chloroethyl esters of the phenoxyacetic acids A 15 ml test tube containing 2 ml of the benzene extract, 1 ml of 2-chloroethanol and 100 µl sulphuric acid are shaken for 1 min, and allowed to stand in a water bath at 50 °C for 30 min. Then 10 ml of distilled water are added, shaken, and centrifuged. The benzene phase is injected into the gas chromatograph and the result compared with a standard treated in the same way.

Renberg[47] recommends the use of γ-BHC (lindane), DDE, or DDT as a gas chromatographic internal standard. The relative retention times of the derivatives corresponding to these internal standards are shown in Table 76. The detection limits for the different substances in 10 grams of an organic tissue or soil are 0.1–1 ppb and for one litre of a water sample 0.0001–0.1 ppb.

Gas chromatography–mass spectrometry

This technique has been used to measure the occurrence of pentachlorophenol and hexachlorophene in water and sewage[48] and chlorophenols in spent bleach liquor trade effluents from sulphate plants.[49,50]

Ingram et al.[51] used a mass spectrometric isotope dilution technique to determine approximately 0.2 µg pentachlorophenol in water with a relative standard deviation of 8%.

Hoben et al.[52] has described a gas chromatographic technique for determining 0.1 ppb pentachlorophenol in water and fish tissues. Confirmation of the identity of the chlorophenol was provided by gas chromatography–mass spectrometry. In this method the pentachlorophenol is extracted from the acidified sample with n-hexane and then re-extracted into a borax solution. It

Table 76 Retention times relative to γ-BHC, p,p-DDE, and p,p-DDT

	OV-17	QF + SF 96
γ-BHC	*7 min at 160 °C*	*9 min at 150 °C*
Methyl ethers of		
2,4,6-trichlorophenol	0.096	0.14
2,3,4,6-tetrachlorophenol	0.25	0.32
pentachlorophenol	0.64	0.70
γ-BHC	*1 min at 200 °C*	*3 min at 180 °C*
p,p-DDE	2.93	3.02
Methyl esters of		
2,4,-dichlorophenoxyacetic acid	1.24	0.94
2,4,5-trichlorophenoxyacetic acid	1.83	1.29
2-Chloroethyl esters of		
2,4-dichlorophenoxyacetic acid	1.70	2.65
2,4,5-trichlorophenoxyacetic acid	2.95	4.00
Methyl ether of		
2,hydroxy-2',4,4'-trichlorodiphenyl ether	2.93	3.27
p,p-DDT	SF 96	
	2 min at 200 °C	
Dimethyl ether of		
hexachlorophene	3.42	

Reprinted with permission from Renberg.[47] Copyright (1974) American Chemical Society.

is then acetylated by extracting with n-hexane containing acetic acid anhydride and pyridine. The resulting pentachlorophenyl acetate is analysed by gas chromatography using an electron capture detector.

Method

Equipment Gas chromatograph: aerograph 600 equipped with an electron capture detector. Column: 1 m × 1.5 mm i.d. glass column packed with 5% QF-1 on Varaport
30, 100–120 mesh
Injection temperature: 160 °C
Column temperature: 150 °C
Detector temperature: 170 °C
Carrier gas: nitrogen 25 ml min^{-1}.

The above conditions give a retention time of 3 min for pentachlorophenyl acetate.
Glassware: Graduated conical centrifuge tubes 12 ml Pasteur capillary pipettes 2 ml.

Reagents All chemicals used should be analytical grade.
(1) Sulphuric acid, 6 M: 340 ml concentrated sulphuric acid are mixed with water, cooled, and diluted with water to 1000 ml.

(2) *n*-Hexane.
(3) Extraction mixture: 50 ml of isopropanol and 250 ml of *n*-hexane are mixed together.
(4) Borax, 0.1 M: 38 g of $Na_2B_4O_7 \cdot 10H_2O$ are dissolved in water and diluted to 100 ml.
(5) Acetylation reagent: 2 ml of pyridine and 0.8 ml of acetic anhydride are mixed in a dry 5 ml injection vial, and the vial is capped with an injection septa and stored cold. The reagent must be prepared fresh every day.
(6) Pentachlorophenol.

Pentachlorophenyl acetate standard   About 500 mg of pure pentachlorophenol are dissolved in 5 ml of pyridine, 1 g of acetic acid anhydride is added and the mixture is kept at about 50 °C for half an hour. The mixture is cooled and 5 ml of water added. The mixture is then extracted with 10 ml of ethyl ether. The aqueous phase is discarded and the ethyl ether phase washed twice with 2 ml of borax. The ethyl ether is evaporated. The pentachlorophenyl acetate melted at 152–153 °C and gave only one peak in the gas chromatogram.

The retention times of the standards and of some other acetates of halogenated phenols are given in Table 77. The retention time for aldrin and dieldrin are included for comparison.

Table 77   Relative retention times of halogenated phenyl acetates and some related compounds (pentachlorophenyl acetate = 3 min)

Compound	Relative retention time
Pentachlorophenyl acetate	1.00
2,3,4,6-Tetrachlorophenyl acetate	0.50
2,4,6-Trichlorophenyl acetate	0.22
2,4-Dichlorophenyl acetate	0.18
2,4,6-Tribromophenyl acetate	0.71
Aldrin	0.96
Lindane	0.74

Reprinted with permission from Hoben *et al.*[52] Copyright (1976) Springer Verlag, NY.

Procedure (water samples)

A 100 ml volume of the sample is acidified with 2 ml of concentrated sulphuric acid in a separatory funnel and extracted with 10.0 ml of *n*-hexane for 1 min. The hexane layer is transferred to a graduated centrifuge tube with conical bottom and the volume read. The hexane phase is extracted twice with 2 ml of borax for 1 min and centrifuged if necessary. The aqueous phase is transferred with a Pasteur capillary pipette to a 5 ml test tube. The extraction is repeated with 2 ml of borax and 0.50 ml of *n*-hexane and 40 µl of the acetylation reagent are added to the combined aqueous extracts and shaken for 1 min. The hexane phase is analysed in the gas chromatograph.

The extract procedure gave 91–98% recovery of pentachlorophenol from water samples and an 83–91% recovery from fish. The method was used successfully to determine pentachlorophenol at the 3–9 µg l^{-1} level in water and the 0.15–3 mg kg^{-1} level in fish. Confirmation of identity of the chlorophenol was established by a combined gas chromatographic–mass spectrometric analysis. To establish whether the peaks obtained corresponded to pure pentachlorophenol or not, samples of authentic pentachlorophenol acetate were subjected to a similar analysis. The sizes of these samples were chosen so as to provide peaks in the gas chromatogram having approximately the same heights as those obtained from the water and fish samples. The intensities of the molecular peaks in the mass spectrograms were about the same, indicating that the material analysed as pentachlorophenol in the gas chromatograph could be considered as pure pentachlorophenol.

Gas chromatography coupled with mass spectrometry has also been used by Hoben et al.[52] for the determination of pentachlorophenol.

High performance liquid chromatography

Column chromatographic techniques have been described[53] for the determination of pentachlorophenol and other chlorophenols, but these methods are not suitable for the determination of low levels. The best method for determining pentachlorophenol is conversion into the methyl ether followed by analysis using gas chromatography with an electron capture detector, or gas chromatography coupled with mass spectrometry.[52] Both of these methods require an extensive amount of pretreatment and highly trained personnel for the operation of the equipment.

Ervin and McGinnis[53] attempted to overcome this problem by developing a high performance liquid chromatographic method for determining in water low concentrations of pentachlorophenol and chlorinated impurities that occur in the technical grade material such as 2,3,4,6-tetrachlorophenol, mono-, di-, and trichlorophenols, octa-, hepta-, and hexachlorodibenzo-*p*-dioxins, and a variety of other polychlorinated aromatic compounds.

The method involves chloroform extraction of acidified waste water samples and rotary evaporation without heat. After redissolving in chloroform the samples were analysed directly by high performance liquid chromatography on a microparticulate silica gel column. A number of solvent combinations are possible and 98:2 cyclohexane–acetic acid (92:2 v/v) is preferred. The minimum detectable concentration is 1 ppm (without sample concentration) and the coefficient of variation is 1–2%. The type of separation achieved with a microparticulate silica gel column is shown in Figure 113. The first peak as determined by gas chromatographic–mass spectrometric analysis, consisted of a complex mixture of polychlorinated compounds, including octa-, hepta-, and hexachlorodibenzo-*p*-dioxins as well as a mixture of products including 2,4,6-trichlorophenol. The third peak was mainly 2,3,4,6-tetrachlorophenol and the fourth peak was pentachlorphenol.

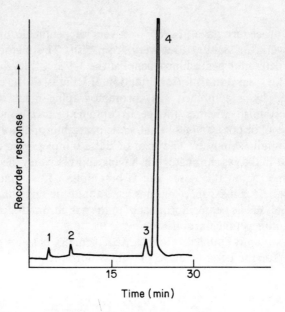

Figure 113 Separation of technical pentachlorophenol using cyclohexane–acetic acid (98:2) as the eluting solvent. Peaks: 1, mixture of dioxins and other polychlorinated products; 2, mixture of chlorinated phenols including trichlorophenol; 3, tetrachlorophenol; 4, pentachlorophenol. Reprinted with permission from Ervin and McGinnis.[53] Copyright (1980) Elsevier Science Publishers

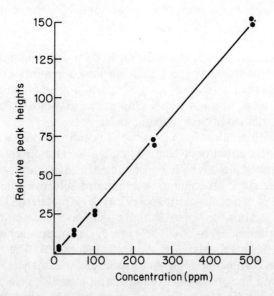

Figure 114 Relationship between peak areas and concentration of pentachlorophenol. Reprinted with permission from Ervin and McGinnis.[53] Copyright (1980) Elsevier Science Publishers

A linear calibration curve was obtained when peak heights were plotted versus concentrations of pentachlorophenol using a fixed wavelength detector (254 nm) (Figure 114). The minimum concentration of pentachlorophenol which can be detected without concentrating the sample is 1.0 ppm. For repeated injections of the same sample, the precision (coefficient of variation) was 1-2% using a standard of pentachlorophenol.

The results in Table 78 show the good agreement obtained between results obtained by this method and by a gas chromatography–mass spectrometric method.[52]

Table 78   Typical analysis of pentachlorophenol from wood treating plants

No.	Source of sample*	Concentration (h.p.l.c.) values (ppm)	Concentration (m.s.) values (ppm)
(1)	Treated waste water	72.2	75.0
(2)	Untreated waste water	17.5	32
(3)	Untreated waste water	13.5	18.0
(4)	Treated water	9.9	9.7
(5)	Treated waste water	9.1	6.4
(6)	Untreated waste water	8.2	17
(7)	Treated waste water	4.7	4.4
(8)	Treated waste water	4.6	4.0
(9)	Treated waste water	4.4	3.9
(10)	Treated waste water	4.3	4.1
(11)	Treated waste water	0.33	0.29
(12)	Treated waste water	0.29	0.16
(13)	Incoming water	0.25	0.14
(14)	Treated waste water	0.027	0.20
(15)	Incoming water	0.111	>0.010

*Samples analysed within 24 hours of collection.
Reprinted with permission from Ervin and McGinnis.[53] Copyright (1980) Elsevier Science Publishers.

A variety of other solvent combinations were found which could also be used to separate the components in technical grade phentachlorophenol. All of these solvent systems could be used to separate pentachlorophenol, the major component of technical pentachlorophenol. The relative retention times of pentachlorophenol using these solvent systems are given in Table 79.

Thin-layer chromatography

Various workers[54-57] have used this technique for determining pentachlorophenol and other chlorophenols in water samples. Thielemann and Luther[57] separated the chlorophenols on Kieselgel G plates with benzene as solvent. The spots were located by spraying with diazotized sulphanilic acid solution or with a mixture (1:1) of 15% ferric chloride solution and 1% $K_3Fe(CN)_6$ solution.

Table 79 Pentachlorophenol retention times with different eluting solvents

Solvents	Proportions (%)	Retention time (min)
Cyclohexane–acetic acid	98:2	24:5
Cyclohexane–methylene chloride*	57:43	9:7
Hexane–methylene chloride*	90:10	17:7
Hexane–methylene chloride*	80:20	10:0
Hexane–methylene chloride*	60:40	8:3
Hexane–acetic acid	98:2	22:7
Hexane–acetic acid	96.7:3.3	19:2
Hexane–acetic acid	95:5	12:1
Hexane–chloroform	95:5	18:6
Cyclohexane–chloroform	80:20	15:0

*Contains 1% acetic acid.
Reprinted with permission from Ervin and McGinnis.[53] Copyright (1980) Elsevier Science Publishers.

Miscellaneous techniques

Pentachlorophenol has been determined in water in amounts down to 0.3 ppm by differential pulse polargraphy.[58] Boyle et al.[59] have studied the degradation of pentachlorophenol in a simulated lentic environment.

Chlorophenols in materials other than water

Stark[60] has described a gas chromatograph method for the determination of pentachlorophenol as the trimethyl silylether in amounts down to 0.5 $\mu g \, g^{-1}$ in fish and soil and 0.01 $\mu g \, l^{-1}$ in water. Rudling[61] determined pentachlorophenol in fish and water by an electron capture gas chromatographic method. In this method a sample of fish tissue (1 g) in water is transferred to a centrifuge tube with 5 ml water, 6 M sulphuric acid (1 ml) is added and the tube is left for 10 min. Isopropyl alcohol–hexane (1:5) (5 ml) is added and the tube shaken for 1 min. The tube is centrifuged and cooled in ethanol–solid carbon dioxide. The organic layer is decanted and extracted with 0.1 M $Na_2B_4O_7$ (2 ml). If necessary, the extract is centrifuged, the aqueous phase separated and the extraction repeated with 0.1 M $Na_2B_4O_7$. Hexane (0.5 ml) and fresh acetylation reagent (pyridine (2 ml) plus acetic anhydride (0.8 ml) stored in the cold) (40 μl) are added to the combined aqueous extracts and shaken for 1 min. The hexane phase is analysed by gas chromatography on a glass column (1 m × 1.5 mm) packed with 5% of QF-1 on Varaport 30 (100–120 mesh) operated at 150 °C with nitrogen as carrier gas (25 ml min^{-1}).

Baird[62] utilized gas chromatography in his study of the biodegradability of chlorinated phenols in sewage sludge.

HALOFORMS

Numerous articles have appeared in the technical and popular press referring to the presence of carcinogens in water and questioning the safety of chlorine when used as a disinfectant of water supplies.[65-67] The impetus for this relatively sudden development was the release of a study by the Environmental Protection Agency (EPA) of the presence of potentially toxic organic substances in the New Orleans Water Supply[63] and an epidemiological study of the implications of cancer-causing substances in the Mississippi river water by the Environmental Defence Fund.[64] This latter document is of some importance in that it suggests a relationship between the above-average incidence of cancer in certain communities and the Mississippi derived water supply. Subsequent to the passage of the Safe Drinking Act (PL93-523) in 1974 the United States Government Environmental Protection Agency sponsored two finished water studies, the National Organics Reconnaissance Survey (NORS) and the National Organics Monitoring Survey (NOMS). These studies confirmed the widespread occurrence of chloroform as well as the other trihalomethanes, bromodichloromethane ($CHBrCl_2$), chlorodibromomethane ($CHBr_2Cl$), and bromoform ($CHBr_3$). Since this work, there have been numerous other reports of the presence of chlorinated and brominated haloforms in river and potable water including studies on East Texas water supplies,[68] Vienna Water supplies,[69] Ontario water supplies,[70] Jowe City Reservoir supplies,[71] San Fransisco Bay water supplies,[72] Ohio river water,[73] West Netherlands water supplies,[74] New York water supplies,[77], North American surface and drinking water supplies.[76,78-81] A survey of 80 cities by the United States Environmental Protection Agency[75] indicated that the occurrence of trihalomethanes in drinking water was widespread and results directly from the chlorination process. The concentration of the four trihalomethanes surveyed (chloroform, bromodichloromethane, chlorodibromomethane, and bromoform) were found to range from high parts per billion (ppb) to sub-ppb with chloroform present in the highest concentration and bromoform in the lowest. Average levels of chloroform, bromodichloromethane, and chlorodibromomethane were found respectively to be 21, 6 and $1.2\,\mu g\,l^{-1}$. European surface water and drinking water supplies[82,83] and Japanese water supplies[85,85] have also been studied.

Brett and Calverley[86] have reported on a 1-year survey conducted in the USA of trihalomethane concentration changes in a water distribution system. The authors were concerned with trihalomethane levels at the point of consumption, not at the treatment plant. A wide range of temperatures, weather conditions, and raw water conditions at all stages in a distribution system were sampled. It was found that where water contained a chlorine residual, a trihalomethane precursor, or both, trihalomethane levels do increase significantly within a distribution system and that higher temperatures and longer detention times exacerbated the problem, reinforcing the view that sampling for potable standards must relate to the samples taken at the consumer taps.

Smith et al.[87] have conducted a study of temporal variations of trihalomethane concentrations in drinking water from a system drawing primarily from surface water. The 24-hour variations in concentration of total trihalomethanes, chloroform, dichlorobromomethane, dibromochloromethane, and bromoform in drinking water were assessed in a dynamic system using a solvent extraction electron capture gas chromatographic procedure. Measurements were made at 4-hourly intervals over a 1-week period. Spectral analysis and replicate harmonic regression were used to evaluate the temporal patterns in fluctuation and trihalomethanes. Data show that within the 24-hour and day-to-day samples, variations were found to be present from which the authors conclude that the time of sampling is an important factor for the reliable monitoring of trihalomethane concentrations in drinking water.

Of the 66 substances reported by the US Environment Protection Agency as being present in the New Orleans Water Supply chlorinated hydrocarbons have aroused most concern. Chloroform in particular has received much attention since it is present at much higher levels than the other substances reported (maximum 133 mg l^{-1} at that time). However, a wide variety of other chlorinated and brominated aliphatics have been identified in drinking water supplies. Research by Rook[82,83] and others has shown that chloroform and related chlorohydrocarbons are formed during the disinfection of water with chlorine. Rook[82,83] suggested that the chlorine reacts with organic substances of natural origin such as humic acids to produce these chlorinated impurities. Work carried out by the Environmental Protection Agency[88] has confirmed the production of halohydrocarbons during chlorination but has tentatively postulated ethanol as a substance for reaction with chlorine. Since both precursors could be of natural origin the implication is that chlorinated substances in the water supply are not necessarily due to industrial or domestic pollution. Although the major concern at the present time involves the role of chlorination during water treatment, the widespread chlorination of sewage effluents prior to discharge to water courses in the US is receiving increased attention.

As well as chlorohydrocarbons a wide range of bromohydrocarbons and bromochlorohydrocarbons such as bromoform, bromodichloromethane, and dimonochloromethane have been identified in water supplies, Luong et al.[89] have recently drawn attention to the role of bromide in water supplies in the formation of brominated trihalomethanes with reference to its interaction during the chlorination process with humic material present in natural waters. Changes during water treatment were examined and subsequent trihalomethane formation, on chlorination of those waters evaluated. For bromide levels in lowland waters of up to 120 μg l^{-1} brominated trihalomethanes were shown to account for up to 54% of the total trihalomethanes formed on treatment.

Chloroform apparently results from reaction between hypochlorite and any of several types of organic precursors in the chlorinated raw water. The brominated and mixed brominated/chlorinated trihalomethanes are presumed to be formed from the reaction of hypobromite and hypochlorite with the same precursors; the hypobromite is formed from the oxidation of bromide by hypochlorite. If iodide salts are also present in the water being chlorinated, an analogous reaction with hypochlorite results in trihalomethanes containing iodine. Bunn et al.[90] detected all ten possible mixed and single halogen-containing trihalomethanes of chlorine, bromine, and iodine when salts containing fluoride, bromide, and iodide were added to a river water sample before chlorination (no fluorinated trihalomethanes were detected). Glaze et al.[91] identified seven trihalomethanes containing chlorine, bromine, and iodine in a tap water sample. Also found in drinking water but not necessarily formed during the chlorination process, are compounds such as methylene chloride, dichlorobenzene, hexachlorobutadiene,[92] tetrachloroethylene,[93] trichloroethylene,[92,93] carbon tetrachloride,[92,94] and 1,2-dichloroethane.[93,94]

The formation of volatile organohalogen compounds by the chlorination of waters containing organic contaminants[95] has received wide attention.[96-98] Investigation carried out by Tardiff and Dunzer[98] confirmed the presence of six halogenated compounds (viz., chloroform, bromoform, bromodichloromethane, dibromochloromethane, tetrachloromethane, and 1,2-dichloroethane) in drinking waters with concentrations varying up to 100 μg l^{-1}.

As a consequence of the concern regarding possible adverse effects of minute quantities of trihalomethanes in drinking water the US Environment Protection Agency[99,100] in 1978 drew up an amendment to US National Interim Primary Drinking Water Regulations designed to protect the public from exposure to undesirable amounts of trihalomethanes (including chloroform) in drinking water. A maximum contaminant level of 0.10 mg l^{-1} has been prescribed for total trihalomethanes, applicable initially to community water supplies for populations in excess of 75,000 where disinfection is practised as part of the treatment process. Monitoring arrangements for total trihalomethane levels in supplies to populations between 10,000 and 75,000 are also envisaged in the regulations, which also require the water utilities to submit design proposals for a granular activated carbon adsorption system designed to achieve the recommended maximum contaminant level for trihalomethanes in their treated water. This proposal became mandatory on November 20, 1979, and water systems will be required to monitor for total trihalomethanes starting during the next 3 years, the exact commencement date depending on their community size.

Quimby et al.[101] used gas chromatography with microwave emission detection to identify the aqueous chlorination and bromination products of fulvic and humic acids in water.

Schnoor et al.[102] have determined the apparent molecular weight range of trihalomethane precursor compounds in the Iowa river and a reservoir near Iowa City. Soluble organics were size fractionated by gel permeation chromatography and the fractions were chlorinated and analysed for trihalomethane yields by electron capture gas chromatography. Of the trihalomethanes formed, 75% were derived from organics of molecular weight less than 3000 and 20% from those of molecular weight less than 1000.

The determination of very low concentrations of haloforms in water presents a challenging problem to the analytical chemist. Due to the low concentrations involved direct examination by gas chromatography has received little attention and most of the published work is concerned with incorporating into the method a suitable preconcentration technique performed prior to gas chromatography. Such techniques include solvent extraction, gas sparging, head-space analysis, and adsorption on resins. These various approaches are discussed below.

Haloforms in drinking water

Direct injection gas chromatography

There is very little published work on this and most of it was carried out prior to 1976. Direct injection methods analysis, although less time-consuming than methods involving a preconcentration step are relatively insensitive and lack selectivity between volatile and non-volatile components and are complicated by reactions that may occur at heated injection parts. Lower detection limits of 100 μg l^{-1} have been claimed with an electron capture detector and 5 μg l^{-1} with a gas chromatograph–mass spectrometer system.[103,104]

Kissinger and Fritz[105] used the technique to determine chloroform, bromodichloromethane, and bromoform in chlorinated drinking water. They observed that the concentrations of these substances increased with the time of storage of the sample.

Nicholson and Meresz[106] directly injected the drinking water sample into a gas chromatograph equipped with a scandium tritide electron capture detector. Glass columns (4 ft × ¼ in) packed with Chromosorb 101 (60–80 mesh) were used for the analysis.

The conditions for operating the instruments were as follows.

Varian 2400	Injector temperature:	230 °C
	Detector temperature:	230 °C
	Oven temperature:	130 °C
	Nitrogen flow rate:	50 ml min^{-1}
Varian 2100	Injector temperature:	220 °C
	Detector temperature:	225 °C
	Oven temperature:	150 °C
	Nitrogen flow rate:	60 ml min^{-1}

The detection limits for the trihalogenated compounds listed in Table 80 are all below the 10 μg l^{-1} level. Figure 115 illustrates typical calibration curves for $CHCl_3$, $CHBrCl_2$, and $CHBr_2Cl$. The percentage standard deviation of five injections was ±1%. All measurements were made using peak heights; thus, the detection limits do not directly reflect the response of the electron capture detector for these compounds.

The detection limits for dichloropropane is only 60 μg l^{-1}, and the dichlorobenzenes cannot be detected below 500 μg l^{-1}. This analysis is not, therefore, suitable for detecting trace levels of some of the dichlorinated hydrocarbons in water.

Direct injection techniques have been reported on by various other workers.[107-114]

Solvent extraction–gas chromatography

Liquid–liquid extraction techniques have been shown to be both convenient and accurate by several workers but have been criticized for lack of selectivity

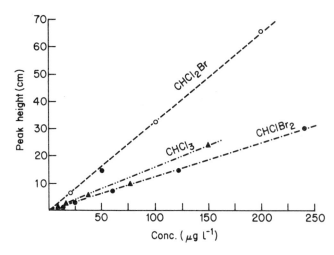

Figure 115 Typical calibration curves: o CHCl$_2$Br, △ CHCl$_3$, ●CHCl$_2$Br. Reprinted with permission from Nicolson and Meresz.[106] Copyright (1975) Springer Verlag, NY

Table 80 Detection limits of some halogenated compounds

Compound	Retention time (min)	Column temperature (°C)	Detection limit (μg l^{-1})
Bromochloromethane	2.8	130	1
Chloroform	3.0	130	3
1,1-Dichloroethane	4.0	130	90
1,2-Dichloroethane	4.0	130	150
Carbon tetrachloride	4.0	130	3
Dibromomethane	6.0	130	0.6
1,2-Dichloropropane	6.8	130	60
Tetrachloroethylene	13.0	130	8.0
Chlorodibromomethane	15.4	130	5
Bromoform	4.5	150	2
Tetrachloroethane	5.2	150	7
p-Dichlorobenzene	9.8	150	500
o-Dichlorobenzene	11.4	150	500

Reprinted with permission from Nicolson and Meresz.[106] Copyright (1975) Springer Verlag, NY.

between volatile and non-volatile components, erratic efficiency, and susceptibility to solvent interference. The earliest references date to 1973–76.[115–117,119–121]

In a relatively early method published by the Water Research Centre (UK)[118] 100 ml of water sample was extracted with 10 ml of petroleum ether (redistilled from potassium hydroxide) in a 250 ml separating funnel, by shaking for 5 minutes. Of this 10 ml extract 5 μl were injected directly into the gas chromatograph. If required, the sample size can be reduced with a corresponding reduction in extractant volume. A blank determination of the solvent to be used was carried out for each batch of analyses.

The gas chromatographic conditions were as follows.

Pye '104 series' gas chromatograph, model 134.
Detector: Electron Capture, 10 mCi ^{63}Ni as radioactive source.
Mode: Pulse mode 150 μS.
Detector oven temperature: 350 °C
Column oven temperature: 100 °C (isothermal).
Column: Glass 2.7 m × 6 mm o.d., 3 mm i.d., 5% FFAP on Chromosorb HPW (100–120 mesh) (other columns of similar polarity, e.g. QF-1, would probably be satisfactory).
Carrier gas: N_2 40 ml min^{-1} (oxygen-free nitrogen).

Retention data for four halomethanes are given below.

Component	Time (from injection)
Chloroform	1 min 30 s
Dichlorobromomethane	2 min 12 s
Dibromochloromethane	3 min 48 s
Bromoform	7 min 12 s

The linear ranges (ng on column) for these substances were 0.1 (dichlorobromomethane), dibromochloromethane, and bromoform) and 1 (chloroform). The detection limits achievable were 0.5 μg l^{-1} (chloroform and bromoform) and 0.02 μg l^{-1} (dichlorobromomethane and dibromochloromethane). Recoveries of all four substances at the 0.05–1.5 μg l^{-1} level were between 83 and 97%. Using this method, levels of haloforms found before chlorination were less than 0.1 μg l^{-1} (dichlorobromomethane, dibromochloromethane, bromoform) and after chlorination 6–80 μg l^{-1} (chloroform), 6–76 μg l^{-1} (dichlorobromomethane), 40 μg l^{-1} (dibromochloromethane) and less than 0.1–8 μg l^{-1} (bromoform). Figure 116 shows a typical chromatogram of a chlorinated water, indicating the four trihalomethanes quantified and also the presence of dichloroiodomethane, which was found on a few occasions. Figure 117 indicates the retention times of some other trihalomethanes sometimes formed on chlorination.

Since some of the halogenated methanes are volatile, a loss in concentration between sampling and analysis could be expected. However, there is increasing evidence that an increase can occur in practice. The increase is due, at least in part, to excess chlorine reacting with organic material (humic acids) in the sample. In practice, therefore, when taking samples of chlorinated treated water, it is important to bear in mind that the reaction producing the halogenated methanes could be incomplete. The addition of ascorbic acid (7 ppm) has been reported to stabilize the levels of halogenated methanes.[122] Fielding et al.[118]

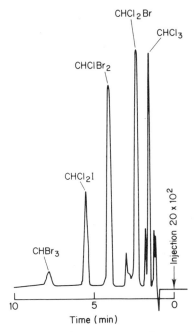

Figure 116 Gas chromatogram of treated water extract. Reprinted with permission from Fielding et al.[118] Copyright (1977) Water Research Centre

showed that addition of ascorbic acid (about 7 mg l^{-1}) at the time of sampling reduced the further formation of halogenated methanes but did not eliminate it. Figure 118 indicates the variation in the levels of halogenated methanes in samples with and without addition of ascorbic acid and with storage of samples in the dark at about 6 °C. Curve A shows the typical behaviour of a sample without attempted reduction of residual chlorine, while curve C shows the effect of adding ascorbic acid. Curve B is the hypothetical curve which could be anticipated if all the chlorine residual were destroyed and no reaction intermediates existed. Curves A and B appear to vary somewhat from sample to sample. This overall effect must be taken into consideration before analysis. Either samples are extracted more or less on site and analysed as soon as possible or standardization of the time (and storage) conditions), between sampling and analysis is strictly observed. Ideally the reaction should be terminated immediately after sampling, but in the relatively few experiments tried this has not been achieved.

Von Rensburg et al.[123] have described a rapid sensitive semi-automatic method for determination of chlorinated organic compounds formed during chlorination of water, and of volatile halogen compounds appearing in surface and other waters. The method is based on a rapid liquid-liquid extraction process using hexane, and gas chromatographic analysis using an electron capture detector. The linear detection range was from 0.3 to 300 µg l^{-1} with a lower limit of detection of 0.1 µg l^{-1}.

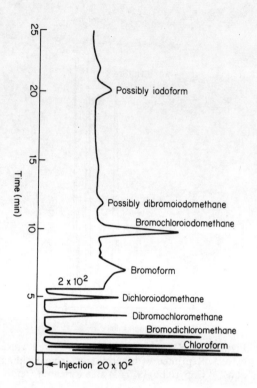

Figure 117 Gas chromatogram of a chlorinated water extract. Reprinted with permission from Fielding et al.[118] Copyright (1977) Water Research Centre

Method

The extraction apparatus (Figure 119) consisted of a 10 or 20 ml all glass syringe (1) with a Luer slip fitting, a connector (2) prepared from PTFE shrink tubing (4.5 mm diameter), a stopper (3), transfer tube (4), and a glass 'needle' (5). Stoppers, transfer tubes, and 'needles' were all prepared from borosilicate glass tubing, 4 mm o.d. × 2 mm i.d. Additional apparatus used was one 500 μl syringe (S.G.E. 500 A-Fn), 2 ml serum vials (6) with aluminium crimp caps (8) and glass vial inserts (7) (200 μl volume). All glassware was thoroughly cleaned in chromic acid and baked at 300 °C for 1 hour.

The glass 'needle' was fitted to the syringe by means of the PTFE connector and a 10 ml water sample drawn from a sample contained (with minimum air space and Teflon-lined screw cap) into the syringe. A small amount of air (1–2 ml) was also drawn into the syringe to facilitate shaking. With the syringe tip uppermost, the glass 'needle' was removed and 250 μl of hexane (BDH extra pure grade) was added through the tip of the syringe. The tip was stoppered and the mixture shaken for 3 minutes. The syringe was then left in the tip uppermost position for about 5 minutes to allow the two phases to separate.

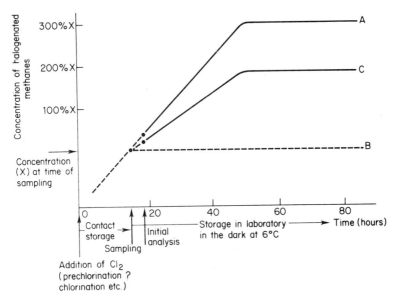

Figure 118 Variation with time of concentrations of halogenated methanes in treated water. Curve A: no attempted reduction of residual chlorine. Curve B: hypothetical effect of complete termination of reaction at time of sampling. Curve C: effect of addition of ascorbic acid to terminate reaction. Reprinted with permission from Fielding et al.[118] Copyright (1977) Water Research Centre

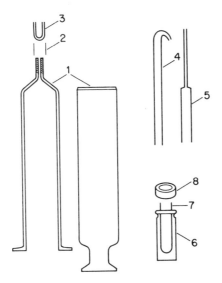

Figure 119 Extraction apparatus. Reprinted with permission from Von Rensberg et al.[123] Copyright (1978) Pergamon Press

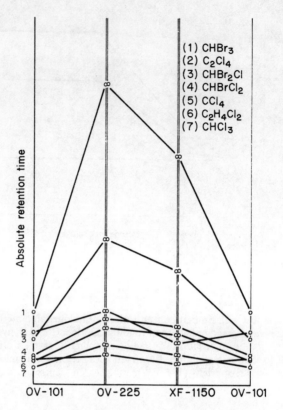

Figure 120 Comparative graphs of absolute retention times of seven volatile organohalogen compounds chromatographed on various stationary phases. Reprinted with permission from Von Rensberg et al.[123] Copyright (1978) Pergamon Press

On separation the transfer tube was fitted to the syringe tip and the syringe clamped above a labjack, with the labjack supporting the syringe plunger. The solvent was carried over to the glass insert on raising the syringe plunger by means of the labjack. The insert was then immediately placed into the serum vial, sealed with aluminium foil followed by a silicone rubber disc, and finally by an aluminium crimp cap.

Gas chromatographic conditions

Apparatus Hewlett-Packard 5710: a gas chromatograph mounted with a Hewlett-Packard 5671 A autosampler.

Detector: A ^{63}Ni electron capture detector.

Column: a 3 m × 2 mm i.d. stainless steel or 3 m × 6.2 o.d. × 2.5 mm i.d. glass column packed with 15% OV 225 on Chromosorb W-AW, 80–100 mesh.

Column and inlet temperature 90 °C, detector temperature 250 °C; carrier gas argon–methane mixture (95%:5%) at a flow rate of 20 ml min^{-1}.

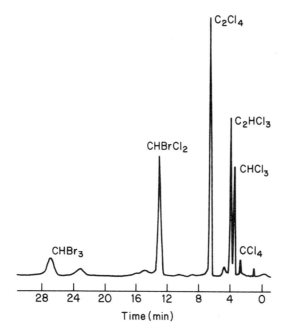

Figure 121 A chromatogram of a typical water sample using OV-225 as stationary phase. Reprinted with permission from Von Rensberg et al.[123] Copyright (1978) Pergamon Press

The autosampler was controlled by a Hewlett-Packard 3352 B data system and programmed to wash the syringe 10 times with solvent, then to rinse the syringe with the sample a further 10 times and finally to inject a 5 μl aliquot on to the column.

Examination of Figure 120 indicates that OV-225 or XF 1150 were the preferred mobile phases for separation of the seven haloforms examined. Separation of these compounds at 90 °C was obtained within 28 min (Figure 121). Components were identified by retention time data. Quantitative results were obtained from peak areas, calibrated by extraction of a standard water solution. The standard solution was prepared from a solution containing 0.5 ml each of: carbon tetrachloride, chloroform, tetrachloroethylene, bromo-dichloromethane, dibromochloromethane, and bromoform in 50 ml dichloro-ethane. This mixture was diluted 2×10^6 times with glass distilled water by using microsyringes. The concentration by mass of each compound was calculated by means of their respective densities.

Recoveries obtained from three different solvent–water ratios are listed in Table 81 which shows that recoveries of 69% (chloroform) and higher were obtained by using a 50:1 water–hexane ratio.

The detection range of the seven compounds examined is shown in Table 82. The linearity of the system extended over approximately three orders of magnitude. The linear range was adapted by varying the solvent–water ratio.

Table 81 Percentage recovery of various compounds by liquid-liquid extraction using various solvent–water ratios

Compound	% Recovery Solvent–water ratio		
	1:100	1:50	1:25
Tetrachloromethane	100	111	103
Chloroform	60	68	78
Dichloroethane	104	100	100
Tetrachloroethylene	102	100	100
Bromodichloromethane	64	80	90
Chlorodibromomethane	79	86	97
Bromoform	100	100	100

Reprinted with permission from Von Rensberg et al.[123] Copyright (1978) Pergamon Press.

Table 82 Detection range of seven compounds analysed

Compound	Lower detection* limit (ng)	System determination lower limit ($\mu g\ l^{-1}$)	Linear range ($\mu g\ l^{-1}$)
Tetrachloromethane	0.015	0.10	0.16–300
Chloroform	0.02	0.10	0.15–450
Dichloroethane	1.40	10.0	12.5–30.000
Tetrachloroethylene	0.02	0.13	0.16–375
Bromodichloromethane	0.025	0.15	0.20–325
Chlorodibromomethane	0.030	0.20	0.25–600
Bromoform	0.025	0.22	0.30–700

*Concentration that produces a signal three times greater than that produced by hexane background.
Reprinted with permission from Von Rensberg et al.[123] Copyright (1978) Pergamon Press.

Table 83 Results from an experimental water reclamation plant

Sample	Concentration (ng l^{-1})						
	CCl_4	$CHCl_3$	$C_2H_4Cl_2$	$CHBrCl_2$	C_2Cl_4	$CHBr_2Cl$	$CHBr_3$
Sand filter effluent							
1	122	749	—	63	52	83	211
2	19	563	—	—	88	84	154
After chlorination							
1	685	33623	—	82	21926	14547	1723
2	308	40647	—	—	24527	15068	3711
Final water*							
1	152	4068	—	—	192	407	175
2	20	572	—	—	63	165	135

*Obtained by passing the water through an activated carbon filter.
Reprinted with permission from Von Rensberg et al.[123] Copyright (1978) Pergamon Press.

Between the lower linear limit and the detection limit, quantitative results were erratic possibly due to adsorption or dissociation effects or both. The reproducibility of the system was found to vary from 1.59 to 6.83% relative standard deviation for the various compounds with an average of 4.17 over the concentration range 2–5 μg l^{-1}. The reproducibility of the autosampler, determined over 10 samples of equivalent concentrations was found to be 4.02% relative standard deviation.

Results of haloform determinations obtained on samples from a water reclamation plant are shown in Table 83.

Dressman et al.[124] have compared the precision and accuracy of three liquid-liquid extraction methods for the determination of trihalomethanes in water. These workers compared liquid-liquid extraction procedures developed by Henderson et al.,[125] Mieure,[126] and Richard and Junk.[127] All of these methods involve the extraction of a small volume of water with an even smaller volume of organic solvent, followed by a gas chromatographic analysis of the extract with an electron capture detector. They then applied the preferred method of these three[126] and the purge and trap method described by Symons[128] and Bellar and Lichtenberg[129] to a range of municipal water supplies.

Analyses were performed using a Model 5730A Hewlett-Packard gas chromatograph with a linear, ^{63}Ni electron capture detector. A 4 mm i.d. × 183 cm glass column was packed with 10% squalane on 80–100 mesh Chromosorb W-AW. The chromatographic conditions included an injection port temperature of 100 °C, an oven temperature of 67 °C (isothermal), and a 95% argon and 5% methane carrier gas at a flow rate of 20 ml min^{-1}. The typical retention times in this study were 1.84 min for chloroform, 4.24 min for dichlorobromomethane, 9.60 min for dibromochloromethane, and 20.72 min for bromoform.

Details of the three liquid-liquid extraction methods are outlined below.

Mieure method[126]

One millilitre of methylcyclohexane is pipetted into a 9 ml screw-cap vial, 5 ml of the water sample is pipetted into the vial and the vial sealed with an open-top cap and a PTFE lined septum. The sample is shaken vigorously for 1 min and allowed to stand 30 s before analysing the organic phase by piercing the septum with a syringe and withdrawing a measured amount for injection into the gas chromatograph. The results are compared to a gas chromatographic standard containing a similar concentration of the trihalomethane mixture in methylcyclohexane.

Richard and Junk method[127]

A 10 ml volume of water sample is pipetted into a 15 ml glass-stoppered, calibrated centrifuge tube and 1 ml of pentane or iso-octane (2,2,4-trimethylpentane) is pipetted into the tube. The sample is shaken for 30 s and allowed to stand for at least 1 min before analysing the organic phase by removing the stopper and withdrawing a measured amount for injection into the gas chromatograph. The results are compared to a gas chromatographic standard containing a similar concentration of the trihalomethane mixture in pentane or iso-octane.

Henderson et al. method[125]

A 120 ml serum bottle is filled to overflowing with the water sample so as to exclude all head-space. The bottle is capped with a PTFE-lined septum and sealed by crimping an aluminium retainer over the bottle top. To the sample is added 5 ml of pentane by piercing the septum with the needles of two 10 ml syringes, one syringe containing 5 ml pentane and the other empty. As the pentane is added to the sample, the displaced water is collected in the empty syringe. The sample is shaken for 15 min at 500 rpm on a gyratory platform shaker and then allowed to stand at least 1 min before analysing the organic phase by piercing the septum with a syringe and withdrawing a measured amount for injection into the gas chromatograph. The results are compared to a gas chromatographic standard containing a similar concentration of the trihalomethane mixture in pentane.

Details of the purge and trap method are outlined below.[128,129] The sample (at 25 °C) is transferred to a valve-equipped 5 ml syringe by pouring it into the back of the barrel with the plunger removed until barrel is completely full. The plunger is replaced, the valve opened, and the plunger depressed to the 5 ml mark. Then the valve is closed and the syringe mounted on the needle of the purge assembly. The trap filled with porous polymer adsorbent is vented and mounted on the gas exit of the purge assembly. The purge gas flow is started, the valve of the syringe is opened, and the plunger depressed to expel the sample into the purge chamber; then the valve is closed. The purge is continued for 11 min at 20 ml min^{-1} for the 5 ml water sample. After the purge is complete, the vent is removed and disconnected from the trap (the water remaining in the purge chamber can be removed at this time with the needle and syringe used for introducing the sample). With the Tenax gas chromatographic column at ambient temperature, the trap is inserted into the modified heated (200 °C) injection port of the gas chromatograph, with the plug fitting removed. The desorbing gas service is immediately connected to the trap and desorption performed at a gas flow rate of 20 ml min^{-1} for 3 min at 220 °C. The trap is removed and the injection port plugged with the prepared fitting. The chromatograph oven is immediately heated to 95 °C and held at 95 °C for 8 min, then programmed at 4 °C min^{-1} up to 205 °C. The results are compared to aqueous standards, analysed by the entire purge–trap–desorb procedure.

Typical retention times using this procedure with a 2 mm × 183 cm Tenax column and a carrier flow of 20 ml min^{-1} were 15.9 min for chloroform, 21.20 min for dichlorobromomethane, 25.9 min for dibromochloromethane

Table 84 Richard and Junk method: percent recoveries* of halomethanes from 1 ml water extracted with 1 ml iso-octane

Dose (μg l^{-1})	CHCl$_3$	CHBrCl$_2$	CHBr$_2$Cl	CHBr$_3$
200	70.7 ± 1.0	77.7 ± 4.3	78.2 ± 1.0	88.2 ± 1.5
100	75.5 ± 1.6	79.2 ± 1.9	75.5 ± 0.3	79.7 ± 3.0
50	72.8 ± 6.0	70.1 ± 1.2	63.7 ± 1.8	66.4 ± 2.6
10	72.5 ± 1.6	76.1 ± 4.3	82.5 ± 1.9	95.9 ± 2.1
1	Solvent interference	70.6 ± 3.3	75.9 ± 2.1	77.2 ± 7.0
Average†	73.5 ± 4.5	75.6 ± 6.4	74.0 ± 7.7	77.7 ± 10.9

*Average of five replicates at each concentration, showing the relative standard deviation.
†Overall average of the 25 results obtained over the concentration range of 1-200 μg l^{-1}, showing the relative standard deviation.
Reprinted from *Journal AWWA*, Vol. 71, No. 7 (July 1979) by permission.
Copyright © 1979, The American Water Works Association.

Table 85 Richard and Junk method: percentage recoveries* of halomethanes from 10 ml water extracted with 1 ml pentane

Dose (μg l^{-1})	CHCl$_3$	CHBrCl$_2$	CHBr$_2$Cl	CHBr$_3$
200	92.5 ± 2.0	92.7 ± 2.2	91.4 ± 1.2	97.0 ± 1.0
100	73.4 ± 2.7	82.9 ± 4.1	95.6 ± 2.9	92.5 ± 2.8
50	84.9 ± 4.1	85.7 ± 5.1	91.8 ± 3.6	96.8 ± 1.5
10	72.5 ± 1.6	76.1 ± 4.3	82.5 ± 5.6	95.9 ± 2.5
1	84.6 ± 8.0	78.4 ± 7.8	76.9 ± 7.3	81.0 ± 1.7
Average†	80.3 ± 8.4	81.8 ± 7.2	87.5 ± 8.0	92.8 ± 6.0

*Average of five replicates at each concentration, showing the relative standard deviation.
†Overall average of the 25 results obtained over the concentration range of 1-200 μg l^{-1} showing the relative standard deviation.
Reprinted from *Journal AWWA*, Vol. 71, No. 7 (July 1979), by permission.
Copyright © 1979, The American Water Works Association.

Table 86 Mieure method: percentage recoveries* of halomethanes from 5 ml water extracted with 1 ml methycyclohexane

Dose (μg l^{-1})	CHCl$_3$	CHBrCl$_2$	CHBr$_2$Cl	CHBr$_3$
200	85.9 ± 3.9	93.4 ± 2.9	93.6 ± 3.8	94.2 ± 2.4
100	85.3 ± 5.2	86.2 ± 1.8	87.5 ± 5.1	91.9 ± 6.5
50	85.8 ± 6.0	89.4 ± 2.4	89.6 ± 0.7	90.1 ± 3.2
10	88.0 ± 1.2	85.3 ± 2.5	90.0 ± 1.9	90.9 ± 2.2
1	88.2 ± 0.9	83.5 ± 4.4	85.5 ± 3.8	93.5 ± 4.2
Average†	86.4 ± 3.7	87.9 ± 4.5	89.5 ± 4.2	92.2 ± 3.7

*Average of five replicates at each concentration, showing the relative standard deviation.
†Overall average of the 25 results obtained over the concentration range of 1-200 μg l^{-1} showing the relative standard deviation.
Reprinted from *Journal AWWA*, Vol. 71, No. 7 (July 1979), by permission.
Copyright © 1979, The American Water Works Association.

Table 87 Henderson method: percentage recoveries* of halomethanes from 115 ml water extracted with 5 ml pentane

Dose ($\mu g\, l^{-1}$)	$CHCl_3$	$CHBrCl_2$	$CHBr_2$	$CHBr_1$
200	63.4 ± 1.9	65.3 ± 1.4	76.5 ± 3.1	75.6 ± 1.1
100	75.3 ± 2.7	76.9 ± 1.5	84.4 ± 1.7	82.9 ± 4.6
50	77.7 ± 3.8	80.3 ± 3.4	83.9 ± 3.9	84.1 ± 1.3
10	71.3 ± 0.8	70.0 ± 1.1	74.2 ± 2.3	85.5 ± 1.3
1	72.3 ± 0.7	70.4 ± 0.7	71.0 ± 2.0	84.2 ± 1.7
Average†	72.0 ± 5.4	72.6 ± 5.6	78.0 ± 6.0	82.5 ± 4.3

*Average of five replicates at each concentration, showing the relative standard deviation.
†Overall average of the 25 results obtained over the concentration range of 1–200 $\mu g\, l^{-1}$ showing the relative standard deviation.
Note: specially prepared water low in organic content* was dosed with the halomethane mixture at concentrations of 200, 100, 50, 10, and 1 $\mu g\, l^{-1}$. Aliquots of the dosed water were then proportioned to the appropriate sample size. Blanks of all the solvents and water were analysed to ensure their purity. All of the high-purity solvents were further cleaned by redistillation and passage through a 2.5 cm i.d. glass column of activity −1 basic alumina.
Reprinted from *Journal AWWA*, Vol. 71, No. 7 (July 1979), by permission.
Copyright © 1979, The American Water Works Association.

Table 88 Minimum detectable concentrations of trihalomethanes

| Trihalomethane | Concentration ($\mu g\, l^{-1}$) | | | |
	Purge and trap*	Mieure†	Richard and Junk†	Henderson et al.
$CHCl_3$	0.1	0.2	0.1	0.1
$CHBrCl_2$	0.1	0.2	0.1	0.1
$CHBr_2Cl$	0.1	0.4	0.2	0.2
$CHBr_3$	0.2	0.8	0.5	0.5

*Electrolytic conductivity detector, halogen-specific mode.
†Linear ^{63}Ni electron capture detector.
Reprinted from *Journal AWWA*, Vol. 71, No. 7 (July 1979), by permission.
Copyright © 1979, The American Water Works Association.

and 30.4 min for bromoform. An electrolytic conductivity detector in the halogen-specific mode was used on the gas chromatograph.

Data obtained on the precision and accuracy of the liquid-liquid extraction methods are presented in Tables 84–87. Individual averages of the five results reported for each compound at each concentration and their accompanying relative standard deviation measure accuracy and precision under the least variable conditions of analysis; namely, replicate same-day analysis. The average reported at the bottom of each column and its accompanying relative standard deviation measure accuracy and precision under the least variable conditions of analysis; namely, replicate same-day analysis. The average reported at the bottom of each column and its accompanying relative standard deviation measures the accuracy and precision on a day-to-day basis. This average is particularly useful because it reflects the effect of a wide range of solute concentrations on the overall accuracy of the method.

For comparison the average percentage recovery of the trihalomethanes obtained by the purge and trap method[128,129] in the 1–200 mg l^{-1} range was 75–95%, much the same as in the liquid-liquid extraction methods.

The minimum detection limits of the trihalomethanes by the liquid-liquid and purge and trap methods are compared in Table 88.

The one feature that most clearly distinguishes one liquid-liquid extraction method from another is the extraction vessel. The Mieure[126] method, which provided the best overall precision, employs the small screw-cap vial as the extraction vessel. The use of this vial is believed to be directly responsible for the greater precision. Those factors that influence precision, such as variable loss of the sample, solvent extract, and head-space gas during the sample manipulation, are well controlled by the screw-cap vial because it remains completely sealed throughout the analysis. In fact, the vial may be conveniently sent to the field for sampling with both the extracting solvent and a chlorine-reducing agent to prevent trihalomethane formation after sampling—in which case, the seal need never be broken after sampling. No special tool is required to seal the vial.

In Table 89 is shown a comparison of results obtained by the Mieure[126] method and the purge and trap method.[128,129] The liquid-liquid extraction results are corrected to account for the recovery efficiencies presented in Table 86. The purge and trap results are not corrected because the standards are purged in the same manner as the sample, making arithmetic correction unnecessary.

As shown in Table 89, the precision of both liquid-liquid extraction and purge and trap methods is equally good when the five replicate field test samples are analysed simultaneously. However, the overall precision of the Mieure method, about ±4% for each compound (Table 86) suggests that the field test results obtained by liquid-liquid extraction may be more accurate than those obtained by purge and trap. By way of comparison, replicates analysed by purge and trap on a day-to-day basis show an overall precision of ±5–20% from the lower boiling chloroform to the higher-boiling bromoform.

The results obtained by the liquid-liquid extraction and purge and trap methods are comparable to the extent that they are of the same order of magnitude on the same sample. Except for several samples having concentrations below 5 μg l^{-1}, the results generally do not vary by more than ±15%.

In addition to these quantitative investigations Dressman et al.[124] also investigated the qualitative accuracy of liquid-liquid extraction and purge and trap methods. They conclude that a greater change of error exists in liquid-liquid extraction methods. This was because the electron capture detector used in a liquid-liquid analysis was specific for all electron capturing moieties, not just the halogens. Thus, in the absence of confirmatory data, the results from a purge and trap analysis are more reliable.

The US Environmental Protection Agency has published two methods for trihalomethane analysis:[130] methods 501.1[131] and 501.2.[132] Method 501.1 is a procedure for the analysis by the purge and trap technique; method 501.2 is a liquid-liquid extraction technique.

Table 89 Trihalomethanes detected in drinking water from seven cities by purge and trap and LLE (Mieure) Methods*

	CHCl$_3$		CHBrCl$_2$		CHBr$_2$Cl		CHBr$_3$	
City	Purge and trap	LLE	Purge and trap	LLE	Purge and trap	LLE	Purge and trap	LLE
1	9.5 ± 0.2	9.5 ± 0.8	0.9 ± 0.03	1.6 ± 0.1	<0.1†	0.3 ± 0.05	‡	‡
2	98.6 ± 3.1	89.1 ± 2.8	17.0 ± 0.9	15.2 ± 0.9	2.6 ± 0.9	2.0 ± 0.3	‡	‡
3	2.1 ± 0.1	3.3 ± 0.2	‡	0.2 ± 0.05	‡	‡	‡	‡
4	40.4 ± 1.8	34.7 ± 1.2	0.9 ± 0.1	0.8 ± 0.1	<0.1†	‡	‡	‡
5	37.0 ± 1.7	36.0 ± 0.9	1.8 ± 0.1	1.9 ± 0.1	<0.2†	‡	‡	‡
6	141.3 ± 1.7	128.0 ± 7.7	65.3 ± 0.8	74.7 ± 3.3	35.9 ± 0.3	36.8 ± 3.4	5.7 ± 0.3	5.3 ± 0.1
7	139.1 ± 6.4	131.3 ± 4.2	34.7 ± 1.4	36.3 ± 1.5	6.3 ± 0.5	6.4 ± 0.4	0.2‡	‡
8	46.2 ± 1.6	37.7 ± 0.8	24.8 ± 0.8	31.6 ± 2.9	16.8 ± 0.6	19.4 ± 1.2	0.5 ± 0.1	1.5 ± 0.6

*for cities 1–7 an average of five replicates for each analysis is reported: for city 8 three replicates.
†At least one sample of the replicate series was equal to or less than this concentration.
‡Not detected.
Reprinted from *Journal AWWA*, Vol. 71, No. 7 (July 1979) by permission.
Copyright © 1979, The American Water Works Association.

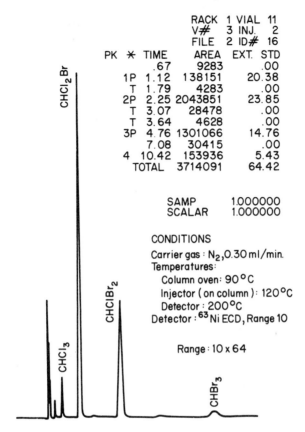

Figure 122 Gas chromatogram of trihalomethanes in finished drinking water. Reprinted with permission from Kirschen.[133] Copyright (1980) Varian AG, Zug, Switzerland

Kirschen[133] applied the liquid-liquid extraction method to community drinking water samples from the San Francisco Bay area, a local commercial bottled water, a well water, and an Environmental Protection Agency quality control sample. Instrumentation used was the Varian Model 3700 Gas Chromatograph with ^{63}Ni electron capture detector; CDS-111 Chromatography Data System, Model 8000 AutoSampler and Model 9176 Recorder. The column was 200 cm × ¼ inch glass, 10% Carbowax 20M on Chromosorb W-HP (80–100).

Procedure

A 10 ml sample of water is extracted in a 14 ml septum capped vial with 2 ml of hydrocarbon solvent. The extract is then analysed for total trihalomethanes on one of several suggested columns using the external standard technique. The limits of detection of the trihalomethanes using the Model 3700 were: chloroform

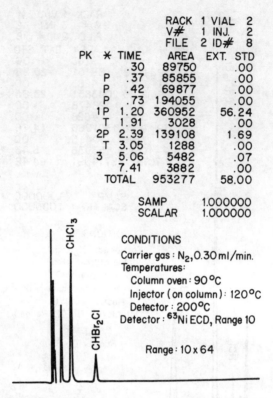

Figure 123 Gas chromatogram of trihalomethanes in finished drinking water. Reprinted with permission from Kirschen.[133] Copyright (1980) Varian AG, Zug, Switzerland

0.12 μg l^{-1}, dichlorobromomethane 0.015 μg l^{-1}, dibromochloromethane 0.029 μg l^{-1}, and bromoform 0.17 μg l^{-1}.

Table 90 and Figures 122 and 123 show the results obtained in analyses of finished water from three utilities. The higher levels of bromine-constituents in the CCCWD and CMWD finished waters are due to the increased bromide ion in the raw untreated water.[134]

Norin and Renberg[135] have investigated the determination of trihalomethanes in water using high efficiency solvent extraction.

Head-space analysis

The static head-space method is based on the fact that when a water sample that contains organic compounds is sealed in a vial, organics will equilibrate between the water and vial head-space. Distribution of compounds between the two phases depends on temperature, vapour pressure for each compound, sample matrix influences on compound activity coefficients, and ratio of head-space to liquid volume in the vial. A major advantage of this method is that only relatively volatile water-insoluble compounds tend to partition into the

Table 90 Trihalomethane analysis of finished water samples

Source	Trihalomethane ($\mu g\ l^{-1}$)				
	$CHCl_3$	$CHBrCl_2$	$CHBr_2Cl$	$CHBr_3$	Total
CCCWD					
Raw	0.2	0.10	—	—	0.3
Partially treated	6.7	7.60	4.66	0.51	19.4
Finished	19.4	23.7	14.8	5.55	63.5
EBMUD	56.3	1.7	0.10	—	58.0
CMWD	22.9	17.9	10.3	1.64	52.8
Commercial bottled water	—	—	—	2.05	2.05
Well water	—	—	—	—	—

Reprinted with permission from Kirschen.[133] Copyright (1980) Varian AG, Zug, Switzerland.

head-space; therefore, a form of sample clean-up is provided. Also, since only gaseous samples are injected into the gas chromatograph, column and detector contamination are prevented and chromatographic interferences are minimized. The earliest application of head-space analysis to the determination of the haloforms is probably that of Rook[136] and Kaiser and Oliver.[137] Kaiser and Oliver[137] have described a head-space method for the determination of chloroform, dichlorobromomethane, dibromochloromethane, bromoform, and carbon tetrachloride in water. This method is based on the equilibration of the dissolved compounds in water with a small volume of gaseous head-space under reduced pressure at elevated temperature. Head-space samples, so equilibrated, are directly injected into conventional gas chromatograph inlets for rapid quantification of the volatile compounds present. With a ^{63}Ni electron capture detector, quantitative determinations of chloroform and similar compounds in the $0.1-10\ \mu g\ l^{-1}$ range in water samples of less than 60 ml are performed in approximately 0.75 h.

Method

A normal laboratory separatory funnel with Teflon stopcock (2 mm bore) 60 ml volume, was filled with the (cold) water sample to leave an air space of approximately 2 ml after closing with the stopper. The funnel was then inverted and the small air space was quickly evacuated to approximately 10 Torr by the suction of a water vacuum pump and the stopcock closed again. The entire funnel was then submersed in the upside-down position into a thermostat-controlled water bath at 30, 50, 70, or 90 °C. Small gas bubbles developed in the sample and rose to the surface. After sufficient time, usually 30 min, had elapsed, the head-space was quickly returned to atmospheric pressure by turning the stopcock once for a half turn. Immediately following, a 5 ml sample of this head-space was withdrawn with an appropriate gas-tight syringe by inserting the syringe needle through the opened stopcock into the head-space after which the stopcock was closed again. Several consecutive head-space samples withdrawn by the same technique allowed multiple determinations of the same water sample.

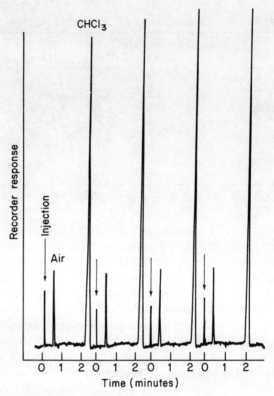

Figure 124 Gas chromatograms of four 5 μl head-space injections of water samples containing 10 μg l^{-1}. Chloroform (attenuation × 4). Reprinted with permission from Kaiser and Oliver.[137] Copyright (1976) American Chemical Society

For the gas chromatographic analysis of the gas samples, a Tracor 550 gas chromatograph with a 1.8 m, 6 mm o.d. and 4 mm i.d. glass column with 10% OV-1 Gas Chrom Q, 80–100 mesh and ^{63}Ni electron capture detector was used. The column temperature was kept at 50 °C, the injection port at 185 °C, and the detector at 330 °C, respectively. A Tracor recorder with 1-mV full scale deflection was used.

Figure 124 shows the gas chromatograms of four consecutive injections of the head-space of a solution of 10 μg l^{-1} chloroform in water after equilibration at 70 °C. At an attenuation × 4, the peak heights of the four chloroform peaks ranged from 196 to 211 mm with a mean of 205 mm and a mean deviation of ± t mm or ± 3%. These samples had been equilibrated for approximately 1 h and showed good reproducibility.

Figure 125 shows the effect of equilibration time on chloroform concentration in the head-space. Evidently, 30 minutes' equilibration time was adequate in this case. Kaiser and Oliver[137] undertook measurements to determine the influence of equilibration temperature. For this purpose, solutions of 1–10 μg l^{-1}

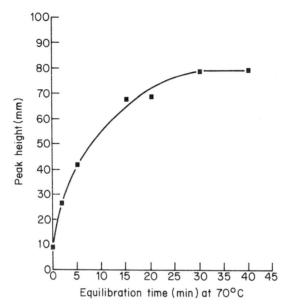

Figure 125 Plot of chloroform peak height *vs* equilibration time; at 10 μg l^{-1} chloroform. Reprinted with permission from Kaiser and Oliver.[137] Copyright (1976) American Chemical Society

chloroform in distilled water were analysed by the head-space technique at equilibration temperatures of 30, 50, 70, and 90 °C. From the results given in Figure 126 a strong dependence of the chloroform concentration in the vapour phase on the bath temperature is evident. Obviously the distribution of compound between water and gas phase depends on several parameters, such as solubility in water and the vapour pressure of the compound at the equilibration temperature, provided there is no saturation in either phase. The chloroform concentration in the head-space at 30 °C was approximately one-tenth of that at 90 °C. Although the equilibrations at 90 °C resulted in the highest sensitivity of the method, most of the subsequent investigations were performed at 70 °C as sufficient sensitivity was obtained at this temperature with good reproducibility (Figure 124). Even at an equilibration temperature of 90 °C only a small percentage of the haloforms in the sample will move into the vapour phase. The ratio of the concentration of chloroform in the vapour phase to the concentration of chloroform in solution is 0.4 at 30 °C, 0.6 at 50 °C, 1.0 at 70 °C, and 3.7 at 90 °C.

At the 0.1 μg l^{-1} concentration level blank determinations are very important. Laboratory distilled water supplies might contain up to 1 μg l^{-1} chloroform. Treatment of the blank water with a Millipone Super Q system reduces the chloroform blank to 0.1 μg l^{-1}. Volumes of head-space injected over the sample can range from 5 to 100 μl.

In addition to chloroform several other halogenated methane derivatives could easily be determined by the head-space technique. Figure 127 shows a gas

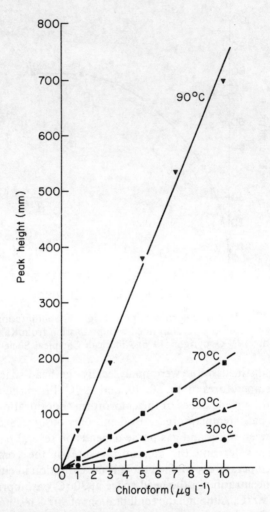

Figure 126 Plot of chloroform peak height *vs* chloroform concentration at several equilibration temperatures. Reprinted with permission from Kaiser and Oliver.[137] Copyright (1976) American Chemical Society

chromatogram of the head-space of a lake water after treatment with 10 mg l^{-1} chlorine for 24 h. By comparison with standard solutions of authentic compounds, the following residues were observed: $CHCl_3$, 43 µg l^{-1}; CCl_4, 0.3 µg l^{-1}; $CHBrCl_2$, 15 µg l^{-1}, and $CHBr_2Cl$ 20 µg l^{-1}. Quantitative measurements of standard solutions of 10 µg l^{-1} each of CCl_4, $CHBrCl_2$, $CHBr_2Cl$, and $CHBr_3$ showed a marked decrease in the overall sensitivity from CCl_4 to $CHBr_3$ as measured by the peak height of the detector response. This, however, could be expected and was in part rectified by a strong temperature programming (20 °C min^{-1} up to 100 °C) of the gas chromatograph

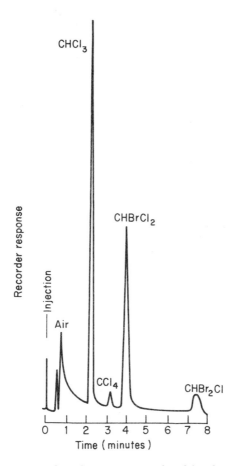

Figure 127 Gas chromatogram of 5 μl head-space of chlorinated lake water (attenuation × 8). Reprinted with permission from Kaiser and Oliver.[137] Copyright (1976) American Chemical Society

immediately following the injection or by isothermal g.c. operation at 100 °C.

Table 91 gives relative retention times and responses of five halogenated methane derivatives at g.c. column temperatures of 50 and 100 °C. These data were obtained from head-space analysis of solutions of 10 μg l^{-1} chlorocarbon each with equilibration at 70 °C. The relative sensitivities in Table 91 therefore, reflect overall sensitivity of the method and detector. As is shown, bromoform does not elute from the column at 50 °C but produces a sharp peak at 100 °C. Also, the response to $CHBr_2Cl$ improved markedly at the higher column temperature.

Head-space gas chromatography has been used to determine 0.1 ppb to 1 ppm chloroform, carbon tetrachloride, trichloroethylene, and tetrachloroethylene in

Table 91 Retention times (RT) and relative detector responses (DT) for halogenated methanes at g.c. column temperatures of 50 and 100 °C

	g.c. temperature			
	50 °C		100 °C	
Compound	RT (s)	DT*	RT (s)	DT*
$CHCl_3$	130	1	57	1
CCl_4	190	5	72	6
$CHBrCl_2$	240	0.8	79	1.4
$CHBr_2Cl$	450	0.1	119	0.3
$CHBr_3$	...	...	187	0.1

*Arbitrary units of peak heights ($CHCl_3 = 1$) values obtained from head-space samples of equipment aqueous solutions at 70 °C.
Reprinted with permission from Kaiser and Oliver.[137] Copyright (1976) American Chemical Society

drinking, natural, and industrial waters.[138] These workers carried out a systematic study of the parameters affecting the accuracy and precision of results obtained by this technique.

Method

Gas chromatographic equipment Head-space analyses were conducted using a Tracor 222 gas chromatograph equipped with a [63]Ni electron capture detector which was operated in the pulsed linearized mode. Separations were effected with a 10 ft × 4 mm i.d. glass column containing 20% SP-2100/0.1% Carbowax 1500 on 100–200 mesh Supelcoport. The column was operated at 85 °C using 90:10 argon–methane carrier gas (65 ml min^{-1}) which was prepurified with a molecular sieve filter and an oxygen trap. The detector was maintained at 275 °C and purged with 40 ml min^{-1} of 90:10 argon–methane. The injection port was maintained at 130 °C and was fitted with Microsep-138 septa. Chromatographic data were recorded using a Hewlett-Packard 3380A printer-plotter integrator using a chart speed of 1 cm min^{-1} and a slope sensitivity of 1.0. Under these conditions, respective retention times for air, $CHCl_3$, CCl_4, C_2HCl_3, and C_2Cl_4 were 0.88, 3.64, 5.12, 6.15, and 12.88 min.

Head-space samples were injected into the gas chromatograph using Precision Sampling Corp. syringes (Pressure-Lok, series D) of 5, 2, and 1 ml capacity and Hamilton syringes (1000 series) of 500, 250, and 100 µl capacity.

Purge trap analyses were conducted using the above chromatograph and column; however, a Tracor 700 Hall electrolytic conductivity detector was used. The detector was operated at 900 °C with a hydrogen reaction gas flow of 40 cm^3 min^{-1}. The electrolytic fluid flow was set at 1.2 ml min^{-1}; detector conductivity range was 10. The chromatographic column employed a nitrogen flow of 60 ml min^{-1} and was operated at 80 °C. The injector was 90 °C and the column to detector transfer line was maintained at 120 °C. Liquid sample concentration was provided by a Tekman LSC-1 (all Tenax trap). The Tekmar trap effluent port was directly interfaced to the gas chromatograph injection port with 0.03 in i.d. (1/16 in o.d.) stainless steel capillary tubing. The desorption heater was a modified replacement heater (P/N 12082) which provides a rapid temperature ramp (180 °C in 40 s). The Tekmar unit was operated using a 15 min sample purge of 20 ml min^{-1} of nitrogen, a 4 min desorption at 200 °C, and 5 ml samples. The purge-trap method was calibrated using aqueous standards (and dilutions thereof). Integration of detector signals was conducted as described above. Observed retention times (from initiation of desorption) for the respective components $CHCl_3$, CCl_4, C_2HCl_3, and C_2Cl_4 were 4.5, 6.2, 7.2, and 14.6 min. Single or multiple component primary aqueous standards were prepared by dissolving 2 ml of a stock methanol solution in 2 l of water. To obtain reproducible standards, all stock methanol solutions were prepared by appropriate dilution of methanol solutions which contained 2 g of halocarbon per 100 ml of methanol. The stock methanol solutions, when tightly sealed and stored at 4 °C, were stable for at least 6 months. The primary aqueous standard contained the following halocarbon concentrations: 50 ppb of chloroform, 2 ppb of carbon tetrachloride, 20 ppb of trichloroethylene, and 20 ppb of tetrachloroethylene. This standard produced head-space chromatographic responses which were midrange on the electron capture detector linearity curves. Aqueous standards with other halocarbon concentrations were prepared by diluting primary aqueous standards. Dilutions up to 100 were employed.

From a primary aqueous standard, ten serum vials were filled, sealed, and stored at 4 °C. Vials containing standard solutions were stable for at least 28 days.

Reagents

Water purification Water for preparing standards and blanks was prepared from distilled water that was purified by passage through a mixed-bed ion-exchange cartridge, then through an adsorptive carbon cartridge. To remove all traces of volatile interferents, this water was sparged for at least 1 h with ultrapure nitrogen. Suitability of each batch of water was checked by the head-space analysis method.

Analysis procedure Twelve hours prior to analysis samples and several standards were taken from storage (4 °C) and placed in the temperature stabilizer chest. From each sample vial and one of the standard vials, 60 ml of water was poured off into a graduated cylinder. Each vial was quickly resealed, vigorously shaken (by hand) for about 1 min, and then returned to the chest. Just prior to head-space sampling of any vial, it was agitated for 15 s.

A 2 ml injection of ultrapure nitrogen was made into the gas chromatograph at the beginning of each day to confirm syringe cleanliness; then 2 ml injections of head-space from the standard vial were made. The halocarbon responses, which usually exhibited a precision of about 2%, were used to calibrate the integrator for subsequent calculations of sample data.

Samples and standards cannot be considered stable once the vial septum is pierced; therefore, replication of sample analyses should be conducted as quickly as possible. Each sample was analysed by starting with a 2 ml head-space aliquot. Samples that produced responses outside the detector linearity range were reanalysed using smaller or larger (5 ml to 100 μl) head-space aliquots without sacrificing analytical accuracy as a result of differences in syringe construction (see Table 92). Component concentrations were calculated by comparing signal responses from samples to those obtained from standards as in the following equation:

$$\text{ppb of } Y = F \times \text{ppb concn of } Y \text{ in std } X$$
$$(\text{area of } Y \text{ in spl.}/\text{area of } Y \text{ in std})$$

where F equals 2 ml divided by the injected aliquot volume.

After conducting sample analyses for approximately 90 min, an injection from the standard vial was made. If the standard response for chloroform exhibited less than 5% deviation from the original calibration, then sample analyses were continued. When a change exceeding 5% was observed, 60 ml of solution were removed from another standard vial which was then equilibrated and used for subsequent analyses.

Periodically a low-level standard containing 1.0 ppb of chloroform, 0.04 ppb of carbon tetrachloride, 0.4 ppb of trichloroethylene, and 0.4 ppb of carbon tetrachloride was tested to provide an estimate of detection limits, confirm methodology, and assure detector linearity.

Dietz and Singley[138] observed that analytical accuracy is influenced by the effects of sample matrices on chromatographic responses. Several matrix conditions were examined experimentally. The presence of organics in the water

Table 92 Accuracy of injection volumes*†

Syringe‡	Injection vol (cm^3)	Rel. std concn	Area response
PS 5 ml	2	1.00	9136
PS 2 ml	1	2.00	9684
H 1 ml	0.5	4.00	10383
H 1 ml	0.4	5.00	10386
H 250 μl	0.2	10.00	10623
H 100 μl	0.1	20.0	9652
H 100 μl	0.05	40.0	10895
		Average response	10108
		R.s.d.	6%

*A 25 ppb standard of C_2HCl_3, used for the test.
†Samples were prepared so that each injection would introduce an equal amount of C_2HCl_3.
‡PS is Precision Sampling Corp. and H represents Hamilton.
Reprinted with permission from Dietz and Singley.[138] Copyright (1979) American Chemical Society.

Table 93 Effect of NaCl on g.c. response

% salt	Relative response			
	$CHCl_3$	CCl_4	C_2HCl_3	C_2Cl_4
0	1.00	1.00	1.00	1.00
0.5	1.00	1.00	1.02	1.01
1.0	1.03	1.04	1.06	1.04
2.0	1.06	1.06	1.12	1.09
5.0	1.27	1.23	1.32	1.28
10.0	1.58	1.45	1.65	1.57
20.0	2.50	2.01	2.48	2.26

Reprinted with permission from Dietz and Singley.[138] Copyright (1979) American Chemical Society

sample might upset halocarbon phase equilibration. However, no effects (2% r.s.d.) from either methanol or acetone were observed up to a 2% (v/v) concentration. This is an important result, because aqueous standards contain 0.1% methanol. Methanol is observed in the gas chromatograms at a 2 min retention time but does not interfere in the halocarbon analyses. Sample pH also does not affect (2% r.s.d.) chromatographic results. The effect of salt is presented in Table 93. The data clearly show that sodium chloride concentrations greater than 1% will significantly increase chromatographic responses. These increased responses reflect an increase in halocarbon activity coefficients (salting out).

As the sample temperature increases, the head-space gas chromatographic response for each halocarbon increases. By assigning a relative response of 1.0 to chloroform, carbon tetrachloride, trichloroethylene, and tetrachloroethylene at 0 °C relative responses of 3.8, 3.0, 4.2, and 3.9 were obtained at 25.5 °C and 5.8, 3.8, 6.4, and 5.6 were observed at 39.0 °C for each respective compound (r.s.d. of ~4%). The observed increase in response as temperature is increased can be advantageously used for lowering analytical detection limits. Day-to-day temperature differences are not important since analytical accuracy is based on standards that are analysed just prior to sample analyses. However, for precise results, daily temperature fluctuations must be kept within ±0.5 °C.

Dietz and Singley[138] also showed that equilibration of the aqueous sample with the head-space with no agitation is very slow. Even after 2 h, equilibration is far from complete when vials are not shaken. In Table 94, group B, the efficacy of a short-term vigorous agitation is demonstrated. Clearly, 1 min of agitation is sufficient for sample phase equilibration. Table 94, group C, points out that equilibrium is completely attained by rapid hand agitation which is comparable to a long-term agitation on a mechanical shaker.

Four factors were identified which determine how closely the gas chromatographic responses for a standard will replicate during a several-hour interval. A difference due to poor sample injection (plugged needle, etc.) can occur. The sample temperature may change slightly, in which case all component responses will increase or decrease. Halocarbons, especially tetrachloroethylene, may be

Table 94 Effects of agitation and equilibration time on gas chromatographic responses

Conditions	Group	Relative responses			
		$CHCl_3$	CCl_4	C_2HCl_3	C_2Cl_4
No agitation; after 0.5 h standing	A	1.00	1.00	1.00	1.00
No agitation; after 1.5 h standing		1.62	2.16	1.62	1.98
No agitation; after 2.0 h standing		1.82	2.93	1.81	2.48
After vigorous hand shaking for 2 min		2.51	11.30	2.93	7.52
After vigorous hand shaking	B	1.00	1.00	1.00	1.00
After 1 min, 20 s vigorous hand shaking		1.05	1.03	1.06	1.03
After 2 min, 20 s vigorous hand shaking		1.05	1.05	1.04	1.02
After 3 min vigorous hand shaking	C	1.00	1.00	1.00	1.00
After 30 min shaking on wrist action shaker		1.00	1.00	1.01	1.02

Reprinted with permission from Dietz and Singley.[138] Copyright (1979) American Chemical Society

lost from the vial once the septum is pierced. Component responses will decrease owing to halocarbon depletion due to multiple injections. In the later case, however, the chloroform response should be reproducible while the carbon tetrachloride response should decrease. For this reason, only the trichloroethylene response was monitored for purposes of confirming that the calibration data generated from the first few injections of standard are valid. When significant response differences due to temperature change are noted, then the calibration data is updated using a fresh standard vial. Table 95 presents data which demonstrate the above-mentioned situations. The first group of data shows a typical series of checks on the standard responses. The overall precision of about 2.5% indicates very good temperature stability. The second group of results represents a standard that experienced an unusual temperature increase (malfunction of heating–cooling system causing a 20 °F change in laboratory temperature) and sample loss. Note that the tetrachloroethylene response decreased while the other component responses increased (vial A). This indicates loss of tetrachloroethylene. Confirmation of this fact is given by results for a new standard (vial B) which exhibits increased responses for all components when compared to the original analysis of vial A. The final set of data is typical for a very small temperature rise in the sample as evidenced by increases in all component responses.

When very carefully conducted, analytical precision of approximately 3% (r.s.d.) can be achieved by this method. For routine analyses of many samples, a precision of from 5 to 10% (r.s.d.) is readily attained. Table 96 presents a study in which ten lake water samples were collected during a 1 h period. These samples were collected, transported, and stored for 30 days at 4 °C, then analysed. The first five samples (Table 96A) were analysed very carefully; a precision of ~ 3% (r.s.d.) was observed. The second set of samples (Table 96B) was analysed using less stringent attention to detail. In this case, a precision of about 5% (r.s.d.) was noted. The analytical method thus provides excellent precision even for rapid screening of samples. That these samples gave such

Table 95 Reproducibility of responses from standards under various conditions

Time	Relative response			
	$CHCl_3$	CCl_4	C_2HCl_3	C_2Cl_4
	Vial with stable temp. and no losses			
0800	1.00	1.00	1.00	1.00
0930	1.01	1.00	1.00	1.00
1100	1.04	1.00	1.03	1.03
1200	0.98	0.95	0.98	0.97
(R.s.d.)	(2.5%)	(2.5%)	(1.6%)	(2.4%)
	Vial with temperature increase and component loss			
(Vial A) 0800	1.00	1.00	1.00	1.00
(Vial A) 0930	1.17	1.10	1.11	0.92
(Vial B) 1000	1.23	1.17	1.29	1.34
	Vial with temperature increase			
0800	1.00	1.00	1.00	1.00
1000	1.07	1.03	1.03	1.07

Reprinted with permission from Dietz and Singley.[138] Copyright (1979) American Chemical Society.

Table 96 Results from analyses of ten lake water samples

Sample	Results (ppb)			
	$CHCl_3$	CCl_4	C_2HCl_3	C_2Cl_4
	Group A			
1	56.1	11.8	11.4	7.9
2	54.8	12.0	11.0	7.8
3	54.6	12.2	11.0	7.9
4	55.1	12.0	11.2	7.9
5	53.3	11.5	11.7	8.2
Average	55.2	11.9	11.3	7.9
R.s.d.	1.1%	2.2%	2.7%	2.0%
	Group B			
1	56.2	13.0	12.7	10.8
2	55.8	13.1	11.8	10.1
3	59.1	14.2	13.0	11.4
4	59.1	14.3	12.9	11.3
5	57.6	13.5	12.8	11.4
Average	57.6	13.6	12.6	11.0
R.s.d.	2.7%	4.5%	2.3%	5.1%

Reprinted with permission from Dietz and Singley.[138] Copyright (1979) American Chemical Society

good precision even after 30 days of storage, indicates sample integrity in well-sealed serum vials.

Table 97 illustrates the results obtained in spiking experiments and confirms the reliability of the method.

Table 97 Spiked drinking water sample

	Results (ppb)			
	$CHCl_3$	CCl_4	C_2HCl_3	C_2Cl_4
Town of Grand Island drinking water				
Vial 1	20.1	0.3	8.5	ND
Vial 2	20.7	0.4	8.5	ND
(Average)	(20.4)	(0.4)	(8.5)	(0)
Amount of spike	9.9	0.4	4.0	4.3
Town of Grand Island drinking water with spike				
Vial 1	30.8	0.8	13.0	4.3
Vial 2	31.7	1.2	13.2	4.4
(Average)	(31.3)	(1.0)	(13.1)	(4.4)
Theory	30.3	0.8	12.5	4.3

Reprinted with permission from Dietz and Singley.[138] Copyright (1979) American Chemical Society

Otson et al.[139] have compared dynamic head-space, solvent (hexane) extraction, and static head-space techniques utilizing Tenax gas chromatographic columns, a ^{63}Ni electron capture detector, and a Hall electrolytic conductivity detector for the determination of trihalomethanes in water. The relative standard deviation between trihalomethane values obtained by the three techniques ranged 9–10% for chloroform, 3–24% for dichlorobromomethane, and 13–61% for dibromochloromethane. Although the dynamic head-space technique was the most sensitive the solvent extraction technique gave comparable precision while the static head-space technique showed relatively poor precision and sensitivity. The solvent extraction technique is recommended for routine monitoring since it requires no special equipment and allows more analyses per hour than the dynamic head-space technique.

The dynamic head-space or gas sparging technique entails purging of a water sample with inert gas, collection of purged trihalomethanes on an adsorbent (e.g. Tenax GC), followed by thermal desorption. The static head-space technique, involving equilibration of trihalomethanes between the water sample and air space in a closed vessel, allows an aliquot of the air space to be analysed for trihalomethanes, liquid-liquid extraction of trihalomethanes from a water sample into a small volume of appropriate organic solvent, and analysis of an aliquot of the organic phase.

Methods

Apparatus Analyses were done using a dual column gas chromatograph (Hewlett-Packard 5837) equipped with 2.2 mm i.d. stainless steel columns, 90 cm and 150 cm in length, packed with Tenax GC 60–80 mesh (Applied Science Laboratories Inc.) and respectively connected to a ^{63}Ni electron capture detector (ECD) and a Hall electrolytic conductivity detector (Tracor 310).

The detector was maintained at 300 °C and both injection ports were at 200 °C. The Hall detector conditions were: 2-propanol–water, 50:50 at 0.12 ml min^{-1}, as electrolytic solvent; furnace temperature at 840 °C; operated in the prolytic mode. The Hall detector signal was integrated by means of an external integrator (Hewlett-Packard 3370B).

A four-port, two-way valve connected to a 5 ml sparging device permitted collection of purged volatile organics on the 150 cm Tenax column when the valve was in position 1. When the valve was in position 2 the sparge/carrier gas bypassed the device and directly entered the injection port of the gas chromatograph.

Culture tubes (32 ml capacity) equipped with screw caps with predrilled, 0.5 cm diameter centred holes and Teflon-coated, silicone rubber disks (Tuf-Bon, Pierce Chemical Co.) were used throughout. All tubes were inverted during storage at 25 °C to ensure formation of an additional liquid seal by the contents. Culture tubes containing 2.5 mg of sodium thiosulphate pentahydrate as residue from evaporation of aqueous thiosulphate at 70 °C, were filled completely with aqueous samples.

Reagents

Methanol and hexanes were distilled-in-glass quality and were checked for trihalomethane before use. Distilled, deionized water was boiled for 8 h to give trihalomethane-free water. Nitrogen gas was passed through molecular sieve and Tenax traps before use.

Procedures—dynamic head-space or gas sparging

Technique Aliquots (5.0 ml) of spiked aqueous trihalomethane solution prepared by injecting 12.8 μl aliquots of the appropriate stock methanolic solution of trihalomethanes into sealed culture tubes filled with trihalomethane-free water and tap water samples were purged with nitrogen at a rate of 40 ml min^{-1}, and volatile organics were collected on the Tenax column held at 50 °C. After 10 min the four-port, two-way valve was switched to bypass the sparging device, and the trapped organics were eluted using temperature programming and nitrogen as carrier gas (34 ml min^{-1} at 185 °C). The column oven temperature was raised at a rate of 30 °C min^{-1} to 130 °C, at 5.0 min the programme was continued at 30 °C min^{-1} to 185 °C, at 9.5 min cooling was begun, and at approximately 14 min purging of a fresh aliquot was initiated.

Static head-space technique

Tap water samples and spiked aqueous trihalomethane solutions which had been used in the gas sparging technique analyses were immediately prepared for static head-space analyses by removing and discarding an additional 5.0 ml of the liquid contents of the culture tubes. After replacing the removed liquid with air, the tubes were quickly sealed and were stored overnight at 25 °C. Aliquots (50 µl for electron capture detector, then 1.0 ml for Hall detector) of the gas phase were then withdrawn through the disk seal and were injected on to the appropriate Tenax column. The gas chromatographic conditions were: nitrogen carrier gas, Hall, 34 ml min^{-1} measured at 185 °C, EC, 54 ml min^{-1} measured at 185 °C; the column oven was held at 140 °C for 2 min, then programmed at 30 °C min^{-1} to 185 °C, at 6.2 min cooling to 140 °C was begun, and at approximately 10 min analysis of a fresh aliquot was initiated.

Liquid-liquid or solvent extraction technique

Aliquots (2.0 ml) of liquid were removed from sealed culture tubes filled with tap water samples or spiked aqueous trihalomethane solutions and were replaced with aliquots (1.0 ml) of hexane. The sealed tubes were agitated (wrist action shaker) for 30 min and were then stored overnight at 25 °C. Aliquots of 1.0 and 5.0 µl of the hexane layer were then respectively analysed by the electron capture detector and the Hall detector. Chromatographic conditions were identical with those used for static head-space analyses.

All analyses were performed in triplicate. Concentrations in tap water samples were determined by the use of calibration curves plotted from the data for spiked aqueous trihalomethane solutions analysed by the same technique. The calibration curves were prepared by plotting mean peak area against trihalomethane concentration.

Otson et al.[139] found that improved sensitivity for trihalomethanes was achieved by operating the Hall detector in the pyrolytic mode. Typical detection limits and retention times for the trihalomethanes are summarized in Table 98. As expected, the electron capture detector gave considerably lower method detection limits than the Hall detector.

Table 98 Detection limits (DL) and retention times (RT)* for trihalomethane

Anal system	Anal† technique	$CHCl_3$ DL (µg l^{-1})	RT (min)	$CHBrCl_2$ DL (µg l^{-1})	RT (min)	$CHBr_2Cl$ DL (µg l^{-1})	RT (min)	$CHBr_3$ DL (µg l^{-1})	RT (min)
Hall‡	GS	0.5	4.0	0.5	5.9	0.5	7.1	0.5	8.4
	SE	5	1.4	1	2.9	0.5	3.9	1	5.1
	HS	10	1.4	2	2.9	4	3.9	8	5.1
EC§	SE	<0.5	0.6	<0.1	1.4	<0.1	2.7	<0.1	3.6
	HS	<0.5	0.6	1	1.4	2	2.7	3	3.6

*Retention time measured from injection and/or start of temperature programme.
†GS, gas spurging; SE, solvent extraction; HS, head-space.
‡Hall detector with 150 cm × 0.22 cm Tenax column.
§ECD with 90 cm × 0.22 cm Tenax column.
Reprinted with permission from Otson et al.[139] Copyright (1979) American Chemical Society

Table 99 Comparison of relative standard deviation for trihalomethane (THM) determination in spiked aqueous solutions

Spiked aq trihalomethane (THM) soln	Anal technique*	Rel s.d. (%)								
		CHCl₃			CHBrCl₂			CHBr₂Cl		
		Theory	Hall	EC	Theory	Hall	EC	Theory	Hall	EC
1	GS-x	20	10		1.6	11		0.40	ND†	
	GS	20	20		1.6	36		0.40	ND	
	SE	20	1	4	1.6	7	5	0.40	ND	11
	HS	20	11	5	1.6	ND	25	0.40	ND	ND
11	GS-x	40	12		3.2	3		0.80	5	
	GS	40	3		3.2	1		0.80	5	
	SE	40	8	5	3.2	13	8	0.80	37	8
	HS	40	6	11	3.2	20	8	0.80	ND	ND
111	GS-x	80	2		6.4	5		1.6	3	
	GS	80	9		6.4	25		1.6	69	
	SE	80	8	8	6.4	11	9	1.6	3	10
	HS	80	13	63	6.4	13	32	1.6	ND	ND
1V	GS-x	120	2		9.9	2		2.4	1	
	GS	120	4		9.9	12		2.4	11	
	SE	120	5	2	9.9	7	3	2.4	13	6
	HS	120	10	14	9.9	8	12	2.4	ND	ND

*GS-x, gas sparging, solution prepared in sparging device; GS gas sparging, solution in culture tubes; SE, solvent (hexanes) extraction; HS, head-space.
†ND, not detected.
Bromoform not found.
Reprinted with permission from Otson et al.[139] Copyright (1979) American Chemical Society

No interfering peaks were produced by methanol, hexane, or trihalomethane-free water. However, a constant trace peak for chloroform ($<0.5\,\mu g\,l^{-1}$), presumably due to retention of chloroform on the Tenax column from previous analyses, was observed for the gas sparging and static head-space techniques when using the electron capture detector. The Hall detector was not sufficiently sensitive to give a detectable peak at this concentration.

An estimate of the precision of the three analytical techniques is given in Table 99 which shows the relative standard deviation obtained for triplicate analyses of spiked aqueous trihalomethane solutions. The precision of the gas sparging and solvent extraction techniques was comparable but the static head-space technique gave relatively poor precision. The precision of solvent extraction and static head-space techniques was essentially independent of the choice of detectors. Table 99 also includes the results for the two types of calibration solutions which were analysed by the gas sparging technique. Spiked aqueous trihalomethane solutions prepared directly in the sparger (type GS-x) gave essentially the same ($\pm 3\%$) chloroform peak area values as aliquots of solutions, with identical trihalomethane concentrations, transferred from culture tubes. Type GS-x, which has commonly been used for the gas sparging technique, gave slightly better precision than solutions transferred from culture tubes, but the latter procedure duplicates the transfer step for tap water analyses.

Otson et al.[139] also studied the effects on reported results of delays of up to 30 hours between taking the sample and carrying out the analysis. Bulk aqueous solutions containing 20–120 μg l^{-1} chloroform, 1.6–9.9 μg l^{-1} dichlorobromomethane, and 0.4–2.4 μg l^{-1} dibromochloromethane showed respectively concentration losses of 12 ± 3, 16 ± 5 and 21 ± 4%. Table 100 shows the precision and trihalomethane concentrations for tap water samples analysed by the three techniques. The gas sparging results are generally lower and the static head-space results are higher in value than the three technique mean. The precision for chloroform and dibromochloromethane determinations in tap water samples is good for all three techniques. The precision for dibromochloromethane was relatively poor, since levels of this compound were near the detection limit. Sample analyses done by means of the solvent extraction technique and using the electron capture detector (Table 101) show excellent agreement with those obtained by means of the Hall detector (Table 100).

Table 100 Values and precision of trihalomethane (THM) concentration in tap water samples as determined by three techniques and using the Hall detector

Sample	Anal.* technique	THM concn (μg l^{-1}) and rel. s.d. (%)					
		CHCl$_3$		CHBrCl$_2$		CHBr$_2$Cl	
		μg l^{-1}	%	μg l^{-1}	%	μg l^{-1}	%
A	GS-x	29	2	2.0	4	<0.5	
	GS	26	2	2.1	4	<0.5	
	SE	31	3	2.1	6	<0.5	
	HS	29	2	3.1	50	ND†	
	Mean	29	9	2.3	24	<0.5	
B	GS-x	58	3	5.0	1	0.5	11
	GS	61	3	5.4	1	0.5	11
	SE	66	2	5.5	3	0.5	31
	HS	73	4	5.7	8	ND	
	Mean	67	9	5.5	3	0.6	13
C	GS-x	79	1	6.9	1	1.7	1
	GS	85	1	7.6	1	1.9	1
	SE	97	3	8.9	6	1.0	14
	HS	104	5	8.2	8	ND	
	Mean	95	10	8.2	8	1.5	44
D	GS-x	164	2	8.5	3	2.4	13
	GS	136	2	8.9	3	2.5	13
	SE	162	2	11.3	1	1.0	8
	HS	153	4	14.0	37	ND	
	Mean	150	9	10.1	17	1.8	61

*Gas-x, gas sparging, calibration solution prepared in sparging device; GS, gas sparging calibration solution prepared in culture tubes; SE, solvent extraction; HS head-space; Mean, of GS, SE, and HS values.
†ND, not detected.
Bromoform not found.
Reprinted with permission from Otson et al.[139] Copyright (1979) American Chemical Society.

Table 101 Values and precision of trihalomethane (THM) concentration in tap water samples as determined by the solvent extraction techniques and using the electron capture detector

	THM concn (μg l^{-1} and rel. s.d. (%))					
	CHCl$_3$		CHBrCl$_2$		CHBr$_2$Cl	
Sample	μg l^{-1}	%	μg l^{-1}	%	μg l^{-1}	%
A	31	1	2.2	1	0.1	3
B	60	2	4.9	1	0.5	2
C	98	3	8.4	12	1.0	6
D	168	3	11.1	3	2.6	2

Bromoform not found.
Reprinted with permission from Otson et al.[139] Copyright (1979) American Chemical Society.

Otson et al.[139] concluded that the three analytical techniques gave comparable trihalomethane values for tap water samples. Although the gas sparging technique was the most sensitive, the solvent extraction technique gave comparable precision. The static head-space technique showed relatively poor precision and inferior sensitivity. The sensitivity of the static head-space and solvent extraction techniques can be improved by increasing the volume of the aliquot injected into the gas chromatograph, but overloading of the electron capture detector by organhalides must be avoided.

Both the static head-space and solvent extraction techniques allowed processing of 6 samples per h, whereas only 2.5 samples per h could be analysed by the gas sparging technique. In view of its precision, relative accuracy, simplicity, and speed of analysis the solvent extraction technique was judged the most suitable for monitoring trihalomethanes in water.

Varma et al.[140] carried out a comparative study of the determination of trihalomethanes in water. They compared the results of chloroform extraction using six liquid-liquid extraction solvents (pentane, methylcyclohexane, iso-octane, hexane, n-heptane, and n-nonane) with the vapour space extraction method. The vapour space method yielded the poorer results.

Friant[141] has described a direct head gas analysis procedure for the isolation of chloroform from aqueous environmental samples. The technique included gas chromatography and mass spectrometry. This worker carried out fundamental studies of the partitioning of organic compounds between the aqueous and vapour phases and systematically examined effects of variations in operating parameters. Possible errors and limitations of this method are discussed.

Gomella and Belle[142] studied the determinations of volatile organhalogen compounds in water by the head-space technique.

Montiel[143] applied the head-space analysis technique to the determination of the halomethane content of chlorinated water samples. He considered the effects of operating variables on the sensitivity of the method and the impact of a number of interfering substances (surface active agents and soluble salts). Particular attention is paid to the operation of the electron capture detector system.

Suffet and Radziul[144] applied various techniques including head-space analysis and solvent extraction to the screening of volatile organics including chloroform and *bis*(2-chloroethyl) ether in water supplies.

Bush et al.[145] developed a method for the determination of halogen-containing organic compounds using measurement of peaks in head-space vapour with an electron capture detector and a chromatograph system. The method was used in a screening survey of New York State drinking water with the objectives of determining significant seasonal variations in halo-organic concentrations of the compounds in chlorinated ground water and surface water, and the frequency of occurrence of halo-organic compounds in chlorinated water. It is shown that chloroform and bromodichloromethane occur most frequently in chlorinated water.

Dynamic head-space techniques for the determination of trihalomethanes have been studied by Symons et al.,[146] Keith,[147] and workers at the Health and Welfare Department, Canada[148] and static head-space techniques have been studied by Keith,[147] Bush et al.,[149] and Morris and Johnson.[150]

Gas purging methods

Analytical methods for determination of trace amounts of volatile, relatively insoluble organic compounds in water generally require a preconcentration step. The target species can be either enriched by concentrating the gases in the head-space on adsorbent traps[151-154] or by purging the aqueous phase with a stream of gas followed by trapping.[155-163]

This section deals with gas purging methods which have been shown to be applicable to determining the following halocarbons in water:

Bromoform
Bromodichloromethane
Bromomethane
Carbon tetrachloride
Chlorobenzene
2-Chloroethylvinyl ether
Chloroform
Chloromethane
Dibromochloromethane
1,2-Dichlorobenzene
1,3-Dichlorobenzene
1,4-Dichlorobenzene
Dichlorodifluoromethane
1,1-Dichloroethane
Chloroethane

1,2-Dichloroethane
1,1-Dichloroethene
trans-1,2-Dichloroethene
1,2-Dichloropropene
*cis*1,3-Dichloropropene
trans-1,3-Dichloropropene
Methylene chloride
1,1,2,2-Tetrachloroethane
Tetrachloroethene
1,1,1-Trichloroethane
1,1,2-Trichloroethane
Trichloroethene
Trichlorofluoromethane
Vinyl chloride

In one gas purging method the volatile compounds are extracted from the water sample by passing pure gas (e.g. nitrogen) through the water sample and collecting the volatile compounds on a small adsorption column. The compounds are introduced into a chromatograph by heating the adsorption column. The determination is carried out by temperature programmed gas chromatography using a halogen-specific detector (Electrical Conductivity Detector) or a more generalized detector (flame ionization or mass spectrometry).[164,165] This technique is quite sensitive ($>0.5\ \mu g\ l^{-1}$).

Bellar and Lichtenburg[165] used a purge and trap method in conjunction with temperature programmed gas chromatography to resolve 23 volatile organohalides all of which have been identified at one time or another in various water samples. This method entails pumping of the water sample with an inert gas, collection of the purged trihalomethane on an adsorbent (e.g. Tenax GC) followed by the thermal desorption. Dressman[124] is of the opinion that purge and trap methods are very amenable to gas chromatographic–mass spectrometric confirmation of the identity of volatile compounds and at levels lower than can be obtained with liquid-liquid extraction methods. The lower minimum detectable concentrations are attainable because virtually all of the trihalomethanes purged from the sample are transferred to the gas chromatographic column and the detector—without the relatively non-volatile interferences coextracted by liquid-liquid extraction methods and without introducing solvent-related interference.

The liquid-liquid extraction methods rarely provide a compound sufficiently concentrated for gas chromatography–mass spectrometry and it may be necessary to resort to a solvent enrichment technique. Enriching the solvent by evaporation commonly results in interference from concentrated solvent impurities to the extent that mass-spectral analysis cannot be performed, or in loss of the more volatile components such that actual enrichment is not achieved.

Quimby and Delaney[166] determined trihalomethanes in drinking water by gas chromatography with an atmospheric pressure microwave emission detector. The organics are isolated by a purge and trap technique. This detector, in addition to being very sensitive, is also element selective, distinguishing between chlorine, bromine, and iodine.

The gas chromatograph, interface microwave components, and spectrometer used by Quimby and Delaney have been described by Quimby et al.[167] The effluent from a gas chromatograph (Varian 2440) is split to direct one-third of the effluent to a flame ionization detector and the balance through a heated transfer line into a small auxiliary oven. The oven contains a high temperature valve which in one position allows the solvent peak (or any other large peaks which contain sufficient material to extinguish the plasma) to be vented, and in the other position directs the column effluent to the plasma. The oven butts against the back of the TM_{010} cavity (RKB Products, Inc., Lexington, Mass.) and maintains the valve and other interface components at the desired temperature independent of the column temperature. The plasma (viewed axially) is imaged on to the entrance slit of a prototype Spectraspan III echelle grating monochromator (Spectrametrics Inc., Andover, Mass.). The photocurrent from the photomultiplier tube (R446) is monitored with a picoammeter, and displayed on a strip chart recorder.

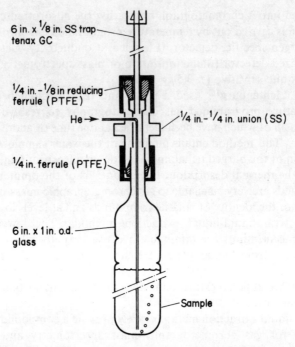

Figure 128 Purge and trap apparatus. Reprinted with permission from Quimby and Delaney.[166] Copyright (1979) American Chemical Society

A ⅛ in o.d. × 10 ft stainless steel Tenax GC column was employed. The helium flow through the analytical column was 30 ml min^{-1} and the injector, detector transfer line, and auxiliary oven were maintained at 210 °C.

The purge and trap apparatus is illustrated in Figure 128. The water sample is contained in a 6 in long glass tube, 1 in o.d. The neck at the top of the tube is ¼ in o.d. and is held in a vertically positioned ¼ in × ¼ in stainless steel union (Parker CPI) with a ¼ in PTFE ferrule. A 6 in × ⅛ in stainless steel trap containing Tenax GC is attached at the other end of the union with a ¼ in × ⅛ in PTFE reducing ferrule. A small hole is drilled through the side of the union, and ¹/₁₆ in o.d. stainless steel tubing is positioned through the hole and silver-soldered so that the end of it extends to the bottom of the glass sample tube. Helium or nitrogen purge gas, which has passed through a trap containing silver oxide heated at 200 °C to oxidize any organic contaminants is bubbled into the water sample through the ¹/₁₆ in tubing.

The organics purged from the water sample are desorbed from the Tenax GC trap into the gas chromatograph by attaching the end of the trap to a helium line and inserting the entire length of the trap into the injection port of the gas chromatograph. The trap is sealed to the inlet using the septum nut and a septum with a ⅛ in hole through it to serve as a gasket. A three-way valve located after the flow controller of the gas chromatograph allows the carrier gas to be directed through the trap and through the analytical column, or through the injection port in the normal fashion when no trap is in place (the hole through the septum gasket is then plugged with a solid ⅛ in o.d. stainless steel rod).

Procedure

To analyse solutions by the purge and trap technique, 10 ml of the standard or sample were placed in the purging vessel with a glass syringe. The vessel was immediately sealed to the union assembly (to which the Tenax GC trap had already been affixed) by tightening the nut around the PTFE ferrule finger tight and the sample purged for 11 min at a flow rate of 20 ml min^{-1}. After sparging the trap was removed and attached to the helium line with the three-way valve positioned so that the carrier gas was directed through the liquid inlet. The plug was removed from the septum, and after inserting the trap into the injection port and sealing it by tightening the injector nut, the valve was immediately turned to redirect the carrier gas through the trap. The temperature of the analytical column was maintained at 100 °C until insertion of the trap was completed and then was programmed from 100 to 220 °C at 10 °C min^{-1}.

One artefact of this particular purge and trap arrangement is the introduction of nitrogen from the sparging and air into the chromatographic column when the trap is inserted into the injection port. This plug of air was often found to be sufficient to extinguish the plasma; thus the high temperature valve was positioned so that the air peak was vented. After elution of the air, the valve was returned to the position where the column effluent entered the plasma.

The detection limits and linearities observed with the microwave emission detector are well suited for its use with the purge and trap technique. For 5 ml samples, detection limits of *ca*. 0.1 ppb are obtained for the four bromine- and chlorine-containing trihalomethanes, and the linear range extends to above the 100 ppb level. Similar results are also obtained for the other organohalides in Table 102. Chlorine-, bromine-, and iodine-selective chromatograms for a standard aqueous solution containing several organohalides at the 25–30 ppb level are given in Figure 129. Since the g.c.–microwave detector system was a single channel instrument, the chromatograms were obtained by performing three replicates on 10 ml aliquots of the standard, during which the monochromator was set to a wavelength corresponding to each of the three elements in turn. The high temperature valve was positioned so that the column effluent was vented until the large air peak eluted and repositioned to direct the analytes into the plasma. This point is indicated by 'v.o.' in the chromatograms.

From Figure 129, the ability to distinguish among the three halogens greatly simplifies the chromatographic requirements of the analysis. With a halogen-selective detector, the sample components would not be completely resolved. Although the single channel system requires replicate runs at the different wavelengths and thus increased analysis time, the necessity to resolve all of the halogenated components in the sample when using a halogen selective detector

Table 102 Relative responses with the microwave emission detector per mole of halogen

	Relative response
Cl(II) 481.0 nm	
CHCl$_3$	1.00
CHBrCl$_2$	1.01
CHBr$_2$Cl	1.00
CH$_2$Cl$_2$	0.96
CHClCCl$_2$	0.99
C$_6$H$_5$Cl	0.95
CHCl$_2$CHCl$_2$	0.98
CCl$_4$	0.95
Values given relative to CHCl$_3$	
Br(II) 470.5 nm	
CHBr$_3$	1.00
CHBr$_2$	1.03
CHBrCl$_2$	1.07
Values given relative to CHBr$_3$	
I(I) 206.2 nm	
CH$_3$I	1.00
CH$_3$CH$_2$I	1.06
CH$_3$CH$_2$CH$_2$I	0.96
Values given relative to CH$_2$I	

Reprinted with permission from Quimby and Delaney.[166] Copyright (1979) American Chemical Society.

would generally also result in increased analysis time. Use of a multichannel spectrometer to monitor the three analytical wavelengths simultaneously would eliminate the need for replicate runs.

The chromatograms for analysis of a tap water sample are presented in Figure 130. The negative responses that appear immediately after the valve is opened are caused by the water peak, which is sufficiently large to lower momentarily the background emission at the analytical wavelength. Comparisons of the responses obtained for the sample with those for a standard solution indicate that methylene chloride is present at around 0.2 ppb, chloroform at 20 ppb, bromodichloromethane at 1 ppb, and chlorodibromomethane at 0.1 ppb. As indicated by the iodine-selective trace, no evidence exists for organo iodine compounds in the sample.

The identity of a given analyte is more firmly established by an element-selective detector than a total halogen detector. When a non-selective detector such as flame ionization is employed, only the retention time serves to identify a compound eluting from the gas chromatograph. Specific halogen-selective detection increases the reliability of the analysis in that both retention time and evidence of halogen content may be used in compound identification. Knowledge of which particular halogen(s) is present in a peak at a given retention time

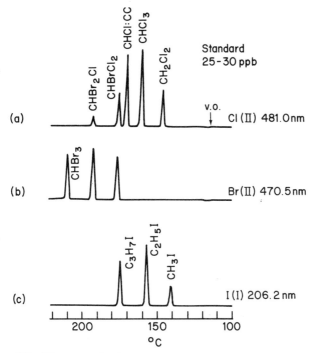

Figure 129 Element selective chromatograms from sparged standard solution. (a) Cl channel, (b) Br channel, (c) I channel. Reprinted with permission from Quimby and Delaney.[166] Copyright (1979) American Chemical Society

provides yet further confirmation of the identity of the sample component. This is illustrated by bromodichloromethane and chlorodibromomethane in Figure 130 where both chlorine and bromine content are clearly evident. In view of the large number of halogenated compounds which have been detected in various water samples, this additional qualitative information can be very useful.

While the majority of analyses performed with the microwave emission detector have dealt with samples of drinking water, the system has also been applied to the analysis of chloroform in secondary effluents. A chlorine-selective chromatogram obtained from such a sample is given in Figure 131 where the chloroform concentration was found to be at the 0.7 ppb level. Also detected were methylene chloride at about 0.5 ppb, carbon tetrachloride at 0.5 ppb, and trichloroethylene at 0.1 ppb.

Kroner[168] has described a microprocessor controlled gas chromatograph monitoring system for determining purgeable carbon tetrachloride, haloforms, and halomethane at concentrations of $0.1~\mu l~l^{-1}$ in river water. Chloroform was found in 70% of the samples analysed; other chlorinated solvents found regularly were trichloroethane, trichloroethylene, tetrachloroethylene, methylene chloride, and carbon tetrachloride.

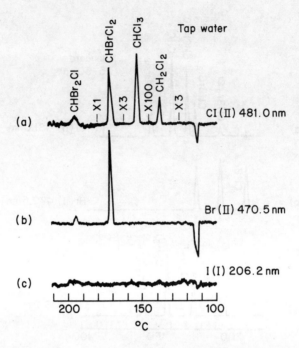

Figure 130 Element selective chromatograms for sparged tap water sample. (a) Cl channel, (b) Br channel, (c) I channel. Reprinted with permission from Quimby and Delaney.[166] Copyright (1979) American Chemical Society

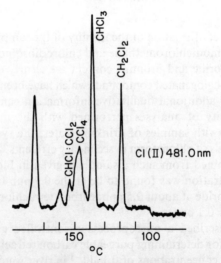

Figure 131 Chlorine selective chromatogram from sparged sample of secondary effluent. Reprinted with permission from Quimby and Delaney.[166] Copyright (1969) American Chemical Society

Kirschen[169] has investigated the Environmental Protection Agency standard purge and trap method[170] for determining trihalomethanes and other halogenated volatiles in water. Kirschen[169] used a Varian Model 3700 gas chromatograph with Model 700A Hall Electrolytic Conductivity Detector VISTA 401 Chromatography Data System. The column comprised 200 cm × 2 m glass column containing 1% Sp-1000 on Carbopack-B (60–80 mesh). The instrument parameters were:

Hall detector parameters

Reactor temperature 750 °C
Hydrogen flow rate 30 ml min^{-1}
n-Propanol flow rate: 0.6 ml min^{-1}
Detector base temperature: 230 °C
Range: 100
G.c. conditions:
  Carrier gas: helium, 30 ml min^{-1}
  Temperatures: column oven: 45 °C for 3 min, 8 °C min^{-1}
  to 220 °C, hold for 2 min; injector 150 °C

This method was used to check the concentrations of four halomethanes, chloroform, bromoform, dichlorobromomethane and dibromochloromethane in potable water samples (Table 103, Figure 132). Relative % standard deviations were normally below 5%.

A purging method has been described for the specific determination of low levels of methyl bromide fumigant in water.[171] The analysis was performed with a Packard Purgatrator, mounted on a P-T-C (packed-trap-capillary) MDSS module, with two electron capture detectors. The module was used in a Packard Model 433 gas chromatograph. The essential instrumentation for methyl bromide determination consists of: a Purgatrator, a single packed column and an electron capture detector (ECD). This configuration is compatible with the Packard 429, 430, and 433 gas chromatographs. A flow diagram for the Purgatrator/MDSS configuration is shown in Figure 133. The four different modes are given in Figures 134 A–D. This valve switching sequence can be entered on a time base programme, so that the complete analysis is automated.

Table 103 THM analysis of finished water samples (μg l^{-1})

Source	$CHCl_3$	$CHBrCl_2$	$CHBr_2Cl$	$CHBr_3$	Total THM
Utility I	23.9	23.4	21.8	1.7	70.8
Utility II	48.7	22.0	8.8	1.8	81.3
Utility III	44.8	39.9	24.8	2.6	112.1
Commercial deionized water	21.5	17.7	8.6	—	47.8
Well water	—	—			

Reprinted with permission from Kirschen.[169] Copyright (1981) Varian AG, Zug, Switzerland.

Conditions

Columns	Column 1	Column 2
Material	Glass	Glass
Length	1 m	25 m
Diameter	i.d. 2 mm	i.d. 0.25 mm
Stationary phase	10% SP-1000	FFAP, df = 0.45 u
Support	Supelcoport (80–100)	....
Carrier gas	$N_2:P_1 = 117$ kPa	$N_2:P_2 = 50$ kPa
Selectable restrictions	R_x = no. 8	R_y = no. 6
Detectors	Detector 1: ECD att 7 range 4	Detector 2: ECD
Temperatures	Detector : 300 °C Injector : 250 °C Auxiliary : 175 °C Oven initial: 50 °C (18.5 min) Rise : 10 °C min^{-1} Oven final : 160 °C	
Purge time	12 min	
Purge flow	40 ml min^{-1} N_2	
Purge volume	25 ml	
Sorbent trap	i.d. 2 mm length 15 cm Sorbent Tenax® (mesh size 80–100)	

® Tenax is a registered trademark of AKZO/Holland.

Figure 135 shows the gas chromatogram obtained for a sample containing 10 µg l^{-1} methyl bromide and other halogenated organics.

Nicolson et al.[113] have described a convenient easily automated method for the analysis of haloforms and some other volatile organohalides in drinking water. This direct aqueous injection method has a detection limit at or below 1 µg l^{-1} for haloforms. Simultaneous analysis of finished drinking water samples with the direct aqueous injection and the gas sparging method revealed hitherto unknown aspects of water treatment chemistry. While the gas sparging technique measures only the free haloforms present in the drinking water, they showed that the direct aqueous injection method quantitates the total potential haloforms that can form after chlorination.

Earlier studies by Nicholson and Meresz[106] had indicated that a gas chromatograph equipped with a Porapak Q column and an electron capture detector could be used for the determination of haloforms in dilute aqueous solution by direct aqueous injection. It was shown that the determination of the haloforms (chloroform, bromodichloromethane, and chlorodibromomethane)

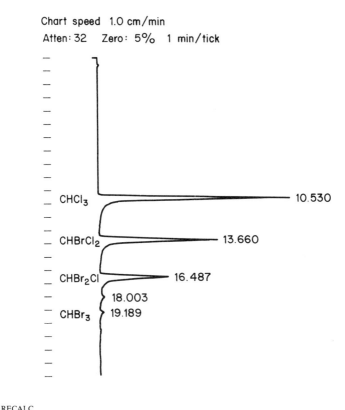

Figure 132 Gas chromatogram of trihalomethanes found in finished drinking water. Reproduced with permission from Kirschen.[169] Copyright (1981) Varian AG, Zug, Switzerland

as well as carbon tetrachloride, trichloroethylene, and tetrachloroethylene could be achieved near the 1 µg l^{-1} level using 9 µl injection volumes. Since no preconcentration was required, this technique was easily automated, allowing the analysis of up to 60 samples per day.

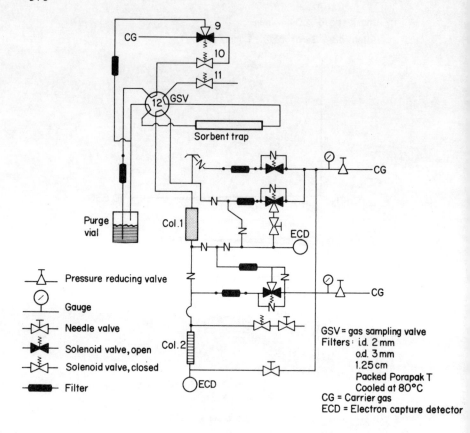

Figure 133 System shown in SFL mode, and stop flow position

However, before adopting this method for routine monitoring, it was essential to compare it with the published gas sparging procedure. Analysis of standard aqueous haloform solutions showed excellent agreement between the two methods. However, when the two methods were compared on samples of chlorinated water obtained from treatment plants a different picture emerged. The direct aqueous injection method gave haloform values consistently higher than those obtained by gas sparging. The differences in data were consistent with the location of sampling sites. A study of this problem, as discussed below, revealed that the two methods are measuring different parameters. Namely, the gas sparging technique quantitates only the free haloforms while the direct aqueous injection method measures the total potential haloform concentration.

Details of the direct aqueous injection and gas sparging methods used by Nicholson et al.[113] are given below.

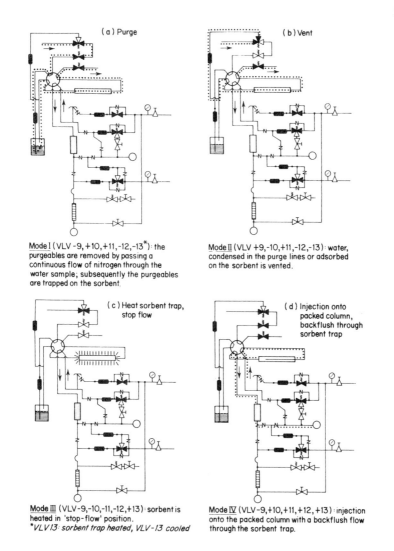

Figure 134 Purge method for determination of organohalogen compounds

Direct aqueous injection method

A Varian 2400 gas chromatograph equipped with a scandium tritide electron capture detector was used. The 6 ft × ¼ in (2 mm i.d.) glass column was packed with Chromosorb 101 (60–80 mesh). The injector and detector were operated at 250 °C. The column oven was operated isothermally at either 130 or 150 °C, depending on the volatility of the compounds being analysed. The flow rate of the nitrogen carrier gas was 30 ml min^{-1}. The analysis was automated by connecting a Varian model 8000 autosampler to the gas chromatograph, then interfacing it with an Autolab System 1VB electronic

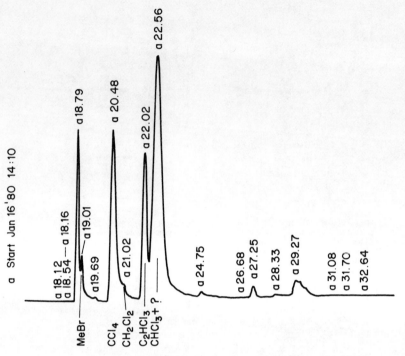

Figure 135 Gas chromatogram of organohalogen compounds, pumped from water

integrator. In all cases, a 9 µl sample volume and 20 min analysis time were used.

Gas sparging method

Figure 136 shows the modified version of the 'Bellar and Lichtenberg'[78] apparatus used for gas sparging. A Varian 2100 gas chromatograph, containing a 6 ft × ¼ in glass column (2 mm i.d.) was used for the analysis. The sample was injected into the system with a 5 ml glass hypodermic syringe (A) (fitted with a Luer lock) through a Hamilton valve (B). This valve was attached to a 19 gauge needle which pierced the septum of the stripper (C). The cylindrical stripper was 14 mm in diameter and 100 mm long. Helium carrier gas was passed through a coarse sintered glass disk, through the sample and the chromatographic column. The three-way valve (D) was attached to a helium line via a four-port valve (F) and to a vacuum line (E). The sample after being purged at 20 ml min^{-1} for 10 min, was isolated from the system by switching the valve (F). The sample was then withdrawn from the stripper by switching the three-port valve (E) to vacuum. The volatile organics, which were trapped on the head of the Chromosorb 101 column, were chromatographed by heating the column oven at 20 °C min^{-1} to 150 °C and programming it from 150 to 200 °C at 4 °C min^{-1}. The effluent from the column was analysed by a Hall electrolytic conductivity detector, operating in the chlorine mode. The furnace of the Hall

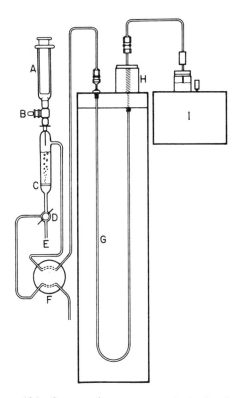

Figure 136 Gas sparging apparatus. A, 5 ml syringe, B, Hamilton 2-way valve; C, glass sparger; D, 3-way stopcock; E, line to vacuum; F, 4-port valve; G, 6 ft Chromosorb 101 column; H, furnace; I, Hall electrolytic conductivity detector. Reprinted with permission from Nicholson et al.[113] Copyright (1977) American Chemical Society

detector was mounted directly over the detector base, and the quartz pyrolysis tube attached to the base in place of the flame tip. The hydrogen was introduced at the detector base at a rate of 40 ml min^{-1}. The pyrolysis oven was held at 800 °C. The electrolyte solution used was 10% isopropane and 90% water by volume.

To achieve reasonable standard curves and reproducibility, several modifications to the Hall detector were made. Figure 137 is a schematic of the flow system used. It was found that the needle valve controlling electrolyte flow and the small ion-exchange cartridge were the main causes of the erratic performance of the detector. The ion-exchange cartridge was doubled in length from 7.5 cm to 15 cm. The ion-exchange resin used was Duolite ARM 381 obtained from Diamond Shamrock Chemical Company. Since the needle valve in the flow system tended to become clogged, it was replaced by a Nupro 5 μm filter placed in the flow system before the ion-exchange cartridge and 45 cm

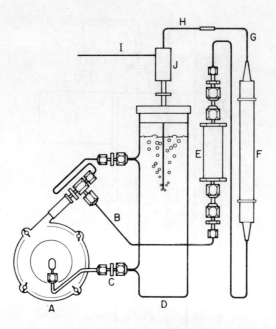

Figure 137 Modifications to the Hall detector. a, pump; b, outlet; c, inlet; d, reservoir, e, 5 μm fritted disk; f, ion-exchange column; g, 0.01 inch capillary tubing; h, ^{1}/₁₆ inch Teflon tubing; i, gas inlet from furnace; j, hall electrolytic conductivity detector. Reprinted with permission from Nicolson et al.[113] Copyright (1977) American Chemical Society

Table 104 Retention time and detection limits for selected organohalides

Compound	Oven temperature (°C)	Directed aqueous injection*		Gas sparging	
		Retention time (min)	Detection[†] limit ($\mu g\ l^{-1}$)	Retention time (min)	Detection[†] limit ($\mu g\ l^{-1}$)
Methylene chloride	130	1.7	500	6.5	0.1
Chloroform	130	3.0	1	8.1	0.1
1,2-Dichloroethane	130	4.0	90		0.2
1,1-Dichloroethane	130	4.0	150		0.2
Carbon tetrachloride	130	4.0	0.1		10
Trichloroethylene	130	4.7	2	9.0	0.3
Bromodichloromethane	130	6.0	0.5	9.6	0.5
Tetrachloroethylene	130	13.0	0.5	11.2	2
Chlorodibromomethane	130	15.4	1	12.7	2
Bromoform	150	4.5	2	16.9	1.5
Tetrachloroethane	150	5.2	7		
p-Dichlorobenzene	150	9.8	500		
o-Dichlorobenzene	150	11.4	500		

*Chromosorb 101 column.
[†]Detection limit = 2 × noise level.
Reprinted with permission from Nicholson et al.[113] Copyright (1977) American Chemical Society.

Table 105 Relative peak areas of standards in water and ethanol

Standard	Peak area in water / Peak area in ethanol
Chloroform	0.74
Carbon tetrachloride	0.11
Trichloroethylene	0.52
Bromodichloromethane	0.67
Chlorodibromomethane	0.68

Reprinted with permission from Nicholson et al.[113] Copyright (1977) American Chemical Society.

of 0.01 inch i.d. capillary tubing between the ion-exchange cartridge and the conductivity cell. These two restrictors reduced the flow of the conductivity solution to 0.5 ml min^{-1} through the cell.

Because all of the compounds listed in Table 104 dissolve readily in ethanol, which has retention characteristics similar to water when chromatographed on Chromosorb 101, standard solutions of the haloforms chloroform, bromodichloromethane chlorodibromomethane, trichloroethylene, and carbon tetrachloride at a concentration of 1 mg l^{-1} were made up in ethanol. From this stock solution, two sets of standard curves ranging in concentration from 10 to 200 ng l^{-1} were made, one in water and one in ethanol. Special care had to be taken in making up the aqueous standards to ensure complete dissolution of the haloforms. One ml of the ethanol stock was added to 900 ml of water (distilled and purified with a Millipore Super Q system) contained in a 1 l volumetric flask. The solution was shaken vigorously by hand for at least 3 min. The solution was then made up to exactly 1 l with water and aliquots of this stock were diluted further with water to the desired concentration. Six point standard curves were recorded using the direct aqueous injection for both the water solutions and the ethanol solutions. The ratio of the areas at each concentration were calculated and averaged.

Table 105 lists the ratio of the peak areas obtained using the two solvents. The higher peak areas of the standards made up in ethanol more accurately reflect the true concentration of the volatile organohalides than do the solutions made up in water. Since ethanol could not be used as solvent for the standards when using the gas sparging technique the ratio of the areas obtained for each standard curve by the direct aqueous injection methods was used to correct the gas sparging results.

Typical chromatograms obtained when analysing a standard aqueous solution containing 50 µg l^{-1} of each of selected volatile organohalides by the direct aqueous injection and gas sparging method are shown in Figures 138 and 139 respectively. Examination of these and of the data presented in Table 104 reveals the characteristic sensitivity of each method.

The detection limits for the determination of the tri- and tetrahalogenated compounds by direct aqueous injection (except for tetrachloroethane) are all

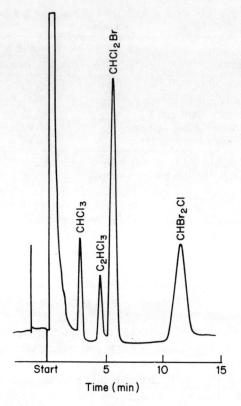

Figure 138 Typical chromatogram obtained by direct aqueous injection. Reprinted with permission from Nicolson et al.[113] Copyright (1977) American Chemical Society

below $2 \mu g \ l^{-1}$ when using a scandium tritide detector. One of the disadvantages of using this detector in the d.c. mode, is that the column oven must be operated isothermally to prevent excessive baseline drift. Since the volatility and polarity of the organohalides of interest cover a wide range, the column oven must be operated successfully at two temperatures (130 and 150 °C for Chromosorb 101) involving two separate injections for the analysis of the whole spectrum of compounds. In practice, however, the lower temperature scan is sufficient for the monitoring of drinking waters. Bromoform has been found in less than 20% of the locations sampled in the US EPA 80-city survey[88] and tetrachloroethane has been detected rarely. The detection limits for the determination of the dihalogenated compounds such as methylene chloride, the dichloroethane, dichloropropanes, and dichlorobenzenes, are all above the $90 \mu g \ l^{-1}$ level making the direct aqueous injection method unsuitable for monitoring these compounds at lower concentrations.

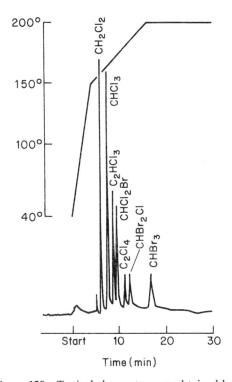

Figure 139 Typical chromatogram obtained by gas sparging. Reprinted with permission from Nicolson et al.[113] Copyright (1977) American Chemical Society

The gas sparging method complements the direct aqueous injection method in that, when using an electrolytic conductivity detector, the detection limits for the determination of the dihalogenated organics mentioned above are all below the 1 μg l^{-1} level. Because of the specificity of the detector, the column oven may be programmed with relatively little shift in the baseline, allowing the determination of the whole range of volatile, chlorinated organics listed in Table 104 with one injection.

The gas sparging method has a detection limit for carbon tetrachloride about 100 times greater than that obtained by direct aqueous injection. The electrolytic conductivity (Hall) detector is much more so. The haloforms were easily pyrolysed and gave reasonable responses regardless of the flow rates of carrier gas, hydrogen, or electrolyte. The determinations of trichloroethylene and carbon tetrachloride were found to be very sensitive to changes in conditions. High concentrations of isopropanol in the electrolyte reduced the sensitivity of the Hall detector to all compounds investigated. The high helium and hydrogen flow rates suggested by the manufacturer were found to be unsuitable for the analysis of carbon tetrachloride and trichloroethylene. The residence time in the pyrolysis furnace for these compounds was critical, with a maximum total

Table 106 Dependence of haloform concentration on sample temperature using gas sparging

Temperature (°C)	Apparent concentration (μg l^{-1})	Apparent bromodichloromethane concentration (μg l^{-1})
15	178	12.8
25	205	20.5

Reprinted with permission from Nicolson et al.[113] Copyright (1977) American Chemical Society

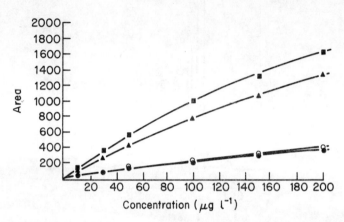

Figure 140 Typical direct aqueous injection calibration curves: ● chloroform, ○ trichloroethylene, ▲ chlorodibromomethane, □ bromodichloromethane. Reprinted with permission from Nicolson et al.[113] Copyright (1977) American Chemical Society

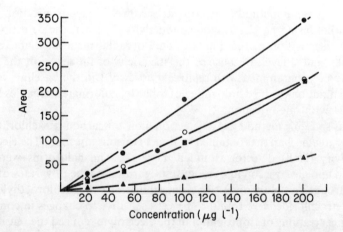

Figure 141 Typical gas sparging calibration curves: ● chloroform, ○ trichloroethylene, ▲ chlorodibromomethane, □ bromodichloromethane. Reprinted with permission from Nicolson et al.[113] Copyright (1977) American Chemical Society

flow of 75 ml min^{-1} giving the greatest sensitivity for all the compounds of interest. It was found that measurement of peak areas gave much more reproducible results than peak heights when temperature programming was used.

The temperature of the samples when introduced into the sparging apparatus was found to critically affect the results. In Table 106 are listed the apparent haloform concentrations in a sample which was injected into the sparger at 15 and at 25 °C. All standards were introduced at 25 °C comparing the chloroform and dichlorobromomethane results. It is apparent that a decrease in temperature has a greater effect on the less volatile components. Both methods produced non-linear calibration curves for most compounds, as shown in Figures 140 and 141. Whereas the slope of the calibration curves obtained using the gas sparging method and electrolytic conductivity detector generally increased with increasing concentration, the direct aqueous injection method with the electron capture detector showed a reverse trend. The curves in Figures 140 and 141 are non-linear least square plots which have been forced through the origin.

The precision of both methods was very good when using an integrator. Listed in Table 107 are the relative standard deviations when using a solution containing 50 μg l^{-1} of each of the organohalides. Use of the peak height measurements for quantitation in the gas sparging method was found to be very imprecise. A relative standard deviation of greater than 10% was common using this measuring technique. Because the gas sparging method is slow, very sensitive to chromatographic conditions and difficult to automate, the direct aqueous injection method is recommended, despite its slightly lower sensitivity, for routine testing of drinking waters for organohalides.

Table 107 Precision (σ) of the direct aqueous injection and gas stripping techniques

Method	$CHCl_3$	C_2HCl_3	σ (%) $CHCl_2Br$	$CHBr_2Cl$
DAI[a]	1.5	7.5	3.3	3.1
GS[b]	1.5	5.9	8.5	9.3

[a] 5 injections in a 24-h period.
[b] 5 injections in an 8-h period.
Reprinted with permission from Nicolson et al.[113] Copyright (1977) American Chemical Society

Comparison of determinations of haloforms, particularly chloroform, in drinking water revealed that direct aqueous injection results were consistently higher than those obtained by gas sparging (Figure 142). A similar trend was shown for chlorodibromomethane. In general, the direct aqueous injection method indicated chloroform and bromodichloromethane concentrations 1.5 and 2.2 times higher respectively, than the values obtained by gas sparging. Analysis of standards by both methods during calibration runs confirmed that purging inefficiency could not account for differences in results. To ascertain that no special matrix effects were involved, standards were made up in raw

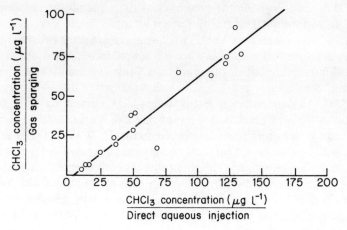

Figure 142 Correlation of results obtained for the concentration of chloroform in treated natural water by direct aqueous injection and gas sparging. Reprinted with permission from Nicolson et al.[113] Copyright American Chemical Society

drinking waters (collected before the prechlorination stage) and analysed by both methods. The results are listed in Table 108 and show only slightly lower values for the gas sparging procedure which can be explained by some inefficiency in the purging process. The results of successive purging experiments with field samples are summarized in Table 109. These data further confirmed that purging can account for only a fraction of the differences observed during analysis of field samples for haloforms by both methods.

These results suggested to Nicholson et al.[113] that actual chlorinated natural water samples, as opposed to synthetic standards, contain non-volatile haloform precursors which are injected in determinations by direct injection but not by gas sparging. If the additional quantities of haloforms observed by the direct aqueous injection method are produced during analysis from non-volatile halogenated organic compounds, then in a given sample the difference should be measurable by analysing prepurged samples by direct aqueous injection. This was proved to be the case as shown by the data summarized in Tables 110 and 111

Table 108 Comparison of spiked raw water samples

| | Concentration ($\mu g\ l^{-1}$) | | | | | | | | |
| | $CHCl_3$ | | | $CHCl_2Br$ | | | $CHBr_2Cl$ | | |
Sample location	Calcd	DAI	GS	Calcd	DAI	GS	Calcd	DAI	GS
Trent River	35	44	36	40	43	38	40	43	32
Trent River	70	74	64	80	80	69	80	82	70
Sudbury	35	37	37	40	43	41	40	41	33
Sudbury	70	72	69	80	79	78	80	81	82

Reprinted with permission from Nicolson et al.[113] Copyright (1977) American Chemical Society

Table 109 Percentage purging efficiency of spiked raw water standards and samples at 25 °C*

Sample no.	$CHCl_3$	$CHCl_2Br$
1817	93	75
1818	95	84
1819	94	88
2033	92	79

*Purging efficiency measured by repeatedly stripping the sample until no haloforms remain. Total concentrations were obtained as the sum of the readings.
Reprinted with permission from Nicholson et al.[113] Copyright (1977) American Chemical Society.

Table 110 Effect of purging on chloroform concentration

	$CHCl_3$ concentration (μg l^{-1})		$CHCl_3$ concentration (μg l^{-1}) after 30 min purge	
Location	Direct aqueous injection	Gas stripping	Direct aqueous injection	Gas stripping
Cayuga, Ontario	129	63	65	3
Cayuga, Ontario	119	61	60	3
Grand Bend	28	24	3	0
Union WTP	13	9	4	0
Elgin, Ontario	20	18	2	0

Reprinted with permission from Nicholson et al.[113] Copyright (1977) American Chemical Society

Table 111 Effects of purging on bromodichloromethane concentration

	$CHCl_2Br$ concentration (μg l^{-1})		$CHCl_2Br$ concentration (μg l^{-1}) after 30 min purge	
Location	Direct aqueous injection	Gas stripping	Direct aqueous injection	Gas stripping
Elgin WTP	10	7	2	0
Union WTP	10	8	1	0
Windsor WTP	13	12	1	0
Amherstburg	12	12	1	0

Reprinted with permission from Nicholson et al.[113] Copyright (1977) American Chemical Society.

which clearly shows that after purging, haloform precursors remain in solution which can only be converted to haloform under the conditions of direct aqueous injection.

As a further confirmation of this mechanism, water samples treated with sodium thiosulphate to neutralize any residual chlorine, were heated in sealed

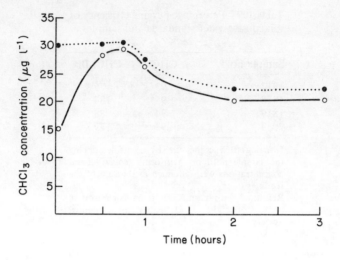

Figure 143 Change of concentration of chloroform on heating sample; ● direct aqueous injection; ○ gas sparging. Reprinted with permission from Nicholson et al.[113] Copyright (1977) American Chemical Society

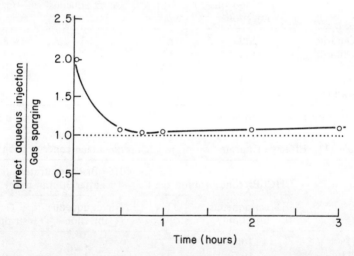

Figure 144 Change in ratio of the results obtained by direct aqueous injection and gas sparging for chloroform upon heating sample. Reprinted with permission from Nicholson et al.[113] Copyright (1977) American Chemical Society

(Teflon-lined septa, no head-space) hypovials. Vials were removed from the boiling water bath at intervals, cooled to 25 °C and analysed by both the direct aqueous injection and gas sparging methods. The results are shown in Figure 143. The values obtained by the direct aqueous injection method show little change over the first hour, whereas the values obtained by the gas sparging method increase dramatically during this period and approach the same concentration as indicated by direct aqueous injection. The apparent decrease in concentration of chloroform, after heating for 1 h, reflected by both techniques, is probably due to either decomposition of the chloroform or losses caused by leakages. Figure 144 shows the ratio of the values obtained by the two methods vs time. The fact that this ratio remains slightly greater than 1 even after extensive heating, is probably due to stripping inefficiency. These results indicate that when using the direct aqueous injection method, compounds which are precursors of chloroform and the other haloforms decompose at high injector temperatures to produce the haloforms.

In conclusion, it can be stated that the direct aqueous injection technique will produce a value which is the maximum haloform concentration which can be reached while the water is in the distribution system. When the free haloform content of the sample has been determined using both methods, the difference in the free and total potential haloform results gives an indication of the amounts of higher molecular weight haloform intermediates which cannot be detected by the gas sparging method.

Pfaender et al.[172] have also compared purge and trap and direct injection gas chromatographic techniques for the determination of chloroform in drinking water and found evidence for the presence of non-volatile precursors. The direct aqueous injection technique employing a bypass valve to vent water and electron capture detector, gave consistently higher values for chloroform than the purge method. Comparable results were obtained if the direct injection value after a 30 min purge was subtracted from the before-purged value. The nature of the residual measured by direct injection after purging was investigated and shown to be due to non-purgeable intermediates that decompose within the injection port of the gas chromatograph to chloroform. The residual varied depending on the source of the water sample examined and the specific configuration of the chromatograph employed. The results indicated the need for caution in the interpretation of chloroform and other trihalomethane values, especially haloform potentials, generated by direct aqueous injection.

These workers also used the Bellar and Lichtenberg[78] gas purging technique. For analysis of chloroform by direct aqueous injection, a Perkin Elmer Model 900 gas chromatograph equipped with a ^{63}Ni electron capture detector was utilized. A four-way valve was installed in the manifold between the column outlet and the detector inlet. The four fittings of the valve were connected to:

(1) analytical column outlet,
(2) make-up column outlet,
(3) the electron capture detector inlet, and
(4) vented to the atmosphere.

The valve permitted the effluent from the analytical column to be vented while the water peak was emerging and then redirected to the detector. During venting the effluent of the make-up column was connected to the detector. Prior to the installation of the valve, it had been observed that water injections of 1 μl caused no appreciable change in the performance of the electron capture detector but injection of larger quantities caused a loss of detector sensitivity. Installation of the valve and venting of the water permitted larger volumes of water to be injected with no loss in detector sensitivity. Chloroform analyses were carried out using 0.3 cm × 1.8 m stainless steel columns for both analysis and make-up, packed with 50–80 mesh Porapak Q. Operating parameters of the gas chromatograph were: column temperature, 150 °C; injector temperature, 160 °C; manifold temperature, 200 °C; detector temperature, 250 °C; and a nitrogen flow rate of 80 ml min^{-1} through both columns. One to ten microlitres of the natural chlorinated water samples (some preserved with thiosulphate) were injected directly into the chromatographic column. After the initial injection the water samples were purged with zero-gas grade nitrogen for 30 min at room temperature to remove any purgeable organics. The water was then reinjected and any residual chloroform measured.

The results of both analyses are presented in Table 112. Water samples from both sources gave consistently higher chloroform values for the initial direct aqueous injection than was measured by the purge method. After 30 min of purging, some apparent residual chloroform was detected by direct aqueous injection as found by Nicholson and Meresz.[106] Additionally, the amount of residual that remains after purging differs depending on the source of the water supply. Table 113 presents the residual after 30 min of purging as a percentage of the original aqueous injection value for several types of water. Thus, the conditions that contribute to the residual measurement vary from source to source. The chloroform value obtained by difference (subtracting the residual value from the initial injection value) appears to represent the chloroform that can be purged from water and is comparable to the chloroform as quantitated by the purge method.

Table 112 Comparison of chloroform analysis by direct aqueous injection and purge methods

		Chloroform (μg l^{-1})			
Sample source	S$_2$O$_3$ added when collected	Purge method	Direct injection (A)[b]	After purging (B)[c]	Difference (A − B)
Cincinnati tap water	+	95 ($\pm$5)[d]	117 ($\pm$9)	15 ($\pm$4)	102 ($\pm$4)
Cincinnati tap water	−	109 ($\pm$2)	128 ($\pm$7)	10 ($\pm$9)	109 ($\pm$16)
Durham tap water	+	143 ($\pm$8)	182 ($\pm$3)	39 ($\pm$2)	143 ($\pm$1)
Durham tap	−	146 ($\pm$4)	177 ($\pm$1)	45 ($\pm$6)	131 ($\pm$4)

[a]Average of triplicates.
[b]DAI method. No purge.
[c]DAI method. 30 min purge.
[d]Standard deviation.
Reprinted with permission from Pfaender et al.[172] Copyright (1978) American Chemical Society

Table 113 Residual chloroform found by direct aqueous injection method after purging for 30 min

Water source	% Residual*
Cincinnati, Ohio, tap water	14.0
Durham, NC, tap water	23.7
Chapel Hill, NC, tap water	26.5
Distilled water	0.0

*Expressed as a percentage of the total chloroform measured by DAI before purging.
Reprinted with permission from Pfaender et al.[172] Copyright (1978) American Chemical Society.

The direct aqueous injection method was also used to detect the other commonly observed trihalomethanes, including dichlorobromomethane, chlorodibromomethane, and bromoform. The minimum detectable amount of chloroform was on the order of 0.01 ng. The lower detection limit of the technique is dependent on the volume of water injected. With the valving arrangement described above, in conjunction with the electron capture detector, it is possible to inject up to 20 μl of water, which would make chloroform at the 0.5 ppb level detectable by direct aqueous injection. The direct aqueous injection technique requires no special glassware or apparatus, only an electron capture detector, and, maybe most significantly, requires little or no handling of water samples. The presence of non-purgeable material that is detected as chloroform after injection, however, raises questions about the usefulness of the technique. Only by determining the difference between direct aqueous injection determined concentrations before and after purging are values comparable to the purge method obtained.

Pfaender et al.[172] went on to examine the nature of the non-purgeable haloform precursors present in chlorinated natural waters. They started on the premise that these substances are formed as a result of water chlorination processes that are intermediates in the formation of chloroform, are not purgeable, but do decompose within the gas chromatograph to produce chloroform. Alternatively, chloroform may associate with less volatile organic compounds present in water and thus not be easily removed by purging but still be detectable as chloroform in the gas chromatograph.

In one experiment, the time and conditions used to purge chloroform from the water were investigated. The results, shown in Table 114, indicate that purging alone, even for extended periods, does not remove all the material measured as chloroform by direct aqueous injection. Treatment with acid, base, or boiling of the water, however, all result in essentially complete elimination of the residual. Although these results can be explained by an increased hydrolysis rate of an intermediate, these experiments do not completely resolve the nature of the residual. Acid, base, or heat treatment could also free chloroform complexed with the other organic material.

Table 114 Effect of purging conditions on chloroform residuals in Durham, NC, tap water

Water treatment	Purge time* (min)	Direct aqueous injection — chloroform ($\mu g \; l^{-1}$)
Tap water	0	312
Tap water + S_2O_3	0	283
Tap water	30	75
Tap water + S_2O_3	30	82
Tap water	90	74
Tap water + S_2O_3	90	65
Tap water	240	38
Tap water + S_2O_3, pH 2.0	30	3
Tap water + S_2O_3, pH 12.4	30	Trace[†]
Tap water + S_2O_3, pH 2	60	5
Tap water + S_2O_3, pH 12.4	60	Trace
Tap water + S_2O_3, 100 °C	30	ND[‡]

*Purging time with zero-gas N_2.
[†]Trace represents less than 0.5 $\mu g \; l^{-1}$.
[‡]ND, none detected.
Reprinted with permission from Pfaender et al.[172] Copyright (1978) American Chemical Society.

During these above experiments, when chloroform-free water was injected immediately following the direct injection of a chlorinated water sample, a small chloroform peak occurred. This ghost peak rarely represents more than 1–3 $\mu g \; l^{-1}$, as chloroform. When distilled water solutions of chloroform were injected, no ghost peak was ever observed on subsequent injections of chloroform-free water. A second injection of chloroform-free water usually produced no ghost peak, but if the residual chloroform in the previous chlorinated sample had been large, it might require two or more chloroform-free water injections before the ghost peak was eliminated.

It is unlikely that any chloroform would be retained on the injector liner or front of the column and be washed off by subsequent water injections at the 160 °C injector temperature used. The presence of these ghost peaks indicates that some chlorinated intermediates become trapped somewhere in the injection system of the gas chromatograph and are then hydrolysed to chloroform by subsequent water injections. While the small quantities measured do not account for the amount of residual observed, they do indicate that intermediates capable of producing chloroform are present. This would tend to substantiate the hydrolysis explanation.

A second experiment was conducted to test whether chloroform binds to organic constituents present in water, thus diminishing its purgeability. Known concentrations of chloroform were added to raw water samples and to deionized controls. Samples were then analysed by the purge method, and in two cases by direct aqueous injection after purging. Table 115 presents the results of this evaluation. It is apparent from the results that essentially all the chloroform

Table 115 Removal by purging of chloroform added to raw water[a]*

Sample	Control	$CHCl_3$ ($\mu g\ l^{-1}$) added	CHCl₃ measured by purge method Peak heights (in)[b]	CHCl₃ measured by purge method Response (mv)[b]	Direct aqueous injection method[c†] $CHCl_3$ ($\mu g\ l^{-1}$)
Durham raw water		75	7.35	3.68	
	Deionized water	75	7.35	3.68	
Durham raw water		7.5	6.00	1.20	
	Deionized water	7.5	6.20	1.24	
Durham raw water		0	0	0	
	Deionized water	0	0	0	
Durham raw water		236			0.0
	Durham raw water	0			0.0

[a]Untreated water from water treatment plant.
[b]Attenuation changed for different concentrations.
[c]Water purged with nitrogen for 30 min.
Reprinted with permission from Pfaender et al.[172] Copyright (1978) American Chemical Society.

added to unchlorinated water is removed by purging. In addition, no ghost peaks were observed when chloroform-free water was injected following the injection of purged, chloroform-treated raw water. These observations lead to the conclusion that the binding of chloroform to organics, if it did occur, has no measurable effect on chloroform purgeability and is an unlikely explanation for the residual observed after purging by direct aqueous injection.

In an attempt to obtain further information regarding the identity on the non-purgeable chloroform precursor Pfaender et al.[172] investigated by means of gas chromatography–mass spectrometry whether a possible precursors, particularly chloral (trichloroacetaldehyde), decomposes quantitatively to chloroform when subjected to direct aqueous injection. All commercial chloral tested produced three peaks, one identified as chloral, one as chloroform, and the third as dichloroacetaldehyde. The chloroform and dichloroacetaldehyde appeared to be contaminants in the chloral, resulting from decomposition or impurities from manufacturing. When purged with nitrogen, the chloroform peak disappeared from the solution while the chloral and dichloroacetaldehyde remained. The injection of the purged (chloroform-free) chloral solution yielded peaks for chloral and dichloroacetaldehyde only. There was no indication that chloral in aqueous solution decomposed to chloroform when directly injected into a gas chromatograph. Chloral does appear to decompose to chloroform during storage and if present in water or formed during chlorination processes, may ultimately add to the chloroform levels present but cannot account for the residual chloroform observed with the direct aqueous injection technique. This contradiction to the results observed and reported by Nicholson et al.[106,113] in addition to the observation of ghost peaks (see above) in this study but not by Nicholson et al., indicates that the decomposition of chlorinated intermediates in the injection port of a gas chromatograph may not always be complete.

The size of the residual measured by direct aqueous injection may therefore depend on the materials and configuration of the injection port of the chromatograph used. This raises doubts as to the usefulness of the residual haloforms measured by direct aqueous injection as an estimate of haloform formation potential. The relationship of the potential as determined by direct aqueous injection and the haloforms formed during distribution of a finished drinking water is still largely unknown.

Resin adsorption–gas chromatography

Two approaches to this analysis are possible. In one, the water[173] sample is purged with purified nitrogen or helium and the purged trihaloforms are trapped on a column of macroreticular resin such as Amberlite XAD-2 or XAD-4. The trapped trihaloforms are then desorbed from the column with a small column of a polar solvent prior to gas chromatography. In the second approach the water sample is contacted directly with the resin and then the trihaloforms desorbed as before. Alternatively the resin can be injected directly into the gas chromatograph injection port.[174] Both methods provide a very useful built-in concentration factor which improves method sensitivity.

Renberg[175] has reported a resin adsorption method for the determination of trihalomethanes and chloroethanes and dichloroethane in water. In this method halogenated hydrocarbons are determined by adsorption on to XAD-4 polystyrene resin and eluted with ethanol. The extract is analysed by gas chromatography and is sufficiently enriched in hydrocarbon to be suitable for other chemical analysis or biological tests. Volatile hydrocarbons yielded recoveries of 60–95%. By using two series-connected columns Renberg[175] was able to study the degree of adsorption and the chloroethane were found to be more strongly adsorbed than the haloalkanes.

Method

Materials Amberlite XAD-4 (Rohm and Haas Company), acetone (pesticide grade), methanol and ethanol (spectroscopic grade) were used. The purity of the methanol and ethanol was checked by the gas chromatography.

Polymeric resin clean-up Amberlite XAD-4 was placed in a column with a glass filter. Five bed volumes of acetone were passed through the column with a flow rate about 0.1 bed volume per minute. Then water was passed upflow through the column at a rate sufficient to expand the bed by about 50% to remove the smallest particles. The resin was then extracted by means of a Soxhlet apparatus with methanol for at least 6 h, then replaced into the column and eluted with 20 bed volumes of purified water and finally stored under purified water. The purity of the resin was checked in a blank procedure.

Purified water Deionized water was distilled through an all glass apparatus. The first 10% of the distillate was discarded and the 10–50% portion was passed through a column containing 10 ml XAD-4 at a flow rate of about 30 ml min^{-1}. If the resin was taken directly from the Soxhlet extraction and thus contained methanol, the first 200 ml of the eluate was discarded.

Adsorption and desorption steps Two glass columns, 15 × 1.1 (i.d.) cm with Teflon stopcocks, were each filled with 5 ml of XAD-4 resin. Purified water was added to the top of the columns and the beds were stirred to release air bubbles. The columns were connected in a series and the water sample was allowed to pass through the columns at a flow rate of 30 ml min^{-1}. The columns were allowed to run dry and disconnected. To each column, ethanol (3 ml) was added, the beds were stirred and the resins were allowed to swell for at least 20 min. The beds were stirred once more to release air bubbles and the columns were eluted separately with ethanol (flow rate 1 ml min^{-1}) until 20 ml of eluate were collected. The eluates were then analysed by gas chromatography.

Gas chromatography system A Varian 3700 gas chromatographic equipped with an electron detector was used. Injector and detector temperatures were 250 °C. The ethanol extracts were injected into a 170 × 0.18 (i.d.) cm glass column, packed with Chromosorb 101 80–100 mesh which was kept at 145 °C. Alternatively, ethanol extract (1 ml) was shaken with purified water (5 ml) and hexane or pentane (1 ml). The organic phase was, after centrifugation, injected into a 50 m OV-17 coated capillary column at a split ratio of 1:10. After injection, the column was held 5 min at 80 °C, then programmed at a rate of 2 °C min^{-1} until 120 °C was reached.

The results of recovery experiments for different volatile organohalogens are summarized in Table 116. Similar results were also obtained with spiked drinking water. It should be pointed out that the flow rates used here for both the adsorption and desorption step (360 bed vol h^{-1} and 12 bed vol h^{-1}, respectively) are much higher than those recommended by the manufacturer (4–16 bed vol h^{-1} and 1–4 bed vol h^{-1}, respectively). Such high flow rates speed up the time of analyses while still maintaining high recoveries.

To study the adsorption character of substances described here, as well as other compounds, Renberg[175] connected two columns in a series and after the passage of the water sample the columns were eluted separately. The recovery of the substances of the second column, compared to those of the first one, will indicate the leakage of substances which is dependent on, for example, water flow rate and polarity of the substances.

The ratio (a_2/a_1) of the amounts found in the second (a_2) and the first columns (a_1) respectively, will predict the degree of adsorption, which can be regarded as a measurement of the lipophilicity of the substances studied. The values of the ratios found in the recovery experiments, shown in Table 116, indicate tetrachloroethane and 1,2-dichloroethane as the most and the least lipophilic of the substances tested. As long as the desorption step is carried out

Table 116 Results of recovery experiments from five spiked samples

Substance	Amount of spiked substance (μg 250 μl^{-1})	Recoveries* Mean value (%) $\pm$ rel. s.d.	Ratio a_2/a_1[†] $\pm$ s.d.
Chloroform	2.9	85 $\pm$ 4.1	0.27 $\pm$ 0.03
Bromodichloromethane	0.63	86 $\pm$ 9.6	0.22 $\pm$ 0.08
Chlorodibromomethane	1.3	91 $\pm$ 6.7	0.12 $\pm$ 0.03
1,2-Dichloroethane	120	95 $\pm$ 9.0	0.44 $\pm$ 0.09
Trichloroethane	2.5	74 $\pm$ 3.6	0.07 $\pm$ 0.03
Tetrachloroethane	1.4	60 $\pm$ 4.5	0.06 $\pm$ 0.02

*Sum of the amounts, found in the two series-connected columns ($a_1 + a_2$).
†Ratio of the amounts found in the second and the first column, respectively.
To prepare solutions purified water (250 ml) was transferred to a 250 ml separatory funnel and 250 μl of an ethanol solution containing chloroform, bromodichloromethane, chlorodibromomethane, 1,2-dichloroethane, and tri- and tetrachloroethane was added. The funnel was shaken and the spiked water allowed to pass the columns. The flow was stopped when the leve just reached the top of the upper bed. In order to rinse the glass walls from substances adsorbed, additional water (3 $\times$ 250 ml) was passed through the funnel columns and the columns were eluted as described. The separatory funnel was shaken with ethanol (25 ml) for the determination of substances adsorbed on the glass walls.
Reprinted with permission from Renberg.[175] Copyright (1978) American Chemical Society.

quantitatively, an approximation of the original concentration can be calculated, using the values from the two individual columns as the first two terms in an infinite geometrical series, giving a finite sum (s) as long as $a_2/a_1 < 1$ according to the formula:

$$s = a_1^2/(a_1 - a_2)$$

Thus using Table 116 the calculated true values for, e.g. chloroform and chlorodibromomethane, are both 92%. However, for tri- and tetrachloroethane, the values are still 74 and 60%, respectively, in spite of their low a_2/a_1 values, which indicates that these compounds are strongly adsorbed, but not so easily desorbed. The elution profiles (see Figure 145) show that this is especially pronounced for tetrachloroethene. It seems that this strong adsorption, compared to, e.g. the trihalomethane, is not only dependent on the non-polar character of the chloroethenes, but may also depend on interference of the π electrons with the polystyrene matrix. The recoveries of the chloroethenes are increased if either a larger elution volume or shorter columns are used. However, the technique described is a compromise between relatively small elution volumes and satisfactory recoveries for compounds with quite different water solubilities.

The method was applied to incoming and outgoing water from a water treatment plant (see Figure 146) and on waste water from a poly(vinyl chloride) plant for the determination of trihalomethanes and 1,2-dichloroethane.

Kissinger[176] has reported a high resolution capillary column gas chromatographic procedure for the determination of trihaloforms in drinking water.

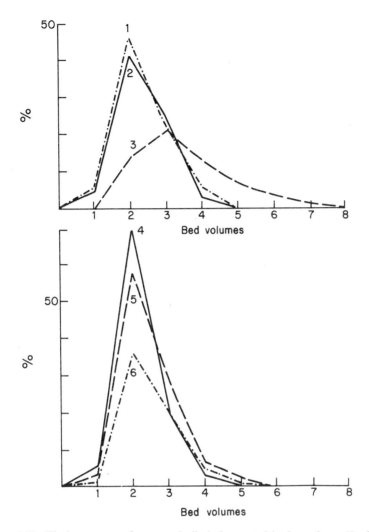

Figure 145 Elution curves of some volatile halogenated hydrocarbons (5 ml XAD-4 column) (1) Chloroform, (2) 1,2-dichloroethane, (3) tetrachloroethene, (4) bromodichloromethane, (5) chlorodibromomethane, (6) trichloroethene. Reprinted with permission from Renberg.[175] Copyright (1978) American Chemical Society

The trihaloforms were determined by sorption on to columns of acetylated XAD-2 resin after the removal of the pyridine solvent. Detection limits for the four chloro- and bromo-compounds of the haloform type in a 100 ml sample of water were below 1 μg l^{-1}.

Various other workers have discussed resin adsorption methods for the determination of trihaloforms in water.[177-182, 184] Kissinger and Fritz[183] have described a method based on adsorption of trihaloforms on acetylated XAD-2 resin, followed by elution with pyridine.

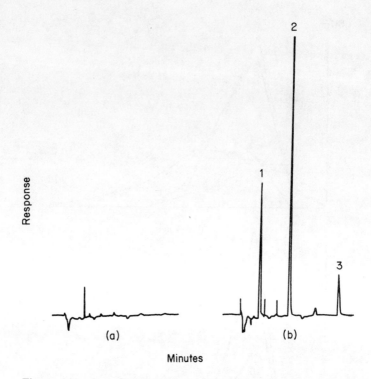

Figure 146 Gas chromatogram of water extracts from a water treatment plant (OV-17 capillary column). (a) Incoming water, (b) outgoing water. (1) Chloroform, 5.0 μg l^{-1}; (2) bromodichloromethane, 2.5 μg l^{-1}; (3) chlorodibromomethane, 0.54 μg l^{-1}. Reprinted with permission from Renberg.[175] Copyright (1978) American Chemical Society

Capillary column gas chromatography

Eklund et al.[185,186] have developed a method for the determination of down to 1 μg l^{-1} volatile organohalides in waters which combines the resolving power of the glass capillary column with the sensitivity of the electron capture detector. The eluate from the column is mixed with purge gas of the detector to minimize band broadening due to dead volumes. This and low column bleeding give enhanced sensivitity. Ten different organohalides were quantified in tap water, sea water, and industrial effluent from a pulp mill. Using this technique these workers detected bromoform in sea water for the first time.

Method

Apparatus A Perkin Elmer 3920 gas chromatograph equipped with a ^{63}Ni electron capture detector was used. The gas chromatograph was modified in order to minimize band broadening effects due to dead volumes, which is essential in work with capillary columns. The injector was modified to a

Grob-type injector. At the column end a scavenger gas (argon plus 5% methane) was added in order to minimize band broadening in the manifold and in the detector. The construction of this gas chromatograph allows heating of the scavenger gas in the manifold before it is added to the column flow. A constant temperature of the scavenger gas, unaffected by the oven temperature, is necessary to avoid baseline drift during temperature programming. The flow rate of the scavenger gas was 30 ml min^{-1}. Glass capillary columns were prepared according to the procedures of Grob.[187,188] Duran 50 glass tubing, 8 mm o.d. 3 mm i.d. were drawn to 0.3 mm i.d. tubing by a glass drawing machine. The inside surface of the glass capillary was modified by deposition of fine $BaCO_3$ crystals. Deactivation of the glass surface was accomplished by coating with dilute solutions of Carbowax 20M and Emulphor ON 870 followed by rapid heat treatment. For direct injection of water samples a Carbowax 400 column, length 40 m, was used. This column was coated with a 20% solution of the stationary phase in methylene dichloride using the mercury plug method.[189] For injections of pentane extracts two SE-52 columns were prepared; one with a length of 33 m with a film thickness of 0.1 μm and one with a length of 10 m with a film thickness of 1.5 μm. These columns were coated by the static method. Column efficiency for the three columns used, expressed as coating efficiency, ranged from 85 to 93%.

Procedure

For the analysis of halogenated compounds in relatively contaminated waters, e.g. tap water and river water, 100 ml of water was manually shaken with 5 ml pentane, for 5 minutes. To the extraction solvent tribromochloromethane was added as internal standard. Errors resulting from differences in the injected volumes are thereby reduced. A volumetric flask is the most suitable device for the extraction since the narrow neck allows easy sampling of the extract with a microlitre syringe for injection into the gas chromatograph. For determination of halogenated hydrocarbons in relatively pure waters, e.g. lake water, sea water, higher water to pentane ratios were used.

Halogenated hydrocarbons in different waters were identified by comparison with a standard solution. A chromatogram of a standard pentane solution of various volatile organochlorine compounds including trihaloforms is shown in Figure 147. Retention times were measured on two columns with different stationary phases, i.e. SE-52 and Carbowax 400.

Eklund *et al.*[185] determined the extraction efficiency obtained by this procedure for solutions of chloroform, bromodichloromethane, and chlorodibromomethane in tap water.

Extraction

Extraction of 100 ml water with various amounts of pentane ranging from 1 to 15 ml showed that the extraction efficiency was increased when using a lower

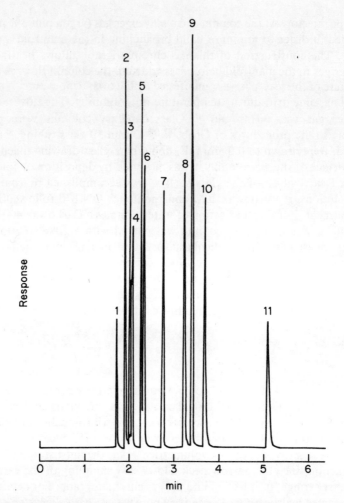

Figure 147 Chromatogram of a standard mixture containing 11 organohalides. Stationary phase, SE-52; 33 m × 0.3 mm i.d., inj. temp. 200 °C, column temp. 50 °C, interface 250 °C, detector temp. 250 °C. Helium carrier gas flow rate, 36 ml s^{-1}, scanvenger gas flow 30 ml min^{-1}. Split ratio 1:20, 1 = CH_2Cl_2, 2 = $CHCl_3$, 3 = CH_3CCl_3, 4 = CCl_4, 5 = $CHClCCl_2$, 6 = $CHBrCl_2$, 7 = $CBrCl_3$, 8 = $CHBr_2Cl$, 9 = CCl_2CCl_2, 10 = $CHCl_2I$, 11 = $CHBr_3$

water to pentane ratio (Figure 148). The extraction efficiency was determined using consecutive extractions of tap water. It is evident that using a water-pentane ratio lower than 20 means a dilution of the extract. With this ratio the extraction efficiency for the above three compounds in tap water was 82, 89, and 84% respectively. The relative standard deviation for the extraction procedure was less than 3%, as determined from five extractions including the relative standard deviation for the split injection. A higher sensitivity is thus achieved using a higher water to pentane ratio. Despite the

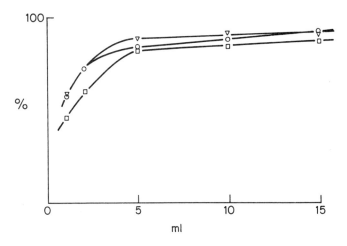

Figure 148 Extraction efficiency for $CHCl_3$ (□), $CHBrCl_2$ (▽) and $CHBr_2Cl$ (○) in 100 ml tap water as a function of amount extraction solvent

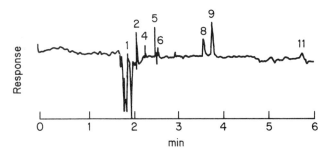

Figure 149 Chromatogram showing the minimum detectable amount for eight organohalides. $0.1\,\mu l$ injected, split ratio 1:70, attenuation ×2. Approximate amounts on column: 1, 200 fg; 2, 20 fg; 4, 1 fg; 5, 10 fg; 6, 3 fg; 8, 4 fg; 9, 4 fg; 11, 8 fg; g.c. conditions and notations as in Figure 147

Table 117 Electron capture detection limits and approximate detection limits of eight organohalides in water

	Electron capture detection limit (fg)	Approximate detection limits of organohalides in water (ng l^{-1})
CH_2Cl_2	500	50
$CHCl_3$	10	1
CCl_4	1	0.1
$CHClCCl_3$	5	0.5
$CHBrCl_2$	2	0.2
$CHBr_2Cl$	2	0.2
CCl_2CCl_2	2	0.2
$CHBr_3$	10	1

lower extraction efficiency, the concentration of the extractable compounds in the extract is higher.

The approximate detection limits of the method for some compounds in water are given in Table 117, together with the gas chromatographic detection limits. A chromatogram showing the gas chromatographic detection limits for eight organohalides is presented in Figure 149. The linear dynamic range of the electron capture detector for these compounds was four orders of magnitude from the detection limit.

Application of the procedure

Tap water analysis A chromatogram from a tap water analysis by the pentane method is shown in Figure 150. The gas chromatographic conditions are given in the figure caption. Quantitative data from this analysis are given in Table 118. A fast analysis of haloforms in tap water on a 10 m glass capillary column is shown in Figure 151. An example of a haloform analysis of tap water by direct aqueous injection is shown in Figure 152. Quantitations of chloroform and bromodichloromethane based on direct water injections on a Carbowax 400 column differ from quantitations based on the pentane extraction procedure. Measured concentrations are 2.8 and 2.6 times higher for chloroform and bromodichloromethane with direct water injections compared to the extraction method. This discrepancy is not due to differences in split ratios for chloroform and bromodichloromethane resulting from the solvent. Equal peak heights were obtained when spiked samples of chloroform and bromodichloromethane in water or pentane were analysed.

Sea water analysis

A chromatogram in the analysis of a sea water sample is shown in Figure 153 and quantitative data are given in Table 118. Bromoform (peak 11) has not been reported in sea water samples before.

Table 118 Concentrations of different organohalides in different waters

	Tap water ($\mu g\ l^{-1}$)	Sea water ($\mu g\ l^{-1}$)	Pulp mill effluent ($\mu g\ l^{-1}$)
CH_2Cl_2	—	—	640
$CHCl_3$	9.5	0.026	760
CH_3CCl_3	0.060	0.046	—
CCl_4	0.017	<0.0005	
$CHClCCl_2$	0.015	0.015	2.8
$CHBrCl_2$	2.2	—	—
$CHBr_2Cl$	0.6	—	—
CCl_2CCl_2	0.008	0.005	1.4
$CHCl_2I$	0.007	—	—
$CHBr_3$	0.016	0.027	—

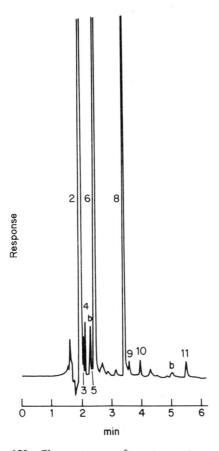

Figure 150 Chromatogram from tap water analysis, 100 ml of water was extracted with 5 ml pentane. 2 μl injected, split ratio 1:20; b = background; g.c. conditions and notations as in Figure 147

Industrial waste water

Effluents from a sulphate pulp mill were investigated for their content of low molecular weight halocarbons. Large amounts of chlorine are used in the bleachery plant resulting in the formation of chlorinated compounds. Results from this analysis are shown in Figure 154 and Table 118.

Trussell et al.[190] used glass capillary gas chromatography for the precise and rapid analysis of trihalomethanes in n-pentane extracts of drinking water. These workers claim that the use of glass capillary gas chromatography has several advantages over conventional gas chromatography. First, the glass capillary column provides better resolution of individual components: the typical packed column has 2000–10,000 theoretical plates whereas capillary columns range from 15,000 to 50,000 theoretical plates. Second, the high-quality resolution allows

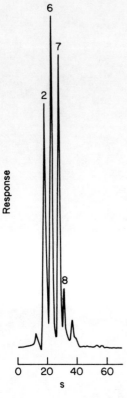

Figure 151 Tap water analysis on a 10 m × 0.3 mm i.d. SE-52 column. Helium carrier gas flow rate 83 ml s^{-1}; g.c. conditions and notations as in Figure 147

shorter gas chromatographic runs without overlapping peaks and thus brings about time savings. Finally, the trihalomethanes can be resolved well on general usage liquid phases, which gave versatility in analytical capability in that a wider range of types of organic compounds can be analysed without requiring time-consuming column changes.

Method

Sampling Water samples are taken in 120 ml bottles constructed of photo-protective amber glass. This precaution is taken because algal growths might alter the organic composition of the sample upon standing and sunlight may cause further production of chlorinated organics by photochemical reaction. The bottles were equipped with pressure-bonded PTFE septa and twist caps with openings for introduction of syringe needles. The bottles were cleaned by standard laboratory procedures, then heated at 400 °C for at least 1 h.

To prevent loss of the volatile trihalomethanes, sample aeration should be minimized during collection, whether sampling is by dipping from a tank or opening a sample tap. A sample tap must be opened and allowed to flow free for several minutes prior to sampling so as to remove particulates that might have settled in the pipe and debris that may have accumulated on the valve seat.

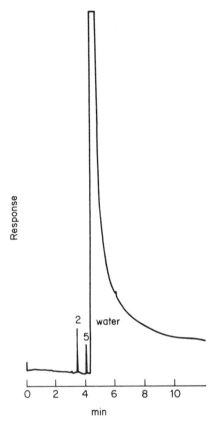

Figure 152 Injection of tap water on a 40 m × 0.3 mm i.d. Carbowax 400 column. Column temp. 90 °C Helium carrier gas flow rate 22 ml s^{-1}. 1 µl injected, split ratio 1:30. Notations as in Figure 147

High flow velocities should be avoided unless this purging process is continued for a considerable length of time as high velocities can dislodge material that ordinarily might not be removed. If the sampling tap is remotely located, it may be necessary to let it run for 15 min or more to obtain a representative sample. The sample bottles were completely filled, allowing no head-space, in order to prevent partitioning of the trihalomethanes between aqueous and gaseous phases.

The trihalomethanes are extracted by the addition of an organic solvent to the sample bottle and the subsequent partitioning of the trihalomethanes into the organic phase in which they are more soluble. The solvent, *n*-pentane, is added to the sealed sample bottles utilizing two 10 ml syringes, one for injecting pentane and the other for removing an equivalent amount of water. The needles of both syringes are inserted through the septum, and 5 ml of pentane are injected. The water displaced is collected in the second syringe and discarded

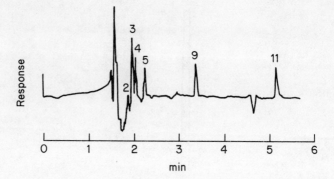

Figure 153 Chromatogram from sea water analysis. 100 ml water was extracted with 2 ml pentane. 2 μl injected, split ratio 1:20; g.c. conditions and notations as in Figure 147

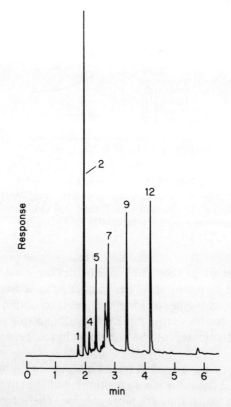

Figure 154 Chromatogram from industrial waste water analysis, 100 ml water was extracted with 5 ml pentane. 2 μl injected, split ratio 1:20. 12 = unidentified. g.c. conditions and notations as in Figure 147

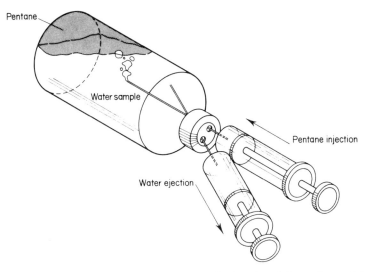

Figure 155 Addition of pentane to water sample. Reprinted from *Journal AWWA*, Vol. 71, No. 7 (July 1979) by permission. Copyright © 1979, The American Water Works Association

(Figure 155). Extraction is performed by 20 min of mechanical shaking at 400 rpm on a gyrorotatory platform shaker. Then a 1 μl aliquot of the upper organic phase is removed with a syringe for subsequent injection.

The extraction of trihalomethanes into an organic solvent is not quantitative. The percentage recovery of trihalomethanes is governed by their partition coefficients, which are functions of the properties of the liquid phases and the trihalomethanes as well as such other physical parameters as temperature and ionic strength. Partition coefficients for several currently used extraction solvents are given in Table 119. (The larger the partition coefficient the greater the percentage recovery.) In the case of pentane the partition coefficient for chloroform corresponds to approximately 80% extraction recovery.

Table 119 Partition coefficients for trihalomethane extraction solvents

Solvent	Partition coefficients				Purity	Boiling point (°C)
	Chloroform	Dichloro-bromomethane	Dibromo-chloromethane	Bromoform		
Methylcyclo-hexane	14	12	14	20	+	101
Cyclohexane	29	26	40	36	+	81
Hexane	18	42	104	116	−	69
Iso-octane	36	24	83	80	+	99
Petroleum ether	54	47	125	134	+	30–60
Pentane	41	43	118	161	0	36
Toluene	17	20	35	28	−	111

Reprinted from *Journal AWWA*, Vol. 71, No. 7 (July 1979) by permission. Copyright © 1979, The American Water Works Association

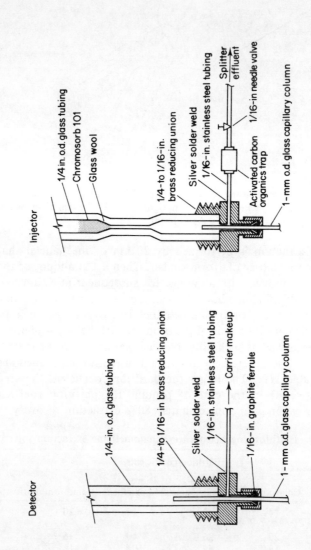

Figure 156 Injector and detector adapters for split mode capillary gas chromatography. Reprinted from *Journal AWWA*, Vol. 71, No. 7 (July 1979) by permission. Copyright © 1979, The American Water Works Association

Chromatographic analyses

The modification to a standard gas chromatograph required for capillary column work is shown in Figure 156. Provision is made for modifying the injection port and the column connections and for supplying a purge gas at the end of the capillary column. The setup involves an injection system that splits the material introduced into the injection port so that a fraction of the injected sample reaches the capillary column. As pictured in Figure 156 approximately 30 mg of Chromosorb 101 assist mixing in the injection port splitter so that a representative portion enters the column. The separation is performed by a 30 mm × 0.25 mm coiled glass column internally coated with a 0.4–0.5 micron film of OV-101. The splitting ratio of the nitrogen carrier gas entering the injection port to that entering the capillary column is 100:1. The 1 ml min effluent from the column is swept into the electron capture detector with an additional 29 ml min^{-1} of nitrogen, giving a total flow of 30 ml min^{-1} to the detector. During the analysis the injection port is maintained at 150 °C and the detector at 300 °C; the column is run isothermally at 80 °C. One microlitre injection of the pentane extracts yield chromatograms with excellent resolution.

Concentrations of the trihalomethanes were calculated with the internal standard method by the daily injection of standards prepared as outlined above. The internal standard, 1,2-dibromoethane, is added to the pentane prior to extraction. The use of the internal standard allows correction for variation in size of injections. The relative responses of the trihalomethanes were calculated by extraction and injection of aqueous standards of the trihalomethanes. These relative responses can be calculated manually by peak area (or peak height) or electronically by the use of an integrator.

Trussell et al.[190] tested the precision of their technique by repeated analysis of water samples prepared in single batches so as to contain the same concentration of trihalomethanes. Accuracy was checked by spiking water samples of known trihalomethane concentrations so that the concentration was approximately doubled. The water samples of these tests were prepared with water containing a natural complement of humic substances (2.5 mg l^{-1} TOC) to simulate more accurately routine monitoring conditions. The three different concentration ranges — low, medium, and high — correspond to total trihalomethane levels of 40, 85, and 160 μg l^{-1} respectively. The distribution of trihalomethane concentrations tested was ($CHCl_3$) > (Cl_2BrCH) > (Br_2ClCH) > ($CHBr_3$); this is seen in the vast majority of samples in the absence of appreciable quantities of bromide ion. Minimum detection levels range from 0.1 to 1.0 μg l^{-1}. The outcomes of these tests are shown in Tables 120–122. The results demonstrate that both variations of this liquid-liquid extraction technique have excellent repeatability and accuracy.

Table 120 Trihalomethane analysis with packed column chromatography*

Parameter	Chloroform concentration ($\mu g\ l^{-1}$)			Dichlorobromomethane concentration ($\mu g\ l^{-1}$)			Dibromochloromethane concentration ($\mu g\ l^{-1}$)			Bromoform concentration ($\mu g\ l^{-1}$)		
	Low	Medium	High	Low	Medium	High	Low	Medium	High	Low	Medium	High
Sample 1	21.4	39.9	78.9	17.7	29.9	50.9	4.2	8.3	17.6	1.8	4.1	7.9
Sample 2	21.3	41.6	77.7	32.4	49.9	4.2	8.8	17.3	2.0	4.3	7.8	
Sample 3	22.1	40.8	78.1	18.1	31.9	53.3	4.3	8.6	17.5	1.8	4.3	7.9
Sample 4	20.0	41.5	77.7	17.0	32.3	47.5	4.1	8.5	17.4	1.9	3.8	7.8
Sample 5	21.3	40.0	78.5	17.5	30.2	50.8	4.2	8.5	17.4	1.9	3.9	7.5
Sample 6	20.9	41.7	79.5	17.4	31.4	48.4	4.1	8.6	17.2	1.7	4.0	7.2
Sample 7	20.1	41.4	79.4	16.9	31.2	49.5	4.1	8.5	17.5	2.0	4.1	7.8
Mean	21.0	41.0	78.5	17.5	31.3	50.0	4.2	8.5	17.4	1.9	4.1	7.8
Relative standard deviation (%)	(3.4)	(1.8)	(0.9)	(3.6)	(3.1)	(3.7)	(1.8)	(1.8)	(0.8)	(5.8)	(4.5)	(1.8)
Actual value		42.2	85.6		31.0	54.1		8.5	17.3		3.9	7.8
Relative error		(3.3)	(8.3)		(1.0)	(7.6)		(0.0)	(0.6)		(5.1)	(0.0)

*2 m × 2 mm i.d. glass columns containing 10% squalane on Chromosorb W-AW (80–100 mesh). The analysis is run isothermally at 67 °C with an injection port temperature of 90 °C and a detector temperature of 300 °C on a dual column gas chromatograph equipped with a ^{63}Ni electron capture detector. Nitrogen at 30 ml min^{-1} is used as the carrier gas; no make-up flow to the detector is supplied. Standard of the trihalomethanes is prepared gravimetrically by micropipetting the trihalomethanes into absolute methanol. The trihalomethanes are added to a tared 50 ml volumetric flask approximately half filled with the alcohol and the weight of each addition was recorded. The volumetric flask was filled to volume and the contents mixed to provide a homogenous standard of known concentration. Aliquots of this standard were sealed in glass ampoules and refrigerated for use within 2 weeks. One 2 ml aliquot for immediate use is stored in a 2 ml vial with a PTFE septum and open topped twist cap. Appropriate dilutions of this primary standard in alcohol were made as required and spiked into trihalomethane-free water with a volumetric pipette. The final concentration of the trihalomethanes in the water standards is set at 10, 50, and 100 $\mu g\ l^{-1}$.

Table 121 Trihalomethane analysis with capillary column chromatography*

Parameter	Chloroform concentration ($\mu g\ l^{-1}$)			Dichlorobromomethane concentration ($\mu g\ l^{-1}$)			Dibromochloromethane concentration ($\mu g\ l^{-1}$)			Bromoform concentration ($\mu g\ l^{-1}$)		
	Low	Medium	High	Low	Medium	High	Low	Medium	High	Low	Medium	High
Sample 1	21.1	42.8	80.4	13.6	27.1	54.5	4.3	8.7	17.0	1.9	3.9	7.5
Sample 2	23.4	46.5	81.0	12.2	27.4	53.4	3.9	8.7	16.7	1.8	4.0	7.4
Sample 3	21.1	45.7	80.9	12.4	27.0	53.8	4.0	8.9	16.9	1.8	4.1	7.6
Sample 4	20.3	45.3	79.1	11.7	27.7	51.2	3.8	8.8	16.5	1.8	4.0	7.4
Sample 5	20.5	46.9	82.9	12.5	28.1	56.3	4.0	9.0	17.3	1.8	4.1	7.6
Sample 6	21.3	44.5	80.5	12.5	27.1	54.8	4.0	8.7	17.2	1.8	4.0	7.6
Sample 7	21.3	42.6	82.6	12.6	28.3	54.3	4.0	8.9	17.5	1.8	4.1	7.9
Mean	21.3	44.9	81.0	12.6	27.5	54.0	4.0	8.8	17.0	1.8	4.0	7.8
Relative standard deviation (%)	(4.7)	(3.8)	(1.6)	(4.6)	(1.9)	(2.9)	(3.8)	(1.4)	(2.0)	(2.1)	(1.9)	(2.2)
Actual value	42.8	85.6		27.1	54.2		8.7	17.4		3.9	7.8	
Relative error (%)		(4.9)	(5.3)		(1.5)	(0.4)		(1.1)	(2.3)		(2.5)	(2.6)

*30 m × 0.25 mm i.d. glass column internally coated with a 0.4–0.5 micron film of OV-101.
Reprinted from *Journal AWWA*, Vol. 71, No. 7 (July 1979), by permission. Copyright © 1979, The American Water Works Association.

Table 122 Comparison of precision and accuracy obtainable with packed and capillary column chromatography

Trihalomethane	Average relative standard deviation (%)		Relative error (%)	
	Packed column	Capillary column	Packed column	Capillary column
Chloroform	2.0	3.4	5.8	5.1
Dichlorobromomethane	3.5	3.1	4.3	1.0
Dibrochloromethane	1.5	2.4	0.3	1.7
Bromoform	4.1	2.1	2.6	2.6
Average total trihalomethanes	2.8	2.8	3.3	2.6

Reprinted from *Journal AWWA*, Vol. 71, No. 7 (July 1979), by permission. Copyright © 1979, The American Water Works Association.

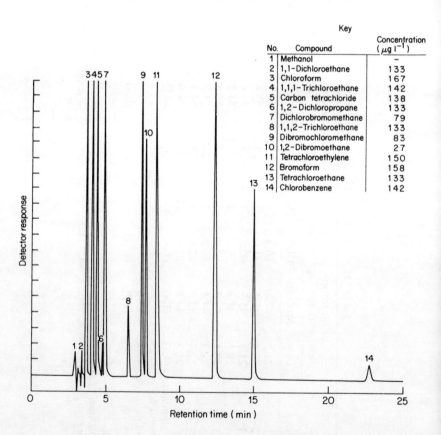

Figure 157 Resolution of halogenated organics by glass capillary chromatograph. Reprinted from *Journal AWWA*, Vol. 71, No. 7 (July 1979) by permission. Copyright © 1979, The American Water Works Association

The chromatogram obtained from the analysis of a water sample spiked with 14 volatile organic compounds is shown in Figure 157. Only two of these compounds were not well resolved on the glass capillary column. 1,2-Dichloroethane appears as a shoulder on the 1,1,1-trichloroethane peak and trichloroethylene is masked by the dichlorobromomethane peak. Trussell et al.[190] point out that although capillary gas chromatography appears to offer no significant advantage in terms of precision and accuracy there are several ways in which it is superior. The sharp resolution and improved separation resolve low molecular weight, volatile halogenated compounds that might ordinarily appear as a single peak, e.g. dichlorobromomethane and 1,2-dichloropropane. If the raw water itself contains interfering compounds, these are more easily resolved as separate peaks by capillary gas chromatography. On occasion, water samples that have shown only the four trihalomethanes on a packed column showed 20 or more compounds on a glass capillary column. The narrow peak widths, the accurately reproducible retention times, the separation of interfering raw water compounds, and the resolution of chlorinated compounds similar to the trihalomethanes all contribute to making more positive identification. This is important because the electron capture detector utilized in liquid-liquid extraction techniques is relatively non-specific, responding to many electron-capturing materials in addition to the halogen.

Gas chromatography–mass spectrometry

Fujii[191-193] has carried out extensive work on the application of direct aqueous injection–mass fragmentography–gas chromatography–mass spectrometry to the analyses of various organhalogen compounds including trihaloforms in water at concentrations below the parts per billion level (1 in 10^9). In one method[191] a sample volume of 100 μl was injected into a precolumn with diglycerol as stationary phase, on which water is retained. The halogenated compounds, which are not retained by the precolumn, are separated by a packed analytical column.

Method

Apparatus Fujii[191] used a Finnigan 3300F gas chromatograph–mass spectrometer equipped with a multiple ion detector and operated in the electron impact mode. The interface between the gas chromatograph and the mass spectrometer was an all-glass jet-type enrichment device. The mass spectrometer was set to unit resolution (10% valley between adjacent nominal masses). The resulting ion currents were recorded on a multichannel strip chart recorder. Other conditions held constant throughout the analysis were: helium carrier gas flow rate, 30 ml min^{-1}; temperature of the gas chromatograph injection port, 200 °C; pressure in the mass spectrometer, 6×10^{-6} Torr); ionizing voltage, 70 eV; emission current, 320 μA. The gas chromatograph precolumn contained 10% diglycerol on 60–80 mesh Chromosorb W NAW (Johns-Manville) and the main column 5% SE-30 on 60–80 mesh Chromosorb W AW DMCS to allow each organohalide to appear before an overload water peak due to aqueous injection and to achieve the required separation.

Although all examined organohalides eluted in less than 6.5 min, this analysis normally required about 60 min, as the column was maintained at 100 °C for at least 50 min so that water and less volatile materials would be removed before the next analysis. The short- and long-term stability of the gas chromatography–mass spectrometry system was good throughout the analysis period. Figure 158 illustrates typical mass fragmentograms of specific ions of $CHCl=CCl_2$, CH_3CCl_3, and $CH_2ClCHClCH_3$ in certain tap water, showing no interference from other organics.

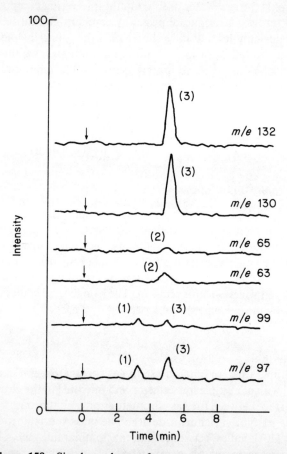

Figure 158 Six-channel mass fragmentograms obtained from a 100 μl sample of the tap water collected near Tokyo. The gas chromatographic effluent is analysed by the mass spectrometer operated in the mass fragmentography mode. The two specific ions in the mass spectrum of each organohalide were monitored. Peaks are: (1) CH_3CCl_3, 0.5 ppb; (2) $CH_2ClCHClCH_3$, 0.8 ppb; $CHCl=CCl_2$, 0.9 ppb. Reprinted with permission from Fujii.[191] Copyright (1977) Elsevier Science Publishers

Table 123 Organohalides detected in the tap water sample

Organohalides	Retention time (min)	Column temperature (°C)	Masses monitored	Detection limit (ppb)
CH_2Cl_2	2.4	50	84 86	0.2
$CHCl_3$	4.1	50	83 85	0.1
$CHCl_2Br$	1.8	70	83 85 127 129	0.2
$CHClBr_2$	3.1	70	127 129	0.1
$CHBr_3$	6.5	70	171 173	0.2
CCl_4	4.3	50	117 119	0.8
$CHCl=CCl_2$	5.0	55	130 132	0.2
$CCl_2=CCl_2$	3.4	70	164 166	0.1
CH_2ClCH_2Cl	4.8	50	62 64 98 100	0.5
CH_3CCl_3	3.2	55	97 99	0.4
$CH_2ClCHClCH_3$	4.8	55	63 65 97 99	0.8

Reprinted with permission from Fujii.[191] Copyright (1977) Elsevier Science Publishers.

Table 123 summarizes the examined organohalides, the gas chromatograph–mass spectrometer conditions, the retention times, and the detection limit of each (concentration of single substance producing a peak three times higher than noise level). The standard deviation precision of the method at the 1 ppb concentration of dichlorobromomethane level was 0.14 and the coefficient of variation 4.2%.

Table 124 summarizes the substances determined in the drinking tap water of five locations near Tokyo. Chloroform was the major component with concentrations in the low ppb range which represented a very large peak in the mass fragmentograms. The profiles obtained from various tap water samples showed similar patterns, although variations in the total concentration were large.

In further work Fujii[192,193] applied direct injection gas chromatography–mass spectrometry with single ion detection to the determination of 1–100 µg l^{-1} of chloroform, bromodichloromethane, dibromochloromethane, bromoform, and

Table 124 Organohalide concentrations (ppb) in Japanese tap water samples (11 December 1976)

Organohalides	Tokorosawa	Fussa	Tsuchiura	Urawa	Hanamuro
CH_2Cl_2	0.3	—	—	—	—
$CHCl_3$	10.2	2.6	13.0	2.7	17.2
$CHCl_2Br$	6.4	1.6	10.5	2.3	4.0
$CHClBr_2$	3.2	0.6	4.0	1.4	0.6
$CHBr_3$	0.5	—	0.4	0.3	—
CCl_4	1.2	—	—	—	—
$CHCl=CCl_2$	0.9	—	0.7	—	—
$CHCl=CCl_2$	0.6	0.2	0.2	0.2	0.2
$CCl_2=CCl_2$	0.9	—	—	—	—
CH_3CCl_3	0.5	—	—	—	—
$CH_2ClCHClCH_3$	0.8	—	—	—	—

Reprinted with permission from Fujii.[191] Copyright (1977) Elsevier Science Publishers.

carbon tetrachloride in chlorinated water supplies. This method uses a liquid diglycerol phase to make the organic peaks appear before the large water peak.

Bertsch et al.[194] have carried out trace analysis of trihaloforms in chlorinated drinking water by gas chromatography–mass spectrometry using glass capillary columns. A simple, glass apparatus is described, with a diagram, for concentrating traces of volatile organic materials in water by gas-phase stripping and adsorption on to a porous polymer. The adsorbent tubes are fitted into the inlet system of a gas chromatograph and the sample is transferred by a one-step elution process into a glass capillary column for gas chromatograph separation.

Haloforms in sewage

De Leer[195] has described a procedure for the determination of trihaloforms (chloroform, bromodichloromethane, dibromochloromethane, bromoform) in secondary and tertiary treated sewages and in chlorinated sewage. The haloforms are extracted into pentane followed by gas chromatographic separation and quantification by electron capture detection. The sensitivity of this procedure was more than adequate (less than $2\,\mu g\,l^{-1}$) for these types of sample.

Miscellaneous volatile organohalogen compounds

Lovelock et al.[196,197] determined methyl fluoride, methyl chloride, methyl bromide, methyl iodide, and carbon tetrachloride in the Atlantic Ocean. This shows a global distribution of these compounds. Murray and Riley[198,199] confirmed the presence of carbon tetrachloride and also found low concentrations of chloroform and tri- and tetrachloroethylene in Atlantic surface waters.

Determination of organohalogen compounds as halogens

In water

Wegman and Greve[200] have described a microcoulometric determination of extractable organic halogens, particularly chlorine, in surface waters and sediments.

Fritschi et al.[201] have described a method for the determination of combined volatile and non-volatile organochlorine compounds in water. Non-volatile compounds were extracted either by hexane or di-isopropyl ether, while volatile compounds were vaporized by a stream of nitrogen and passed through a furnace, the resulting chloride being measured coulometrically.

In fish

White[202] developed methods for the detection of trace amounts of organic halogens in organisms. Three fractions (lipid-soluble material, cationic water-soluble molecules, and large macromolecules) were assayed for each organism, the inorganic halogens being monitored by means of radioactive chloride and radioactive iodide. Fractions were burned in an oxygen combustion tube and the resulting adsorbed halides were assayed on two automatic analytical instruments, chloride and bromide on one instrument and iodide on the other.

Linde et al.[203] determined organohalogen compounds, as halogen, in fish samples. The samples were steam distilled with cyclohexane for halogen-containing non-polar compounds and hexane extracts of oils from all species were treated with concentrated sulphuric acid. Total amounts of halogens in the original oils, in the volatile compounds in the cyclohexane distillate, and in the sulphuric acid-treated hexane extracts were determined by neutron activation analysis. The total level of organic chlorine ranged from 30 to 240 ppm: 2-10% of the chlorinated compounds were volatile, and from 5 to 50% of the chlorinated compounds remained after acid treatment. This chloride exceeded chlorine in polychlorinated biphenyl by a factor of 1.5-5, and most of the chlorine in untreated and acid-treated lipids could not be accounted for as known compounds.

REFERENCES

1. Hassler, J. and Rippa, F. *Vyskummy Ustai Vodneko Hospodarstra; Veda a Vyskum Praxi*, No. 50, 50 pp (1977).
2. Glaze, W., Henderson, J. E., Bell, J. E. and Van Wheeler, A. *J. Chromatog. Sci.*, **11**, 580 (1973).
3. Zitko, V. *J. Chromat.*, **81**, 152 (1973).
4. Zitko, V. *J. Ass. Off. Anal. Chem.*, **57**, 1253 (1974).
5. Friedman, D. and Lombardo, P. *J. Ass. Off. Anal. Chem.*, **58**, 703 (1975).
6. Hollies, J. I., Pinnington, P. F., Handley, A. J., Baldwin, M. K. and Bennett, D. *Anal. Chim. Acta*. **111**, 201 (1979).
7. Kaiser, K. L. E. and Oliver, B. G. *Anal. Chem.*, **48**, 2207 (1976).
8. Hrivňak, F., Siskupič, P. and Hässler, J. *Vodni Hospodarstivi. Series B*, **28**, 195 (1978).
9. Dowty, B. J., Carlisle, D. R. and Laster, L. J. *Environ. Sci. Technol.* **9**, 762 (1975).
10. Dawson, R., Riley, J. P. and Tennant, R. H. *Mar. Chem.*, **4**, 83 (1976).
11. Parejko, R. and Keller, R. *Bull. Environ. Contam. Toxicol.*, **14**, 480 (1975).
12. Solomon, J. *Anal. Chem.*, **51**, 186 (1979).
13. De Leon, I. R., Maberry, M. A., Overton, E. B., Roschke, C. K., Remele, P. C., Steele, C. F., Warren, V. L. and Leister, J. L. *J. Chromatog. Sci.*, **18**, 85 (1980).
14. Murray, A. J. and Riley, J. P. *Anal. Chim. Acta*, **65**, 261 (1973).
15. Murray, A. J. and Riley, J. P. *Nature (London)*, **242**, 37 (1973).
16. Novak, J., Zluticky, J., Kubulka, V. and Mostecky, J. *J. Chromat.*, **76**, 45, (1973).
17. Kleopfer, R. D. and Fairless, B. T. *Environ. Sci. Technol.*, **6**, 1036 (1972).
18. Jensen, S., Jernelov, A., La ge, R. and Palmork, K. H. Food and Agriculture Organisation of the United Nations. Report No. FIR:MP 170/E-88 November (1970).
19. Dietz, F. and Traud, J. *Vom Water*, **41**, 137 (1973).
20. Deetman, A. A., Demeulemeester, P., Garcia, M., Hauck, G., Hauck, G., Hollies, J. I., Krockenberger, D., Palin, D. E., Prigge, H., Rohrschneider, L. and Schmidthammer, L. *Anal. Chim. Acta*, **82**, 1 (1976).

21. Guam, C. S. and Wong, M. K. *J. Chromat.* **72**, 283 (1972).
22. Kummert, R., Molnar-Kubica, E. and Giger, W. *Anal. Chem.*, **50**, 1637 (1978).
23. Stozek, A. and Beumer, W. *Korrespondenz Abwasser*, **26**, 632 (1979).
24. Dilling, W. L., Tefertiller, O. and Kallos, G. *J. Environ. Sci. Technol.*, **9**, 833 (1975).
25. Simonov, V. D., Popova, L. N. and Shamsutdinov, T. M. *Dokl. Neftehim. Sekt. bashkirsk. resp. Provl. vses. khim. Obshch. Mendeleeva* (**6**) 357 (1971). Ref. *Zh. Khim.* 19GD Abstract No. 59302 (1972).
26. Renberg, L. *Anal. Chem.*, **50**, 1836 (1978).
27. Burgasser, A. J. and Calaruotolo, J. F. *Anal. Chem.*, **49**, 1588 (1977).
28. *Determination of Vinyl Chloride Monomer in Aqueous Effluents*, Analytical Chemistry Branch. Southeast Environmental Research Laboratory, Environmental Protection Agency, Athens, Ga. USA (1974).
29. Alberti, J. and Jonke, B. *Zeitschrift für Waisser and Abwasser Forschung*, **8**, 140 (1975).
30. Bellar, T. A., Lichtenberg, J. J. and Eichelberger, J. W. *Environ. Sci. Technol.*, **10**, 926 (1976).
31. Fujii, T. *Anal. Chem.*, **49**, 1985 (1977).
32. Wolen, R. L. and Pierson, H. E. *Anal. Chem.*, **47**, 2068 (1975).
33. Rivera, J., Cuberes, M. R. and Albaiges, J. *Bull. Environ. Contamin. Toxicol.*, **18**, 624 (1977).
34. Kuchl, D. W., Kopperman, H. L., Veith, G. D. and Glass, G. E. *Bull. Environ. Contam. Toxicol.*, **16**, 127 (1976).
35. Kapila, S. and Aue, W. A. *J. Chromatog. Sci.*, **15**, 569 (1977).
36. Johnsen, J. L., Stalling, D. L. and Hogen, J. *Bull. Environ. Contam. Toxicol.*, **11**, 393 (1974).
37. Bjorseth, A., Carlberg, G. E. and Moller, M. *Science of the Total Environment*, **11**, 197 (1979).
38. Chau, A. S. Y. and Coburn, J. A. *J. Ass. Off. Anal. Chem.*, **57**, 389 (1974).
39. Chriswell, C. and Cheng, R. *Anal. Chem.*, **47**, 1325 (1975).
40. Rudling, L. *Water Research*, **4**, 533 (1970).
41. Junk, G. A., Richard, J. J., Grieser, M. D., Willek, J. L., Argquello, M. D., Vick, R. and Svec, H. J. *J. Chromat.*, **99**, 745 (1974).
42. Farrington, D. S. and Mundy, W. *Analyst (London)*, **101**, 639 (1976).
43. Morgade, C., Barquet, A. and Pfaffenberger, C. D. *Bull. Environ. Contamin. Toxicol.*, **24**, 257 (1980).
44. Stanley, C. W. *J. Agric. Food Chem.*, **14**, 321 (1966).
45. Shafik, T. M., Sullivan, H. C. and Enos, H. F. *J. Agric. Food Chem.*, **21**, 295 (1973).
46. Ashiya, K., Otani, H. and Kajino, K. *Japan Water Works Association*, No. 536, **12**, (1979).
47. Renberg, L. *Anal. Chem.*, **46**, 459 (1974).
48. Buhler, D. R., Rasmussan, M. E. and Nakaue, H. S. *Environ. Sci. Technol.*, **7**, 929 (1973).
49. Lindstrom, K. and Nordin, J. *J. Chromat.*, **128**, 13 (1976).
50. Carlberg, G. E., Gjos, N., Maller, M., Gustavsen, K. O., Tveten, G. and Renberg, L. *Science of the Total Environment*, **15**, 3 (1980).
51. Ingram, L. I., McGinnis, G. D. and Porikl, S. V. *Anal. Chem.*, **51**, 1077 (1979).
52. Hoben, H. J., Ching, S. A., Casarett, L. J. and Young, R. A. *Bull. Environ. Contam. Toxicol.*, **15**, 78 (1976).
53. Ervin, H. E. and McGinnis, G. D. *J. Chromat.*, **190**, 203 (1980).
54. Henshaw, B. G., Morgan, J. W. W. and Williams, N. *J. Chromat.*, **110**, 37 (1975).
55. Zigler, M. G. and Phillips, W. F. *Environ. Sci. Technol.*, **1**, 65 (1967).
56. Wolkoff, A. W. and Larose, R. H. *J. Chromat.*, **99**, 731 (1974).

57. Thielemann, H. and Luther, M. *Pharmazie*, **25**, 367 (1970).
58. Wade, A. L., Hankridge, F. M. and Williams, H. P. *Anal. Chim. Acta*, **105**, 91 (1979).
59. Boyle, T. P., Robinson-Wilson, E. F., Petty, J. D. and Weber, W. *Bull. Environ. Contam. Toxicol.*, **24**, 177 (1980).
60. Stark, A. *J. Agric. Food Chem.*, **17**, 871 (1969).
61. Rudling, L. *Water Research*, **4**, 533 (1970).
62. Baird, R. B., Kuo, C. L., Shapiro, J. S. and Yanko, W. A. *Archs Environ. Contam. Toxicol.*, **2**, 165 (1974).
63. Analytical Report — *New Orleans Area Water Supply Study*, EPA 906/10-74-002, Surveillance and Analysis Division, USEPA, Region VI, Dallas, Texas Mimeo, 30pp (1974).
64. *The Implications of Cancer — Causing Substances in Mississippi River Water.* Environmental Defence Fund, Washington DC (1974).
65. Tardiff, R. G. and Dunzer, M. *Toxicity of Organic Compounds in Drinking Water.* Environmental Protection Agency, Cincinnati, Ohio, USA (1974).
66. *Report on the Carcinogenesis Assay of Chloroform.* US Natl. Cancer Inst. Bethesda, MD (1976).
67. Cantor, R. H., Mason, T. J. and McCabe, L. J. *Association of Cancer Mortality Rates and Trihalomethane Levels in Municipal Drinking Water Supplies.* Proc. 10th Ann. Mtg. Soc. Epidemiol. Res. (1977).
68. Glaze, W. H. and Rawley, R. *Am. Water Wks Ass.* **71**, 509 (1979).
69. Boltzer, W. *Gas, Wasser, Wärme*, **33**, 80 (1979).
70. Nicholson, A. A. and Meresz, O. *Bull. Environ. Contam. Toxicol.*, **11**, 453 (1975).
71. Schnoor, J. L., Nitzschke, J. L., Lucas, R. D. and Veenstra, J. N. *Environ. Sci. Technol.* **13**, 1134 (1979).
72. Kirshen, N. A. *Varian Instrument Applications*, **14**, 10 (1980).
73. Kroner, R. C. *Public Works*, **111**, 81 (1980).
74. *Packard Technical News* No. 7528, Feb. (1981).
75. Symons, J., Bellar, T. A., Carswell, J. K., DeMarco, J., Kropp, K. L., Dobeck, G. G., Seeger, D. R., Slocum, C. J., Smith, B. J. and Stevens, A. A. *J. Am. Water Works Assoc.* **67**, 634 (1975).
76. Kopfler, F. C., Melton, R. G., Lingg, R. D. and Coleman, W. E. *GC/MS Determination of Volatiles for the NORS of Drinking Water.* Ed. Keith, L. H. Ann Arbor Science, Michigan. Identification and analysis of organic pollutants in water, Chapter 6, 87 (1976).
77. Bush, B., Narang, R. S. and Syrotynski, S. *Bull. Environ. Contam. Toxicol.*, **18**, 436 (1977).
78. Bellar, T. A., Lichtenberg, J. J. and Kroner, R. C. *J. Am. Water Works Ass.*, **66**, 703 (1974).
79. Bunn, W. W., Haas, B. B., Deane, E. R. and Kleopfer, R. D. *Environ. Letters*, **10**, 205 (1975).
80. Symons, J. M., Beller, T. A., Carswell, J. K., DeMario, J., Kropp, K. L., Robeck, G. G., Seeger, D. R., Slocum, C. J., Smith, B. L. and Stevens, A. A. *J. Am. Water Works Ass.*, **67**, 634 (1975).
81. Foley, P. D. and Missingham, G. A. *J. Am. Water Works Ass.*, **68**, 105 (1976).
82. Rook, J. J. *Water Treat. Exam.*, **23**, 234 (1974).
83. Rook, J. J. *J. Am. Water Works Ass.*, **68**, 168 (1976).
84. Fujii, T. *Bull. Chem. Soc. Japan*, **50**, 2911 (1977).
85. Fujii, T. *J. Chromat.*, **139**, 297 (1977).
86. Brett, R. W. and Calverley, R. A. *J. Am. Water Works Ass.*, **71**, 515 (1979).
87. Smith, V. L., Cech, I., Brown, J. H. and Bogdan, G. F. *Environ. Sci. Technol.*, **14**, 190 (1980).
88. Symons, J. M. *National Organics Reconnaissance Survey for Halogenated Organics in Drinking Water*, Environmental Protection Agency. USA (1967).

89. Luong, T., Peters, C. J., Young, R. J. and Perry, R. *Environ. Technol. Letters*, **1**, 299 (1980).
90. Bunn, W. W., Haas, B. B., Deane, E. R. and Kleopfer, R. D. *Environ. Letters*, **10**, 205 (1975).
91. Glaze, W. H., Henderson, J. E. and Smith, G. Private communication.
92. Rook, J. J. *Water Treat. Exam.*, **21**, 259 (1972).
93. Dowty, B. J., Carlisle, D. R. and Laseter, J. L. *Environ. Sci. Technol.*, **9**, 762 (1975).
94. New Orleans Area Water Supply Study (Draft Analytical Report), Lower Mississippi River Facility, USEPA (1974).
95. Stevens, A. A., Slocum, C. J., Seiger, D. R. and Robeck, G. G. *Conf. Environ. Impact of Water Chlorination*, Oak Ridge National Lab., Oak Ridge, Tennessee (1975).
96. Stevens, A. A. and Symons, J. M. *Conf. Environ. Impact of Water Chlorination*, Oak Ridge National Lab., Oak Ridge, Tennessee (1975).
97. Symons, J. M., Bellar, T. A., Carswell, J. K., Demarco, J., Kropp, K. L., Roebeck, G. G., Seegar, D. R., Slocum, C. J., Smith, B. L. and Stevens, A. A. *National Organics Reconnaissance Survey for Halogenated Organics in Drinking Water*, EPA, Cincinnati, Ohio (1975).
98. Tardiff, R. G. and Dunzer, M. *Toxicity of Organic Compounds in Drinking Water*, EPA, Cincinnati (1973).
99. US Environment Protection Agency *Federal Register* No. 28, **43**, 5756 (1978).
100. USEPA Control of Organic Chemical Contaminants in Drinking Water. *Federal Register* **44**, 5755 et seq. (Feb. 9, 1978).
101. Quimby, B. R., Delaney, M. F., Uden, P. C. and Barnes, R. M. *Anal. Chem.*, **52**, 259 (1980).
102. Schnoor, J. L., Nitzschke, J. L., Lucas, R. D. and Veenstra, J.N. *Environ. Sci. Technol.*, **13**, 1134 (1979).
103. Harris, L. E., Budde, W. L. and Eichelberger, J. W. *Analyst (London)*, **46**, 1912 (1974).
104. Nicholson, A. A. and Meresz, O. *Bull. Environ. Contam. Toxicol.*, **14**, 4 (1975).
105. Kissinger, L. D. and Fritz, J. S. *J. Am. Water Works Ass.*, **68**, 435 (1976).
106. Nicholson, A. A. and Meresz, O. *Bull. Environ. Contam. Toxicol.*, **14**, 453 (1975).
107. *National Survey for Halomethanes in Drinking Water*. Health and Welfare, Canada, 77-EHD-9 (1977).
108. Smillie, R. D., Nicholson, A. A., Meresz, O., Duholke, W. K., Rees, G. A. V., Roberts, K., and Fung, C. *Organics in Ontario Drinking Waters, Part II. A Survey of Selected Water Treatment Plants*, Ontario Ministry of the Environment, April, 1977.
109. Bush, B., Narang, R. S. and Syrotynski, S. *Bull. Environ. Contam. Toxicol.*, **13**, 436 (1977).
110. Morris, R. C. and Johnson, L. G. *J. Am. Water Works Ass.*, **68**, 492 (1976).
111. Richard, J. J. and Junk, G. A. *J. Am. Water Works Ass.*, **69**, 62 (1977).
112. Mieure, J. P. *J. Am. Water Works Ass.*, **69**, 60 (1977).
113. Nicholson, A. A., Meresz, O. and Lemyk, B. *Anal. Chem.*, **49**, 814 (1977).
114. Hammerstrand, K. *Chloroform in Drinking Water*, Varian Instrument Application, 10:2:2 (1976).
115. Dietz, F. and Traud, J. *Vom Wasser*, **41**, 137 (1973).
116. Grob, K., Grob, K. and Grob, G. *J. Chromat.*, **106**, 299 (1975).
117. Goldberg, M. C., Delong, L. and Sinclair, M. *Anal. Chem.*, **45**, 89 (1973).
118. Fielding, M., McLoughlin, K. and Steel, C. *Water Research Centre Enquiry Report ER 532 August 1977*. Water Research Centre, Stevenage Laboratory, Elder Way, Stevenage, Herts UK (1977).

119. Henderson, J. E., Peyton, G. R. and Glaze, W. H. *Recommended Procedure: A Convenient Liquid-Liquid Extraction Method for the Determination of Chloroform and Other Halomethanes in Water.* Rpt. Trace Analysis Lab., Inst. Applied Sci., North Texas State Univ., Denton, Texas (1976).
120. Webb, R. G. *Isolating Organic Water Pollutants: XAD Resins, Urethane Foams, Solvent Extraction.* USEPA-660/4-75 003 pp. 17–18 (1975).
121. Trussell, A. R. *Microextraction of Organic Pollutants: Application to Trihalomethanes in Drinking Water.* President's Fellowship Rpt. Univ. of California, San Diego (1976).
122. Kissinger, L. D. and Fritz, J. S. *J. Am. Water Works Ass.*, No. 8, **68**, 435 (1976).
123. Von Rensburg, J. F. T., Van Huyssteen, J. T. and Hassett, A. J. *Water Research*, **12**, 127 (1978).
124. Dressman, R. C., Stevens, A. A., Fair, J. and Smith, B. *J. Am. Water Works Ass.*, **71**, 392 (1979).
125. Henderson, J. E., Peyton, G. R. and Glaze, W. H. *A Convenient Liquid-Liquid Extraction Method for the Determination of Halomethane in Water at the Parts per Billion Level. Identification and Analysis of Organic Pollutants in Water* (L. H. Keith, editor) Ann Arbor Science, Ann Arbor, Mich, **105**, 195 (1976).
126. Mieure, J. P. *J. Am. Water Works Ass.*, **69**, 62 (1977).
127. Richard, J. J. and Junk, G. A. *J. Am. Water Works Ass.*, **69**, 62 (1977).
128. Symons, J. M. National Organics Reconnaissance Survey for Halogenated Organics in Drinking Water, *J. Am. Water Works Ass.*, **11**, 634 (1975).
129. Bellar, T. A. and Lichtenberg, J. J. *J. Am. Water Works Ass.*, **12**, 66 739 (1974).
130. USEPA Part III, Appendix C. Analysis of trihalomethanes in drinking water. *Federal Register*, **44**, No. 231, 68672 (Nov. 29, 1979).
131. Method 501.1, *The Analysis of Trihalomethanes in Finished Waters by the Purge and Trap Method*, USEPA, EMSL, Cincinnati, Ohio 45268 (November 6, 1979).
132. Method 501.2, *The Analysis of Trihalomethanes in Drinking Water by Liquid/Liquid Extraction*, USEPA, EMSL, Cincinnati, Ohio 45268 (November 6, 1979).
133. Kirschen, N. A. *Varian Instrument Applications*, **14**, 10 (1980).
134. Lange, A. L. and Kawczynski, E. *J. Am. Water Works Ass.*, **70**, 653 (1978).
135. Norin, H. and Renberg, L. *Water Research*, **14**, 1397 (1980).
136. Rook, J. J. *J. Water Exam.*, **21**, 259 (1974).
137. Kaiser, K. L. E. and Oliver, B. G. *Anal. Chem.*, **48**, 2207 (1976).
138. Dietz, E. A. and Singley, K. F. *Anal. Chem.*, **51**, 1809 (1979).
139. Otson, R., William, D. T. and Bothwell, P. D. *Environ. Sci. Technol.*, **13**, 936 (1979).
140. Varma, M. M., Siddique, M. R., Doty, K. T. and Machis, H. *J. Am. Water Works Ass.*, **71**, 389 (1979).
141. Friant, S. L. Thesis Drexel University, University Microfilms Ltd, London, 468 pp., 29162 (1977).
142. Gomella, C. and Belle, J. P. *Techniques et Sciences Municipales*, **73**, 125 (1978).
143. Montiel, A. *Tribune de Cabedeau*, **32**, 422 (1979).
144. Suffet, I. H. and Radziul, O. *J. Am. Water Works Ass.*, **68**, 520 (1976).
145. Bush, B., Narang, R. S. and Syrotynski, S. *Bull. Environ. Contam. Toxicol.*, **18**, 436 (1977).
146. Symons, J. M., Bell, T. A., Carswell, J. K., DeMarco, J., Kropp, K. L., Robeck, G. G., Seegar, D. R., Slocum, D. J., Smith, B. L. and Stevens, A. A. *J. Am. Water Works Ass.*, **67**, 643 (1975).
147. Keith, L. H. *Identification and Analysis of Organic Pollutants in Water*, Ann Arbor Science Publishers, Inc., Ann Arbor, Mich. (1976).
148. *National Survey for Halomethanes in Drinking Water*, Health and Welfare Canada, 77-EHD-9 (1977).

149. Bush, B., Narang, R. S. and Syrotynski, S. *Bull. Environ. Contam. Toxicol.*, **13**, 436 (1977).
150. Morris, R. L. and Johnson, L. G. *J. Am. Water Works Ass.*, **68**, 492 (1976).
151. Rook, J. J. *Water Treat. Exam.*, **21**, 259 (1972).
152. Rook, J. J. *Water Treat. Exam.*, **23**, 234 (1974).
153. Lovelock, J. E. *Nature (London)*, **256**, 193 (1975).
154. Lovelock, J. E., Maggo, R. J. and Wade, R. J. *Nature (London)*, **241**, 194 (1973).
155. Zlatkis, A., Lichtenstein, A. and Tuchbee, A. *Chromatographia*, **8**, 67 (1973).
156. Mieure, J. P. and Dietrich, M. M. *J. Chromatog. Sci.*, **11**, 559 (1973).
157. Dowty, B. and Laseter, J. L. *Anal. Lett.*, **8**, 25 (1975).
158. Murray, A. J. and Riley, J. P. *Anal. Chim. Acta*, **65**, 261 (1973).
159. Bellar, T. A. and Lichtenberg, J. J. *J. Am. Water Works Ass.*, **66**, 739 (1974).
160. Grob, K. *J. Chromat.*, **84**, 255 (1973).
161. Grob, K. and Grob, G. *J. Chromat.*, **90**, 303 (1974).
162. Grob, K., Grob, K. Jnr. and Grob, G. *J. Chromat.*, **106**, 299 (1975).
163. Bertsch, W., Andersson, E. and Holzer, G. *J. Chromat.*, **112**, 701 (1975).
164. Stevens, A. A. and Symons, J. M. *Environ. Impact of Water Chlorination*. Oak Ridge National Lab., Oak Ridge, Tennessee (1975).
165. Bellar, T. A. and Lichtenberg, J. J. *J. Am. Water Works Ass.*, **66**, 739 (1974).
166. Quimby, B. D. and Delaney, M. F. *Anal. Chem.*, **51**, 875 (1979).
167. Quimby, B. D., Uden, P. C. and Barnes, P. C. *Anal. Chem.*, **50**, 2112 (1978).
168. Kroner, R. C. *Public Works*, **111**, 81 (1980).
169. Kirschen, N. A. *Varian Instrument Applications*, **15**, 2 (1981).
170. *Federal Register* (US) 69464, **44**, No. 223 Dec. 3rd (1979).
171. *Packard Technical News* 1., No. 7528 February (1981).
172. Pfaender, F. E., Jones, R. B., Stevens, A. A., Moore, L. and Hass, J. R. *Environ. Sci. Technol.*, **12**, 438 (1978).
173. *Anonymous Chemical Engineering News*, **54**, 35 (1976).
174. Ligon, V. V. and Johnson, R. L. *Anal. Chem.*, **48**, 481 (1976).
175. Renberg, L. *Anal. Chem.*, **50**, 1836 (1978).
176. Kissinger, L. D. US National Technical Information Service, Springfield, Va Report No. 1S-T-845. 161 pp. (32411) (1979).
177. Keith, L. H. *Identification and Analyses of Organic Pollutants in Water*, Ann Arbor Science Publishers Inc., Ann Arbor, Michigan, USA (1976).
178. Brodtmann, N. V. *J. Am. Water Works Ass.*, **67**, 558 (1975).
179. Coburn, J. A., Valdmanis, I. A. and Chau, A. S. Y. *The Extraction of Organochlorine Pesticides and P.C.B.'s from Natural Waters with XAD-2*. Water Quality Branch, Canada Centre of Inland Waters, Burlington, Ontario, L7R 4A6 Canada.
180. Musty, P. R. and Nickless, G. *J. Chromat.*, **89**, 185 (1974).
181. Junk, G. A., Rickard, J. J., Grieser, M. D., Witlak, D., Witlak, J. L., Arguello, M. D., Vick, R., Svec, H. J., Fritz, V. S. and Calder, G. V. *J. Chromat.*, **99**, 745 (1974).
182. Coburn, J. A., Valdmanis, I. A. and Chau, A. S. Y. *J. Ass. Off. Anal. Chem.*, **60**, 224 (1977).
183. Kissinger, L. D. and Fritz, J. S. *J. Am. Water Works Ass.*, **68**, 435 (1976).
184. Junk, G. A., Chriswell, D. C., Chiang, R. C., Kissinger, L. D., Richard, J. J., Fritz, J. S., Svec, H. J., Fresenius, Z. *Anal. Chem.*, **282**, 331 (1976).
185. Eklund, G., Josefsson, B. and Roos, C. *J. High Res. Chromat. Chromat. Commun.*, **1**, 34 (1978).
186. Eklund, G., Josefsson, B. and Roos, C. *J. Chromat.*, **142**, 575 (1977).
187. Grob, K. and Grob, G. *J. Chromat.*, **125**, 471 (1970).
188. Grob, K., Grob, G. and Grob, K. Jnr. *Chromatographie*, **10**, 181 (1977).
189. Schomburg, G. and Husmann, H. *Chromatographia*, **8**, 517 (1975).

190. Trussell, A. R., Umphres, M. D., Leong, L. Y. C. and Trussell, R. R. *J. Am. Water Works Ass.*, **71**, 385 (1979).
191. Fujii, T. *J. Chromat.*, **139**, 297 (1977).
192. Fujii, T. *Anal. Chim. Acta*, **92**, 117 (1977).
193. Fujii, T. *Bull. Chem. Soc. Japan*, **50**, 2911 (1977).
194. Bertsch, W., Anderson, E. and Holzer, G. *J. Chromat.*, **112**, 701 (1975).
195. Le Leer, E. W. B. H_2O, **13**, 171 (1980).
196. Lovelock, J. E., Maggs, R. J. and Wade, R. J. *Nature (London)*, **241**, 194 (1973).
197. Lovelock, J. E. *Nature (London)*, **256**, 193 (1975).
198. Murray, A. J. and Riley, J. P. *Nature (London)*, **242**, 37 (1973).
199. Murray, A. J. and Riley, J. P. *Anal. Chim. Acta*, **65**, 261 (1973).
200. Wegman, R. C. C. and Greve, P. A *Science of the Total Environment*, **7**, 235 (1977).
201. Fritschi, E., Fritschi, G. and Kussmanl, H. *Z. für Wasser und Abwasser Forsch.*, **11**, 165 (1978).
202. White, R. H. Thesis, Illinois University, University Microfilms Ltd., Tylers Green Penn, Bucks., 232 pp. (24232) (1968).
203. Linde, G., Gether, J. and Steinnes, E. *Ambio*, **5**, 180 (1976).

Chapter 5
Miscellaneous Compounds and Ozonization Products

SQUOXIN PISCICIDE (1,1′METHYLENEDI-2-NAPHTHOL)

Kiigemagi et al.[1] developed a method for the determination of residues of this substance in water and aquatic organisms. They used derivatization gas chromatography and spectrophotometric methods to determine down to 0.1 ppm and to 2 ppm, respectively, of this substance in fish tissues and water samples.

Gas chromatographic method

Reagents

Diazomethane was prepared in ether solution by the method of Fieser and Fieser.[2]

Reagent grade hexane was passed through a column of activated Florisil prior to use. Reagent grade benzene was refluxed over metallic sodium for a minimum of 12 h and distilled. Reagent grade diethylether was refluxed over metallic sodium for 48 h and distilled. Sodium sulphate was dried at 400–450 °C overnight and stored in a sealed bottle. Florisil was activated at 400–450 °C overnight and stored at 100 °C.

Extraction

Fish were homogenized by grinding with dry ice according to the method of Benville and Tindle.[3] The grinding is done in a 1 quart container adapted to fit a 1.5 hp Waring Blendor. The pulverized mixture is poured into a plastic bag which is lightly sealed and placed in a −10 °C freezer overnight to allow the carbon dioxide to sublime. The samples were stored in this condition prior to extraction.

Five-gram samples of the finely pulverized fish tissue are weighed into a 90 ml Sorvall Omnimixer cup and blended with 10 ml of distilled benzene. The benzene is then recovered by suction filtration and the extraction is repeated with two additional portions of benzene. The combined benzene filtrate plus approximately 15 ml of benzene used in the transfer are then evaporated to 3-5 ml.

The benzene extract from above contained in a 50 ml beaker is treated with an excess of ethereal diazomethane (yellow colour of diazomethane persists on swirling). After standing at ambient temperature for about 3 min with occasional swirling, the mixture is heated gently on a steam bath until the yellow colour disappears. Samples are stored at 0 °C until the clean-up procedures are begun.

The methylated material, volume less than 5 ml, is taken up in 100 ml of reagent grade acetonitrile and transferred to a 250 ml separatory funnel. The sample is extracted three times with 50 ml portions of purified hexane. Emulsions which result occasionally can be broken by allowing them to stand for about 16 h. The acetonitrile layer is then flooded with 400 ml of distilled water and the mixture extracted with four 100 ml portions of purified diethylether. Addition of a small amount of solid sodium chloride aids in the delineation of layers.

Florisil column clean-up and g.c. analysis

A 22 mm o.d. column containing 2.5 cm of anhydrous sodium sulphate above 5 cm of activated Florisil is washed with 50 ml of purified hexane. The sample from the previous step is evaporated to 1-2 ml and introduced into the column with 100 ml of purified hexane. The column is then eluted with 50 ml of benzene followed by 100 ml of 10% diethylether in benzene. These are discarded because they contain lipid material which interferes with subsequent analysis. The Squoxin dimethylether is eluted with 250 ml of diethylether. The ether eluate, evaporated to 3-6 ml, is ready for gas chromatographic analysis.

Samples were analysed with a Varian Aerograph Model 204-B gas chromatograph equipped with a ^{3}H foil electron capture detector. The 0.25×120 cm glass column was packed with 4% SE 30 on Chromosorb G, 80-100 mesh. Temperatures were: column, 235 °C; injector, 230 °C; and detector, 220 °C. The nitrogen flow rate was 60 ml min^{-1}. After packing, the column was conditioned for 48 h at 250 °C. Injection of Squoxin dimethylether standards at this point resulted in broad, badly tailing peaks of low sensitivity. Sensitivity was increased about 20- to 40-fold by massive injections of the standard. Further improvement of sensitivity and peak shape occurred as additional fish samples were injected during the course of analysis. The retention time for the dimethylether was 6.25 min and the response of the detector was linear in the range 0.5-15 ng. Typical chromatograms of a standard and untreated and fortified fish samples are shown in Figure 159. As little as 0.5 ng of the dimethylether can be detected by this method permitting the use of 5 g samples and an overall sensitivity of 0.1 ppm of Squoxin. Squoxin recoveries of $85 \pm 5\%$ were obtained from fish samples by this procedure.

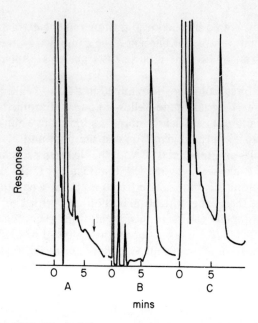

Figure 159 Typical chromatograms from Squoxin analysis as Squoxin dimethylether; (A) untreated rainbow trout; (B) 9.5 ng of Squoxin dimethylether standard; (C) treated rainbow trout containing 1.26 ppm of Squoxin measured as the dimethylether. Reprinted with permission from Kiigemagi et al.[1] Copyright (1975) American Chemical Society

Spectrophotometric method

Reagents

Diazo Blue B reagent was prepared by mixing 2 parts of 1% aqueous Diazo Blue B (*o*-dianisidine, terazotized, zinc chloride complex, Sigma Chemical Co., St Louis, Mo.) with 5 parts of 5% aqueous sodium lauryl sulphate. This reagent must be prepared every 2 days and stored in a refrigerator. Phosphate buffer (pH 8) was prepared by mixing 5.3 ml of NaH_2PO_4 solution (27.8 g l^{-1}) with 94.7 ml of $NaHPO_4 \cdot H_2O$ solution (53.65 g l^{-1}) and diluted to 200 ml. All other chemicals were reagent grade and were used as received.

Procedure

A calibration curve is prepared from standards containing 2–10 μg of Squoxin in 1 ml of ethanol and 3.2 ml of pH 8 buffer and 0.8 ml of Diazo Blue B reagent are added. The absorbance at 552 nm is measured within 10 min.

The 1500 ml water sample in a 200 ml separatory funnel is acidified with 1 ml of 6 M hydrochloric acid and 50 ml of 2-propanol and 25 ml of carbon tetrachloride are added. The mixture is shaken for 2 min, the layers allowed to separate, and the lower layer removed. The aqueous layer is re-extracted with 15- and 10-ml portions of the solvent and the extracts combined and filtered if necessary. A suitable aliquot (2–10 μg of Squoxin) is taken and the carbon tetrachloride removed by evaporation. The residue is dissolved in 1 ml of ethanol, and 3.2 ml of pH 8 buffer added followed by 0.8 ml of the Diazo Blue reagent. The absorbance is measured at 552 nm within 10 min.

Squoxin rapidly breaks down in water to a wide range of decomposition products. None of these products interferes sufficiently with the spectrophotometric method to hinder its usefulness for monitoring purposes.

COPROSTANOL

For the past 50 years the determination of the sanitary quality of water has been based on the enumeration of indicator microorganisms (e.g. coliform bacteria). Recently, the adequacy of coliform enumeration methods for this purpose has been questioned.[4] The current trend of year-round disinfection of waste water effluents and the increasing discharge of both toxic substances and heat from industrial outfalls cast further doubt on the accuracy of biological indicator systems.[4,5]

A number of chemical tests have been suggested to be used concurrently with microbiological monitoring of pollution/contamination of waterways.[6,7] Some of these analyses, such as the measurements of nitrites, free ammonia, chloride, and uric acid, have been shown to be non-specific in character and are no longer considered to be of practical value in the determination of the sanitary quality of water.

Recently, the use of 5β-cholestan-3β-ol (coprostanol) as a molecular marker of faecal pollution of water has been suggested.[8–11] It has been shown that this saturated sterol satisfies the criteria for an indicator of faecal contamination of water.[5,12]

5β-Cholestan-3β-ol satisfies the generally accepted criteria of a good indicator of faecal pollution. It is believed that the only source of this compound is the faeces of higher animals including man. It is biodegradable and can be removed from domestic sewage by adequate treatment. Furthermore, it has been unambiguously proved that the concentration of coprostanol is highest in the overtly faecal polluted water and there is a progressive decrease in the concentration of this compound in the lesser polluted waters. Since its isolation and identification is unaffected by chlorination or by heat and toxic substances discharged from industrial outfalls the advantage of using a molecular rather than a biological indicator of faecal pollution is further demonstrated. Such a characteristic of coprostanol is especially significant in the current trend promoting disinfection of raw and treated waste water. Because of this unique property, coprostanol might also be a useful indicator in monitoring the source, course, and extent of faecal pollution in the ocean or brackish waters where bacteriological evidence is often doubtful.

Dawson and Best[14] investigated the use of coprostanol as an internal tracer for faecal contamination in water by investigating the method of analysis and the effects of sewage treatment on the concentration of coprostanol, and by marine surveys at sewage outfalls. Analysis time is greater than for faecal bacteria but immediate analysis is not necessary and the results are less subject to error.

Singley et al.[13] have described a gas chromatographic method for the analysis of coprostanol. This technique was used in extensive field studies, degradation studies, and studies on treatment plant efficiency, and was also used as a standard for evaluating a colorimetric method that was developed and shown to be capable of determining coprostanol in polluted water at levels of $1~\mu g~l^{-1}$. It was shown that there was good correlation between coprostanol and Biochemical Oxygen Demand, Chemical Oxygen Demand, and total organic carbon (TOC).

Wun et al.[15] used XAD-2 resin for the analysis of coprostanol in river and lake water and secondary sewage treatment plants. They showed that extraction of coprostanol from water by adsorption on a column of Amberlite XAD-2 resin is as efficient as conventional liquid-liquid extraction. Maximal recovery depends on the pH value of the sample, flow rate, resin mesh size, and concentration of coprostanol.

Method

Three litre water samples were collected and 3 ml of concentrated sulphuric acid added as a preservative. Generally, the coprostanol content was analysed on the same day of sample collection. The saponification was not performed. Only the free alcohol was analysed. This was done since various workers[16-18] have reported that omission of the saponification and trimethylsilyl ether derivative preparation steps have no significant effect on the isolation, identification, and quantitation of coprostanol.

Liquid-liquid extraction

The faecal sterols were extracted by liquid-liquid partitioning and/or XAD-2 column extraction. When solvent extraction was performed, the procedure used by Dutka et al.[18] was followed with a slight modification: 2 ml of concentrated hydrochloric acid and 5 ml of 20% (w/v) sodium chloride were added to each litre of water sample. The sample was extracted with vigorous mixing three times with 100 ml each of hexane for 30 min. The combined extract was washed with two 50 ml portions of acetonitrile (saturated with hexane) followed by two 50 ml portions of 70% ethanol. The hexane was then brought to dryness on a rotary evaporator under reduced pressure. The sample was redissolved in 100–200 μl of carbon disulphide and 1–5 μl of the solution was used for gas chromatographic analysis.

Column extraction

Columns packed with Amberlite XAD-2 neutral resin (mesh sizes 40–50 or 60–120) which was prewashed with distilled water, methanol, acetone, methanol, and water in the above order were employed.

Water samples (250 ml to 1 l) were passed through the column at a flow rate of 3 ml min^{-1}. The column was then washed with 20 ml of distilled water and eluted with 15–30 ml acetone depending on column size. Acetone eluant was dried and redissolved as above and 1–5 µl of the solution was used for gas chromatographic analysis as described below.

A Perkin Elmer 900 gas chromatograph equipped with flame ionization detectors was used for gas chromatographic analyses. A 1.43 m × 0.3 o.d. glass column packed with 3% OV-225 coated Chromosorb Q (AW/DMCS), 60–80 mesh was used. The column temperature was 250 °C, injection port and detector temperature were maintained at 260 °C and nitrogen carrier gas pressure, hydrogen gas pressure to the detector, and the air pressure were kept at 45 psi, 25 psi, and 35 psi, respectively.

The presence and amount of coprostanol in water was identified and quantified by comparing the retention time and peak height with those of the reference standard. Prior to each series of analyses a series of known coprostanol standards were analysed for calibration purposes.

The results in Table 125 indicate that the extraction efficiency was pH dependent. At pH 2 the extraction efficiencies for coprostanol were only 70 and 72% at a flow rate of 5 and 3 ml min^{-1}, respectively. Alkaline conditions also affected extraction. A pH of between 6.5 and 8 is satisfactory at a flow rate of 3 ml min^{-1}.

The effect of high pH was more pronounced at a higher flow rate than a 3 ml min^{-1}. The adsorption efficiency was affected by the concentration of faecal sterol in the water samples (Table 126). About 30–35% of the cholesterol was not adsorbed when the concentration of sterol in water was high

Table 125 Effect of pH on faecal sterol extraction by XAD-2 resin

Flow rate	pH of sample	Coprostanol (µg)		Recovery (%)
		Added	Recovered	
3 ml min^{-1}	2.0	50	36	72
	6.5	50	48.5	97
	7.0	50	47.0	94
	8.0	50	48	96
5 ml min^{-1}	2.0	50	35	70
	6.5	50	45	90
	7.0	50	44	88
	8.0	50	37	74

*Average of 6–8 experiments.
Reprinted with permission from Wun et al.[15] Copyright (1976) Pergamon Press.

Table 126 Efficiency of adsorption in relation to concentration of sterol in water samples

Water sample	Washing (200 ml) dist. H$_2$O each	Cholesterol recovered (dpm/200 ml)	(µg/200 ml)	Cholesterol leaked out (%)
A: 1 µCi [$_{14}$C]cholesterol	0	93,540	0.85	4.5
= 7 µg	1st	20,860	0.19	0.9
+ 13 µg cholesterol	2nd	Background		
Total = 20 µg	3rd	Background		
B: 1 µCi [^{14}C]cholesterol	0	166,920	4.30	7.5
+ 50 µg cholesterol	1st	25,740		
Total = 57 µg	2nd	Background		
	3rd	Background		
	4th	Background		
C: 1 µCi [^{14}C]cholesterol	0	739,040	103.10	33.6
+ 300 µg cholesterol	1st	65,280	9.10	3.0
Total = 307 µg	2nd	Background		
	3rd	Background		
	4th	Background		

Reprinted with permission from Wun et al.[15] Copyright (1976) Pergamon Press.

(1.5 µg ml^{-1}). However, the percentage of cholesterol leaking out diminished gradually when its concentration was lowered. The amount of cholesterol retained by columns A and B was not due to the capacity of XAD-2 resin in adsorbing the sterol but solely to the concentration of the compound in water. Results also indicate that once the sterol was adsorbed onto the resin, it could not be eluted by water.

In Table 127 are given coprostanol values for influent and effluent taken at a sewage treatment plant. It is evident that a significant amount of the faecal sterol was removed in the process of treatment. Results also indicate that a high concentration of the faecal sterol occurred in river water at points close to treatment plant discharge (Mill River, Amherst, and Old Deerfield River, Old Deerfield). It was further demonstrated that the coprostanol concentration decreased significantly at points further down river from the effluent outfall (Lake Warner).

In order to compare the efficiency of XAD-2 column extraction with that of solvent extraction, water samples collected from a heavily polluted sewage treatment plant and from a lake remote from the sewage outfall were divided into two 1 l aliquots. One aliquot of each sample was analysed by the liquid-liquid partitioning, the other by the XAD-2 column extraction procedure. Results, presented in Table 128 indicate a slightly higher recovery rate of coprostanol was obtained by the column extraction method. It was observed that the particulate matter present in the field samples tended to block the water passage in the column and a constant flow rate could not be maintained. Applying pressure or stirring the column reduced the recovery rate to about 50–70%. This difficulty was overcome by filtering the water sample before it reached the resin. The particulate matter accumulated on the filter paper was washed with chloroform–methanol (2:1) and the washing was combined with the acetone eluant for analysis.

Table 127 Coprostanol and faecal coliform content of water samples

Location	Coprostanol ($\mu g\ l^{-1}$)	Faecal coliform per 100 ml
Amherst sewage treatment plant, effluent	85–504	$12-22 \times 10^3$
Mill River, Amherst	130–180	9×10^4
Mill River, Millers Falls	7	9×10^2
Lake Warner	0.45	11
Old Deerfield sewage treatment plant, influent	170	
Old Deerfield sewage treatment plant, effluent	4.6–6.8	7×10^2
Old Deerfield River	1.2	1.3–2.5
Chicopee River, Ludlow	3.6	12
Turners Falls	0.9	1
Connecticut River, Hadley	Non-detectable	3–5
Connecticut River, Sunderland	Non-detectable	4–6

Hexane extraction procedures and conditions for g.l.c. analysis are outlined in the text.
Reprinted with permission from Wun et al.[15] Copyright (1976) Pergamon Press.

Table 128 Recovery of coprostanol from water samples by the two extraction methods

	Coprostanol ($\mu g\ l^{-1}$)	
Location	Hexane extraction	Column extraction
Amherst sewage treatment plant, effluent	90	106
Lake Water	0.45	0.57

Hexane extraction = procedure as described in the Water Extraction Section. Column extraction = XAD-2 (60–120 mesh) 10×2.2 cm i.d. column. Flow rate was maintained at 3 ml min^{-1}. Water sample passed through a filter before reaching the resin.
Reprinted with permission from Wun et al.[15] Copyright (1976) Pergamon Press.

In further work Wun et al.[19] improved the efficiency of the extraction of coprostanol using an XAD-2 resin column by decreasing the extraction time using a 'closed' column technique and by determining the effects of sample pH on adsorption processes.

Coprostanol was strongly adsorbed to polystyrene XAD-2 adsorbents, at pH 2, with 100% retention, and the adsorbed sterol was easily eluted with acetone adjusted to pH 8.5–9.0 with ammonium hydroxide. It was also shown that with a closed column method, large volumes of fresh or sea water can be extracted in a relatively short time and with higher sensitivity than that of the liquid-liquid partitioning procedure.

Method

XAD-2 column extraction method

Amberlite XAD-2 resin was ground and sieved to obtain the required (60–120) mesh size. Columns (15 mm × 50 cm). An adjustable plunger was inserted into the top of the column after the desirable height was achieved. Water samples were delivered to the columns at flow rates of 15–18 ml min^{-1} employing a metering pump. The column was then washed with 100 ml of distilled water and eluted with 100 ml of acetone at a flow rate of 3 ml min^{-1}. The samples were adjusted to pH 2 with 2 M hydrochloric acid. An accumulation of suspended solids in the column can result in reduced adsorption efficiency. Samples were filtered through a filter paper before introducing the samples into the columns.

The earlier work of Wun et al.[15] had shown that using 'open' XAD-2 resin columns a flow rate of 3 ml min^{-1} was optimum for maximum coprostanol recovery. In this 'open' column system, the resin can only be packed loosely in the column. At high flow rates, channelling of water through the interparticular spaces probably caused insufficient contact of the sorbate and the adsorbent which resulted in an unsatisfactory extraction. This phenomenon was also observed when coarse mesh size Amberlite neutral resins (40–50) was used as a packing material.

The more recent work[19] showed that 'close' columns packed with fine mesh size XAD-2 (60–120) resins could be used for the extraction of faecal sterols. Tightly packed columns were obtained by the application of pressure. Employing these columns, the flow rate could be increased to 15–18 ml min^{-1} without diminishing the extraction efficiency as the data in Table 129 indicate.

Table 129 XAD-2 resin column extraction of sterol from water

Adsorbent	Cholesterol Added (μg l^{-1})	Added (dpm*)	Eluted (dpm)	Eluted (μg l^{-1})	% recovery
XAD (60–120)	100	1.16×10^6	1.09×10^6	94	94
	30	1.14×10^6	1.10×10^6	28.9	96
	0.7	2.2×10^5	2.2×10^5	0.7	100

Water sample: 1 litre; pH 6.5. Flow rate: 15 mm × 20 cm packed with 60–120 mesh resin.
*dpm = disintegrations per min.

Table 130 Effects of pH of water sample on extraction of faecal sterol by XAD-2 resin

pH	Water sample (dpm)	Radioactivity Recovered in H$_2$O passed through column (dpm)	Retained by column (dpm)	Retained by column (%)	Elution from column (dpm)	Elution from column (%)
2	2.29×10^6	Background	2.29×10^6	100	1.78×10^6	78
6.5	1.32×10^6	1.21×10^5	1.2×10^6	91	1.2×10^6	100
7	1.50×10^6	1.69×10^5	1.33×10^6	89	1.29×10^6	97
9	2.01×10^6	1.39×10^5	1.87×10^6	93	1.87×10^6	100

Water sample: 1 litre. Flow rate of sample: 17 ml min^{-1}. Eluant: Acetone: 100 ml at a flow rate of 3 ml min^{-1}. Concentrations of cholesterol: range from 5 to 100 ppb.

Table 131 Isolation of sterol from water by XAD-2 columns

Added ($\mu g\ l^{-1}$)	(dpm)	Recovered in H_2O passed through column (dpm)	Cholesterol retained by column (dpm)	(%)	Acetone elution (dpm)	(%)
100	1.86×10^6	Background	1.86×10^6	100	1.75×10^6	94
5.5	1.69×10^6	Background	1.69×10^6	100	1.60×10^6	95

Water samples: 1 litre, pH 2. Column: 15 mm × 20 cm, packed with 60–120 mesh resin. Flow rate: 17.5–18 ml min^{-1}. Eluant: acetone (100 ml); pH was adjusted to 8.5–9 by adding a few drops of ammonium hydroxide (14.5 M). Eluted at 3 ml min^{-1}.

These workers demonstrated that acidification (pH 2) of the samples favoured the adsorption of sterol (Table 130). Unfortunately, the sterol was found to be so strongly adsorbed onto the resin at pH 2, that removal of the adsorbate by acetone was incomplete (up to 78%). Although the columns retained 90–93% of the added cholesterol from high pH samples, 97–100% of this was routinely recovered by acetone elution. Table 131 lists the results of experiments in which water samples adjusted to pH 2 were extracted by XAD-2 columns which were subsequently eluted with acetone adjusted to pH 8.5–9.0 with concentrated ammonia. As indicated, the columns retained 100% of the added cholesterol at low and high concentrations at a flow rate of 17.5–18 ml min^{-1}. By simply adjusting the acetone eluant to pH 8.5–9.0, over 94% of the adsorbed sterol was recovered. These experiments were conducted using ^{14}C-labelled cholesterol. Labelled coprostanol was not available.

^{14}C-labelled cholesterol was also used to test the recovery of 5–100 μg of faecal sterols from sea water (labelled coprostanol not being available). The radioactivity of the samples and eluates was measured by a two-channel liquid scintillation counter. Percentage recovery was calculated on the basis of the amount of labelled material recovered in the acetone eluant. The results (Table 132) indicate that column extraction efficiency is not adversely affected by the salinity of the water samples.

Wun et al.[19] do not discuss the lower detection limit of their method. They do, however, report 92% recovery of coprostanol when 10 litres of dilute sewage containing 187.5 μg coprostanol is passed through the XAD-2 column.

Table 132 Extraction of radioactive cholesterol from sea water

Water sample (1 litre)	Cholesterol Added ($\mu g\ l^{-1}$)	(dpm)	Acetone elution (dpm)	(% recovery)
Synthetic sea water	5	1.6×10^6	1.52×10^6	95
	100	1.16×10^6	1.09×10^6	95
Natural sea water	50	1.48×10^6	1.43×10^6	97

Column: 15 mm × 20 cm, packed with XAD-2 (60–120) resin. Flow rate: 18 ml min^{-1}. Eluant: Acetone: 100 ml at 3 ml min.$^{-1}$.

FLUORESCENT WHITENING AGENTS

These substances can occur in river waters originating from their extensive use as ingredients of synthetic detergents for washing and cleaning operations. Abe and Yoshima[20] used thin-layer chromatography to study fluorescent whitening agents in river waters.

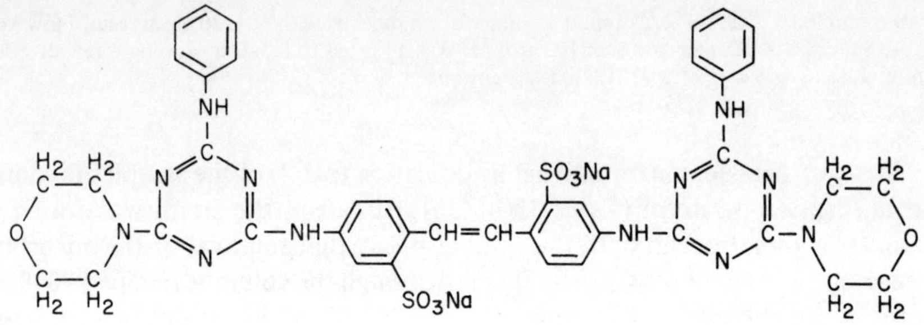

FB 260

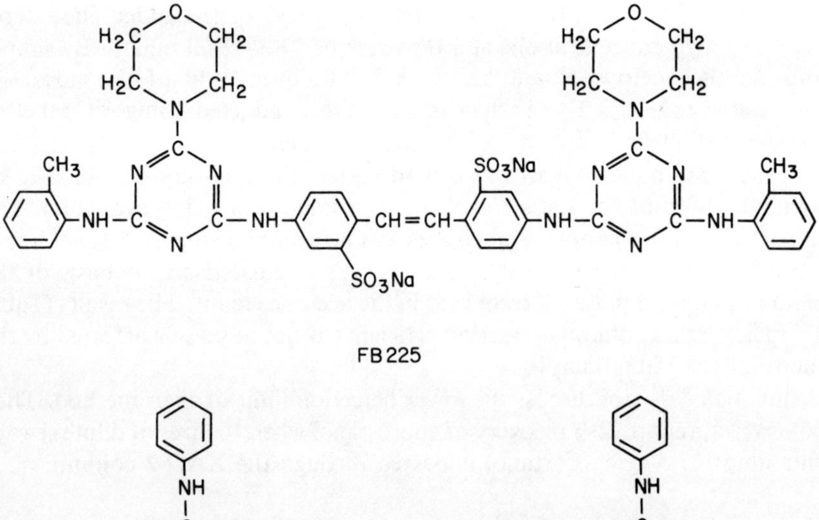

FB 225

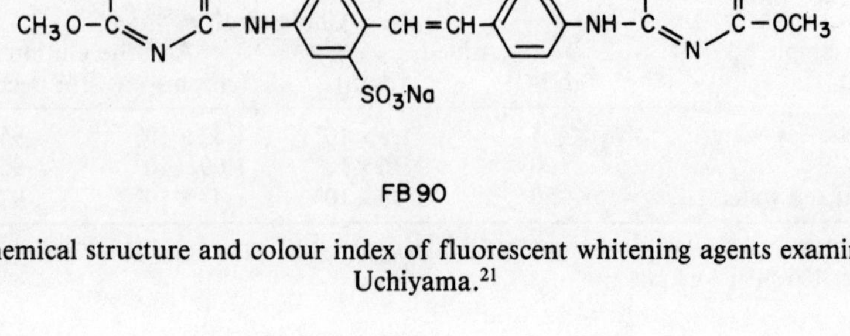

FB 90

The chemical structure and colour index of fluorescent whitening agents examined by Uchiyama.[21]

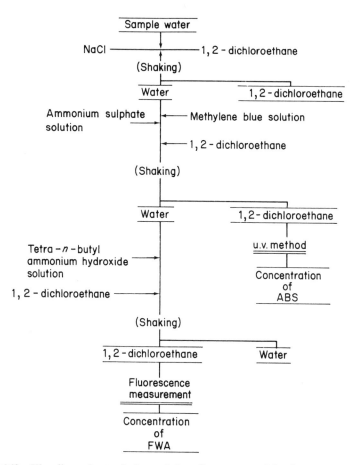

Figure 160 The flow sheet of determining fluorescent whitening agent and alkyl benzensulphonate. Reprinted with permission from Uchiyama.[21] Copyright (1979) Pergamon Press

Uchiyama[21] has given details of a procedure he developed down to 0.2 ppm for the isolation and determination of fluorescent whitening agents in water and bottom deposits. The fluorescent whitening agents were sodium salts of a sulphonated stilbene derivative and this was measured by fluorescence with the use of tetra-n-butyl ammonium hydroxide.

Method

Apparatus

All fluorescence excitation and emission spectra were measured on a Hitachi MPF-3 spectrofluorometer.

Procedure

The flow sheet of the analytical procedure is shown in Figure 160. To a 100 ml of sample water in a separatory funnel are added 10 g sodium chloride and 20 ml 1,2-dichloroethane. The funnel is shaken for 5 min and allowed to settle. The organic phase is discarded. Then 5 ml methylene blue aqueous solution (0.025% w/v), 10 ml ammonium sulphate solution (13% w/v), and 20 ml 1,2-dichloroethane are added to the water phase. The funnel is shaken for 5 min and left to settle. The organic phase is used for analysing alkyl benzenesulphonate by an ultraviolet method. To the water phase are added 3 ml tetra-n-butyl ammonium hydroxide aqueous solution (10% v/v) and 20 ml 1,2-dichloroethane. The funnel is shaken for 5 min and left to settle. The organic phase is used for fluorescent measurement to determine the concentration of fluorescent whitening agent. The wavelength of excitation and emission is 370 and 405 nm respectively for all three whitening agents studied (see formulae above). Figure 161 shows the excitation and emission spectra of FB 260.

Uchiyama[21] showed that pre-extracting the water sample with 1,2-dichloroethane was a very effective means of removing interfering fluorescent materials such as oil which would otherwise interfere in the determination of fluorescent whitening agent. Excellent calibration curves were obtained for the three optical whitening agents studied (Figure 162). Alkyl benzenesulphonates interfere in the procedure.

Uchiyama[21] applied this method to the determination of fluorescent whitening agents and alkyl benzenesulphonates and also methylene blue active substances in water and bottom mud samples taken in a lake. The muds were filtered off with a suction filter and frozen until analysed. About 20 g of wet

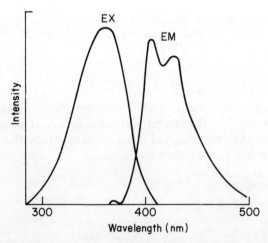

Figure 161 The excitation (EX) and emission (EM) spectra of fluorescent whitening agent (FB 260). Reprinted with permission from Uchiyama.[21] Copyright (1979) Pergamon Press

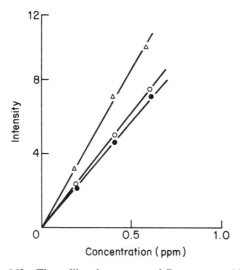

Figure 162 The calibration curves of fluorescent whitening agent. △, FB 225; ○, FB 260; ●, FB 90. Wavelength: excitation, 370 nm, emission, 405 nm. Reprinted with permission from Uchiyama.[21] Copyright (1979) Pergamon Press

Table 133 Analytical results of MBAS, ABS, and FWA in water

Station no.	Methylene blue active substances (ppm)	Alkyl benzene sulphonates (ppm)	Fluorescent whitening agents (ppm)
1	1.46	0.60	0.007
2	0.94	0.47	0.005
3	0.97	0.47	ND
4	0.84	0.47	ND
5	0.73	0.40	ND
6	0.31	0.40	ND
7	0.31	0.27	ND
8	0.29	ND	ND

ND, not detected.
Reprinted from Uchiyama.[21] Copyright (1979) Pergamon Press.

Table 134 Analytical results of MBAS, ABS, and FWA in bottom muds

Station no.	Methylene blue active substances (μg g^{-1} dry mud)	Alkyl benzene-sulphonate (μg g^{-1} dry mud)	Fluorescent whitening agent (μg g^{-1} dry mud)
2	377.8	73.2	1.35
3	288.3	96.3	0.83
4	190.3	37.5	0.55
5	281.5	26.9	0.55
6	117.0	28.1	0.25
7	303.5	17.3	0.70
8	107.3	16.9	0.25

Reprinted from Uchiyama.[21] Copyright (1979) Pergamon Press.

bottom mud was extracted three times with a methanol–benzene (1:1) mixture. After the solvent was evaporated using a water bath, the residue was dissolved in hot water and this solution used for analysis. Tables 133 and 134 show the analytical results for methylene blue active substances (MBAS), alkyl benzenesulphonate (ABS), and fluorescent whitening agent (FWA) in water and bottom muds. In this analysis, FB 260 and dodecyl benzenesulphonate were used as the standard materials.

MESTRANOL (17α-ETHYNYL-3-METHOXYESTRA-1,3,5(10)-TRIEN-17β-OL) AND ETHYNYLOESTRADIOL

Okunu and Higgins[22] have described a procedure for determining residual levels of mestranol animal damage control chemisterilant and its 3-hydroxy homologue, ethynyloestradiol in plant, soil, and water samples. The lower limits of detection were 0.05 ppm for foliage, 0.1 ppm for soils, and 0.01 ppm for water samples. After extraction in acidic medium, samples are cleaned up by Florisil column chromatography. Water samples can then be analysed by gas-liquid chromatography. Soil and foliage samples are further cleaned up on a gel permeation chromatographic column so that the ethynyloestradiol fraction can be analysed by gas chromatography. The mestranol fraction is again cleaned up by gel separation and a Florisil column. Thin-layer chromatography was used to confirm the results obtained by gas chromatography.

Method

Materials and equipment

Reagents
Reference standards of mestranol and ethynyloestradiol (EED) were supplied by Syntex Corporation. All organic solvents used were Burdick and Jackson 'distilled in glass' grade. Carbon tetrachloride was redistilled before use, and the chloroform contained commercially added 1% ethanol as a preservative.

Kuderna-Danish evaporative concentrator The unit consists of a graduated 10 ml concentrator tube, a 250 ml flask, and a three-ball Snyder column.

Florisil columns Florisil, 60–100 mesh, is heated at 130 °C for 24 hours, then deactivated with 2–3% water and allowed to stand for at least 24 hours. To check the activity, 0.25 g of the Florisil is added to 0.5 ml of a chloroform solution containing 100 ppm mestranol and the concentration of mestranol in the liquid phase compared with that in the original solution by gas chromatographic analysis. At optimum activity, the ratio of the two concentrations (with Florisil/original) should be 0.30 ± 0.02. For a 10 g column, 10 g of Florisil are added to a 400×22 mm i.d. chromatographic column. The column is tapped lightly so that the Florisil packs uniformly, and 10 g of anhydrous sodium sulphate are added over the Florisil. For a 5 g column, 5 g of Florisil are similarly packed into a 200 × 10 mm i.d. glass column with a 125 ml reservoir, and 10 g of anhydrous sodium sulphate added.

Gel permeation chromatographic column The gel column apparatus consists of a 250 × 14 mm i.d. glass column with a 250 ml reservoir at top and a glass wool plug and Teflon stopcock at bottom. The top of the reservoir is fitted with an 18/9 ball joint to facilitate connection to an air pressure source. The gel (Biobeads S-X2 Gel, 200–400 mesh, Bio-Rad Laboratories) is swelled overnight in carbon tetrachloride, then poured as a slurry through a funnel into the column and allowed to pack by gravity until the length of the gel bed is 23 cm. Solvent should always be maintained in the column bed. A glass wool plug, approximately 1 cm in length, is placed on top of the gel. At least 50 ml of solvent should be pumped through the column until the carbon tetrachloride does not contain any substance that will interfere in the gas chromatographic analysis when evaporated from 10 to 0.1 ml. A constant flow rate between 2 and 3 ml min^{-1} is maintained by applying air pressure to the solvent head by means of an air pump or gas cylinder. The column should be flushed with at least 20 ml of solvent between runs.

Gas chromatography A Beckman GC-5 gas chromatograph was used, equipped with a Model 5671000 Electrometer, flame ionization detector, and a 4 foot × ¼ inch o.d. aluminium column packed with 3% OV-17 on 80–100 mesh Gas Chrom Q. The electrometer was operated at a sensitivity of 4×10^{-11} A mV^{-1}. Flow rates were about 60 ml min^{-1} for hydrogen, 300 ml min^{-1} for air, and 60 ml min^{-1} for the nitrogen carrier gas. Operating temperatures were 260 °C for the column, 275 °C for the inlet, and 290 °C for the detector.

Sample preparation

Vegetation samples

Ten grams of air-dried and homogeneously ground sample are weighed into a glass-stoppered 300 ml flask and shaken for 30 minutes with 100 ml of acetonitrile containing 5 ml of 1.2 M hydrochloric acid. The extract is decanted and filtered through a No. 1 Whatman filter paper into another 300 ml flask. The extraction is repeated twice with 50 ml each of acetonitrile. A Teflon boiling stone is added to the flask, a three-ball Snyder column is connected to the flask and a water-aspirator to the column, and the flask evaporated under vacuum just to dryness on a steam bath.

Soil samples

Ten grams of air-dried and pulverized soil sample are weighed with 50 ml each of 1.2 M hydrochloric acid and acetone, decanted and filtered through a No. 1

Whatman filter paper into another 300 ml flask, and rinsed twice with 50 ml each of acetone. A three-ball Snyder column is connected to the flask which is placed on a steam bath to evaporate off the acetone. After cooling, 10 ml of ethyl ether and 50 ml of chloroform are added and the flask shaken for 2 minutes. The contents and two 50 ml of chloroform rinses are transferred to a 250 ml separatory funnel and shaken for 5 minutes. The lower organic layer is drawn off and passed slowly through 10 g of granular anhydrous sodium sulphate into a 300 ml flask, then the sodium sulphate is rinsed with 10 ml of chloroform and the rinse added to the sample. A Teflon boiling stone is added to the flask, a three-ball Snyder column is attached to the flask which is then evaporated just to dryness on a steam bath.

Water samples

A 200 ml volume of water sample is shaken with 10 ml of 1.2 M HCl, 10 ml of ethyl ether, and 50 ml of chloroform in a 500 ml separatory funnel for 10 minutes. The lower layer is passed through 10 g of sodium sulphate into a Kuderna-Danish unit and evaporated to about 5 ml on a steam bath. The 10 ml concentrator tube is placed in a tube heater (60 °C) and a gentle stream of air is passed over the sample to concentrate it just to dryness.

Sample clean-up and analysis

10 g Florisil column

The sample extract from the vegetation or soil sample is dissolved in 20 ml of chloroform, using an ultrasonic cleaner to facilitate solution. This and a 10 ml chloroform rinse are added to the dry chromatographic column and eluted with an additional 110 ml of chloroform. All the eluate is collected in a Kuderna-Danish unit and evaporated just to dryness, then the first 5 g Florisil column clean-up procedure is followed.

First 5 g Florisil column

The extract (from water samples) or the concentrate (from the 10 g Florisil clean-up of vegetation and soil samples) is dissolved in 1 ml of chloroform and this and a 1 ml chloroform rinse are added to the dry Florisil column, then eluated. The remainder is collected in a Kuderna-Danish unit and concentrated just to dryness. The gel column is then used for further clean-up of concentrates from vegetation and soil samples. The concentrate from water samples is dissolved in 100 μl of carbon tetrachloride and analysed for mestranol by gas chromatography. If the sample contains interfering material and further clean-up is needed, the gel column procedure is used.

Gel column

The stopcock is opened, air pressure is applied to the reservoir and enough solvent removed until its surface is level with the top of the glass wool plug. The air pressure line is disconnected, the stopcock closed, and a 10 ml Kuderna-Danish concentrator tube placed under the column. The sample from the first 5 g Florisil clean-up is dissolved in 1 ml of carbon tetrachloride, the sample solution withdrawn with a 1 ml syringe and transferred to the gel column. The stopcock is opened, air pressure applied, and the sample allowed to sink just to the top of the glass wool plug. The transfer is repeated with another 1 ml of carbon tetrachloride, then 80 ml of carbon tetrachloride are added to the reservoir and six 10 ml fractions collected in Kuderna-Danish concentrator tubes. Fraction 4 contains mestranol and fractions 5 and 6 are concentrated in Kuderna-Danish units (without Snyder columns) to about 3 ml on a steam bath, then evaporated to just dryness in the tube heater under a stream of air. Fraction 4 from water samples, and the composite of fractions 5 and 6 from vegetation, soil, and water samples, are clean enough for gas chromatographic analysis. For fraction 4 from vegetation and soil samples, the gel column clean-up and concentration procedure is repeated, retaining only fraction 4 for additional clean-up by Florisil chromatography.

Second 5 g Florisil column

Fraction 4 from the second gel clean-up is added to a 5 g Florisil column and eluted with 30 ml of chloroform. Only the second 10 ml of eluate is collected. This fraction is evaporated to dryness and dissolved in 100 μl of carbon tetrachloride for analysis by gas chromatography. If early emerging materials interfere with the measurement of mestranol, the second 5 g Florisil column clean-up step is repeated until they are removed.

G.l.c. analysis

The sample extracts from the clean-up procedures are dissolved in 100 μl of carbon tetrachloride so that the sample equivalence is 100 mg μl^{-1} for water samples. The amount of sample injected for gas chromatographic analysis will depend on the concentration of mestranol or ethynyloestradiol in the sample solution and the sensitivity desired. The retention time is about 9 minutes for mestranol and 11.5 minutes for ethynyloestradiol (Figure 163), but at least 40 minutes should be allowed between injection of mestranol sample extracts so that late emerging substances will not interfere with subsequent injections. The ethynyloestradiol fraction is relatively free of these late interfering substances.

T.l.c. analysis

Thin-layer chromatography is used for qualitative and semi-quantitative confirmation. An appropriate amount of sample solution is spotted on a silica

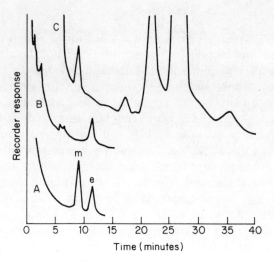

Figure 163 Gas chromatographic analysis of extracts from clean-up of a grass foliage sample fortified with 0.1 ppm each of mestranol and EED: A, mixed reference, 2 μl containing 10 ppm each of mestranol (m) and EED (e); B, composite of fractions 5 and 6 from first gel clean-up, 4 μl (sample equivalence, 100 mg μl^{-1}); C, fraction 4 from second gel clean-up after two separations on second 5 g Florisil column, 4 μl (sample equivalence, 100 mg μl^{-1}). Reprinted with permission from Okuno and Higgins.[22] Copyright (1977) Springer Verlag, NY

gel thin-layer sheet (Eastman Chromatogram, Type R 301R2) and developed with chloroform, allowing the solvent front to migrate about 10 cm. After drying, the sheet is sprayed with a 1:1 mixture of methanol and sulphuric acid and warmed for 5–10 minutes at 60 °C.[23] Mestranol ($R_f = 0.75$) and ethynyloestradiol ($R_f = 0.25$) will appear as red spots under visible light and orange under ultraviolet light (154 nm). The lower limit of detection is about 100 ng for each compound.

In the above method, the 10 g Florisil column is used initially to separate mestranol and ethynyloestradiol from most of the coextracted materials that are more polar. The first 5 g Florisil column is used to remove additional polar materials that were not removed in the 10 g column. Recoveries of both compounds added as standard solutions to the 10 g column were essentially quantitative (Table 135). On the 5 g column, mestranol and ethynyloestradiol recoveries averaged 103% and 92%, respectively, in the eluate collected after the first 10 ml (Table 135). On the second 5 g Florisil column, used as an additional clean-up step for mestranol only, recovery averaged only 91% because only the second 10 ml fraction is retained (Table 135). About 9% is sacrificed to exclude interfering substances, including chlorophyll and xanthophyll from plant samples, that appear after 20 ml of chloroform have eluted. For additional

Table 135 Recoveries of mestranol and ethynlyoestradiol (EED) added as standard to Florisil and gel permeation chromatographic columns

Column	ml eluant	μg added	N	Compound	Eluate retained	% Recovery Mean	s.d.
Florisil							
10 g	140	10	6	Mestranol	All	100	3.0
				EED	All	103	3.6
1st 5 g	70	1	6	Mestranol	All but first 10 ml	103	2.5
				EED	All but first 10 ml	92	6.1
2nd 5 g	30	1	6	Mestranol	2nd 10 ml only	91	7.4
Gel	60	1	5	Mestranol	3rd 10 ml only	86	4.1
				EED	4th 10 ml only	77	5.8

Reprinted with permission from Okuno and Higgins.[22] Copyright (1977) Springer Verlag, NY.

Table 136 Recoveries of mestranol and ethynyloestradiol (EED) from foliage, soil, and water samples spiked with both compounds

Sample	Sample size	μg added	N	Compound	No. column separations 1st 5 g Florisil	Gel	2nd 5 g Florisil	% Recovery Actual Mean	s.d.	Corr* mean
Grass foliage	10 g	1.0	5	Mestranol	1	2	2	40	8.8	65
				EED	1	1	0	49	4.6	69
Mixed foliage	10 g	0.5	2	Mestranol	1	2	4	38	0	75
				EED	1	1	0	56	2.1	79
Soil A	10 g	1.0	2	Mestranol	1	2	2	26	3.5	42
				EED	1	1	0	13	1.4	18
Soil B	10 g	10.0	2	Mestranol	1	2	2	29	4.9	47
				EED	1	1	0	12	2.1	17
Soil C	10 g	10.0	2	Mestranol	1	2	2	30	2.8	49
				EED	1	1	0	34	0.7	48
Water	200 ml	2.0	5	Mestranol	1	0	0	99	7.2	99
				EED	1	0	0	73	7.1	79

*Corrected for clean-up losses on the basis of the following recoveries (Table 135): for mestranol, 100% for 1st 5 g Florisil, 86% for each gel, 91% for each 2nd 5 g Florisil; for EED, 92% for 1st 5 g Florisil, 77% for gel.
Reprinted with permission from Okuno and Higgins.[22] Copyright (1977) Springer Verlag, N.Y.

clean-up, repeated separations with the 5 g Florisil column were more effective than a single clean-up on a larger column. Gel permeation chromatography is used to remove extractables of high molecular weight that are not easily removed on the Florisil column. Substances of high molecular weight are eluted first, followed by mestranol and then ethynyloestradiol. By beginning the mestranol fraction at 30 ml, some mestranol is lost in order to remove as much of the interfering substances as possible. Some ethynyloestradiol is sacrificed in the fourth 10 ml fraction to obtain maximum recovery of mestranol. Beginning the ethynyloestradiol fraction at 40 ml also excludes remnants of interfering material that trail, producing a much cleaner fraction for gas chromatographic analysis. Recoveries from the gel column averaged 86% for mestranol and 77% for ethynyloestradiol (Table 135).

Recoveries of mestranol and ethynyloestradiol from spiked samples varied with sample type and with the number of steps required for clean-up. Recoveries from five replicate analyses of grass foliage samples spiked at 0.1 ppm averaged 40% for mestranol and 49% for ethynyloestradiol, or 65% and 69% after correction for the losses due to the various clean-up steps (Table 136). Recoveries from a mixed foliage sample (field collected grasses and forbs) spiked at 0.05 ppm averaged 38% for mestranol and 56% for ethynyloestradiol, or 75% and 79% after correction.

Recoveries were lower in soil samples, averaging less than 50% even after correction (Table 136). This may have been due to degradation of the compounds by soil microorganisms or to chemical and physical interactions with the soil. Mestranol recoveries averaged 26% from soil A, a podzol type spiked at 0.1 ppm, and 29% and 30% from similar type soils B and C spiked at 1.0 ppm. Recoveries of ethynyloestradiol were even lower, presumably because of its greater chemical reactivity due to the slightly acidic hydrogen in the 3-hydroxy position. The better recovery of ethynyloestradiol from soil C than from A and B was probably due to relatively large amounts of less adsorbing vegetative debris in soil C.

Good recoveries were obtained from water samples. For a field-collected runoff sample spiked at 0.01 ppm, recoveries averaged 99% for mestranol and 73% for ethynyloestradiol, or 99% and 79% after correction (Table 136). The lower ethynyloestradiol recoveries in this case apparently reflect a less favourable partitioning of the compound into the chloroform phase.

CARBOXYMETHOXY SUCCINATE

A method has been described[24] for determining this detergent builder in river water and sewage effluent. The sample is centrifuged, adjusted to 1 M in hydrochloric acid and, after 30 min, recentrifuged. The supernatant solution (1 ml) is heated at 100 °C for 30 min, then treated with 0.5% phenylhydrazine solution in 2 M hydrochloric acid (0.3 ml) and heated for a further 30 min; 10% barium chloride solution (0.5 ml), concentrated aqueous ammonia (0.15 ml), and ethanol (6 ml) are added and the mixture is kept at 4 °C for 20 h. The precipitate is centrifuged off and baked at 130 °C for 30 min, 0.05% 2-naphthol solution in 92.5% sulphuric acid (1 ml) is added and the residue is completely dissolved by vigorous mixing. After heating the solution on a boiling water bath for 1 h, 80% sulphuric acid (3 ml) is added and the mixed solution is set aside at room temperature for 20 min before the extinction is measured at 480 nm against water. As little as 1 µg of carboxymethoxy succinate can be determined.

ISOPROPYLMETHYLPHOSPHONOFLUORIDATE (GB) AND S-2-(DI-ISOPROPYLAMINO)ETHYL O-ETHYLMETHYLPHOSPHONOTHIOATE (VX)

To determine these substances at the parts per trillion (US) level in sea water, Michel et al.[25] mixed the sample (0.1 ml) with 0.1 ml of eel chloinesterase solution in a buffer solution of pH 7.2 (0.1 M in morpholinopropane sulphonic acid, 0.01 M in EDTA and containing 0.1% of gelatin). The mixture is incubated at 25 °C for up to 30 h. To determine residual enzyme activity, 0.1 ml of substrate mixture (2 nM in 5,5′-dithio-*bis*-(2-nitrobenzoic acid) is added and, after 1 h, the extinction of the mixture is measured at 412 nm. Standards are analysed similarly, and a blank determination is carried out without use of the enzyme. To determine VX alone, a 25 ml sample of sea water is extracted with dichloromethane (2 × 5 ml), the combined extracts are mixed with 0.1 ml of 0.1 M hydrochloric acid and the mixture is evaporated under nitrogen at room temperature. The residue is dissolved in water and an aliquot is analysed as described above.

2,3-DICHLORO-1,4-NAPHTHAQUINONE

A spectrophotometric method has been described for determining down to 2 µg of this fungicide in water and soil based on the formation of a coloured reaction product with aniline.[26]

BENETHONIUM SALTS

These compounds have been determined in waste water by a spectrophotometric method based on reaction with tetrabromo-phenolphthalein ethyl ester.[27]

ααα-TRIFLUORO-4-NITRO-*m*-CRESOL

This pesticide has been determined gas chromatographically in fish in amounts down to 0.01 ppm.[28] The fish sample is homogenized with hexane–ethyl ether (7:3). The phenol is extracted into 0.1 M sodium hydroxide, back-extracted into hexane–ether (7:3) after acidification, and methylated with diazomethane. The product formed is analysed by gas chromatography at 140 °C with the use of a glass column (1.8 m × 4 mm) packed with 3% of OV-1 on 80–100 mesh Chromosorb W, with nitrogen as carrier gas (60 ml min^{-1}) and electron capture detection.

PYRETHRINS

Down to 0.2 µg pyrethrine can be determined in water by gas chromatography.[29] A hexane extract of the sample is washed with aqueous sodium chloride solution then following an elaborate working up procedure, it is gas chromatographed on a helical glass column (4 ft × 0.25 in) packed with 5% of SE-30 on AW-DMCS Chromosorb W (60–80 mesh) and operated at 190 °C with nitrogen (40 ml min^{-1}) as carrier gas and flame ionization detection. This method permits the simultaneous determination of pyrethrin (linear response range 0.2–2.2 µg) and of the synergists piperonyl butoxide (range 0.6–5.6 µg) and *N*-(2-ethylhexyl)-norborn-5-ene-2,3-dicarboximide (range 0.6–1.8 µg). Recoveries averaged 93–94%.

TOXAPHENE (CAMPHECHLOR)

Hempel et al.[30] have described a colorimetric method for determining down to 15 parts per 10^9 of this pesticide in water. Toxaphene is first extracted into hexane then the extract washed with concentrated sulphuric acid then chromic acid then water. Hexane is removed *in vacuo* and the residue taken up in acetone prior to evaporating to dryness *in vacuo*. Pyridine is then added and the mixture heated for 30 min at 100 °C, then aqueous ethanol and sodium hydroxide added. The extinction of the coloured reaction product is measured at 550 nm.

Hughes and Lee[31] have used gas chromatography to determine toxaphene in water and fish samples.

CARBOFURAN

Carbofuran and its degradation products carbofuranphenol, 3-ketocarbofuran, 3-hydroxycarbofuranphenol, *N*-hydroxymethylcarbofuran, 3-hydroxycarbofuran, have been determined in water samples and crustacae by thin-layer chromatography of ether extracts of the sample.[32]

GEOSMIN (*TRANS* 1,10-DIMETHYL-*TRANS*-DECALOL)

This is a compound with a musty odour that is a metabolite of some kinds of *Actinomyces*.[33-36] A musty odour occurs in water from polluted rivers and lakes, and Kikuchi and coworkers[37,38] isolated *Actinomyces* from some lakes and, after cultivation, detected geosmin in the culture medium by gas chromatography–mass spectrometry.

Kikuchi et al.[39] identified geosmin in water supplies using gas chromatography–mass spectrometry. The water was chlorinated, treated with aluminium sulphate, and filtered through a layer of sand; it was then passed through a column of activated charcoal to adsorb odoriferous compounds. The charcoal was extracted several times with dichloromethane and the oily material recovered from the extract was distilled with water, the distillate was extracted with dichloromethane, and the solvent was removed by evaporation. The residue was purified by chromatography on silica gel, with pentane and pentane–ethyl ether (99:1) as solvents. The eluate, which contained a musty-smelling oil, was subjected to gas chromatography at 150 °C on a stainless steel column (2 m × 3 mm) packed with 15% of Reoplex 400 on Chromosorb WN AW with nitrogen as carrier gas (30 ml min^{-1}) and flame ionization detection. The identity of geosmin was confirmed by mass spectrometry and by further gas chromatography at 115 °C on a glass column (1.5 m × 2 mm) packed with 25% of PEG 20M Chromosorb W with helium as carrier gas (2 kg cm^{-2}).

Otsuhara and Suwa[40] conducted studies on odorous compounds in reservoir water, filtration plant water, and broth cultivated with actinomycetes and algae by gas chromatographic–mass spectrometric analysis. Odorous compounds were isolated from water by adsorption and extraction methods and identified as geosmin and 12-methylisoborneol. Geosmin and 12-methylisoborneol were found in the surface water of a Japanese reservoir and *Micromonospora* isolated from a filtration plant produced 12-methylisoborneol. It was suggested algae were much concerned in the production of odorous compounds.

Yasuhara and Fuwa[41] have described a quantitative method for determining geosmin in river water using computer controlled mass fragmentography. A JEOL Model JMS-D 100 mass spectrometer was connected with a JEOL Model JGC-20K gas chromatograph and a JEOL Model JMA-2000 mass data analysis system.

Method

A water sample (1 l) was extracted with dichloromethane (50 ml) after dissolution of sodium chloride (300 g). The extracted solution was dried for several hours over anhydrous sodium sulphate (15 g), then concentrated to several hundred microlitres using a Kuderna-Danish concentrator under atmospheric pressure.

A 1 m × 2 mm i.d. column packed with 1% OV-1 on Chromosorb W (80–100 mesh) was used. The column temperature was 100 °C, the injection port temperature 250 °C, and the carrier gas was helium at a flow rate of 40 ml min^{-1} (3 kg cm^{-2}).

Geosmin showed a base peak at m/e 112 whereas organic substances that were present in geosmin-free river water did not show a peak at m/e 112 in the mass fragmentogram under the conditions used. As the dynamic range in the mass spectrometer is very narrow and injection of an exact volume with a microsyringe is not sufficiently accurate, quantitative analysis was carried out using an internal standard method. *n*-Butyl benzoate was selected as the internal standard, as the retention times of geosmin and *n*-butyl benzoate in the gas chromatogram were similar and a base peak at m/e 105 in the mass spectrum of *n*-butyl benzoate was near to the m/e 112 peak of geosmin. *n*-Butyl benzoate in hexane solution was added to a hexane or dichloromethane solution of geosmin at a rate of 1 μg ml^{-1} of solution.

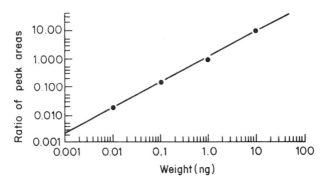

Figure 164 Calibration graph for geosmin. Reprinted with permission from Yasuhara and Fuwa.$_{41}$ Copyright (1979) Elsevier Science Publishers

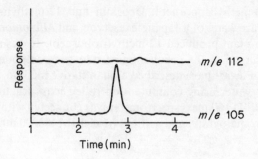

Figure 165 Mass fragmentograms of geosmin (m/e 112, 50 pg) and n-butyl benzoate (internal standard) (m/e 105, 1 ng). Reprinted with permission from Yasuhara and Fuwa.[41] Copyright (1979) Elsevier Science Publishers

In Figure 164 is shown a calibration graph prepared using the ratio of the peak areas of geosmin and n-butyl benzoate. The detection limit of geosmin was 10 ng ml^{-1}, but a more reliable value was over 50 ng ml^{-1}. An example of a mass fragmentogram is shown in Figure 165. The coefficient of variations ranged by 5.4% at the 10 μg ml^{-1} level, 1.9% at 1 μg l^{-1} level, to 18% at the 0.05 μg ml^{-1} level.

OZONATION OF WATER

Since the detection of halogenated organics in drinking water much research effort has been directed toward finding water treatment processes to remove such organics or their precursors and toward finding disinfectants other than chlorine. Great interest has been focused upon ozonation because both disinfection and organic removal can be accomplished with this process. As ozonated end-products will occur in water produced by such processes and these could be potentially toxic and would accumulate in waste water after repeated cycles of use it is necessary to ascertain what end-products occur in water that has been ozonated and subsequently chlorinated.

Kuo et al.[42,43] have studied the response of solutions of 2-propanol, acetic acid, and oxalic acid to ozonation, with and without ultraviolet irradiation, in the laboratory. The effect of ultraviolet irradiation was to enhance the rate of oxidation of these organic compounds; the products of the reaction were identified and the rate of decomposition determined in each case. Chlorination and ozonation of the retentates from the reverse osmosis and ultrafiltration of secondary-sewage effluent was also performed. Methylene chloride, chloroform, carbon tetrachloride, and corresponding bromo-substituted derivatives were identified in the products of chlorination but not in those of ozonation, which appeared to convert the residual organic materials (humic and fulvic acids) to harmless products, leading ultimately to formation of carbon dioxide.

Kuo et al.[42] selected 2-propanol, acetic acid, and oxalic acid for their study of ozonation and ultraviolet irradiation effects because they are low molecular weight organics and are poorly removed in the two advanced water treatment processes, namely reverse osmosis and activated carbon treatment.

Kuo et al.[42] examined the ozonation products gas chromatographically using a Hewlett-Packard 5750B gas chromatograph equipped with dual flame ionization detectors. A 6 ft × ¼ in o.d. glass column packed with 0.2% Carbowax 1500 on 80–100 Carbopack C was used for determining alcohols and ketones. The derivative gas chromatographic procedures of Bethge and Lindstrom's method[44] were followed to determine formic, acetic, and propionic acids. The benzyl esters of acids were separated and determined on a 6 ft × ¼ in o.d. glass column packed with 3% butane-1,4-diol succinate polyester on 100–200 AW Chromosorb W. Oxalic acid was determined on the latter gas chromatographic column after the compound was converted to methyl ester using diazomethane.[45] Glyoxylic acid was determined by a colorimetric method described by Kramer et al.[46] The chromatropic acid method was used to determine formaldehyde.[47] Both the ozonated and the control mixtures were analysed for volatile halogenated organics by a gas-stripping–gas chromatography and gas chromatographic–mass spectrometric method. A solution of 125 ml was stripped at 60 °C and the stripped volatile organics were adsorbed in a Tenax GC column. The adsorbed organics were analysed on a 12 ft × 2 mm i.d. glass packed column packed with 0.2% Carbowax 1500 on 60–80 mesh Carbopack C. The adsorbed compounds were desorbed from the Tenax gas chromatographic column trap at a peak temperature of 250 °C for 6 min with a helium flow of 40 ml min^{-1}. The glass column was held at ambient temperature during this period. The oven temperature was then programmed to increase at 8 °C min^{-1} to 200 °C and held there for the remainder of the run.

2-Propanol was oxidized by ozonation and ultraviolet light irradiation to acetone (Figures 166 and 167), which in turn was oxidized to acetic and oxalic acids (Table 137). Trace amounts of formaldehyde and formic acid were also detected in the ozonated acetone mixtures (Table 137). Table 137 also gives a material balance on the total organic carbon as well as the sum of the organic compounds determined. The discrepancy in organic carbon between them is the intermediates which remained to be determined.

Pyruvic and keto malonic acids may be among the undetermined ozonation products. Ozonation of acetic acid resulted in the formation of glyoxylic acids. Glyoxlyic acid was, however, present in trace amounts (<1 ppm). The organic carbon present during the course of the reaction can be fully accounted for by acetic and oxalic acids resulting from the ultraviolet ozonation of acetic acid. No organics except oxalic acid itself were detected in reaction mixtures resulting from the ozonation of oxalic acid. The total organic carbon of the reaction mixtures could be completely accounted for by the oxalic acid, indicating that oxalic acid was oxidized directly to carbon dioxide under the experimental conditions.

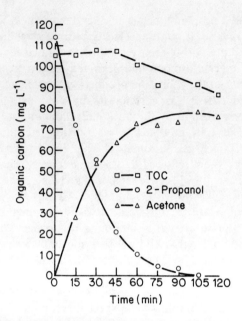

Figure 166 Ozonation of 2-propanol. TOC, total organic carbon. Reprinted with permission from Kuo *et al.*[42] Copyright (1977) American Chemical Society

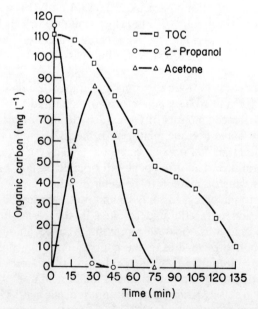

Figure 167 u.v.-ozonation of 2-propanol. TOC, total organic carbon. Reprinted with permission from Kuo *et al.*[42] *Copyright (1977) American Chemical Society*

Table 137 Concentration of determined organic compounds in u.v.-ozonated acetone mixtures

Ozonation time (min)	Acetone	Formaldehyde	Formic	Acetic	Oxalic	ΣOC*	TOC†
0	66	...	...	...	...	66	66
30	19	2	1	9	2	33	49
60	3	1	<1	6	7	17	29

Concentration of organic carbon (mg l^{-1})

*Sum of concentration of organic carbon (ΣOC) of determined organic compounds.
†Total organic carbon (TOC) of ozonation mixtures determined on TOC analyser.
Reprinted with permission from Kuo et al.[42] Copyright (1977) American Chemical Society.

Table 138 Organic removal by ozonation and u.v.-ozonation of 2-propanol and acetic acid

	2-Propanol		Acetic Acid	
pH	7	7	7*	7*
u.v.	No	Yes	No	Yes
Initial TOC (mg l^{-1})	114	116	84	83
Ozonation time (h)	2	2	2	2
% TOC removal	17	82	14	92
TOC removal				
Order	1st	1st	1st	Other
Period (min)	0–120	30–105	0–120	0–60
Rate constant	1.66×10^{-3} min^{-1}	1.29×10^{-2} min^{-1}	1.52×10^{-3} min^{-1}	0.892 mg l^{-1} min^{-1}
Reactant removal				
Order	1st	Other	1st	Other
Period (min)	0–90	0–15	0–120	0–60
Rate constant	1.35×10^{-2} min^{-1}	5.00 mg l^{-1} min^{-1}	3.23×10^{-3} min^{-1}	1.17 mg l^{-1} min^{-1}

*Initial pH.
Reprinted with permission from Kuo et al.[42] Copyright (1977) American Chemical Society.

Table 139 Organic removal by ozonation and u.v.-ozonation of oxalic acid at various pHs

pH	5	7	9	5
u.v.*	Yes	Yes	Yes	No
Initial TOC (mg l^{-1})	44	44	44	38
% TOC removal				
1 h ozonation	100	95	98	39
2 h ozonation	100	100	100	71
Oxalic or TOC removal†				
Order	1st	1st	1st	Other
Period (min)	0–30	0–45	0–60	0–90
Rate constant	0.179 min^{-1} min^{-1}	4.09×10^{-2} min_{-1}	6.15×10^{-2} min^{-1}	0.250 mg l^{-1} min^{-1}

*New u.v. lamp.
†Organic carbon of reaction mixtures was fully associated for by oxalic acid.
Reprinted with permission from Kuo et al.[42] Copyright (1977) American Chemical Society.

Glyoxylic, oxalic, formic, and acetic acids were the common ozonation endproducts prior to their complete oxidation to carbon dioxide. The oxidation of oxalic and formic acids to carbon dioxide is partially responsible for the removal of organic carbon by ozonation. Acetic acid was oxidized through ω-oxidation to yield glyoxylic acid. Glyoxylic acid is an intermediate that appears to have a very rapid rate of oxidation under the experimental conditions because only a very small amount of it was detected during the ozonation of the acetic acid solution. The effect of ultraviolet irradiation in enhancing the removal of organics during ozonation is apparent from the data in Tables 138 and 139. The percentage of organic removal at the end of a 2 h ozonation period was greatly improved with ultraviolet irradiation. The organic carbon removal and reactant removal followed either zero or first-order reaction kinetics. The rate constant for organic carbon removal was improved under ultraviolet irradiation by a factor of approximately eight for 2-propanol. For acetic acid the removal rate was improved by a factor of six.

Glaze et al.[48] analysed ozonation by products produced before and after chlorination in water and waste water using size exclusion chromatography and halogen-specific microcoulometry. Using these procedures, Glaze et al.[48] examined lake water before and after chlorination. Gas chromatography with an electron capture detector and gas chromatography–mass spectrometry of the water showed no detectable amounts of purgeable or volatile extractables before chlorination, with the exception of traces of aliphatic hydrocarbons. Chlorination of the unbuffered lake water with a dose of 20 mg l^{-1} caused the formation of trihalomethanes as determined by liquid-liquid extraction procedure on 25 ml samples using a modification of the procedure described by Henderson et al.[49] The yield of trihalomethanes reached a plateau after 3 days. At a buffered pH of 6.5 the yield of trihalomethanes after this period is 189 $\mu g\ l^{-1}$, consisting of 81% chloroform, 16% dichlorobromomethane, and 3% dibromochloromethane (by weight). Expressed as halogen, this yield is 4.47 μmole l^{-1}.

Gas chromatography–mass spectrometry of the lake water after chlorination

The analytical schemes shown in Figures 168 and 169 were utilized to examine the lake water for volatile and non-volatile organic compounds before and after chlorination. In each case a system blank was performed. Analysis of the base-neutral fraction and the diazomethane derivatized acid fraction were carried out using a Finnigan 3200/6000 gas chromatography–mass spectrometry system. The purge and trap procedure utilized the Tekmar sampler with Tenax/silica gel trap and a 6 ft $\times$ 2 mm i.d. glass column with 0.2% Carbowax 1500 on Carbopac C. A 30 m $\times$ 0.25 mm i.d. glass capillary with SE-30 coating was used.

High performance liquid chromatographic exclusion separations were carried out on a Waters ALC-201 instrument with 6000 A pump and refractive index detector. The column was 25 cm $\times$ 4.6 mm i.d. stainless steel with Partisil 10 (Whatman, 11 μ particle size, 60 Å pore size) deactivated with bonded glycerylpropylsilane (50) and by treatment with Carbowax 20M. Two columns

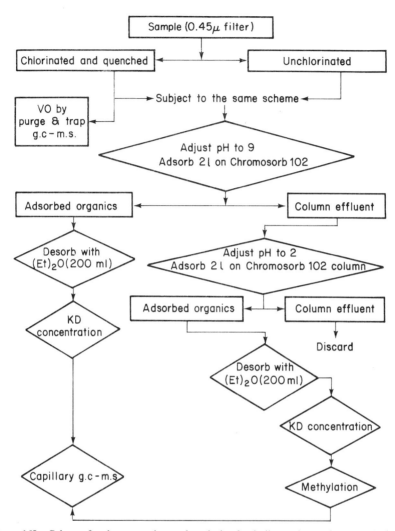

Figure 168 Scheme for the separation and analysis of volatile products of water chlorination

Table 140 Characteristics 11 μ particle size. 60 Å pore size) deactivated with glycerolpropyl silane and treated with carbowax 20M

	Analytical	Preparative
Length (cm)	25	25
i.d. (mm)	4.6	9.4
Number of theoretical plates (N)	2500	3500
HETP	0.1	0.07
Linear velocity (cm sec^{-1})	0.185	0.097
Void volume V_o (ml)	2.25	3.1
Permeation volume V_r (ml)	4.9	7.4
Interstitial volume V_t (ml)	2.65	4.3

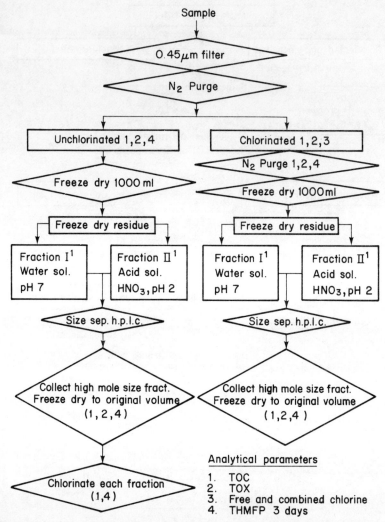

Figure 169 Scheme for the separation and analysis of non-volatile products of water chlorination. TOC = total organic carbon, TOX = total organic halogen, THMFP = trihalomethane

were used, the characteristics of which are shown in Table 140. A calibration curve which fits both columns is shown in Figure 170. The solvent used for this calibration was carbon-filtered deionized water, but 2% isopropanol–water was also used.

Total organic halogen was measured by a modification of the procedure of Glaze et al.[51] using Chromosorb 102 macroreticular resin as the accumulator material. The sample (usually 125 ml) was percolated through a 6 cm × 2 mm i.d. bed of 100–120 mesh Chromosorb 102 (ca 200 mg) at a flow rate of approximately 5 ml min^{-1}. The column was washed with 2 ml of 0.001 M nitric

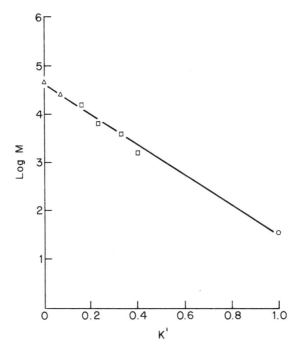

Figure 170 Calibration curve for Partisil 10/glycophase size exclusion columns. △ Proteins (ovalbumin, $m = 45,000$; chymotrypsinogin A, $m = 25,000$). □ Sodium polystyrene sulphonates ($m = 6000, 6500, 4000, 1600$). O Methanol

acid in carbon-purified deionized water. Organics were eluted with a two-step solvent elution:

(1) 1.0 ml of 1 M ammonia in methanol (1 part 2 M ammonia, 1 part methanol) and
(2) 1.0 ml of n-pentane.

Solvents were eluted into a sealed mini-valve containing enough nitric acid to neutralize (with a slight excess) the ammonium hydroxide. The sealed vial was shaken to partition the lipophilic organics (trihalomethanes) into the pentane layer, then each layer was analysed by pyrolysis (800 °C)/microcoulometry using the Dohrmann system. Trihalomethanes were measured in the pentane layer by gas chromatography with ^{63}Ni EC detector,[52] using the scheme shown in Figure 168. This revealed only one unidentified compound in the purgeable fraction other than the trihalomethanes. Analysis of the extractable volatiles resulted in the identification of trichloracetic acid which has been postulated as an intermediate[53] in haloform reactions of natural waters in addition to previously reported chlorinated compounds.

The carbon matrix of the lake water was largely a non-volatile fraction, presumably consisting of a mixture of carbohydrates, fulvic acid, and other

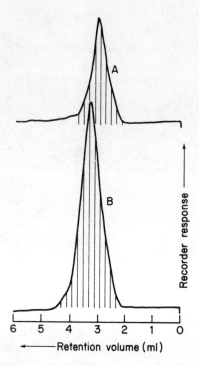

Figure 171 Size exclusion chromatogram of water-soluble fractions of freeze dried Cross Lake water: A, unchlorinated sample, apparent m 10.5×10^3. Samples concentrated × 100 compared to lake water

components. After microfiltration and freeze drying, the residue obtained is mostly soluble in purified water (50% of original TOC) and in dilute nitric acid (24% of TOC). A darkly coloured solid remains which is partially base soluble and which probably represents humic acids and clay particles which are presumably colloidal in size in the original sample. After freeze drying, the water-soluble fraction of the lake water sample was analysed in size exclusion high performance liquid chromatography. Figure 171 shows chromatograms of the water-soluble fraction before and after 20 mg l^{-1} chlorination for 5 days. Apparent average molecular weight has shifted slightly downward upon chlorination as might be expected. Exclusion chromatograms of the reinjected fractions are shown in Figure 172. The data allows an important conclusion; namely, that the THMFP and non-volatile TOX fractions (for key see Figure 169) are evenly distributed throughout the molecular weight range of the polymer. High performance liquid chromatography showed that the average molecular weight of the acid-soluble fraction (Figure 169) is substantially below that of the water-soluble portion. Average values for the chlorinated and non-chlorinated acid-soluble fractions are 3.8×10^3 and 4.2×10^3 respectively. These lower values may be due to the presence of smaller molecules, possibly containing basic nitrogenous compounds.

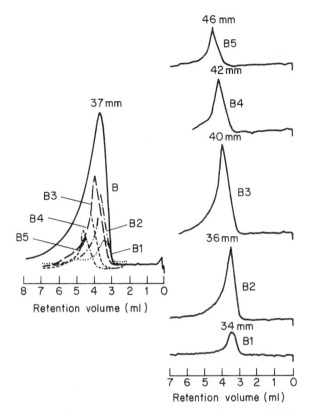

Figure 172 Size exclusion chromatograms of water-soluble fractions of chlorinated freeze dried Cross Lake water. Fractions collected and reinjected (1-5) and composite curve superimposed on original chromatogram

It is interesting to note that the total non-volatile halogen content of the five water-soluble fractions after chlorination is approximately 260 μg Cl l^{-1} or 7.3 μmole Cl l^{-1}. The combined TOC value of the fractions is 4.2 mg C l^{-1} or 350 μmole C l^{-1}. Thus, the average atomic ratio of chlorine to carbon in the polymer is 1:50, then 4.2 mg C l^{-1} would correspond to 8.4 × 10^3 μg polymer l^{-1}. At an average molecular weight of 8.2 × 10^3, this would correspond to a concentration of 1.0 μmole polymer l^{-1} meaning that the average polymer molecule would contain approximately seven chlorine atoms.

Glaze et al.[48] showed that the majority of the organic halogen found in chlorinated lake water is not accounted for by trihalomethanes but is present, to a large extent as 2,4,6,2'4'6'-hexachlorobiphenyl. They studies the ozonation and photolytic ozonation of this compound. The products identified are shown in the reconstructed gas chromatograms (Figures 173 and 174) and in the proposed reaction scheme in Figure 175. Except for the keto-enol pair shown in Figure 175 all products, starting with 2-chloro-3-(2,4,6-trichlorophenyl)maleic

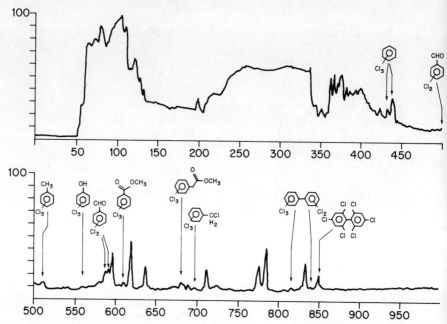

Figure 173 Reconstructed gas chromatogram of ozone, u.v. byproducts of hexachlorobiphenyl to neutral fraction; g.c.-ms. conditions in text

acid and beyond were identified from their mass spectra. Significantly, the scheme proposes that the ring first ruptured is degraded in a series of rapid steps before the second ring is attacked. The repetitive process which destroys these compounds is (1) double bond cleavage resulting in aldehydes which (2) oxidize to carboxylic acids which then (3) decarboxylate. Further cleavage naturally occurs when the system contains additional double bonds, so that organic carbon is converted to carbon dioxide via either intermediate carboxylic acids or the terminal compound in the ozone process, oxalic acid.

Somiya et al.[54] identified and determined products resulting from the ozonation of secondary effluents. Trace amounts of formaldehyde, acetaldehyde, propionaldehyde, methylglyoxylic acid, and pyruvic acid were detected by gas chromatography and colorimetry as their 2,4-dinitrophenylhydrazones. Acetic acid, formic acid, and propionic acid were also detected by gas chromatography for their benzyl esters. Formaldehyde, methylglyoxal, pyruvic acid, and acetic acid were quantified and their behaviour examined during the reaction.

Jolly et al.[55] characterized non-volatile organics produced in the ozonation and chlorination of waste water. The treated effluents were analysed by high pressure liquid chromatography and gas-liquid chromatography as a means of detecting both non-volatile and volatile organic constituents. While chlorinated

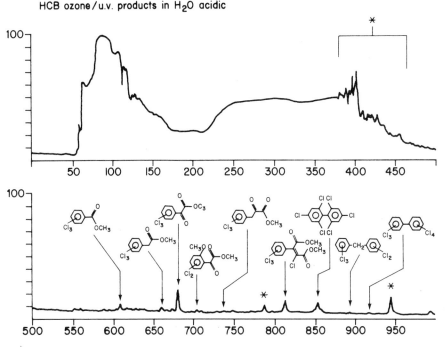

Figure 174 Reconstructed gas chromatogram of ozone, u.v. byproducts of hexachlorobiphenyl to acid fraction after methylation: g.c.-m.s. conditions in text. *Indicates non-chlorinated compound

primary effluent was repeatedly found to be mutagenic towards a particular strain, neither chlorinated, ozonated nor u.v.-irradiated secondly effluents exhibited mutagenic activity even when tested at 10- to 20-fold concentration.

Brunet et al.[56] studied the formation of organic materials during the ozonation of filtered surface water. They found that ozonation leads to a significant decrease in aromatic compounds together with an increase in polar compounds. Ozonation rates of the order of 2 mg l^{-1} appear to cause the degradation of the most reactive polyhydroxylated aromatic haloform precursors at pH 7.0. There was an increase of precursors of the methylketone type which are very reactive with iodine and with chlorine at pH greater than 7. The results, obtained by ultraviolet spectrometry and high pressure liquid chromatography, were confirmed by data obtained by ozonation of dilute aqueous solutions of pure products.

REFERENCES

1. Kiigemagi, U., Burnard, J. and Terriere, L. C. *J. Agric. Food Chem.*, **23**, 717 (1975).
2. Fieser, L. F. and Fieser, M. *Reagents from Organic Synthesis*. Wiley, New York, NY, p.19 (1968).

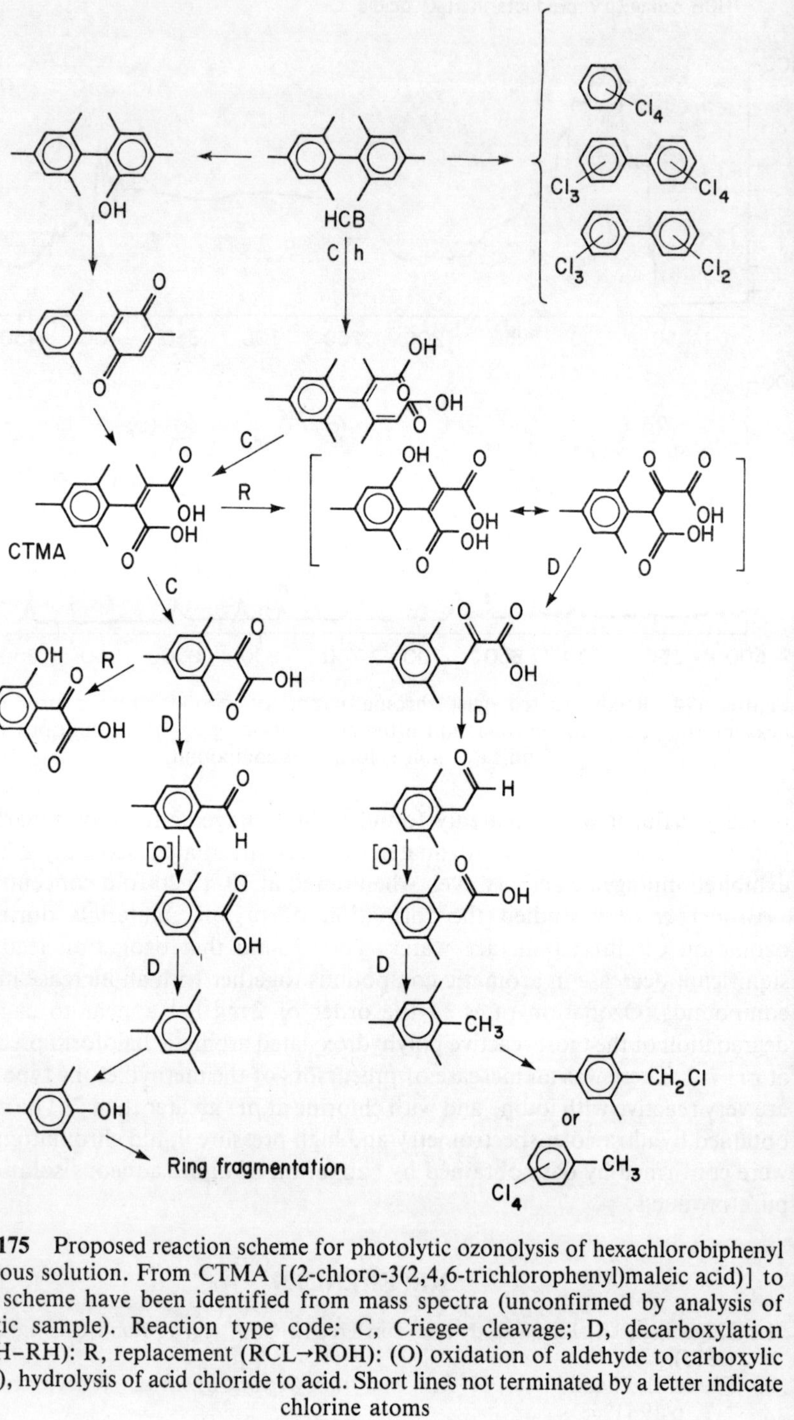

Figure 175 Proposed reaction scheme for photolytic ozonolysis of hexachlorobiphenyl in aqueous solution. From CTMA [(2-chloro-3(2,4,6-trichlorophenyl)maleic acid)] to end of scheme have been identified from mass spectra (unconfirmed by analysis of authentic sample). Reaction type code: C, Criegee cleavage; D, decarboxylation (RCO$_2$H–RH): R, replacement (RCL→ROH): (O) oxidation of aldehyde to carboxylic acid; (h), hydrolysis of acid chloride to acid. Short lines not terminated by a letter indicate chlorine atoms

3. Benville, P. E. and Tindle, R. C. *J. Agric. Food Chem.*, **18**, 948 (1970).
4. Geldreich, E. E. *J. Am. Water Works Ass.* **63**, 225 (1971).
5. Kirchmer, C. J. *5β-Cholestan-3β-ol: an indicator of Fecal Pollution.* University of Florida PhD Thesis (1971).
6. Bunch, R. L. and Ettinger, M. B. *J. Water Pollut. Control Fed.*, **36**, 1411 (1964).
7. Kukchek, G. J. and Edwards, G. P. *J. Water Pollut. Control Fedn*, **34**, 376 (1962).
8. Dutka, B. J., Chau, A. S. Y. and Coburn, J. *Water Research*, **8**, 1047 (1974).
9. Murtaugh, J. J. and Bunch, R. L. *J. Water Pollut. Control Fedn*, **39**, 404 (1967).
10. Smith, L. L. and Gouron, R. E. *Water Research*, **3**, 141 (1969).
11. Tabak, H. H., Bloomhuff, R. N. and Bunch, R. L. *Development in Industrial Biology*, Vol. 13, pp. 296-307. Publication of the Society for Industrial Microbiology, American Institute of Biological Science, Washington, DC (1972).
12. Gould, R. G. and Cook, R. P. *Cholesterol, Chemistry, Biochemistry and Pathology.* (Edited by R. P. Cook), pp. 289-292, Academic Press, New York (1958).
13. Singley, J. E., Kirchmer, C. J. and Miura, R. US Environmental Protection Technology Series EPA-660/2-74-021, US Government Printing Office, Washington, DC, pp. 126 (1974).
14. Dawson, J. P. and Best, G. A. *Proc. Anal. Div. Chem. Soc.*, **12**, 311 (1975).
15. Wun, C. K., Walker, L. W. and Litsky, W. *Water Research* **10**, 955 (1976).
16. Kirchmer, C. J. *5β-Cholestan-3β-ol: an indicator of Fecal Pollution.* PhD Thesis, University of Florida (1971).
17. Tabak, H. H., Bloomhuff, R. N. and Bunch, R. L. *Developments in Industrial Microbiology* Vol. 13, pp. 296-307. Publication of the Society for Industrial Microbiology, American Institute of Biological Sciences, Washington DC (1972).
18. Dutka, B. J., Chau, A. S. Y. and Coburn, J. *Water Research*, **8**, 1047 (1974).
19. Wun, C. K., Walker, R. W. and Litsky, W. *Hlth Lab. Sci.*, **15**, 67 (1978).
20. Abe, A. and Yoshima, H. *Water Research*, **13**, 1111 (1979).
21. Uchiyama, M. *Water Research*, **13**, 847 (1979).
22. Okuno, I. and Higgins, W. H. *Bull. Environ. Contam. Toxicol.*, **18**, 428 (1977).
23. Appelgren, Lars-Erik and Karlsson, R. *Acta Pharmacol. et Toxicol.*, **29**, 65 (1970).
24. Viccaro, J. P. and Ambye, E. L. *J. Am. Oil Chem. Soc.*, **50**, 213 (1973).
25. Michel, H. O., Gordon, E. C. and Epstein, J. *Environ. Sci. Technol.*, **7**, 1045 (1973).
26. Burkat, S. E., Medvedeva, N. Ya. and Ivanov, B. G. *Zh. analit. Khim.*, **24**, 284 (1969).
27. Tsubouchi, M. *Bull. Chem. Soc. Japan* **44**, 1560 (1971).
28. Allen, J. L. and Sills, J. B. *J. Ass. Off. Anal. Chem.*, **57**, 387 (1974).
29. Kawano, Y. and Bevenue, A. *J. Chromat.*, **72**, 51 (1972).
30. Hempel, D., Liebmann, R. and Hellwig, A. *Fortschr. Wasserchem. ihrer Grenzgeb.*, **13**, 181, (1971).
31. Hughes, R. A. and Lee, G. F. *Environ. Sci. Technol.*, **7**, 934, (1973).
32. Ya, C. C., Booth, G. H., Hanson, D. J. and Larsen, L. P. *J. Agric. Food Chem.*, **22**, 431 (1974).
33. Gerber, N. N. and Lechevalier, H. A. *Appl. Microbiol.*, **13**, 935 (1965).
34. Gerber, N. N. *Tetrahedron Lett.*, 2971 (1968).
35. Medsker, L. L., Jenkins, D. and Thomas, J. F. *Environ. Sci. Technol.*, **2**, 461 (1968).
36. Gerber, N. N. *J. Chem. Ecol.*, **3**, 475 (1977).
37. Kikuchi, T., Mimura, T., Itoh, Y., Moriwaki, Y., Negoro, K., Masada, Y. and Inoue, T. *Chem. Pharm. Bull.*, **21**, 2339 (1973).
38. Kikuchi, T., Mimura, T., Harimaya, K., Yano, H., Arimoto, T., Masada, Y. and Inoue, T. *Chem. Pharm. Bull.*, **21**, 2342 (1973).
39. Kikuchi, T., Mimura, T., Masada, Y. and Inoue, T. *Chem. Pharm. Bull., Tokyo*, **21**, 1847 (1973).
40. Otsuhara, M. and Suwa, M. *J. Water and Waste*, **19**, 31 (1977).

41. Yasuhara, A. and Fuwa, F. *J. Chromat.*, **172**, 453 (1979).
42. Kuo, P. P. K., Chian, E. S. K. and Chang, B. T. *Environ. Sci. Technol.*, **11**, 1177 (1977).
43. Kuo, P. P. K., Chian, E. S. K., DeWalle, F. B. and Kim, J. H. *Anal. Chem.*, **49**, 1023 (1977).
44. Bethge, P. O. and Lindstrom, K. *Analyst*, **99**, 137, (1974).
45. Webb, R. G., Garrison, A. W., Keith, L. H. and McGuire, J. J. US Environmental Protection Agency Report No. EPA-R2-277 (1973).
46. Kramer, D. N., Klein, N. and Basellce, R. A. *Anal. Chem.*, **31**, 250 (1959).
47. Houle, M. J., Long, D. E. and Smette, D. *Anal. Lett.*, **3**, 401 (1970).
48. Glaze, W. H., Peyton, C. R., Salch, F. Y. and Huang, F. Y. *Int. J. Environ. Anal. Chem.*, **7**, 143 (1979).
49. Henderson, J. E., Peyton, G. R. and Glaze, W. H. *Identification and Analysis of Organic Pollutants in Water* (L. H. Keith, Ed.), (Ann Arbor Science, Ann Arbor, pp. 105–112 (1976).
50. Regnier, F. E. and Noel, R. J. *Chromatog. Sci.* **14**, 316 (1974).
51. Glaze, W. H., Peyton, G. R. and Rawley, R. *Environ. Sci. Technol.*, **11**, 685 (1977).
52. Henderson, J. H., Peyton, G. R. and Glaze, W. H. in *Identification and Analysis of Organic Pollutants in Water* (L. H. Keith, Ed.), Ann Arbor Sciences, p. 105 (1976).
53. Morris, J. C. and Baum, B. in *Water Chlorination. Environmental Impact and Health Effects* (L. R. Jolly, H. Gorchev and D. H. Hamilton, Eds), Ann Arbor Science, Ann Arbor **2**, 29 (1978).
54. Somiya, I., Yamada, H., Izumi, Y. and Odagaki, M. *Japan J. Water Pollut. Control*, **2**, 181 (1979).
55. Jolly, R. L., Cummings, R. B., Harman, S. J., Penton, M. S., Lewis, L. R., Lee, N. E., Thompson, J. E., Pitt, W. W. and Mashni, C. I. US National Technical Information Service, Springfield, Va. Report No. ORNL/TM-6555 p. 89 (32710) (1979).
56. Brunet, R., Bourdigot, M. M., Legube, B. and Dore, M. Aqua. No. 4 *Scientific and Technical Review* **76** (1980).

Chapter 6
Natural Pigments in Water

HUMIC AND FULVIC ACIDS

Naturally occurring organic compounds are found in significant concentrations in waters throughout the world. A recent survey of 27 Canadian rivers and lakes[1] revealed that the majority of dissolved organic carbon was of natural origin with fulvic acid, humic acid, tannins, and lignins being the major components. In fact, Telang[2] has proposed a method for the determination of humic and fulvic acids in water based on the determination of dissolved organic carbon at a pH of 9 and a pH of 2. Obviously, other dissolved organics in the sample would interfere in this method.

Humic and fulvic acids are among the most widely distributed products of plant decomposition on the Earth's surface, occurring in soil, water, sediments, and a variety of other deposits. They are amorphous, yellow-brown or black, hydrophilic, acidic, polydisperse substances of wide-ranging molecular weight (< 10,000 for the fulvic acid, 10,000–300,000 for humic acid are the usual ranges although some values outside of these ranges have been reported by Rashid and King[3]). Humic and fulvic acids are responsible for the yellow-brown coloration of many lakes and rivers. By definition, humic acid is that fraction of organic material soluble at pH 9 but insoluble at pH 2, while fulvic acid is soluble at both pH 9 and pH 2.[4]

Tannins and lignins are high molecular weight polycyclic aromatic compounds widely distributed throughout the plant kingdom. Pearl[5] defined lignin as 'the incrusting material of plants which is built up of methoxy- and hydroxy-phenylpropane units'. It is not hydrolysed with acids but is readily oxidized and soluble in hot alkali and bisulphite. The exact chemical composition of the lignin-like compounds commonly found dissolved in natural waters is not known. Tannins, according to Geissman and Crout[6] are either polymers of gallic acid linked to carbohydrate residues (hydrolysable tannins) or polymeric flavinoid compounds (condensed tannins). Both tannins and lignins are highly resistant to chemical and biological degradation. How closely these compounds resemble their laboratory counterparts, tannic acid and lignosulphonic acid, is not really known.

Some of the earliest references to the investigation of humic acids in water are those of Miroslav,[7] Moed,[8] and Eberle.[9] Miroslav used ultraviolet spectrophotometry to investigate the pollution of surface water by humic acid and lignosulphonic acid originating from pulp mills. He derived coefficients for the conversion of extinction values into the concentration of these substances and found a correlation between extinction at 280 nm, oxidizability and absolute concentration. Moed[8] showed that aluminium oxide removed about 98% of the unidentified yellow material present in fresh water samples and used ultraviolet spectrophotometry at 270 nm to evaluate the extinction of the samples. Between 66% and 80% of the adsorbed material could be eluted with sodium dihydrogen phosphate (at pH 7), several fractions being eluted with 0.008 M and 0.3 M sodium dihyrogen phosphate.

Eberle[9] extracted humic acids from water with trioctylamine and showed that the salts thus formed are readily soluble in aromatic or chlorinated hydrocarbons and can thus be extracted from aqueous solution. By use of 10% (v/v) trioctylamine solution in chloroform humic acids are extracted from water at pH less than 5 with a partition coefficient of less than 100 and are re-extracted quantitatively at a pH of greater than 9. He describes a procedure for the extraction—spectrophotometric determination of humic acids in water. Lignosulphonic acid behaves like humic acids with respect to extraction by trioctylamine and all these species can be determined simultaneously by calculation from the extinction measured at two wavelengths.

Interest in ultraviolet spectroscopic methods for the determination of humic acid, lignosulphonic acids, tannin, and fulvic acids in water has continued to the present day.

Recent ultraviolet spectroscopic methods

Sontheimer and Wagner[10] determined humic and lignosulphonic acids in water by measuring the differences in the ultraviolet spectra of the sample between 250 and 300 nm. These workers confirmed that they could be used for the approximate determination of these constitutents in water, following a preliminary extraction and concentration procedure. Tests on several natural waters showed that the method could be used for a preliminary assessment of water quality.

Lawrence[11] described a simple ultraviolet spectroscopic method for semi-quantitatively determining fulvic acid, tannic acid, and lignin in natural waters. This method is based on the fact that in a mixture of three components, the concentrations of each component can be determined by measuring the absorbance at three different wavelengths. The method requires the absorption spectra of the three components to be appreciably different from one another, and a knowledge of the extinction coefficients of the three components at each of the three wavelengths chosen. From Figure 176 it is apparent that the spectra for fulvic, tannic, and lignosulphonic acids are sufficiently different for this method to be viable.

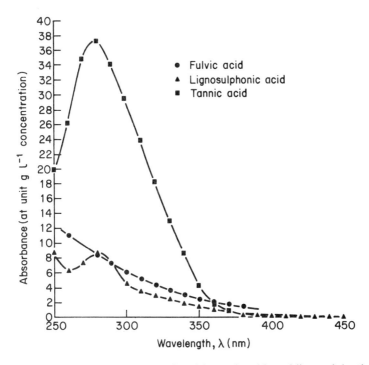

Figure 176 Absorption spectra of fulvic acid, tannic acid, and lignosulphonic acid in phosphate buffer. Reprinted with permission from Lawrence.[11] Copyright (1979) Pergamon Press

Beer's Law for a three component system of fulvic acid (F), tannic acid (T) and lignosulphonic acid (L) states:

$$E_\lambda = \epsilon_{\lambda,F} C_F d + \epsilon_{\lambda,T} C_T d + \epsilon_{\lambda,L} C_L d \tag{4}$$

where E is measured absorbance of wavelength λ ϵs are the extinction coefficients at that wavelength, C the concentrations and d, the optical path length. For simplicity, the path length d can be set to 1 cm and hence disappears from the equation. Since the molecular weights of these compounds are not known, the gram extinction coefficients (calculated for unit g l^{-1}) are used and concentrations are expressed in g l^{-1}.

Combining equation 4 for three different wavelengths and solving for C_F, C_T, and C_L results in the following:

$$C_F = \frac{D(E_{\lambda 3}\epsilon_{\lambda 1} \cdot T - E_{\lambda 1}\epsilon_{\lambda 3} \cdot T) - B(E_{\lambda 3}\epsilon_{\lambda 2} T - E_{\lambda 2}\epsilon_{\lambda 3} \cdot T)}{DA - BC} \tag{5}$$

$$C_T = \frac{F(E_{\lambda 3}\epsilon_{\lambda 1} L - E_{\lambda 1}\epsilon_{\lambda 3} \cdot L) - E(E_{\lambda 3}\epsilon_{\lambda 2} \cdot L - E_{\lambda 2}\epsilon_{\lambda 3} \cdot L)}{ED - BF} \tag{6}$$

$$C_L = A(E_{\lambda3}\epsilon_{\lambda2} \cdot F - E_{\lambda2}\epsilon_{\lambda3} \cdot F) - C(E_{\lambda3}\epsilon_{\lambda1} \cdot F - E_{\lambda1}\epsilon_{\lambda3} \cdot F) \qquad (7)$$

where

$$A = \epsilon_{\lambda3} \cdot F \; \epsilon_{\lambda1} \cdot T - \epsilon_{\lambda1} \cdot F \; \epsilon_{\lambda3} \cdot T$$

$$B = \epsilon_{\lambda3} \cdot L \; \epsilon_{\lambda1} T - \epsilon_{\lambda1} \cdot L \; \epsilon_{\lambda3} \cdot T$$

$$C = \epsilon_{\lambda3} \cdot F \; \epsilon_{\lambda2} \cdot T - \epsilon_{\lambda2} \cdot F \; \epsilon_{\lambda3} \cdot T$$

$$D = \epsilon_{\lambda3} \cdot L \; \epsilon_{\lambda2} \cdot T - \epsilon_{\lambda2} \cdot L \; \epsilon_{\lambda3} \cdot T$$

$$E = \epsilon_{\lambda3} \cdot F \; \epsilon_{\lambda1} \cdot L - \epsilon_{\lambda1} \cdot F \; \epsilon_{\lambda3} \cdot L$$

$$F = \epsilon_{\lambda3} \cdot F \; \epsilon_{\lambda2} \cdot L - \epsilon_{\lambda2} \cdot F \; \epsilon_{\lambda3} \cdot L$$

Hence, measuring the absorbance at three wavelengths and having a total of nine extinction coefficients, C_F, C_L, and C_T can be calculated.

Lawrence[11] carried out all ultraviolet absorbance measurements using 10, 50, or 100 mm quartz cells. Best results were obtained by measuring the absorbance at numerous wavelengths between 280 and 400 nm, computing the concentrations for each combination of three wavelengths, and averaging the results. This multiple wavelength approach had the advantage of detecting the presence of other interfering compounds. If another compound with a significant ultraviolet spectrum was present, it quickly became apparent by the wide variation of concentrations obtained for different wavelength combinations. In the absence of interfering compounds, the variation was usually less than 20%. The gram extinction coefficients for the three components at various wavelengths are listed in Table 141.

Table 141 Gram extinction coefficients for fulvic acid, lignosulphonic acid, and tannic acid

Wavelength (nm)	Fulvic acid	Tannic acid	Lignosulphonic acid
260	11.0	26.0	6.3
280	8.5	37.7	8.6
290	7.3	34.4	7.2
300	6.1	29.8	4.4
310	5.2	24.3	3.5
320	4.4	18.1	2.9
330	3.6	13.8	2.4
340	3.0	8.6	1.9
350	2.5	4.2	1.4
360	2.1	1.6	1.1
370	1.7	0.74	0.72
380	1.4	0.42	0.55

Reprinted with permission from Lawrence.[11] Copyright (1979) Pergamon Press.

Table 142 Measured concentrations of fulvic acid, tannic acid, and lignosulphonic acid (mg l^{-1}) in phosphate buffer. Numbers in parentheses refer to the known gravimetric concentrations

Soln no.	Fulvic acid	Tannic acid	Lignosulphonic acid
1	7.0	6.2	11.1
	(6.7)	(6.7)	(10.7)
2	18.7	1.5	2.0
	(20.0)	(1.9)	(1.9)
3	7.9	1.0	1.8
	(8.0)	(0.8)	(1.3)
4	11.0	1.1	0.5
	(10.0)	(0.95)	(0.95)
5	2.6	0.3	0.1
	(2.0)	(0.2)	(0.2)
6	8.0	0.8	0.7
	(6.7)	(0.6)	(0.6)
7	15.9	0.7	0.7
	(15.0)	(1.0)	(0.8)
8	15.2	0.9	1.2
	(15.0)	(1.0)	(0.8)
9	24.9	1.0	2.0
	(25.0)	(1.5)	(1.0)
10	2.1	0.08	0.04
	(1.6)	(0.08)	(0.08)

Reprinted with permission from Lawrence.[11] Copyright (1979) Pergamon Press.

Table 142 lists some results for solutions of fulvic acid, tannic acid, and lignosulphonic acid in pH 6.88 phosphate buffer. The deviations from the known concentrations (figures in parentheses) are mostly <15% for the fulvic acid, <30% for the tannic acid, and <50% for lignosulphonic acid. The increasing deviations, fulvic<tannic<lignosulphonic acid could be due to the magnitude of the extinction coefficients involved in the calculations or to differences in the nature of the tannin and lignin substances obtained from natural and commercial sources.

Fulvic acid concentrations in natural waters usually fall within the range 1-10 mg l^{-1} which is about the same range as the solutions in Table 142. Attempts were made to preconcentrate the dilute solutions prior to analysis by absorbing the fulvic acid on to XAD-2 resin and eluting with methanolic ammonia but quantitative recovery was not achieved. Rotary evaporation, however, was successfully used to preconcentrate the dilute solutions, but the final results had about the same precision as measurements made directly on the dilute solutions using 100 mm cells.

The spectra of fulvic, tannic, and lignosulphonic acids are very pH dependent and consequently all solutions must be buffered before measuring the absorbance.

Table 143 Measured concentration of fulvic acid, tannic acid, and lignosulphonic acid from different origins

Sample no.	Origin	Gravimetric conc. (mg l^{-1})	Dissolved organic carbon	Conc DOC	Fulvic acid (mg l^{-1})	Tannic acid (mg l^{-1})	Lignosulphonic acid (mg l^{-1})	Fulvic + tannic + lignosulphonic acids (mg l^{-1})	% Recovery
1	Water	30.1	13.3	2.3	19.8	2.0	6.9	28.7	95
2	Sediment	16.3	5.0	3.3	10.8	2.2	—	13.0	80
3	Sediment	20.8	6.1	3.4	14.3	3.2	—	17.5	84
4	Soil	24.3	8.5	2.9	15.8	1.7	3.4	20.9	86
5	Soil	11.7	4.0	2.9	8.0	0.8	1.3	10.1	86

Reprinted with permission from Lawrence.[11] Copyright (1979) Pergamon Press.

Table 144 Concentrations of fulvic, tannic, and lignosulphonic acids in natural water streams

Sample no.	Fulvic acid (mg l^{-1})	Tannic acid (mg l^{-1})	lignosulphonic acid (mg l^{-1})	Fulvic + tannic + lignosulphonic acid (mg l^{-1})	Tannin* lignin (mg l^{-1})	DOC (mg l^{-1})	2.3 × DOC (Dissolved organic carbon)
AG-2-3	1.7	0.4	2.2	4.3	2.7	1.8	4.2
AG-2-4	1.4	0.3	2.4	4.1	2.9	1.4	3.2
AG-4-3	3.5	0.8	4.8	9.1	5.1	5.2	12
AG-4-4	4.2	0.9	6.2	11.3	4.9	14.6	34
AG-13-2	4.5	1.4	4.1	10.0	4.6	5.3	12
AG-13-6	8.7	2.3	14.2	25.2	7.2	12.4	29

*Determined by the sodium tungstate–phosphomolybdic acid method (American Public Health Association)[27]
Reprinted with permission from Lawrence.[11] Copyright (1979) Pergamon Press.

Table 143 shows the measured concentrations of fulvic acid, tannin, and lignin for gravimetrically prepared solutions of fulvic acid extracted from water, lake sediment, and soil. The final column gives the sum of the measured concentrations expressed as a percentage of the prepared concentration. Two things are immediately apparent from this table: first, that extracted fulvic acid, even after clean-up through an XAD column, still contains appreciable amounts of tannins and lignins; second, that sediment and soil fulvic acids do not respond to the analytical techniques in quite the same way as fulvic acid extracted from water. The increased concentration/dissolved organic carbon (DOC) ratio for the soil and sediment fulvic acids indicates that there could be a fundamental difference between the fulvic acids from different origins or that the base-extraction procedures used for soil and sediment results in a modified form of fulvic acid.

Humic acid responds to the test in much the same way was fulvic acid and consequently the measured concentrations of fulvic acid are actually the combined concentrations of fulvic and humic acids. The unit-gram extinction coefficients for humic acid (at pH 6.88) are only 10–50% higher than those for fulvic acid between 260 and 420 nm so it is to be expected that humic acid would respond similarly. It has been claimed[12] that fulvic acid accounts for at least 90% of the dissolved fulvic–humic material in natural waters at near natural pH.

In Table 144 are presented fulvic acid, tannin, and lignin contents obtained by this and other methods from natural water samples. There is reasonable agreement between the sum of the concentrations for fulvic acid, tannin, and lignin measured spectrophotometrically and the total concentration of organics (dissolved organic carbon, DOC, multiplied by Concentration/DOC ratio of 2.3 determined for sample 1 in Table 143) for all samples except AG-4-4. The agreement is surprising in view of the higher fulvic/lignin ratio for sample 1 of Table 143. Some samples, notably those collected from areas in which the predominant vegetation was cedar also yielded concentrations of organics which did not correlate well with $2.3 \times DOC$.

The sum of the concentrations of tannin and lignin agreed reasonably well with the tannin and lignin concentration determined by the photophomolybdic acid method for four out of the six samples. The reason for the discrepancy with samples AG-4-4 and AG-13-6 is not known although a slight precipitate was observed during the phosphomolybdic acid test.

Wagner and Hoyer[13] also used ultraviolet spectroscopy for the simultaneous determination of humic acid and lignosulphonic acids in natural waters. These workers also compared absorption characteristics in the wavelength range 250–300 nm. To eliminate interfering compounds, especially highly polar organic acids, the extraction was performed using a mixture of trioctylamine and chloroform. Equations were derived which enabled the concentration of both the humic and lignosulphonic acids to be determined from measurements of the extinction coefficient at three wavelengths.

Gel filtration

Snow[14,15] applied gel filtration on Sephadex to the fractionation of raw water from a peaty source supplied to a treatment plant. The fractionation of the coloured organic acids was found to be dependent on the elution conditions, particularly pH value and ionic strength of the sample and/or the eluting solution. Five main fractions were identifiable in the elution patterns obtained over a 17-month period, but no correlations were observed with season, rainfall, or raw water quality. Snow investigated the stability of the coloured acids, their association with iron, and the effects of treatment.

Chian and De Walle[16] used gel permeation gas chromatography on Sephadex G-75 and G-200, ultrafiltration, and gas chromatography to characterize soluble organic matter in heavily polluted ground water samples and leachates from landfills. The largest organic fraction consisted of free volatile fatty acids, and the next largest was a fulvic-like material with a relatively high carboxyl and aromatic hydroxyl group density. There was also a small percentage of a high-molecular-weight humic carbohydrate-like complex with which liquids were associated.

The ultrafiltration membranes used in this study were types UM05, UM2, UM10, and XM100A (Amicon, Lexington, Mass.) with nominal molecular-weight cutoffs of 50, 1000, 10,000, and 100,000 respectively. After ultra-centrifugation of one litre of leachate at 30,000 rpm for 30 min, the supernatant was concentrated five-fold with a 500-MW UF membrane. The retentate was then desalted by diluting it to twice its initial volume with distilled water followed by reduction to the original volume. This step was then repeated. The percentage retained was calculated from a material balance. Following the ultrafiltration step, separate aliquots of the ultrafiltration retentate were further separated into components of different molecular weights using Sephadex G-75 (1000–50,000 MW) and G-200 (1000–200,000 MW) columns. After the molecular weight distribution of the 500-MW UF retentate was established by this method, other ultrafiltration membranes with cutoffs at higher molecular weights were employed to obtain sufficient quantities of each of the fractions of different molecular weights for chemical analyses.

The ultrafiltration membrane fractions and Sephadex fractions were characterized by measuring the total organic carbon, COD, carbohydrates, amino acids as lysine, carboxyl groups as acetic acid, carbonyl groups as acetophenone and aromatic hydroxyl groups as tannic acid. Dissolved free amino acids were also measured. Each membrane fraction was also separated according to its polarity by extraction with two solvents; hexane which separates the less polar lipids, hydrocarbons, and fatty acids from the water phase; and after pH adjustment of the aqueous layer to 2, with butanol which separates humic substances from the water phase. Only 104 mg l^{-1} or 0.5% of the volatile residue of 18,403 mg l^{-1} was removed with the ultracentrifuge, indicating that most of the organics were present in the soluble fraction. Subsequent membrane ultrafiltration showed that 27.2% of the initial total organic carbon was retained by the 500-MW

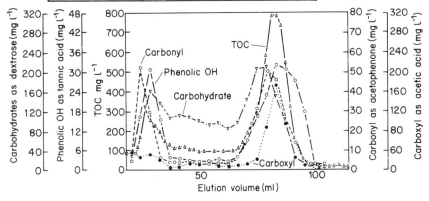

Figure 177 Eluate of 500 MW UF retentate of UI leachate sample separated on Sephadex G-75 column as characterized by total organic carbon, specific organics, and functional groups. Reprinted with permission for Chian and De Walle.[16] Copyright (1977) American Chemical Society

ultrafiltration membrane (fraction 1b). When the 500-MW UF retentate was applied on a Sephadex G-75 column, the total organic carbon distribution in the eluate showed that 22% of the 500-MW ultrafiltration retentate, corresponding to 6% of the original total organic carbon, was excluded from the column as it eluted between 20 and 36 ml (Figure 177), indicating a molecular weight greater than approximately 30,000–50,000. The application of the 500-MW ultrafiltration retentate to a Sephadex G-200 column resulted in essentially the same elution pattern. This fraction, therefore, had a molecular weight considerably greater than 50,000. The results in Figure 177 further showed that the majority of the 500-MW ultrafiltration retentate consisted of organics with a molecular weight less than 1000–3000. Substantial quantities of the different molecular weight fractions were thereafter obtained using ultrafiltration membranes having larger molecular weight cutoffs. An aliquot of the original 500-MW ultrafiltration retentate (fraction 1b) was diluted to its original strength and pH and passed through a 10,000-MW ultrafiltration membrane. The retentate (fraction 3b) represented 21.8% or the original total organic carbon. Fraction 3b, after being diluted to its original strength, was passed through a 100,000-MW ultrafiltration membrane. The membrane retained 11.1% of the original total organic carbon (fraction 5b). The molecular weights obtained by membrane fractionation were several times higher than those estimated from the gel permeation data.

Chian and De Walle[16] used extraction with hexane and butanol to separate the different membrane ultrafiltration fractions according to their polarity. The

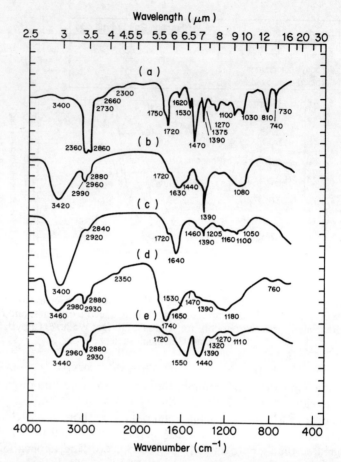

Figure 178 Infrared spectra of residue of hexane and butanol extracts. (a) Hexane extract of 10,000-MW UF retentate; (b) butanol extract of 10,000-MW UF retentate; (c) butanol extract of hydrolysed 10,000-MW UF retentate; (d) butanol extract of 500-MW UF retentate; (e) hexane extract of 500-MW UF retentate. Reprinted with permission from Chian and De Walle.[16] Copyright (1977) American Chemical Society

hexane extraction of the high-molecular-weight fraction (10,000-MW ultra-filtration retentate) represented only 0.5% of the organic carbon, assuming an organic carbon/weight ratio of 0.75 as calculated for palmitic acid (Table 145). The infrared spectrum of the hexane residue of this fraction (Figure 178), while showing the presence of impure mixtures, does suggest that this fraction consists primarily of fatty acids, some of which are esterified. If 50% of the weight of the butanol residue consisted of carbon, 3% of the organic carbon in the 10,000-MW ultrafiltration retentate was comprised of this group (Table 145). The residue of the butanol extract had a dark brown, oily appearance and an acidic and 'mouldy' odour. The infrared spectrum was almost identical to a

Table 145 Effectiveness of hexane and butanol extraction in removing organics from different molecular-weight fractions

Membrane fraction	Hexane extract as percentage of carbon	Extraction of specific organics	Butanol extract as percentage of carbon	Extraction of specific organics
High molecular weight (10,000-MW UF retentate)	0.5	0.8% of carbohydrates extracted with hexane	2.5	4% of carbohydrates and 100% of aromatic hydroxyls extracted with butanol
Intermediate molecular weight (500-MW UF retentate)	0		9	100% of aromatic hydroxyls extracted with butanol
Low molecular weight (500-MW UF permeate)	1.5	0.5% of carbohydrates and 10% of aromatic hydroxyls extracted with hexane	7	65% of aromatic hydroxyls extracted with butanol

Reprinted with permission from Chian and De Walle.[16] Copyright (1977) American Chemical Society.

humic carbohydrate-like, high-molecular-weight fraction previously isolated from the effluent of an activated sludge unit.

The solvent extraction results show that the hexane residue will slightly overestimate the amount of fatty acids and lipids, since free volatile fatty acids and humic fragments are coextracted. The butanol, used to extract the fulvic acid fraction, may also extract some of the free volatile fatty acids. On the other hand, butanol did not extract all of the fulvic-like material in the 500-MW ultrafiltration retentate.

Thus, in conclusion, membrane ultrafiltration, gel permeation chromatography, and specific organic analyses were used to separate and determine the main classes of soluble organics. With ultrafiltration a major fraction of the organics permeated a 500-MW membrane, and further analysis showed that most were present as free volatile fatty acids. The next largest fraction was a fulvic-like material with a relatively high carboxyl and aromatic hydroxyl group density. A small percentage of the organics consisted of a high-molecular-weight, humic carbohydrate-like complex, also characterized by a significant amount of hydrolysable amino acids. Using different model compounds to reflect the actual organics, 75% of organic matter was tentatively identified. Solvent extraction of the different membrane fractions indicated the presence of lipids associated with the high-molecular-weight humic fraction. Butanol preferentially extracted aromatic hydroxyl compounds present in the intermediate-molecular-weight fulvic fraction.

Membrane filtration

Farrah et al.[17] showed that membrane filtration concentrated a material resembling humic acid from tap water samples with an efficiency of about 50%.

Ion-exchange separation

Weber and Wilson[18] used anion and cation exchange resins to isolate fulvic and humic acids from soil and water. Boening et al.[19] also studied the use of ion-exchange resins (styrene–divinylbenzene and acrylic resins) for the removal of humic substances from water.

Activated carbon adsorption

Boening et al.[19] have discussed the use of activated carbon adsorption for the removal of soil and leaf fulvic acids and humic acid from water. These workers used total organic carbon and fluorescence measurements to determine the concentration of humic substances. Although this work was orientated to the removal of humic substances from water on a commercial scale, it may, nevertheless, be of analytical interest.

Divalent metal complexes of humic acids

Buffle et al.[20] measured the complexation properties of humic acid and fulvic acid in natural waters with lead and copper ion-selective electrodes. He obtained values for mean molecular weight of the ligand, the stability constants of the complexes, number of ligands fixed per metal ion, and the dependency of the stability of the complexes on pH.

Stevenson[21-23] used potentiometric titration to study the nature of divalent copper, lead, cadmium, and zinc complexes of humic acids. They showed that at least two major sites were involved in complex formation, and differences in the results using different humic acids were negligible. Stability constants decreased in the order copper, lead, cadmium, zinc.

These methods used a sequential titration procedure in which sequential additions of the metal ion were made to solutions of the humic acid at constant pH. The pH was returned to the initial starting point after each addition, using carbon dioxide-free potassium hydroxide.

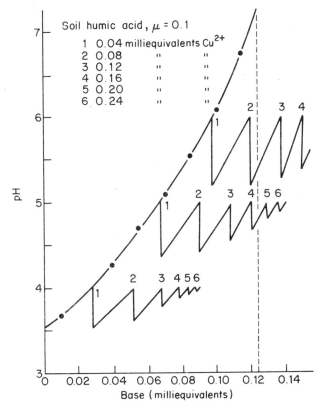

Figure 179 Consumption of base by sequential additions of Cu^{2+} to soil humic acid at pH values of 4, 5, and 6 ($\mu = 0.1$). The pH was returned to the initial starting point after each addition. Reprinted with permission from Stevenson.[21] Copyright (1977) The Williams & Wilkins Co., Baltimore

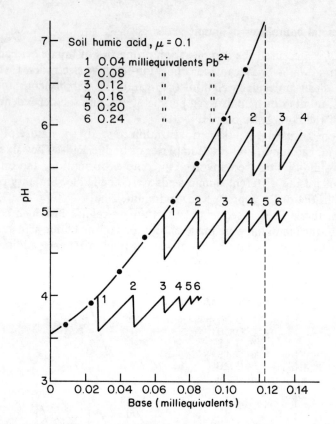

Figure 180 Consumption of base by sequential additions of Pb^{2+} to soil humic acid at pH values of 4, 5, and 6 ($\mu = 0.1$). The pH was returned to the initial starting point after each addition. Reprinted with permission from Stevenson.[21] Copyright (1977) The Williams & Wilkins Co., Baltimore

Results obtained by sequential additions of Cu^{2+}, Pb^{2+}, and Cd^{2+} to the soil humic acid at pH values of 4, 5, and 6 are shown in Figures 179–181. Each addition of metal ion depressed the pH to a lesser extent than the previous addition and less and less base was required to neutralize the liberated protons. As can be seen from the vertical dashed line, not all acidic groups had reacted at pH 4.0, apparently due to competition of protons for the ligand. For Cu^{2+} and Pb^{2+}, more protons were liberated at pH 5 and 6 than could be accounted for by acidic functional groups. This can be accounted for by release of protons from hydration water of the metal ion held in 1:1 complexes. In the case of the reactions carried out at pH 6.0, some of the extra protons undoubtedly resulted from the formation of insoluble oxide hydrates. The fact that the amount of base consumed in the Cd^{2+} experiments did not exceed the amount required for neutralization of acidic functional groups, even at pH 6.0, can be explained on the basis that cadmium oxide hydrate is not formed until the pH approaches 6.8; oxide hydrates of Cu^{2+} and Pb^{2+} are formed at pH values

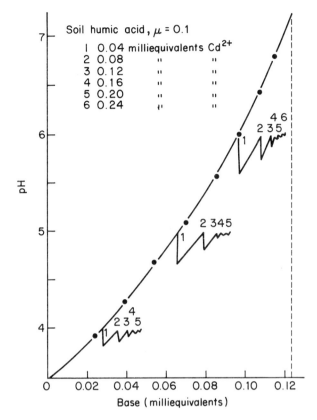

Figure 181 Consumption of base by sequential additions of Cd^{2+} to soil humic acid at pH values of 4, 5, and 6 ($\mu = 0.1$). The pH was returned to the initial starting point after each addition. Reprinted with permission from Stevenson.[21] Copyright (1977) The Williams & Wilkins Co., Baltimore

near 5.3 and 6.0, respectively. Zinc gave results identical to Cd^{2+} except that slightly less base was required for neutralization. In Figure 182, the results for Pb^{2+} and Cd^{2+} at pH values of 4.0 and 5.0 are presented as plots of milliequivalents of base consumed *vs* milliequivalents of metal ion added. For Cd^{2+} at both pH values and for Pb^{2+} at pH 4.0, smooth curves were obtained which fell below the horizontal line representing the content of undissociated acidic groups. For Pb^{2+} at pH 5.0, smooth curves were obtained up to base additions equivalent to about two-thirds of the undissociated acidic group content, after which a pronounced change in slope occurred and more protons were released than could be accounted for by acidic functional groups. Results obtained for Cu^{2+} were similar to those for Pb^{2+}; Zn^{2+} behaved similarly to Cd^{2+} (Figure 183). The anomalous behaviour of Cu^{2+} and Pb^{2+} at pH 5.0 can be explained by the strong tendency of hydration water of 1:1 complexes to dissociate, as follows:

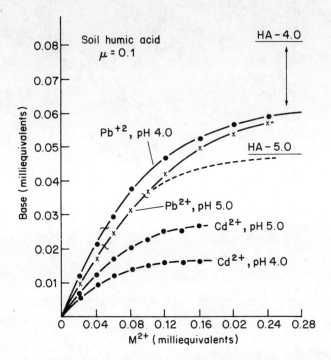

Figure 182 Base consumption *vs* amount of Pb^{2+} and Cd^{2+} added to soil humic acid at pH values of 4.0 and 5.0 ($\mu=0.1$). The horizontal lines indicate the content of undissociated functional groups (HA) at the two pH levels. Reprinted with permission from Stevenson.[21] Copyright (1977) The Williams & Wilkins Co., Baltimore

The first indication of oxide hydrate formation (Cu^{2+} and Pb^{2+} only) coincided with the formation of insoluble M^{2+} humates. It is probable, therefore, that the formation of precipitates enhanced dissociation of hydration water from the metal in 1:1 complexes. Borggaard[24] determined the acidity and pK value of humic acids extracted from soils using a titration procedure.

Miscellaneous

Polargraphy has been used to determine fulvic acids in natural water.[25] Ishibashi[26] applied electron microscopy to the measurement of particle size and other properties of natural organic colouring matter in water.

Combined tannins and lignins can be analysed in water by the sodium tungstate–phosphomolybdic method described by the American Public Health

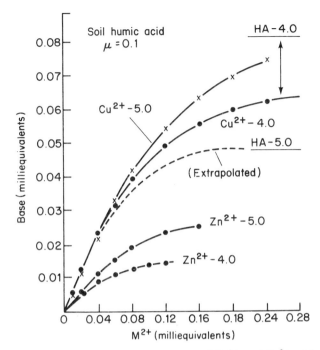

Figure 183 Base consumption vs amount of Cu^{2+} and Zn^{2+} added to soil humic acid at pH values of 4.0 and 5.0 ($\mu = 0.1$). The horizontal lines indicate the content of undissociated functional groups (HA) at the two pH levels. Reprinted with permission from Stevenson.[21] Copyright (1977) The Williams & Wilkins Co., Baltimore

Association.[27] However, this method is not specific for lignin and tannin in as much as other reducing substances respond similarly. Eberle and Schweer[28] have proposed a method for the simultaneous determination of humic acid and lignin by liquid-liquid extraction and ultraviolet spectroscopy. Edeline et al.[29] and Kerndorft and Schnitzer[30] respectively, have studied the nature of humic and fulvic materials in Belgian and Canadian rivers.

Humic and fulvic acids in sewage

Sposito et al.[31] used C^{13} and proton n.m.r. spectroscopy to investigate fulvic acid extracted from sewage sludge. C^{13} n.m.r. spectroscopy was found to be useful for studying the organic structures found in fulvic acids.

Rebhun and Mouka[32] have reported that about 40–50% of the organics in secondary sewage effluents constitute humic substances, the remainder being anionic detergents, carbohydrates, proteins, tannins, lignins, and ether extractables. They extended this work[33] to a more detailed examination of the humic substances in secondary effluents obtained from high-rate trickling filters, the effluents from a stabilization pond, and from an extended aeration activated sludge plant.

Table 146 Composition of soluble organics in secondary effluents as percentage of total COD

Samples of secondary effluent and its COD as mg O l^{-1}	Constituent (%)								
	Proteins	Carbohydrates	Anionic detergents	Tannins and lignins	Ether extractables	Fulvic acid	Humic acid	Hymathomelanic acid	Total
TF COD = 160	18.9	7.1	19.6	1.1	13.4	20.8	13.5	7.1	101.5
TF COD = 180.5	21.4	5.4	14.1	1.6	15.6	30.2	9.7	6.8	104.8
TF COD = 148.3	21.6	6.7	13.3	1.4	10.8	27.5	12.9	7.8	99.9
TF COD = 158.1	24.3	6.0	15.6	1.2	15.6	23.5	10.9	6.4	102.9
TF COD = 172.1	21.7	4.2	20.5	1.4	11.7	25.2	15.7	10.3	107.3
SP COD = 120.4	21.4	8.0	11.2	2.4	10.3	27.9	14.3	6.2	101.7
SO COD = 153.2	20.8	7.5	13.1	1.8	13.2	25.3	15.1	7.2	107.0
AS COD = 105.5	21.0	5.1	15.6	1.3	16.5	16.8	9.6	4.8	90.7
AS COD = 135.7	23.4	5.1	15.3	0.8	16.7	29.6	4.3	5.8	101.0
AS COD = 112.7	24.8	4.6	15.9	1.0	19.9	24.0	7.0	6.6	103.8
AS COD = 167.4	23.0	3.6	17.2	0.9	12.2	25.6	3.6	1.9	98.0

T, trickling filler; SP, stablization pond; AS, extended aeration activated sludge.
Reprinted with permission from Rebhun and Mouka.[32] Copyright (1971) American Chemical Society

The investigation of humic substances included examination of their infrared spectrum, determination of carboxylic and phenolhydroxylic functional acidic groups, and estimation of their molecular weight distribution. The humic substances isolated from secondary effluents were further divided into humic, hymathomelanic, and fulvic acids. The humic and hymathomelanic fractions were then dissolved in 0.5 M sodium hydroxide reprecipitated with 0.5 M hydrochloric acid, washed thoroughly with distilled water and dried in a vacuum desiccator. Each of the isolated fractions (including fulvic acid) was weighed and an aliquot was dissolved in ethylene diamine for carboxyl and phenolic hydroxyl determination. Another aliquot dissolved in 0.25 M sodium hydroxide was used for molecular sieving.

The acidic functional groups were determined by potentiometric titration in non-aqueous medium, using platinum-saturated calomel electrode combination.[34] A 0.1 M solution of sodium aminoethoxide in ethylene diamine, calibrated against benzoic acid, was used as titrant. To avoid interferences of carbon dioxide from the air, the titrations were carried out in a current of nitrogen.

The molecular weight distribution was examined by fractionation on Sephadex gels G-25, G-50, and G-75 having exclusion limits 100, 5000, 500–10,000, and 1000–50,000 (by dextrans), respectively. The gel chromatographic procedure utilized a glass column of 5 mm i.d. filled with about 45 ml of swollen Sephadex. The sample volume used did not exceed 2 ml. The 0.01 M solution of sodium chloride adjusted to pH 8 was applied as eluant. At a flow rate of 12 ml approximately, the effluent was collected in 2 ml fractions. These fractions were evaluated photometrically at 350 nm. The Sephadex columns were calibrated with Blue Dextran having a molecular weight of 2×10^6 excluded from all the gels. The samples of humic acids being examined were first passed through the gel column G-25. The portions of eluate excluded from this gel were collected and reduced in volume in a rotating evaporator. The next fractionation was made on G-50 and on G-75 respectively. For evaluating the molecular weight distribution, the eluation patterns from the three gels were considered. The areas corresponding to the respective ranges of molecular weight were measured and expressed as a percentage of the total area included under the elution pattern from G-25.

The analyses obtained by Mouka et al.[33] are tabulated in Table 146.

The average distribution of the organic groupings and fractions is shown in Table 147.

The infrared spectra of humic and fulvic acids isolated from secondary effluents are fairly similar. A typical infrared spectrum of fulvic acid is presented in Figure 184. The main adsorption bands are in the regions of: 3400—hydrogen-bonded OH; 2980, 2940, 2860, 1460—methylene and methyl stretching; 1730—C=O of carboxyl and ketonic carbonyl; 1630—amide carbonyl or aromatic C=C; 1270, 1130, 1080—carbon-oxygen C-O stretching, which is in agreement with the results of other investigations.[34-37] Analytical results

Table 147 Distribution of organic groupings in secondary effluents—mean values

	Percentage of total COD		
Organic groupings and fractions	Municipal waste water; high rate trickling filter	Municipal waste water; stabilization pond	Domestic waste water; extended acration activated sludge
Proteins	21.6	21.1	23.1
Carbohydrates	5.9	7.8	4.6
Tannins and lignins	1.3	2.1	1.0
Anionic detergents	16.6	12.2	16.0
Ether extractables	13.4	11.9	16.3
Fulvic acid	25.4	26.6	24.0
Humic acid	12.5	14.7	6.1
Hymathomelanic acid	7.7	6.7	4.8

Reprinted with permission from Rebhun and Mouka.[32] Copyright (1971) American Chemical Society

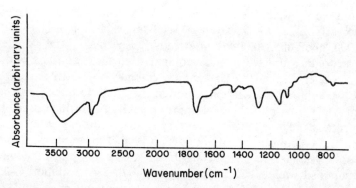

Figure 184 Infrared spectrum of fulvic acid isolated from secondary effluent. Reprinted with permission from Rebhun and Mouka.[32] Copyright (1971) American Chemical Society

showing the content of carboxyl and phenolic hydroxyl groups of humic compounds are presented in Table 148.

The total acidity of the humic compounds arises mainly from the dissociable hydrogen in aliphatic and aromatic carboxyl groups and in phenolic hydroxyl groups. The presence of those acidic groups, especially carboxyls, increases the solubility of humic compounds and enhances their resistance to coagulation, thus increasing their lifetime in the water. This may explain the fact that fulvic acid, having greater acidity than humic and hymathomelanic acids (Table 148), is also more widespread in effluents (Tables 146 and 147). The relatively low content of the phenolic hydroxyl group may be caused by deficiency of lignin residues in the effluent.

The molecular weight distribution of humic substances isolated from secondary effluents is presented in Table 149. The majority of fulvic and

Table 148 Acidic groups of humic compounds present in secondary effluents

Humic compound	Acidity in mEq g^{-1}		
	Total acidity	Carboxyl	Phenolic hydroxyl
Fulvic acid	12.3	9.2	3.1
Humic acid	8.3	8.3	—
Hymathomelanic acid	8.7	8.7	—

Reprinted with permission from Rebhun and Mouka.[32] Copyright (1971) American Chemical Society.

Table 149 Molecular weight distribution of humic substances from secondary effluents

Molecular weight range	Percentage of humic compound present		
	Fulvic acid	Humic acid	Hymathomelanic acid
<500	27.5	17.9	4.5
500–1000	7.8	6.2	12.2
1000–5000	35.7	29.4	48.0
5000–10,000	15.3	7.8	28.0
10,000–50,000	9.4	36.7	7.5
>50,000	4.3	2.0	0

Reprinted with permission from Rebhun and Mouka.[32] Copyright (1971) American Chemical Society.

hymathomelanic acids isolated from secondary effluents have molecular weights ranging from 1000 to 5000, whereas most of the humic acid is present in the range 10,000–50,000. A relatively large proportion of fulvic acid has been found to have a molecular weight of less than 500.

CHLOROPHYLL AND OTHER PLANT PIGMENTS IN WATERS AND ALGAE

Estimates of chlorophyll concentrations in natural waters are frequently required in studies related to primary production. Determination of chlorophyll are also of value in estimating the degree of eutrophication, since the content of this pigment is proportional to the content of nitrogen and phosphorus in the water.[38,39] Probably the most widely used analytical technique depends upon spectrophotometric measurements of extinctions of 90% acetone extracts at three different wavelengths. Solution of appropriate simultaneous equations (trichromatic) provides estimates of chlorophylls *a*, *b*, and *c*. Phaeophytin *a* can also be estimated by this method after acidification of the extract solution. Various improvements to the trichromatic equations have been published by a number of authors; for example, new data for the extinction coefficients of the chlorophylls, including c_1 and c_2, based on highly purified samples of each pigment.[38] The simplest and most widely adopted technique for chlorophylls is that developed by Richards[40] and subsequently modified by others. In this, spectrophotometric measurements are made at selected wavelengths on 90% acetone extracts of the separated phytoplankton. With simultaneous equations[41-43] it is possible to calculate the amounts of chlorophylls a, b, and c and total carotenoids present in the sample. The method, which was the subject of an extensive investigation by a UNESCO working group,[43] is subject to a number of drawbacks:[44]

(1) there is serious overlap between the absorption bands of the chlorophylls, and the precision for chlorophylls b and c is therefore rather poor;
(2) there are considerable differences between the results obtained with the various sets of equations;
(3) chlorophyll degradation products have absorption spectra similar to those of the chlorophylls and seriously interfere in the determination;
(4) the sensitivity is relatively low, necessitating the use of large water samples;
(5) the values found for the total carotenoids are usually only very approximate.

It is possible to overcome the majority of these difficulties by separating the pigments chromatographically before determining them.

Youngman[45] of the Water Research Centre (UK) has studied methods of extracting these pigments from algae and the spectrophotometric determination of chlorophylls in the extracts.

Ryding[46] has carried out an intercalibration study of various methods of determining chlorophyll a. Mainly the results agreed very well although none of the 13 laboratories performed the analyses in an identical manner. The chief differences were the choice of wavelengths for the photometric readings, and in evaluation. In calculating the concentration of chlorophyll a from optical density values, two pairs of alternatives are involved, namely, whether to disregard or allow for the contribution made by degradation products such as phaeophytin, and whether to base the estimation on optical density at one wavelength or at several wavelengths.

Raghi-Atri[47] attempted to ascertain the optimal system for the determination of chlorophyll in nutrient-rich surface waters. Samples were analysed using a number of extraction solvents and filter media. The recommended procedure is based on the use of 90% acetone as the extractant, coupled with a Whatman GF/C glass fibre filter of 25 mm diameter.

Jensen[48] showed that there was no difference between the chlorophyll extracted by either methanol or acetone from fresh diatom material, but with green algae, methanol gave 30% higher results than acetone. Low levels were found in stored filters, and this was most pronounced with methanol. Immediate analysis of the sample is recommended; even storage by freezing should be avoided.

Fluorimetric methods, most of which give crude estimates of chlorophyll a only, have also been used.[49] Loftus and Carpenter[50] have refined the fluorimetric method for the analysis of chlorophylls using the Turner fluorimeter, employing a series of emission filters to improve the selectivity between the emission spectra of each chlorophyll. Their method involves the measurement of the emission through three separate filter combinations and again requires the solution of simultaneous equations to yield the concentrations of each pigment. Acidification of the extract, followed by a further three measurements through the filter combinations yields, in theory, the concentration of the corresponding phaeophytins. However, only phaeophytin a can be estimated with any degree of certainty, and then only if chlorophyll b is not present as a significant component of the mixture. Improved accuracy of chlorophyll b and c determinations compared with spectrophotometric methods is claimed.

The most obvious advantage of a fluorimetric technique is the greatly increased sensitivity compared with spectrophotometry. An increase in sensitivity of two orders of magnitude is easily achieved and, provided that no loss of accuracy is incurred, fluorimetric techniques should provide a much faster method than conventional spectrophotometry for performing chlorophyll analyses in natural water samples. In addition, fluorimetry eliminates any possible interference by the absorption of non-chlorophyllous pigments in the region of the spectrum have 600 μm.

More recently, the applicability has been examined of high performance liquid chromatography to the determination of various chlorophylls.

Chlorophyll in natural waters is frequently estimated by trichromatic spectrophotometry of algal extracts. When no interfering compounds are present, these trichromatic equations are good estimates. The major criticism is that in natural plankton extracts, spectrally similar chlorophyll breakdown products are frequently present.[51-53] Thus, chlorophyll cannot be accurately determined by this method. One way to avoid the problem is to chromatographically separate the breakdown products from the chlorophylls before measurement. Separation of the spectrally similar chlorophylls a and b (as well as degradation products) will result in a more accurate determination of the chlorophylls.

Several chromatographic methods have been published, and most of these utilize thin-layer chromatography to separate the chlorophylls. Thin-layer materials employed have included the following: layers of Kieselguhr G impregnated with triolein, castor oil, or paraffin oil[52,56] layers prepared from glucose shaken with ether,[54] plates coated with Kieselguhr G impregnated with peanut oil dissolved in iso-octane[55] and layers of powdered confectioners icing sugar containing 5% cornflour suspended in light petroleum.[51]

Garside and Riley[57] and Shoaf and Lium[58] have used thin-layer chromatography to achieve a preliminary separation of chlorophylls on solvent extracts of water and algae prior to a final determination by spectrophotometry or fluorimetry. Garside and Riley[57] filtered sea water samples (0.5–5 l) through Whatman GF/C glass fibre coated with a layer, 1–2 mm thick, of light magnesium carbonate. This retains the smallest particles of organic matter and it is easy to extract the pigment from it. The filter is extracted with acetone (3-5 ml) and then with methanol (10 ml) using ultrasonic vibration. The solution is passed through anhydrous sodium sulphate to remove water and then evaporated *in vacuo* at less than 50 °C. The residue is dissolved in ethyl ether-dimethylamine (99:1, 1-2 ml) and this applied as a spot to a plate coated with silica gel PF_{254}. The chromatogram is developed with light petroleum (60-80)-ethyl acetate-dimethylamine (55:32:13) until the centre of the chlorophyll *a* spot has travelled about 10 cm from the origin. The solvent is allowed to evaporate and the plate scanned by reflection of the light passing through an Ilford 601 filter (603 nm). The integration reading for each peak is measured and the *R* values noted relative to chlorophyll *a*. Xanthophylls are identified by scraping off the spots and measuring the absorption spectrum of an extract of the scrapings. Chlorophyll *c* remains at the origin and can be developed in light petroleum-ethyl acetate-dimethylformamide (1:1:2) and scanned as before. Chlorophylls *a*, *b*, and *c*, carotene, xanthophylls, and certain degradation products can be determined. The analysis takes 1 h, the sensitivity for chlorophyll *a* is about 0.12 μg and the precision for most pigments is ±5% or better at the 0.5 μg level.

Shoaf and Lium[58] used thin-layer chromatography to separate algal chlorophylls from their degradation products. Chlorophyll is extracted from the algae with dimethyl sulphide and chromatographed on commercially available thin-layer cellulose sheets, using 2% methanol and 98% petroleum ether as solvents, before determination by either spectrophotometry or fluorimetry.

Method

Algae (both pure laboratory cultures and natural samples) are filtered through a glass fibre capable of retaining particles whose diameter is 0.45 μm or greater. Each filter is then rolled and placed into a glass tissue grinder; 3-4 ml of dimethyl sulphoxide are added, and the sample is ground with a Teflon pestle for 3 min at 500 rpm. The sample is transferred to a 15 ml screw-cap graduated centrifuge tube, and the pestle and grinding vessel are rinsed with dimethyl sulphoxide, which was added to the sample. An equal volume (approximately 6 ml) of diethylether is added, the cap is screwed on, and the sample is shaken vigorously for 10 s. After waiting an additional 10 s, the sample is again shaken vigorously for 10 s. The cap is removed and distilled water equal to 25% of the total volume is added slowly, drop by drop. As the water is added, the dimethyl sulphoxide-diethylether solution separates into two immiscible liquids, the dimethyl sulphoxide-water solution on the bottom and the diethylether layer containing the green chlorophylls above. The tube is capped and shaken well so that all the chlorophyll migrates into the diethylether layer. The sample is centrifuged at 1000 g for 10 min to cleanly separate the two layers and to sediment the glass filter and algal cell debris. After centrifugation the upper diethylether layer containing all the chlorophyll is pipetted off and placed into a 15 ml screw-cap graduated centrifuge tube. An equal volume of distilled water is added, and the tube is shaken vigorously for 10 s. The distilled water is added to remove traces of dimethyl sulphoxide dissolved in the diethylether layer. The addition of the water will isolate the dimethyl sulphoxide in the water phase where it can be removed. If the dimethyl sulphoxide is not removed, serious trailing will result in the thin-layer chromatography step. The sample is again centrifuged at 1000 g for 5 min. After centrifugation the upper diethylether layer is pipetted off and placed in a 15 ml graduated centrifuge tube. The tube is placed in a 25 °C water bath, and the solution is evaporated almost to dryness by blowing nitrogen over the ether surface. When the sample is almost dry, the total volume is brought up to 0.5 ml by the addition of acetone, and the sample is mixed until a uniform solution is obtained. The solvent change from ether to acetone is necessary because of the difficulty in quantitatively removing a given volume from a highly volatile ether solution.

An aliquot of the acetone solution is removed with a microlitre syringe and streaked onto a thin-layer cellulose sheet (Bakerflex. No. 0-4468) 50 × 200 mm, approximately 15 mm from one end and 6 mm from each side. The thin-layer sheet is then placed in a chromatography tank containing 98% petroleum ether (30–60 °C fraction) and 2% methanol and allowed to develop in the dark. The approximate time for the solvent to travel to about 20–30 mm from the top of the sheet is 30 min. The ratio of petroleum ether to methanol is very critical, and this is the factor most likely to affect the migration rate and separation of the chlorophylls. This solution should be prepared fresh daily because the ratio of the solvents will change as a result of evaporation. After development, spots or streaks of chlorophyll are removed by scraping the cellulose containing the chlorophyll into a screw-capped centrifuge tube and adding 3 ml of 90% acetone and mixing. After centrifuging to remove the cellulose, the concentration in the supernatant is determined with a spectrophotometer or spectrofluorimeter.

Typical values for chlorophylls and degradation products are shown in Table 150. Recoveries of pure chlorophylls a and b were 98 and 96% respectively. Thus chlorophylls a and b and their phaeophytins may be readily separated

Table 150 R_f values for chlorophylls and degradation products (relative to solvent)

Compound	R_f	Colour
Phaeophytin a	0.89	Grey
Chlorophyll a	0.76	Blue green
Phaeophytin b	0.61	Greenish yellow
Chlorophyll b	0.34	Yellowish green
Phaeophytin c	0	Yellowish green
Chlorophyll c	0	Yellowish green

Note: Highly purified chlorophylls a and b were purchased from a commercial source. Chlorophyll c was extracted from a fresh water species of *Cyclotella*. The phaeophytins were formed by acidification of the chlorophylls with hydrochloric acid.

and determined by this method. It is not possible to accurately determine chlorophyll c by using this method with natural samples for two reasons. In some samples other degradation products of chlorophyll a and b and phaeophytin a and b (apparently chlorophylides and phaeophorbides; that is, chlorophylls or phaesophytins missing part or all of the phytol tail) do not migrate but remain at the origin with chlorophyll c and phaeophytin c, each of which also lacks a phytol tail. Chlorophyll c is only sparingly soluble in diethylether or acetone, so that total recovery is not possible.

Chlorophyll a and b are accurately determined by this method without interference from degradation products.

Boto and Bunt[59] also used thin-layer chromatography for the preliminary separation of chlorophylls and phaeophytins and combined this with selective excitation fluorimetry for the determination of the separated chlorophylls a, b, and c and their corresponding phaeophytin components. An advantage of the latter technique is that appropriate selection of excitation and emission wavelengths reduces the overlap between the emission spectra of each pigment to a greater extent than is possible with broad band excitation and the use of relatively broad band filters for emission.

The fluorescence studies were performed on 90% acetone solutions with an Aminco-Bowman spectrofluorimeter (J4-8203G Model). Special attention must be given to the slit combinations in the optical system of this instrument. For the degree of resolution considered acceptable and for good sensitivity, the slit combination used was as follows: xenon lamp emission, 1 mm (i.e. 5 nm bandpass): slit slide 1, fully open; excitation monochromator, 2 mm; slit slide 2, fully open; photomultiplier, 2 mm. The calibration and linearity of the instrument were checked at frequent intervals with standard 1.0 and 0.1 μg ml^{-1} quinine/0.1 N sulphuric acid solutions. Also, after every five measurements, the emission response for the 0.01 μg ml^{-1} standard was checked and adjusted if necessary, as a short-term drift of $\pm 5\%$ was sometimes noted.

Solutions of the phaeophytins were prepared from the corresponding chlorophylls by adding 2 drops of 0.1 M hydrochloric acid per 100 ml solution.

Figure 185 shows the excitation and emission spectra for chlorophylls a, b, and c and their respective phaeophytins. The excitation spectra were obtained by holding the emission wavelength at the emission maximum and slowly scanning the excitation wavelength over the required range. The emission spectra shown have not been corrected for changes in photomultiplier response with wavelength. These spectra show that with suitable choice of excitation wavelengths, good selectivity can be achieved. Assuming that a given mixture contains all three chlorophylls and their phaeophytins, a detailed set of equations can be derived for the emission responses at the usual emission maxima of each pigment when the mixture is excited at various chosen wavelengths. After acidification of the mixture to phaeophytins alone, further information can be obtained as indicated below.

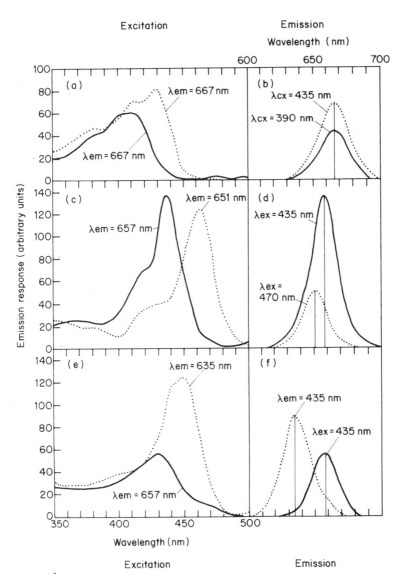

Figure 185 Excitation and emission spectra of: (A and B) chlorophyll a (---) and phaeophytin a (——), conc. of both = 0.134 μg ml^{-1}; (C and D) chlorophyll b (---) and phaeophytin b (——) conc. of both = 0.172 μg ml^{-1}; (E and F) chlorophyll c (---) and phaeophytin c (——), conc. of both = 0.042 μg ml^{-1}. Note that the phaeophytin c curves are shown on a ten-fold higher sensitivity scale to the others. All solutions are in 90% acetone. Reprinted with permission from Boto and Bunt.[59] Copyright (1978) American Chemical Society

In the following equations $H(\lambda_{ex}, \lambda_{em})$ denotes the height of the emission measured at λ_{em} following excitation of λ_{ex}, and $K(\lambda_{ex}, \lambda_{em}, x)$ denotes the coefficient of emission response versus concentration (of species x) at the same λ_{ex} and λ_{em}. C_x is the concentration of chlorophyll x in the mixture while

C_{px} refers to the concentration of pheophytin x. All concentrations are in μg ml^{-1} in the actual extract solution. After acidification $C^1_{px} = C_{px} + C_x$, i.e. the new phaeophytin x concentration is now equal to that originally present (C_{p_x}) plus the amount formed from the chlorophyll (C_x). The equations are as follows.

Before acidification of the solution:

$$H(435,667) = \underline{K(435,667,a)C_a} + K(435,667,b)C_b + K(435,667,c)C_c + K(435,667,P_a)C_{pa} + K(435,667,pb)C_{pb} + K(435,667,pc)C_{pc} \quad (8)$$

$$H(470,651) = \underline{K(470,651,b)C_b} + \underline{K(470,651,c)C_c} + K(470,651,a)C_a + K(470,651,pa)C_{pa} + K(470,651,pb)C_{pb} + K(470,651,pc)C_{pc} \quad (9)$$

$$H(435,635) = \underline{K(435,635,c)C_c} + K(435,635,b)C_b + K(435,635,a)C_a + K(435,635,pa)C_{pa} + K(435,635,pb)C_{pb} + K(435,635,pc)C_{pc} \quad (10)$$

After acidification:

$$H(390,667) = \underline{K(390,667,pa)C^1_{pa}} + K(390,667,pb)C^1_{pb} + K(390,667,pc)C^1_{pc} \quad (11)$$

$$H(435,657) = K(435,657,pb)C^1_{pb} + K(435,657,pc)C^1_{pc} + \underline{K(435,657,pa)\ C^1_{pa}} \quad (12)$$

The terms underlined are the major terms in each equation. Note that in equation 12, as the excitation and emission spectra of phaeophytins b and c overlap greatly, only one equation is obtainable and the concentrations of phaeophytins b and c must be estimated by a reiterative procedure (below). Although the equations appear complex, it must be remembered that, in most cases, the K values in the minor terms are such that many can be neglected at the 2% error level. Further simplification is often possible when one considers that, in most offshore sea water samples, the only primary pigments found are chlorophylls a and c along with phaeophytin a. Chlorophyll b and phaeophytins b and c are either absent or present in minor amounts. Phaeophorbides also are commonly found. However, these would probably have very similar fluorimetric properties to the phaeophytins and, hence, could not be separately determined by any fluorimetric method.

Table 151 shows a full listing of all the K values required for Equations 8–12. H values were measured as the number of recorder divisions obtained with the photomultiplier set on the most sensitive setting, i.e. the 0.1 multiplier scale on this instrument. For purposes of cross-calibration with other instruments, emission peak height H 350ex, 448em) of the 0.01 μg ml^{-1} quinine standard was 2100 divisions on the 0.1 multiplier scale. The response for the standard was always set to this value as noted above. The K values were then calculated ($K(ex,em,x) = H(ex,em)/C_x$) either by the slope of H (ex,em) versus C_x calibration graphs for each pigment, or, in some cases where K was small, by direct calculation from the excitation and emission spectra for each pure pigment in a solution of known concentration. The H versus C calibration graphs were linear, in all cases, in the concentration range 0.003–0.2 μg ml^{-1} with never more than a 2% deviation from linearity for any individual point.

Table 151 The $K(\lambda_{em}, \lambda_{em}, x)$ coefficients required for solution of equations 8 to 12

$K(435,667,a) = 5013$	$K(435,635,c) = 22{,}966$
$K(435,667,b) = 731$	$K(435,635,a) = 401$
$K(435,667,c) = 850$	$K(435,635,b) = 731$
$K(435,667,pa) = 308$	$K(435,635,pa) = 25$
$K(435,667,pb) = 3893$	$K(435,635,pb) = 892$
$K(435,667,pc) = 629$	$K(435,635,pc) = 115$
$K(470,651,b) = 5800$	$K(390,667,pa) = 3463$
$K(470,651,a) = 47$	$K(390,667,pb) = 561$
$K(470,651,c) = 1469$	$K(390,667,pc) = 262$
$K(470,651,pa) = 76$	$K(435,657,pb) = 8111$
$K(470,651,pb) = 282$	$K(435,657,pc) = 1048$
$K(470,651,pc) = 76$	$K435,657,pa) = 185$

Reprinted with permission from Boto and Bunt.[59] Copyright (1978) American Chemical Society

K values obtained by Boto and Bunt[59] are listed in Table 151. It is recommended that any laboratory wishing to consider this modified fluorimetric technique carry out individual calibrations as described above.

Using the K values listed in Table 151 and taking into account the most likely pigment compositions of the phytoplankton in most sea water samples, a simplified set of equations was obtained as shown below. As the quantum efficiency of chlorophyll c appears to be very high its emission spectrum is very little affected by interference from the emissions of other pigments. Therefore, its concentration can be quite accurately calculated and this value then used to give accurate estimates of the others, viz:

Before acidification:
$$H(435,635) = K(435,635,c)C_c \tag{13}$$

$$H(435,667) = K(435,667,a)C_a + K(435,667,c)C_c \tag{14}$$

$$H(470,651) = K(470,651,b)C_b + K(470,651,c)C_c + K(470,651,a)C_a \tag{15}$$

After acidification:
$$H(390,667) = K(390,667,pa)C^1_{pa} \tag{16}$$

$$H(435,657) = K(435,657,pb)C^1_{pb} + K(435,657,pc)C^1_{pc} \tag{17}$$

To solve equation 17, one substitutes $C^1_{pc} = C_c$ and solves for C^1_{pb} and then this value is used to give a better estimate of C^1_{pc} and so on. Usually only two or three iterations are required for convergence. This procedure is, of course, subject to substantial error when one considers the likely errors involved in reading the actual H values. Hence the final estimates of C_{pb} or C_{pc} originally present probably will only represent order-of-magnitude estimates unless either pb or pc is present in relatively large amounts. In all analyses and calibrations, the H values must be corrected for the usually small, but non-negligible solvent fluorescence at the particular wavelength investigated. However, use of a simple computer programme enables greater refinement of the estimates using equations 8-13 with reiterative procedures to apply the small corrections needed. In applying these procedures to pigment analyses from samples of sea water, it was found that this refinement is usually unnecessary at a 5% level although it became necessary for some artificial mixtures where the concentration ratios of each pigment were deliberately adjusted to give analytically unfavourable conditions.

A number of samples containing mixtures of (1) chlorophylls a and b only, (2) chlorophylls a, b, and c, and (3) chlorophylls a and c only were prepared and analysed, with the results shown in Table 152. Also, two actual analyses of 10 ml extracts from sea water are shown in Table 152.

From the results shown in Table 152 it can be seen that for mixtures containing only chlorophylls a and b, very good estimates were obtained, the worst error of +5% for C_a obtained with a solution containing only 0.0176 μg ml^{-1} of a and 0.00954 μg ml^{-1} of b. The C_b/C_a ratio of ca. 0.5 in these analyses represented an unfavourable case where correction of C_a for the C_b term is necessary. On acidification, the estimates of C^1_{pa} and C^1_{pb} ($=C_a$ and C_b respectively in these solutions as no extra pa or pb is present), were also very good although it should be noted that, in mixtures containing a, b, and c, the phaeophytin b estimations would be expected to be less accurate. Analysis of a, b, and c mixtures showed that the estimates of a and c were within ±7% or better while the estimate of b could be subject to fairly large errors (up to 30%) when C_b was very small or if $C_c:C_b$ was 5:1 or greater. Theoretical calculations using equations 8-10 with estimates of the inaccuracies involved in the readings of the emission responses, showed that the estimation of C_b was subject to errors in agreement with those actually found. Also, similar calculations showed that if only a and c were present, and if C_c and/or C_a were of the order of 0.1 μg ml^{-1}, then C_b could be estimated as being 0.006 μg ml^{-1} when none was actually present. Actual analysis of such a mixture (Table 152) confirmed this possibility. Thus such levels of chlorophyll b, if found in real samples, would need to be treated with suspicion.

The sea water sample analysis (Table 152) showed an unusually high concentration of phaeophytin c. In this situation, it could be demonstrated by use of equations 8-12 that the expected order of accuracy would be 10-20%. Therefore the method at least gives a reasonably reliable estimate of phaeophytin c if present at these levels.

The first application of high performance liquid chromatography to plant pigments was by Evans et al.[60] who separated phaeophytins a and b on Corasil II with a mobile phase consisting of a 1:5 (v/v) mixture of ethyl acetate and light petroleum. Eskins et al.[61] have employed two 0.62 m columns of C_{18}-Porasil B for preparative separation of plant pigments by means of programmed stepwise elution with methanol-water-ether. However, the method is of little value for routine application because of the time required and also because the chlorophyll degradation products, other than phaeophytin, are not

Table 152 Results of analyses using the fluorimetric method for artificial mixtures and actual sea water extracts (in 90% acetone)

Sample type	Actual concentrations*			Measured conc.* (errors in parentheses)					
	C_a	C_b	C_c	C_a	C_b	C_c	C^1_{pa}	C^1_{pb}	C^1_{pc}
Prepared mixture	0.176	0.0954		0.171 (−2.8%)	0.0935 (−2.0%)		0.178 (+0.09%)	0.0943 (−1.1%)	
Prepared mixture	0.0176	0.00954		0.0185 (+5.1%)	0.00968 (+1.5%)		0.0172 (−2.4%)	0.0091 (−5.0%)	
Prepared mixture	0.161	0.054	0.0595	0.155 (−3.7%)	0.049 (−9.3%)	0.057 (−4.2%)			
Prepared mixture	0.094	0.014	0.595	0.087 (−7.4%)	0.0135 (−3.6%)	0.059 (−0.8%)			
Prepared mixture	0.021	0.007	0.00774	0.020 (−3.8%)	0.0053 (−2.4%)	0.0076 (−1.8%)			
Prepared mixture	0.115		0.563	0.120 (+4.3%)	0.006	0.057 (+1.2%)	0.110 (−4.3%)	0.0003	0.050 (−11.1%)
Sea water extract†				0.059	0.001	0.0087	0.0086	0.0018	0.0327
Similar collected near same spot†				0.046	0.0002	0.0088	0.090	0.0012	0.0458

*Concentrations in $\mu g\ ml^{-1}$ of the actual extract.
†Results shown here are averages of duplicates for each sea water sample.
Reprinted with permission from Boto and Bunt.[59] Copyright (1978) American Chemical Society

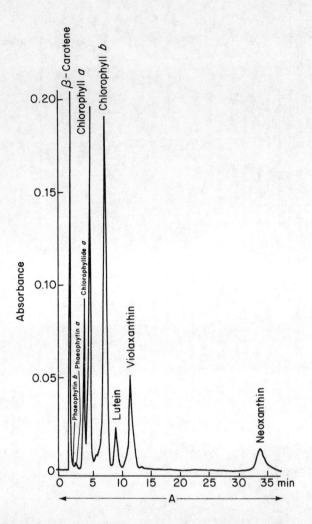

Figure 186 Chromatogram of extract of pigments from *Dunaliella tertiolecta*. Reprinted with permission from Abayachi and Riley.[63] Copyright (1979) Elsevier Science Publishers

separated. Shoaf[62] has used high performance liquid chromatography to separate the chlorophylls *a* and *b* of a pigment extract from which the carotenoids had been previously removed. Good resolution of the two pigments and several of their unspecified degradation products was achieved on a 25 cm column of Partisil PXS 1025 by elution with aqueous 95% methanol; however, chlorophylls were not determined quantitatively.

Abayachi and Riley[63] used high performance liquid chromatography to determine chlorophylls, their degradation products, and carotenoids in phytoplankton and marine particulate matter.

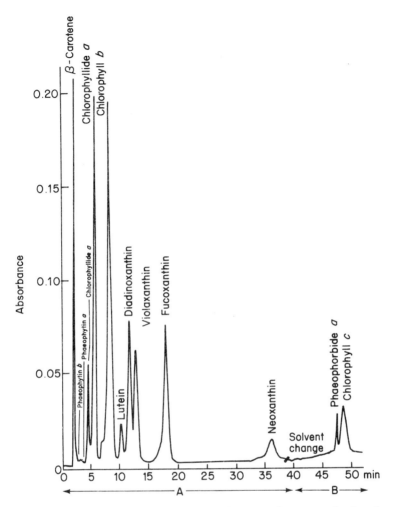

Figure 187 Chromatogram of extract of pigments from a mixed culture of *Phaeodactylum tricornutum* and *Dunaliella tertiolecta*. Reprinted with permission from Abayachi and Riley.[63] Copyright (1979) Elsevier Science Publishers

Pigment extraction is carried out with acetone and methanol. After evaporation of the combined extracts under reduced pressure, the pigments are separated on a Partisil 10 stationary phase with a mobile phase consisting of light petroleum (b.p. 60–80 °C), acetone, dimethyl sulphoxide and diethylamine (75:23.25:1.5:0.25 by volume). When chlorophyll *c* is present, a further development is performed with a similar, but more polar, solvent mixture. Detection is carried out spectrophotometrically at 440 nm. The method has a sensitivity for the chlorophylls of *ca.* 80 ng and for carotene of *ca.* 5 ng. The coefficient of variation of the chromatographic stage of the procedure lies in the range 0.6–1.8%.

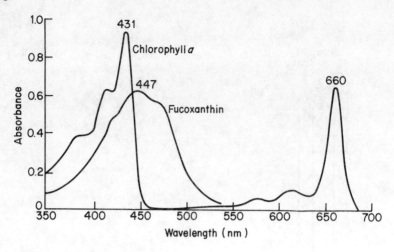

Figure 188 Absorption spectra of chlorophyll *a* and fucoxanthin separated by h.p.l.c. from extract of *Phaeodactylum tricornutum* (4 cm cuvette). Reprinted with permission from Abayachi and Riley.[63] Copyright (1979) Elsevier Science Publishers

The procedure described by Garside and Riley[64] was adopted for the extraction of the pigments of algal cells because of its simplicity. Tests carried out with the high performance liquid chromatographic technique endorsed the claim that it causes negligible degradation of both the chlorophylls and the xanthophylls. A more efficient separation of plant pigments could be achieved on a silica stationary phase than on a C_{18} reversed-phase medium. A 30 cm column packed with Partisil 10 gave an efficient separation of the individual carotenoids, chlorophylls *a* and *b*, and many of the degradation products of the latter pair (Figure 186). The solvent consists of light petroleum (b.p. 60–80 °C), acetone, dimethyl sulphoxide, and diethylamine in the ratio 75: 23.25: 1.5: 0.25 volume. Unfortunately, this solvent is not sufficiently polar to elute phaeophorbide and chlorophyll *c*. With samples containing these pigments, it is necessary to carry out an additional, stepwise, elution with a more polar solvent. Further tests showed that excellent resolution of these compounds could be achieved (Figure 187) by means of a mixture containing light petroleum (b.p. 60–80 °C), acetone, methanol, and dimethyl sulphoxide in the ratio 30:40:27.3 by volume, respectively.

In order to identify the various peaks on the chromatograms, extracts of a range of algae from various classes were injected repeatedly onto the Partisil 10 column operated at a flow rate of 2 ml min^{-1}. The eluates corresponding with individual peaks of known retention times were collected and combined until a suitable quantity of pigment had been accumulated. The combined eluates were then taken rapidly to dryness in a rotary evaporated at less than 30 °C and, after dissolution in appropriate solvents, were identified from their absorption spectra. Chlorophylls and their degradation products were characterized by their spectra in ether and in (1 + 9) water–acetone, based on data from Strickland.[65] Carotenoids were identified by comparison of their wavelengths of maximum absorption in hexane, ethanol, and carbon disulphide with the tabulated values published by Davies.[66] Comparison of the absorption spectra with those of the pure pigments showed that the separated pigments were recovered in a satisfactorily pure state (Figure 188). Identifications were confirmed by thin-layer chromatography on silica gel G.[67] The retention times of the various pigments are shown in Table 153 together with their standard deviations which indicate that the retention times are extremely reproducible.

Since the high performance liquid chromatographic method provides pigments with a high degree of purity, it is relatively easy to standardize the technique. For this purpose, an extract containing the desired pigment is injected into the instrument and the separated pigment is collected. After rotary evaporation at near ambient temperature (which has been shown not to lead to degradation), the pigment is immediately dissolved in a known volume of an appropriate solvent and its concentration is determined photometrically. If the procedure is repeated with several different dilutions of the extract, a calibration curve can be constructed relating the chromatographic peak height to the photometrically determined amount of pigment. Figure 189 shows the calibration curves derived in this way for a number of pigments; the absorptivities shown in Table 154 are used for the evaluation of the pigment concentrations.

Table 153 Retention times and corresponding coefficients of variation for various phytoplankton pigments on a 30 cm Partisil 10 column*

Pigment	Retention time(s)	Coefficient of variation (%)
	First mobile phase	
β-Carotene	107 ± 1	0.9
Echinenone	134 ± 1	0.7
Phaeophytin b	141 ± 1	0.7
Phaeophytin a	183 ± 1	0.5
Chlorophyllide a	237 ± 0.3	0.1
Chlorophyll a	293 ± 2	0.6
Chlorophyll b	424 ± 2	0.6
Diatoxanthin	478 ± 2	0.4
Myxoxanthophyll	488 ± 1	0.6
Lutein	533 ± 1	0.3
Diadinoxanthin	580 ± 3	0.5
Violaxanthin	664 ± 1	0.2
Fucoxanthin	809 ± 5	0.6
Neoxanthin	1773 ± 15	0.9
	Second mobile phase	
Phaeophorbide a	2395 ± 19 (622 in 2nd solvent)	0.8
Chlorophyll c	2497 ± 21 (724 in 2nd solvent)	0.8

*Solvent flow rate, 2 ml min^{-1}.
Reprinted with permission from Abayachi and Riley.[63] Copyright (1979) Elsevier Science Publishers.

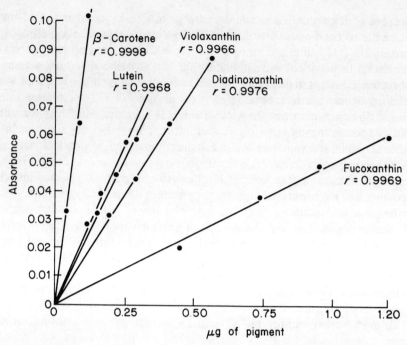

Figure 189 Calibration curves for various pigments. Reprinted with permission from Abayachi and Riley.[63] Copyright (1979) Elsevier Science Publishers

Table 154 Specific absorptivities used in calibration of the high performance liquid chromatographic instrument

Pigment	$E_{1cm}^{1\%}$	Absorption maximum (nm)	Solvent
Chlorophyll a	911	663	90% Acetone
Chlorophyll b	525	645	90% Acetone
Chlorophyll c	1400	446	90% Acetone
β-Carotene	2592	453	Hexane
Echinenone	2153	458	Light petrol
Lutein	2550	445	Ethanol
Diadinoxanthin	2500	446	Ethanol
Violaxanthin	2550	443	Ethanol
Fucoxanthin	1060	449	90% Acetone
Neoxanthin	2243	439	Ethanol
Myxoxanthophyll	2160	478	Acetone

Reprinted with permission from Abayachi and Riley.[63] Copyright (1979) Elsevier Science Publishers.

From Figure 189 it can be seen that the calibration is very closely linear (correlation coefficients ranging from 0.9966 for chlorophyll c to 0.9998 for β-carotene).

The sensitivity of the method varies considerably from one pigment to another. If a peak height of 5 mm is considered to be significant with a full scale deflection of 25 cm corresponding to 0.2 absorbance units, then the detection limit of the method varies from about 5 ng for β-carotene to about 80 ng for chlorophyll *a*. It should be noted however that this is not the maximum sensitivity which could be attained; the wavelength of 440 nm, at which the absorbance is measured, does not in general correspond with that of the absorbance maxima of the pigments, but is a compromise enabling reasonable sensitivity to be achieved for both chlorophylls and carotenoids. This probably reduces the attainable sensitivity by a factor of 2–3.

Abayachi and Riley[63] compared results obtained by the high performance liquid chromatographic method with those obtained by a reflectometric thin-layer chromatographic method and the SCOR/UNESCO polychromatic procedure.[43] The results obtained from the latter were evaluated by the SCOR/UNESCO equations and also by the more recent ones of Jeffrey and Humphrey.[68] The carotenoids were determined collectively from the absorbance of the 90% acetone extract at 480 nm by means of the equations of Strickland and Parsons.[42] The results of these comparative studies (Table 155) show that there is satisfactory agreement for all pigments between the two chromatographic methods. However, although the results for chlorophyll *a* by the polychromatic method were in reasonable accord with those derived chromatographically, many of those for the other chlorophylls were highly discrepant. Obviously, the polychromatic method, in particular, is unsatisfactory with respect to the interference of chlorophyll degradation products, as these are nearly always present in environmental samples.

Method

Liquid chromatography

Solvent was delivered at a rate of 2 ml min^{-1} and a pressure of about 14 kg cm^{-2} by means of a minipump. Pulse damping was achieved by means of two sealed vertical stainless steel tubes (90 cm long, 4.5 mm bore) which were attached by 1 mm bore tubing to branches on the pump outlet. The column, constructed of 0.45 mm bore stainless steel tubing and packed with a 30 cm layer of Partisil 10 at a pressure of 350 kg cm^{-2}, had a plate count of 4600. For precision in sampling, the sample was introduced to the column by means of a syringe-loading sample injection unit fitted with a 20 μl sampling loop. The effluent from the column was fed to an 8 μl flow cell with a 1 cm lightpath in a Cecil CE373 visible spectrophotometer operated at 440 nm and range expansion of 0.2 A. The linear absorbance signal from the spectrophotometer was fed to a Honeywell chart recorder.

Reagents

Magnesium carbonate suspension

About 3 g of finely powdered basic magnesium carbonate are suspended in 100 ml of distilled water and resuspended immediately before use.

Table 155 Comparison between algal pigment analyses carried out by the high performance liquid chromatographic method, by reflectometric t.l.c., and by the photometric method

Pigment	H.p.l.c. μg dl^{-1}	H.p.l.c. % of total carotenoids	T.l.c. μg dl^{-1}	T.l.c. % of total carotenoids	Polychromatic SCOR/UNESCO[43] μg dl^{-1}	Polychromatic Jeffrey and Humphrey[68] μg dl^{-1}
Phaeodoctylum tricornutum						
Chlorophyll a	2.68		2.80		2.9	3.0
Chlorophyll b	0.00		0.00		0.005	−0.34
Chlorophyll c	0.44		0.47		0.59	0.43
β-Carotene	0.44	10.5	0.46	10.5		
Fucoxanthin	2.84	68.0	2.88	69.9	4.6	
Diadinoxanthin	0.90	21.5	1.03	23.6		
Chl. a Chl. c	1:0.17		1:0.16		1:0.2	1:0.14
Dunaliella tertiolecta						
Chlorophyll a	4.53		4.40		4.3	4.5
Chlorophyll b	2.18		2.20		0.04	0.17
β-Carotene	1.10	33.3	1.01	30.2	2.9	
Lutein	0.53	16.0	0.55	16.6		
Violaxanthin	1.27	38.5	1.33	40.2		
Neoxanthin	0.41	12.2	0.43	13.0		
Chl. a; Chl. b	1:0.48	1:0.50			1:0.52	1:0.42
Oscillatoria spp.						
Chlorophyll a	17.07		17.75		15.6	16.11
Chlorophyll b	0.00		0.00		−0.54	−2.27
Chlorophyll c	0.00		0.00		1.89	6.3
β-Carotene	1.97	45.3	2.08	46.8		
Echinenone	1.65	37.9	1.66	37.4	0.8	
Myxoxanthophyll	0.73	16.8	0.70	15.8		

Reprinted with permission from Abayachi and Riley.[63] Copyright 1979 Elsevier Science Publishers.

First mobile phase

From glass-redistilled solvents, a mixture containing light petroleum (b.p. 60-80 °C), acetone, dimethyl sulphoxide, and diethylamine in the ratio 75:23.25:1.5:0.25 by volume is prepared. Before use, the mixture is passed through a porosity 5 sintered glass filter under slight suction.

Second mobile phase

From glass-distilled solvents a mixture of light petroleum (b.p. 60-80 °C), acetone, methanol, and dimethyl sulphoxide in the volume ratio 30:40:27:3 is prepared. Using a slight vacuum, the mixture is filtered through a sintered glass Buchner funnel (porosity 5) before use.

Procedure

To avoid degradation of the pigments, extraction of the sample and processing of the extract must be carried out as rapidly as possible in subdued light. Although it is possible to store the filtered organisms safely for up to a week at $-20\,°C$ in closely stoppered nitrogen-filled tubes, the analysis should be completed without delay once the extraction has been commenced.

Extraction of pigments

A glass filter holder is fitted with a 7 cm Whatman GF/C glass fibre filter and under slight suction coated with a 1-2 mm layer of magnesium carbonate using the suspension. A known volume of the sample is filtered, sufficient to provide *ca.* 5 μg of chlorophyll *a* (normally 2-10 l of sea water or 20-150 ml of cultures) under a vacuum not exceeding 0.3 bar. The filter is washed with a few ml of distilled water and then transferred with its coating quantitatively to a test tube. To the test tube is added 3-5 ml of acetone and the filter gently disintegrated with a glass rod. The test tube is placed in an ultrasonic agitator for 5 min, 10 ml of methanol are added, mixed and the ultrasonic treatment continued for a further 10 min. The extract is decanted through a drawn-out glass tube containing a 2 cm layer of anhydrous sodium sulphate supported on a glass fibre plug. The dry extract is collected in a 25 ml pear-shaped flask. The residue in the test tube and the drying column is washed with about 2 ml of methanol and combined with the extract. If, exceptionally, the cell debris is not practically colourless, it is agitated ultrasonically with 5 ml aliquots of methanol until no further pigment is extracted. The combined extracts are evaporated *in vacuo* in a rotary evaporator at a bath temperature not exceeding 40 °C. Immediately after the last of the solvent has been removed, the residue is taken up in an exactly known volume (0.25-0.5 ml) of the first mobile phase.

Chromatographic separation of pigments

The high performance liquid chromatographic column is equilibrated with the first mobile phase. After equilibration has been achieved, a 20 µl aliquot of the pigment extract is injected via the sampling valve. The chromatogram with the first mobile phase is developed at a flow rate of 2 ml min^{-1} until the last peak has appeared (*ca.* 30 min). The high pressure pump is switched off and the solvent intake tube rapidly transferred to the second mobile phase. The pump is switched on again immediately and the development continued at a flow rate of 2 ml min^{-1} until the elution of chlorophyll *c* is complete. Because of differences in refractive index between the two solvent systems some instrumental noise will be recorded when their interface enters the flow cell; however, a steady baseline should soon be re-established. Before carrying out the next injection the apparatus should be allowed to re-equilibrate with the first mobile phase for a few minutes. If chlorophyll *c* and the phaeophorbides are not to be determined, the analysis can be terminated after the last carotenoid has been eluted, and the next sample can then be injected. However, there will be a tendency for strongly adsorbed materials, e.g. chlorophyll *c* and degradation products, to accumulate on the column. As this will cause its performance to deteriorate gradually, these compounds must be removed by eluting periodically with the second mobile phase. If used with care, the column should perform satisfactorily for several hundred injections.

To calibrate the method, the pigments are extracted from an aliquot of an appropriate species of phytoplankton which has been grown in culture to a reasonably high cell density and 20 µl of the solution of the pigments in the first mobile phase are injected into the liquid chromatograph. The chromatogram is developed, the fraction corresponding with each peak collected and evaporated to dryness rapidly in a rotary evaporator at 30 °C. The residue is immediately dissolved in 2.5 ml of an appropriate solvent (see Table 154) and the absorbance of the solution measured in a 4 cm microcuvette at the wavelength of maximum absorption of the particular pigment. The experiment is repeated with other dilutions of the extract. A calibration curve of chromatographic peak height against the amount of pigment found photometrically is plotted.

Wun et al.[69] have described a method for the simultaneous extraction of the water quality marker algal chlorophyll *a* and the faecal sterol coprostanol from water. This method utilizes a column of Dunberlite XAD-1 for the simultaneous extraction of both markers. Chlorophyll content was determined by the trichromatic method[70] using a double beam spectrophotometer. Gas-liquid chromatographic analyses of faecal sterols were performed with a Perkin Elmer 900 gas chromatograph equipped with flame ionization detectors. Conditions for the gas-liquid chromatograph analyses have been described by Wun.[71]

Wun et al.,[69] using ^{14}C-labelled cholesterol *c*, showed that 100% of this substance is adsorbed from a 500–1000 ml water sample by XAD-1 resin. When the column was subsequently eluted with 50 ml of basic methanol and 30 ml of benzene, up to 97% of the added sterol was recovered in the benzene eluate.

Table 156 Efficiency of the XAD-1 (60–120) resin column for the simultaneous isolation of phytoplankton chlorophyll a and coprostanol

Sample*	Chlorophyll a[†] (μg/sample)	Coprostanol[‡] (μg/sample)
50 ml sewage + 50 ml distilled water	17.01	11.0
50 ml sewage + 450 ml distilled water	16.99	10.0
50 ml sewage + 950 ml distilled water	17.01	10.5
100 ml sewage + 900 ml distilled water	16.78	26.0
200 ml sewage + 800 ml distilled water	16.36	50.0
500 ml sewage + 500 ml distilled water	16.05	125.0

*2.0 ml aliquot of unialgal culture of *Chlorella vulgaris* (total cell count: 9.0×10^7) was added to each of the test samples containing 50–500 ml of sewage. Photosynthetic pigments and coprostanol were extracted by the XAD-1 (60–120) resin column procedure.
†Chlorophyll a content of the 2.0 ml *Chlorella vulgaris*, as determined by the conventional aqueous acetone extraction method, was 11.75 μg. When analysed by the column method the sewage was found to contain approximately 0.64 μg chlorophyll a dl^{-1}. This amount was subtracted from the total chlorophyll a recovered.
‡The coprostanol concentration of the sewage as determined by hexane liquid-liquid partitioning was 219 ppb.
Reprinted with permission from Wun et al.[69] Copyright (1979) D. Reidel Publishing Co.

Table 157 A comparison of column and conventional extraction procedures for the recovery of chlorophyll a and coprostanol

	Extraction method			
	XAD-1 column		90% Aqueous acetone	Hexane partitioning
Sample	Chlorophyll a*	Coprostanol[†]	Chlorophyll a*	Coprostanol[†]
Sewage[‡] + 4.0 ml *Chlorella vulgaris*	33.3	350.0	32.0	330.0
Sewage + 4.0 ml *Chlamydomonas moewuii*	13.5	326.3	9.2	250.5
Sewage + 2.0 ml *Chlorococcum hypnosporum*	16.6	330.00	15.1	270.0
Sewage + 7.0 ml *Oocystis marssonii*	51.4	590.0	17.1	560.0
Sewage + 8.0 ml *Oscillutoria tenius*	26.9	912.00	16.0	912.0

*chlorophyll a = μg per sample.
†coprostanol = ppb.
‡Sewage samples were collected from the sewage treatment plant (a primary treatment plant) on different days. Samples were each divided into two equal aliquots of 500 ml each. Coprostanol content of one aliquot was extracted by hexane partitioning. The other, after adding a known amount of plankton (chlorophyll a content predetermined by 90% aqueous acetone extraction of an equal amount of the algal sample) was extracted by the XAD-1 (60–120) resin column procedure.
Reprinted with permission from Wun et al.[69] Copyright (1979) D. Reidel Publishing Co.

To ascertain the suitability of the column method for the simultaneous extraction of coprostanol and chlorophyll a in actual field testing situations, various unialgal cultures were mixed with sewage samples. The effectiveness of the neutral resin column extraction for these markers was evaluated with that of

Table 158 Recovery of chlorophyll a from field samples by the two extraction procedures

Field samples* (100–400 ml)	Chlorophyll a recovery (μg dl^{-1})		Ratio of extraction (column/ aqueous acetone)
	Column	90% Aqueous acetone	
Sewage lagoon 1	2.42	1.74	1.39
Sewage lagoon 3	4.77	2.98	1.60
Sewage lagoon 4	54.71	38.38	1.43
Sewage lagoon 5	147.54	110.57	1.33
Sewage effluent (chlorinated)	106.99	74.04	1.45
River, 10 ml below outfall	47.18	33.56	1.41

Conditions: extraction procedures as described in footnote‡, Table 157.
*Samples obtained from the oxidation ponds (sewage lagoons) of a secondary sewage treatment plant. Major species of algae found in the samples were *Euglena sp.* and *Chlamydomonas sp.*: plankton cell densities range from 500 to 1.2×10^5 and $600-7.0 \times 10^4$ per ml respectively. Coprostanol concentration in the sewage lagoon 1 sample as determined by gas chromatographic analyses was approximately 0.2 ppb.
Reprinted with permission from Wun et al.[69] Copyright (1979) D. Reidel Publishing Co.

the conventional procedures. Representative results are presented in Tables 156 and 157.

Results presented in Table 156 indicate that the efficiency of the resin column for the simultaneous extraction of coprostanol and phytoplankton chlorophyll a was comparable to or better than the conventional extraction procedures for the respective compounds. Dilution of the samples and the presence of extraneous materials did not affect the recovery efficiency significantly. Data in Table 157 reveal that the column technique was effective in isolating chlorophyll a from various algae. The superiority of the column method was more pronounced when the small green alga (*Oocystis* sp.) and the blue-green alga (*Oscillatoria* sp.) were used as test organisms. The coprostanol extraction efficiency was again shown to be comparable to that of the hexane liquid-liquid partitioning process.

Representative results of field sample analyses are shown in Table 158. From these data, it is apparent that a more complete extraction of phytoplankton chlorophyll a from water samples can be obtained by the column method. Furthermore, the column procedure is much faster; a one litre sample required a processing time of approximately 1–2 h compared to 24 h for the conventional aqueous acetone method. Coprostanol contents of these samples were too low (<0.2 ppb) for meaningful comparisons and are not presented.

COBALAMIN (VITAMIN B$_{12}$)

Beck[72] has described a competitive intrinsic factor binding method for the determination of benzyl alcohol extractable cobalamins in natural waters. He compares results of assays with those determined by high pressure liquid chromatography.

Sharma et al.[73] compared results obtained in the determination of cobalamins in ocean waters by radioisotope dilution and bioassay techniques. These workers showed that the isotopic methods measure both biologically active and inactive cobalamins indiscriminatively when porcine intrinsic factor is used as the B_{12} specific binder.

Beck and Brink[74] have described a sensitive method for the routine assay of cobalamins in activated sewage sludge. The method involves extraction with benzyl alcohol, removal of interfering substances using a combination of gel filtration and chromatography on alumina, concentration of the extract by lyophilization, and direct determination of total cobalamin by high-speed liquid chromatography, in comparison with cobalamin standards.

REFERENCES

1. Lawrence, J. *National Inventory of Natural Organic Compounds—An Interim Report*. Canada Centre for Inland Waters Unpublished Report, Burlington, Ontario.
2. Telang, S. A. *Water Quality and Forest Management*. Report of the Environmental Sciences Centre (Kananaskis) University of Calgary, Canada (1976).
3. Rashid, M. A. and King, L. H. *Chem. Geol.*, **7**, 37 (1971).
4. Schnitzer, M. Khaus, *Humic Substances in the Environment*. Marcel Dekker, New York (1972).
5. Pearl, I. A. *The Chemistry of Lignin* Marcel Dekker, New York (1967).
6. Geissman, T. A. and Crout, D. H. G. *Organic Chemistry of Secondary Plant Metabolism*. Freeman Cooper, San Francisco (1969).
7. Miroslav, M. J. *Water Pollution Central Fed.*, **41**, 1923 (1969).
8. Moed, J. R. *Limnol. Oceanogr.*, **16**, 140 (1971).
9. Eberle, S. H. *Ber. Kernforschungzentrum Karlsruhe*, KFK-1731 UF 18 pp. (1973).
10. Sontheimer, H. and Wagner, I. *Zeitschrift für Wasser and Abwasser Forschung*, **10**, 77 (1977).
11. Lawrence, J. *Water Research*, **14**, 373 (1979).
12. Lawrence, J. *National Inventory of Natural Organic Compounds*. Canada Centre for Inland Waters, Burlington, Ontario (1975).
13. Wagner, I. and Hoyer, O. *Vom Wasser*, **45**, 207 (1975).
14. Snow, M. G. *Proc. Anal. Div. Chem. Soc.*, **12**, 253 (1975).
15. Snow, M. G. *Water Treatment and Examination*, **24**, 297 (1975).
16. Chian, E. S. K. and De Walle, F. B. *Environ. Sci. Technol.*, **11**, 158 (1977).
17. Farrah, S. R., Goyal, S. M., Gerba, C. P., Majahan, V. K., Wallis, C. and Melnick, J. L. *Water Research*, **12**, 303 (1978).
18. Weber, J. H. and Wilson, S. A. *Water Research*, **9**, 1079 (1975).
19. Boening, P. H., Beckmann, D. P. and Snoeyink, V. L. *J. Am. Water Works Ass.*, **72**, 54 (1980).
20. Buffle, J., Greter, F. L. and Haerdi, W. *Anal. Chem.*, **49**, 216 (1977).
21. Stevenson, F. J. *Soil Sci.*, **123**, 10 (1977).
22. Stevenson, F. J. *Environmental Geochemistry* (J. O. Nriagu, ed.), pp. 95–106. Ann Arbor Science Publishers, Inc., Ann Arbor, Mich. (1976).
23. Stevenson, F. J. and Goh, K. M. *Soil Sci.*, **117**, 34 (1974).
24. Borggaard, O. K. *Acta Chem. Scand.*, **28**, 121 (1974).
25. Rubinstein, R. *Sil'chenko Zhurnal Analiticheskoi Khimi*, **12**, 2448 (1975).
26. Ishibashi, T. *Aqua. Sci. Tech. Rev.*, **3**, No. 1 (1980).
27. American Public Health Association, *Standard Methods for the Examination of Water and Wastewater*, 13th edn., p. 346 American Public Health Association Inc., New York (1971).

28. Eberle, S. B. and Schweer, H. K. *Vom Wasser*, **41**, 27 (1973).
29. Edeline, F., Lambert, G., Lorenzi, G. and Fatticcioni, H. *La Tribune du Cebedeau*, No. 385, **28**, 432 (1975).
30. Kerndorft, H. and Schnitzer, M. *Water, Air and Soil Pollution*, **12**, 319 (1979).
31. Sposito, G., Schaumberg, G. D., Perkins, T. G. and Hotzclaw, K. M. *Environ. Sci. Technol.*, **12**, 931 (1978).
32. Rebhun, M. and Mouka, J. *Environ. Sci. Technol.*, **7**, 606 (1971).
33. Mouka, J., Rebhun, M., Mandelbaun, A. and Borginger, A. *Environ. Sci. Technol.*, **8**, 1017 (1974).
34. Black, A. B., and Christman, R. F. *J. Am. Water Works Ass.*, **55**, 897 (1963).
35. Bear, F. E. *Chemistry of the Soil*, p. 224. Reinhold, New York, N.Y. (1964).
36. Felbeck, G. T., Jr. *Advances in Agronomy*, **17**, 327 (1965).
37. Schnitzer, M. *Metal Organic Interactions in Soils and Waters* in *Organic Compounds in Aquatic Environments* (S. D. Faust and J. V. Hunter, eds), pp. 297–315 (1971).
38. Jeffrey, S. W. and Humphrey, G. F. *Biochem. Physiol. Pflanz*, **167**, 191 (1975).
39. Davies, J. and Decasta, J. *Hydrobiologica*, **71**, 19 (1980).
40. Richards, F. A. and Thompson, T. G. *J. Mar. Res.*, **2**, 150 (1952).
41. Creitz, G. I. and Richards, F. A. *J. Mar. Res.*, **14**, 211 (1955).
42. Strickland, J. D. H. and Parsons, T. F. *A Practical Handbook of Seawater Analysis*, Fisheries Research Board of Canada, Ottawa (1968).
43. *UNESCO Monographs on Oceanographic Methodology* No. 1 (1966).
44. Parsons, T. R. and Fish, J. *Res. Board Cam.*, **18**, 1017 (1961).
45. Youngman, R. E. U. K. Report TR.82 Water Research Centre, Medmanham, Marlow, Bucks. (1978).
46. Ryding, S. O. *Vatten*, **31**, 327 (1975).
47. Raghi-Atri, F. *Gesundheits-Ingenieur*, **99**, 380 (1978).
48. Jensen, K. S. *Valten*, **32**, 337 (1976).
49. Holm-Hansen, O., Lorenzer, C. J., Holmes, R. W. and Strickland, J. D. H. *J. Cons., Cons. Int. Explor. Mer.*, **30**, 3 (1965).
50. Loftus, M. E. and Carpenter, J. H. *J. Mar. Res.*, **29**, 319 (1971).
51. Jeffrey, N. W. *Biochim. Biophys. Acta*, **162**, 271 (1968).
52. Daley, R. J., Gray, C. B. J. and Brown, S. R. *J. Chromat.*, **76**, 175 (1973).
53. Daley, R. J. *Arch. Hydrol.*, **72**, 400 (1973).
54. Madgwick, J. C. *Deep-Sea Research*, **12**, 233 (1965).
55. Jones, I. D., Butler, L. S., Gibbs, E. and White, R. C. *J. Chromat.*, **70**, 87 (1972).
56. Riley, J. P. and Wilson, T. R. *J. Mar. Biol. Ass.*, **45**, 583 (1965).
57. Garside, C. and Riley, J. P. *Anal. Chim. Acta*, **46**, 179 (1969).
58. Shoaf, W. T. and Lium, B. W., *J. Res. U.S. Geol. Surv.*, **5**, 263 (1977).
59. Boto, K. G. and Bunt, J. S. *Anal. Chem.*, **50**, 392 (1978).
60. Evans, N., Games, D. E., Jackson, A. H. and Matlin, S. A., *J. Chromat.*, **115**, 325 (1975).
61. Eskins, K., Scholfield, C. R. and Dutton, H. H. *J. Chromat.*, **135**, 217 (1977).
62. Shoaf, W. T. *J. Chromat.*, **152**, 247 (1978).
63. Abayachi, J. K. and Riley, J. P. *Anal. Chim. Acta*, **107**, 1 (1979).
64. Garside, C. and Riley, J. P. *Anal. Chim. Acta*, **46**, 179 (1969).
65. Strickland, D. H. in J. P. Riley and G. Skirrow, *Chemical Oceanography*, Vol. I, p. 494. Academic Press, London and New York (1965).
66. Davies, B. H. in T. W. Goodwin, *Chemistry and Biochemistry of Plant Pigments*, Vol. II, p. 108. Academic Press, London and New York (1976).
67. Wiley, J. P. and Wilson, T. R. *J. Mar. Biol. Ass.*, **45**, 583 (1965).
68. Jeffrey, S. W. and Humphrey, G. H. *Biochem. Physiol. Phanz*, **167**, 191 (1975).
69. Wun, C. K., Rho, J., Walker, R. W. and Litsky, W. *Water, Air and Soil Pollution*, **11**, 173 (1979).

70. American Public Health Association, *Standard Methods for the Examination of Water and Wastewater*, p. 1007. American Public Health Association, Inc., New York, N.Y. (1976).
71. Wun, C. K., Walker, R. W. and Litsky, W. *Water Research*, **10**, 955 (1976).
72. Beck, R. A. *Anal. Chem.*, **50**, 200 (1978).
73. Sharma, G. M., DuBois, Henry, R., Pastore, Albert, T. and Bruno, Stephen F. *Anal. Chem.*, **51**, 196 (1979).
74. Beck, R. A. and Brink, J. J. *J. Environ. Sci. Technol.*, **10**, 173 (1976).

Index

Abietic acids, determination of, 128
Acetic acid
  determination of, 117-118
  g.l.c. of, 105
  ozonation of, 448-459
Acetone, determination of, 131
Aconitic acid, determination of, 117-118
Acrolein, determination of, 130
Acrylic acid, determination of, 128
Acrylamide
  g.l.c. of, 206-210
  in sewage, 210
  in trade effluents, 210
Acrylonitrile, determination of, 214
Adenosine triphosphate
  determination of, 240-246
  in sediments, 244
  in trade effluents, 244-245
Adipic acid, determination of, 118
Air, chlorinated hydrocarbons in, 292-293
Alanine, g.l.c. of, 198-199
Alcohols, determination of, 129-130
Aldehydes, determination of, 130
Algae
  chlorophylls in, 484-485, 496
  organosulphur in, 270
  organotin in, 80
Aliphatic chlorocompounds, determination of, 190-191, 283-292
Alkyl lead salts, determination of, 91-92
Alkyl mercury compounds, determination of, 46
Alloisoleucine, g.l.c. of, 199
Amines
  g.l.c. of, 190
  in trade effluents, 190
Amino acids
  determination of, 192-199
  in sea water, 197-199
  in sediments, 199
Anthaquinone
  determination of, 149
  t.l.c. of, 149
Aromatic amines
  determination of, 191
  in trade effluents, 191
Aromatic chlorocompounds, determination of, 313-328
Aromatic hydrocarbons, g.l.c. of, 167
Argenine, g.l.c. of, 198
Arylphosphate esters, in fish, 246-247
Aspartic acid, g.l.c. of, 199
Atomic absorption spectroscopy of
  organoantimony, 98
  organoarsenic, 49-55
  organogermanium, 98
  organolead, 97-98
  organotin, 61-71
  silicones, 98

Benethonium salts, determination of, 445
Benzene stannoic acid, determination of, 81-90
Benzidine
  determination of, 191
  g.l.c. of, 191
Benzoic acid, determination of, 128
1,4-Benzoquinone
  determination of, 149
  t.l.c. of, 149
Benzthiazole
  determination of, 270-272
  g.l.c.-m.s. of, 270
  in trade effluents, 270-272
Biodegradation of
  N-methyl pyrrolidone, 192

Biodegradation *(continued)*
  pentachlorophenols, 328
  phthalate esters, 143-144
Biogenic materials, total phosphorus in, 253-259
Biological materials, organomercury in, 19
Biota
  phthalate esters in, 144
  salty acids in, 106
Biphenyl, determination of, 131
Bis-(2-chloroethyl) ether, g.l.c. of, 366
Bis-(2-ethylhexyl) adipate, determination of, 131
Bis-(2-ethylhexyl) phthalate, determination of, 131
Bromochloroiodomethane, g.l.c. of, 336
Bromodichloromethane, g.l.c. of, 322
Bromoform
  g.l.c. of, 322-416
  in sea water, 398-402
Butyl benzoate, determination of, 131
Butyl mercaptan, g.l.c. of, 267
Butyric acid, determination of, 117-118
Butyltin compounds
  determination of, 72-81
  g.l.c. of, 72-75
  mass fragmentography of, 72-76
  t.l.c. of, 76-81
Butyltin trichloride, determination of, 74
Butyltin trimethyl, determination of, 75-78

Calcium lignosulphonate, determination of, 276
ϵ-Caprolactam, determination of, 148
Carbofuran
  determination of, 446
  t.l.c. of, 446
Carbohydrates
  determination of, 144-148
  in sea water, 146-147
  in sediments, 147
  in trade effluents, 147-148
Carbon tetrachloride
  g.l.c. of, 292-298, 333, 338-339, 349, 352-361, 371
  g.l.c.-m.s. of, 415-416
β-Carotene, h.p.l.c. of, 498
Carotenoids
  h.p.l.c. of, 494
  in sediments, 494-504
Chlorinated alkylnaphthalenes, g.l.c. of, 314
Chlorinated hydrocarbons
  g.l.c.-m.s. of, 291-292
  head space analysis of, 291

  in air, 292-293
  in fish, 292-293
  in mud, 292-293
  in sea water, 291
  in sediments, 292-293
  in soil, 292, 301
  in tissues, 283
  in trade samples, 292
  t.l.c. of, 284-292
Chlorine, determination of, 416
Chlorobromomethane, g.l.c. of, 333
Chloroethanes
  determination of, 298
  g.l.c. of, 394-397
Chloroform, g.l.c. of, 292-298, 322-416
Chloroolefins, determination of, 298-313
Chlorophenols
  biodegradability of, 328
  g.l.c. of, 314-325
  g.l.c.-m.s. of, 322-325
  h.p.l.c. of, 325-327
  in fish, 319-322
  in sewage, 322
  in soil, 319-322
  in tissue, 319-322
Chlorophenoxybutyric acid, determination of, 128
Chlorophyll *a*, determination of, 484-505
Chlorophyll *b*, determination of, 484-505
Chlorophyll *c*, determination of, 484-505
Chlorophylls
  determination of, 483-505
  h.p.l.c. of, 498
  in algae, 484-485, 496
  in diatoms, 484
  in sediments, 494-504
  t.l.c. of, 485-486
5β-Cholestan-3β-ol, faecal marker, 427-433
Citric acid, determination of, 117-118, 127
Camphetchlor, determination of, 446
Cobalamin
  determination of, 504-505
  h.p.l.c. of, 504
  in sewage, 505
Coprostanol
  determination of, 427-434
  faecal marker, 427-433
  in sea water, 437
o-Cresol, g.l.c. of, 163-167
m-Cresol, g.l.c. of, 163-167
Cresols
  determination of, 170-171
  g.l.c. of, 160-162

Crotonic acid, determination of, 118
Crustaceae, organomercury in, 46
Cyanuric acid, determination of, 227
Cyclohexanol, determination of, 129
Cysteine, g.l.c. of, 199

Diadinoxanthin, h.p.l.c. of, 498
Dialkylmercury compounds, determination of, 19
Diatoms, chlorophylls in, 484
Dibromoiodomethane, g.l.c. of, 336
Dibromomethane, g.l.c. of, 333
Dibutylnitrosamine, g.l.c. of, 204
Dibutyl phthalate
  determination of, 133, 135
  g.l.c. of, 140–144
  in sewage, 139
Dibutyl phthalate-bis(2-2-butoxyethoxy)-methane, determination of, 131
Dibutyltin dichloride, determination of, 74
Dibutyltin dimethyl, determination of, 75–78
Dichlorobenzenes, g.l.c. of, 333, 382
2,5-Dichloro-4-bromophenol, g.l.c. of, 318–319
Dichloroethane, g.l.c. of, 291, 333, 382, 394–397
Dichloroiodomethane, g.l.c. of, 336
Dichloromethane, g.l.c. of, 291
2,3-Dichloro-4,4-naphthaquinone, determination of, 445
Dichlorophenols
  g.l.c. of, 162
  h.p.l.c. of, 325–327
2,4-Dichlorophenol, g.l.c. of, 318–319
Dichloropropane, g.l.c. of, 322–333, 382
1,2-Dichloropropane, g.l.c. of, 291, 333
Diethylene glycol, determination of, 130
Di-2-ethylhexyl phthalate
  determination of, 131, 135
  g.l.c. of, 140–144
  in sewage, 139
Diethyllead dichloride, t.l.c. of, 92
Diethyl phthalate, determination of, 131
Dihydric phenols, g.l.c. of, 162
Dihydroxy phenols, determination of, 160
Diisooctyl phthalate, determination of, 131
S-2-(Diisopropylamino)ethyl-O-ethylmethyl-phosphonothioate
  determination of, 444–445
  in sea water, 444–445
Dimethyl arsinite, determination of, 63–67
Dimethyl dioxan, determination of, 148–149

Dimethyllead dichloride, t.l.c. of, 92
Dimethylnitrosamine, determination of, 199–200
Dimethyl sulphoxide
  determination of, 272–276
  in phytoplankton, 272–275
  in sea water, 272
Dimethyltin hydride, determination of, 68–71
4,6-Dinitro-o-cresol, determination of, 215
Dioctyl phthalate, determination of, 131
Dioxans, determination of, 149
Diphenyltin oxide, determination of, 81–90
Dipropylamine, g.l.c. of, 190
Dipropylnitrosamine, determination of, 204
Dodecyl guaridine, t.l.c. of, 215

Echinenone, h.p.l.c. of, 498
Esters, determination of, 131
Ethylenediaminetetraacetic acid
  determination of, 216–221
  g.l.c. of, 216–221
  in sewage, 217
Ethylene glycol, determination of, 130
Ethylene glycol dinitrate, h.p.l.c. of, 216
N-(2-Ethylhexyl)norborn-5-ene-2, 3-dicarboxamide
  g.l.c. of, 444–445
Ethynyloestradiol
  g.l.c. of, 438–444
  in foilage, 438–444
  in soil, 438–444
  t.l.c. of, 438–444
Ethylmercury compounds, determination of, 37–40, 45
Ethylmercury chloride, 8, 18–19

Fatty acids
  determination of, 105–129
  h.p.l.c. of, 106–117
  in biota, 106
  in sea water, 117–118
  in sediments, 106, 126
  in sewage, 118–126
  in silage, 126
  in trade effluents, 126
  mass spectrometry, 116–117
  t.l.c. of, 106, 116–117
Faecal marker, 427–433
Faecal sterols, h.p.l.c. of, 502
Fish
  aryl phosphate esters in, 246–247
  chlorinated hydrocarbons in, 292–293

511

Fish *(continued)*
  chlorophenols in, 319–322
  haloforms in, 416
  halogen in, 416–417
  hexachlorobenzene in, 313
  methoxymercury compounds in, 19
  1,1'-emthylenedi-2-naphthol in, 424–427
  organoarsenic compounds in, 45, 50–55
  organolead in, 93–98
  organomercury compounds in, 12–16, 44–46
  pentachlorophenol in, 322–325, 328
  phenoxyacetic acid herbicides in, 319–322
  Squoxin, in, 424–427
Fluorescent whitening agents
  determination of, 434–438
  in muds, 436–438
  t.l.c. of, 434
Foliage
  ethynyloestradiol in, 438–444
  mestranol in, 438–444
Formaldehyde, determination of, 128–130, 272
Formic acid, determination of, 128
Fucoxanthin, h.p.l.c. of, 498
Fulvic acid
  determination of, 463–483
  in sediments, 469
  in sewage, 479–483
  IR spectroscopy, 481
  NMR, 479
  UV, 464–469
Fumaric acid, determination of, 118
Furfuraldehyde, g.l.c. of, 130

Gas chromatography of
  acetaldehyde, 458
  acetic acid, 105, 124
  acrylamide, 206–210
  β-alanine, 198–199
  alloisoleucine, 199
  amino acids, 192–199
  argenine, 198
  arylmercury compounds, 19
  aromatic amines, 191, 198
  aromatic hydrocarbons, 167
  aspartic acid, 199
  benzene stannoic acid, 81–90
  benzidine, 191
  bis-(2-chloroethyl) ether, 366
  bromochloroiodomethane, 336
  bromoform, 322–416
  butyl mercaptan, 267

butyltin compounds, 72–75
butyltin oxide, 74
butyltin trichloride, 74
butyltin trimethyl, 75–81
butyric acid, 124
carbohydrates, 145–148
carbon tetrachloride, 333, 338–339, 349, 352–361, 371
chlorinated alkylnaphthalenes, 314
chlorinated hydrocarbons, 283, 291–298
chloroform, 292–298, 322–416
chloroethanes, 394–397
chlorodibromomethane, 333–416
chlorophenols, 314–325
m-cresol, 163–167
o-cresol, 163–167
cresols, 160–162
dialkylmercury compounds, 19
dibromoiodomethane, 336
dibromomethane, 333
dibutylnitrosamine, 204
di-*n*-butyl phthalate, 135, 140–144
dibutyltin dichloride, 74
dibutyltin dimethyl, 75–81
dichlorobenzenes, 322–333, 382
dichlorobromomethane, 267
2,5-dichloro-4-bromophenol, 318–319
1,2-dichloroethane, 333, 382
1,1-dichloroethane, 333
dichloroiodomethane, 336
dichloromethane, 291
dichlorophenol, 162, 318–319
dichloropropane, 322, 382
1,2-dichloropropane, 291, 333
di-2-ethylhexyl phthalate, 135, 140–144
dihydroxyphenols, 160, 162
dimethyldiethyllead, 93–98
dimethylnitrosamine, 200–204
dimethyl sulphoxide, 272–275
diphenyltin oxide, 81–90
dipropylamine, 190
dipropylnitrosamine, 204
esters, 131, 135–139
ethylenediaminetetraacetic acid, 216–221
ethylmercury compounds, 18
N-(2-ethylhexyl)norborn-5-ene-2,3-dicarboximide, 445–446
ethynyloestradiol, 438–444
faecal sterols, 502
fatty acids, 105–126
formaldehyde, 458
furfuraldehyde, 130
geosmin, 446–448
glutamic acid, 199

Gas chromatography *(continued)*
  glycine, 199
  glycals, 130
  haloforms, 331–416
  hexachlorobenzene, 298–301, 313
  hexachlorobutadiene, 292–301
  hexachlorocyclopentadiene, 298–301
  hexachloroethane, 292–298
  hexachlorophene, 322–325
  hexanoic acid, 124
  hexyl mercaptan, 269
  histidine, 199
  hydrogen sulphide, 259–270
  iodoform, 336
  isobutyric acid, 124
  isovaleric acid, 124
  leucine, 199
  2,5-lutidine, 191–192
  lysine, 199
  mercaptans, 259–270
  mestranol, 438–444
  methanol, 129
  methionine, 199
  methoxymercury halides, 199
  methylarsenic compounds, 55–56
  methyl bromide, 373–374, 416
  methyl chloride, 416
  methylene chloride, 371, 382
  1,1′-methylenedi-2-naphthol, 424–431
  methyl fluoride, 416
  methylglyoxal, 458
  methylglyoxylic acid, 458
  methyl iodide, 416
  methyl mercaptan, 268
  methylmercury chloride, 45
  methylmercury compounds, 18, 45
  methyltriethyllead, 93–98
  N-methyl pyrrolidone, 192
  monochlorophenols, 162
  monohydroxyphenols, 160–162
  nitriloacetic acid, 220, 222–223, 226–227
  4-Nitrophenol, 215–216
  nitrosamines, 200–204
  N-nitrosodiethanolamine, 199–202
  N-nitrosodiethylamine, 199–202, 204–205
  N-nitrosodimethylamine, 199–202, 204–205
  N-nitroso-5-methyl-1,3-oxazolidine, 199–204
  N-nitrosomorpholine, 199–202
  N-nitrosopiperidine, 199–202
  N-nitrososarcosinate, 207
  octachloropentene, 298–301

organoarsenic compounds, 55–56, 63
organomercury compounds, 18–20, 45–46
organosulphur compounds, 259–270
orthinine, 199
ozonation products, 449–459
pentachlorobutadiene, 292–298
pentachloroethane, 292–298
pentachlorophenol, 314–325, 328
perchloroethylene, 292–298
phenols, 160–170
phenoxyacetic acid herbicides, 319–322
phenylalanine, 198
phenylmercury compounds, 18
phthalate esters, 144
2-picoline, 191–192
piperonal butoxide, 445–446
propanol, 190
propionaldehyde, 458
propionic acid, 124
propylamine, 190
pyrethrins, 445–446
pyridine, 191–192
pyruvic acid, 458
resorcinol, 160–162
serine, 199
Squoxin, 424–431
syringaldehyde, 131
taurine, 198
tetralkyllead compounds, 90–98
tetrabutyltin, 90
1,1,1,2-tetrachloroethane, 292–298
1,1,2,2-tetrachloroethane, 292–298
tetrachloroethylene, 292, 333, 338–339, 352–361, 375
tetrachloromethane, 291–298
2,3,4,5-tetrachlorophenol, 318–319
2,3,4,6-tetrachlorophenol, 318–319
tetraethyllead, 93–98
tetraethyltin, 90
tetramethyllead, 93–98
tetraphenyltin, 81–90
theonine, 199
toxaphene, 446
tributyltin methyl, 75–81
1,1,1-trichloroethane, 292–298
trichloroethylene, 291–292, 352–361, 371, 375, 384–385
trichloromethane, 291
trichlorophenols, 162
2,3,5-trichlorophenol, 318–319
2,4,5-trichlorophenol, 318–319
$\alpha\alpha\alpha$-trifluoro-4-nitro-m-cresol, 445
trihalomethanes, 394–397

Gas chromatography *(continued)*
  trimethyllead ethyl, 93–98
  triphenyltin hydroxide, 81–90
  urea, 198
  valeric acid, 124
  vinyl chloride, 301–313
  vinylidine chloride, 313
  2,6-xylenol, 163–167
  xylenols, 160–167
Gas chromatography–mass spectrometry
  of benzthiazole, 270
  carbon tetrachloride, 415–416
  chlorinated hydrocarbons, 291–292
  chlorophenols, 322–325
  esters, 131
  geosmin, 446–448
  haloforms, 413–416
  hexachlorobenzene, 313
  2-mercaptobenzthiazole, 270
  nitrosamines, 202, 205
  ozonation products, 452–459
  vinyl chloride, 301–313
Geosmin
  determination of, 446–448
  g.l.c.–m.s. of, 446–448
  mass fragmentography of, 447
Glucose, determination of, 145
Glutamic acid, g.l.c. of, 199
Glycine, g.l.c. of, 199
Glycols
  determination of, 130
  g.l.c. of, 130
Glyphosphate residues, determination of, 246

Haloforms
  determination of, 328–417
  g.l.c. of, 331–416
  g.l.c.–m.s. of, 413–416
  head space analysis of, 348–366
  incidence of cancer, 329
  in fish, 416–417
  in potable water, 328–416
  in sea water, 398–402
  in sewage, 416
  in trade effluents, 398–403
Head space analysis of
  chlorinated hydrocarbons, 291
  haloforms, 348–366
  organic sulphides, 259–265
Heptachlorodibenzo-*p*-dioxins, h.p.l.c. of, 325–327
Heterocyclic compounds, determination of, 191–192

Hexachlorobenzene
  determination of, 313
  g.l.c. of, 298–301
  g.l.c.–m.s. of, 313
  in fish, 313
Hexachlorobutadiene
  determination of, 298
  g.l.c. of, 292–301
Hexachlorocyclopentadiene
  determination of, 298
  g.l.c. of, 298–301
Hexachloroethane, g.l.c. of, 292–298
Hexachlorphene
  g.l.c.–m.s. of, 322–325
  in sewage, 322
Hexachloropropane, determination of, 298
Hexamine, determination of, 191
Hexane 1,6-diamino acetic acid, determination of, 191
Hexyl mercaptan, g.l.c. of, 269
High-performance liquid chromatography of
  acrylamide, 210–212
  acrylic acid, 128
  aliphatic amines, 190
  carotene, 498
  carotenoids, 494
  chlorophenols, 325–327
  chlorophylls, 485, 498
  cobalamin, 504–505
  dichlorophenols, 325–327
  di-2-ethylhexyl phthalate, 133, 135–139
  di-*n*-butyl phthalate, 133, 135–139
  echinenone, 498
  esters, 135–139
  ethylene glycol dinitrate, 216
  fatty acids, 106–117
  fucoxanthin, 498
  heptachlorodibenzo-*p*-dioxins, 325–327
  lutein, 491
  monochlorophenols, 325–327
  myxoxanthophyll, 498
  nevxanthin, 498
  octachlorodibenzo-*p*-dioxins, 325–327
  ozonation products, 459
  pentachlorophenol, 325–327
  phaeophytins, 492–504
  phenols, 172
  tetrachloroethylene, 298
  2,3,4,6-tetrachlorophenol, 325–327
  trichlorophenols, 325–327
  violaxathin, 498
  vitamin B_{12}, 504
  xanthophylls, 496

Humic acids
  determination of, 463–483
  in sewage, 479–483
  IR of, 481
  metal complexes of, 475–478
  UV of, 464–469
Hydrazine, determination of, 191
Hydrogen sulphide, g.l.c. of, 259–270
m-Hydroxybenzoic acid, determination of, 171–172
Hydroxymethylsulphinite salts, determination of, 272
m-Hydroxyphenylacetic acid, determination of, 171–172
m-Hydroxyphenylpropionic acid, determination of, 171–172

Infrared spectroscopy of
  fulvic acid, 481
  humic acid, 481
Inositol phosphate esters, determination of, 246
Iodoform, g.l.c. of, 336
Isopropylmethylphosphorofluoridate
  determination of, 444–445
  in sea water, 444–445

Ketones, determination of, 131

Lactams, determination of, 148–149
Lactic acid, determination of, 117–118, 127
Leucine, g.l.c. of, 199
Lignin,
  in sediments, 469
  UV of, 464–469
Lignosulphonic acid, determination of, 464–469
Lutein, h.p.l.c. of, 498
2,5-Lutidine, determination of, 191–192
Lutidine, g.l.c. of, 199
Lysine, g.l.c. of, 199

Maleic acid, determination of, 117–118
Malic acid, determination of, 127
Malonic acid, determination of, 117–118
Mandelic acid, determination of, 127
Mass fragmentography of
  butyltin compounds, 72–76
  geosmin, 447
  vinyl chloride, 308–311
Mass spectrometry of
  fatty acids, 116–117
  silicones, 98
Melamine, determination of, 227

Mercaptans, g.l.c. of, 259–270
2-Mercaptobenzthiazole,
  determination of, 270–272
  g.l.c. of, m.s. of, 270
  in trade effluents, 270–272
Mercury in
  fish, 12–16
  plants, 12–16
  sediments, 12–16
  trade effluents, 14–16
Mestranol
  determination of, 438–444
  g.l.c. of, 438–444
  t.l.c. of, 438–444
  in foliage, 438–444
  in soil, 438–444
Methacrylic acid, determination of, 128
Methanol
  determination of, 129
  g.l.c. of, 129
  in sewage, 129
Methoxymercury compounds, determination of, 19, 31–32, 37
Methoxymercury halides, determination of, 19
Methionine, g.l.c. of, 199
Methylarsenic compounds, determination of, 55–56, 63
Methylation of organolead compounds, 94
Methyl bromide, g.l.c. of, 373–374, 416
Methyl chloride, g.l.c. of, 416
Methylene chloride, g.l.c. of, 371, 382
1,1′-Methylenedi-2-napthol
  determination of, 424
  g.l.c. of, 424–431
Methyl fluoride, g.l.c. of, 416
Methyl iodide, g.l.c. of, 416
Methyl mercaptan, g.l.c. of, 267
Methylmercury chloride, determination of, 8, 18–19
Methylmercury compounds, determination of, 21–26, 37–40, 45
N-Methyl pyrrolidone
  biodegradation of, 192
  determination of, 192
Methyltin compounds, determination of, 68–72
Methyltin hydride, determination of, 68–71
Monochlorophenols
  determination of, 162, 325–327
  g.l.c. of, 162
  h.p.l.c. of, 325–327
Monohydroxyphenols, g.l.c. of, 160–162

515

Monomethyl arsonate, determination of, 63–67
Monosaccharides, determination of, 145
Mud
  chlorinated hydrocarbons in, 292–293
  fluorescent whitening agents in, 436–438
Myxooxanthophyll, h.p.l.c. of, 498

Naphthol, determination of, 170–171
Neoxanthin, h.p.l.c. of, 498
Neutron activation analysis of organomercury compounds, 20
Nitriles, determination of, 212–215
Nitriloacetic acid
  determination of, 220–227
  g.l.c. of, 220, 222–223
  in sewage, 221–226
Nitrosamines
  determination of, 199–206
  g.l.c.-m.s. of, 202, 205
N-Nitrosodiethanolamine, g.l.c. of, 199–202
N-Nitrosodiethylamine, g.l.c. of, 199–202, 204–205
N-Nitroso-5-methyl-1,3-oxazalidine, g.l.c. of, 199–202
N-Nitrosomorpholine, g.l.c. of, 199–202
N-Nitrosopiperidine, g.l.c. of, 199–202
N-Nitrosopyrrolidone, g.l.c. of, 199–202
N-Nitrososarcosinate, 204
Nitrocompounds, determination of, 215–216
4-Nitrophenol, g.l.c. of, 215–216
Nuclear magnetic resonance spectroscopy of fulvic acids, 479
Nucleic acids
  determination of, 227
  in sea water, 227

Octachlorodibenzo-p-dioxins, h.p.l.c. of, 325–327
Octachloropentene, g.l.c. of, 298–301
Organoarsenic compounds
  determination of, 49–68
  in fish, 45, 50–55
  in sea water, 56–62
Organoantimony compounds, determination of, 98
Organic disulphides
  determination of, 259–270
  g.l.c. of, 259–270
Organogermanium compounds, determination of, 98

Organolead compounds
  determination of, 92–98
  methylation of, 94
  in fish, 93–98
  in sediments, 92
Organomercury compounds
  determination of, 1–49
  in crustaceae, 46
  in fish, 44–46
  in sea water, 20, 26–36
  in sediments, 36–44
  in sewage, 46–47
  in tissue, 46
  in trade effluents, 46–47
  methylation of, 31–32
  t.l.c. of, 46–47
Organic nitrogen
  determination of, 227–240
  in sea water, 228–234
  in sewage, 228
Organic phosphorus, determination of, 247–259
Organic sulphur compounds
  determination of, 259–270
  g.l.c. of, 259–270
  head space analysis of, 259–265
  in algae, 270
  in trade effluents, 259–265, 267
Organic tin compounds
  determination of, 68–90
  in algae, 80
  in sea water, 70–71, 80, 88–90
Orthinine, g.l.c. of, 199
Oxalic acids
  determination of, 117–118
  ozonation of, 448–459
Ozonation of
  acetic acid, 448–459
  oxalic acid, 448–459
  2-propanol, 448–459
  sewage, 458

Pentachlorobutadiene, g.l.c. of, 292–298
Pentachloroethane, g.l.c. of, 292–298
Pentachlorophenol
  g.l.c. of, 314–325, 328
  g.l.c.-m.s. of, 322–325
  h.p.l.c. of, 325–327
  in fish, 322–325, 328
  in soil, 328
  polorography of, 328
  t.l.c. of, 327–328
Perchloroethylene, g.l.c. of, 292–298
Phaeophytins a, determination of, 487–505

Phaeophytin *b*, determination of, 487–505
Phaeophytin *c*, determination of, 487–505
Phaeophytins
 determination of, 487–505
 h.p.l.c. of, 492–504
 in sea water, 492
 t.l.c. of, 488
Phenols
 determination of, 149–183
 g.l.c. of, 160–170
 h.p.l.c. of, 172
 in sewage, 156–159, 162, 177
 in trade effluents, 156–159, 170, 172, 176
 Raman spectroscopy of, 173–176
 t.l.c. of, 170–172
Phenoxyacetic acid herbicides
 in fish, 319–322
 in soil, 319–322
 in tissue, 319–322
Phenylalanine, g.l.c. of, 198
Phenyloglyoxylic acid, determination of, 127
Phenylmercury chloride, determination of, 8
Phenylmercury compounds, determination of, 8, 21–26, 40, 81–90
Phosphorus, determination of, 247–259
Phthalate esters
 biodegradation of, 143–144
 in biota, 144
 in sediments, 139, 143
Phthalic acid, determination of, 118
Phytoplankton, dimethyl sulphoxide in, 272–275
2-Picoline, determination of, 191–192
Piperonal butoxide, g.l.c. of, 444–445
Plant pigments, determination of, 483–504
Plants, organomercury compounds in, 12–16
Polychlorinated styrenes, determination of, 313
Polypeptides
 determination of, 199
 in sewage, 199
Propionic acid, determination of, 117–118
Propanol, g.l.c. of, 190
2-Propanol, ozonation of, 448–459
Propylamine, g.l.c. of, 190
Pyrethrins
 determination of, 445–446
 g.l.c. of, 445–446
Pyridine, determination of, 191–192
Pyruvic acid, determination of, 118

Quinones, determination of, 149

Salicylic acid, determination of, 128
Saturated polychlorocompounds, determination of, 292–298
Sea water
 amino acids in, 197–199
 bromoform in, 398–402
 carbohydrates in, 146–147
 chlorinated hydrocarbons in, 291
 coprostanol in, 433
 S-2-(Diisopropylamino)ethyl-*O*-ethylmethylphosphonothioate in, 444–445
 dimethylsulphoxide in, 272
 fatty acids in, 117–118
 haloforms in, 398–402
 isopropylmethylphosphonofluoridate in, 444–445
 nucleic acids in, 227
 organoarsenic in, 56–62
 organomercury in, 10, 12, 20, 26–36
 organonitrogen in, 228–234
 organotin in, 70, 80, 88–90
 phaeophytins in, 492
Sediments
 adenosine triphosphate in, 244
 amino acids in, 199
 carbohydrates in, 147
 carotenoids in, 495–504
 chlorinated hydrocarbons in, 283–292
 chlorophylls in, 494–504
 fatty acids in, 106, 126
 fulvic acid in, 469
 lignin in, 409
 organolead in, 92–98
 organomercury in, 12–16, 19–20, 24–26, 36–44
 phthalate esters in, 139, 143
 tannin in, 469
 total phosphorus in, 252–259
Serine, g.l.c. of, 199
Sewage
 acrylamide in, 210
 chlorophenols in, 322
 chlorophenoxybutyric acid in, 128
 cobalamins in, 505
 di-*n*-butyl phthalate in, 139
 di-2-ethylhexyl phthalate in, 139
 ethylenediaminetetraacetic acid in, 217
 fatty acids in, 118–126
 formaldehyde in, 130
 fulvic acid in, 479–483
 haloforms in, 416
 hexachlorophene in, 322

Sewage *(continued)*
  humic acid in, 479–483
  methanol in, 129
  nitriloacetic acid in, 221–222, 226
  organomercury in, 10, 46–47
  organonitrogen in, 228
  ozonation of, 458
  phenols in, 156–159, 162, 177
  polypeptides in, 199
  salicylic acid in, 128
Silage, fatty acids in, 126
Silicones, mass spectrometry of, 98
Soil
  chlorinated hydrocarbons in, 292, 301
  chlorophenols in, 319–322
  dimethylnitrosamine in, 199–200, 203
  ethynyloestradiol in, 438–444
  mestranol in, 438–444
  pentachlorphenol in, 328
  phenoxyacetic acid herbicides in, 319–322
Squoxin
  determination of, 424
  g.l.c. of, 424–431
  in fish, 424–427
Stannane, determination of, 68–71
Succinic acid, determination of, 118
Syringaldehyde, g.l.c. of, 131

Tannins
  u.v. of, 464–469
  in sediments, 469
Tartaric acid, determination of, 117–118
Taurine, g.l.c. of, 198
Tetralkyllead
  determination of, 90–93
  t.l.c. of, 91
Tetraalkyltin, determination of, 90
Tetrabutyltin, g.l.c. of, 90
1,1,1,2-Tetrachloroethane, g.l.c. of, 292–298
1,1,2,2-Tetrachloroethane, g.l.c. of, 292–298
Tetrachloroethylene
  determination of, 298, 333, 338–339, 352, 361, 371, 375
  g.l.c. of, 333, 338–339, 352, 361, 371, 375
  h.p.l.c. of, 298
Tetrachloromethane, g.l.c. of, 291
2,3,4,5-Tetrachlorophenol, g.l.c. of, 318–319
2,3,4,6-Tetrachlorophenol
  g.l.c. of, 318–319
  h.p.l.c. of, 325–327
Tetraethyltin, g.l.c. of, 90
Tetramethylthiuram, determination of, 276
Tetramethyltin, determination of, 71
Tetraphenyltin, determination of, 81–90
Theonine, g.l.c. of, 199
Thin-layer chromatography of
  alkyllead salts, 91–92
  anthraquinone, 149
  benzidine, 191
  p-benzoquinone, 149
  1,4-benzoquinone, 149
  butyltin compounds, 76–81
  carbofuran, 446
  chlorinated hydrocarbons, 284–292
  chlorophylls, 485–486, 488
  dodecylguanidines, 215
  ethynyloestradial, 438–444
  fatty acids, 106, 116–117
  fluorescent whitening agents, 434
  mestranol, 438–444
  organomercury compounds, 46–47
  pentachlorophenols, 327–328
  phaeophytins, 448
  phenols, 170–172
  tetraalkyllead compounds, 91
Tin tetrahydride, determination of, 68–71
Tips, chlorinated hydrocarbons in, 292
Tissues
  chlorinated hydrocarbons in, 283
  chlorophenols in, 319–322
  organomercury in, 46
  phenoxyacetic acid herbicides in, 319–322
Thiomersal, determination of, 8
Thiram, determination of, 276
p-Tolyl mercury chloride, determination of, 8
Total phosphorus
  determination of, 247–259
  in biogenic materials, 253–259
  in sediments, 252–259
Toxaphene
  determination of, 446
  g.l.c. of, 446
Trade effluents
  acrylamide in, 210
  adenosine triphosphate in, 244–245
  aromatic amines in, 191
  amines in, 190–191
  benzthiazole in, 270–272
  carbohydrates in, 147–148
  chlorophenols in, 322
  fatty acids in, 126

Trade effluents *(continued)*
  haloforms in, 398–403
  hexachlorophene in, 322
  2-mercaptobenzthiazole in, 270–272
  organomercury in, 14–16, 26, 46–47
  organosulphur in, 259–265, 267
  phenols in, 156–159, 170, 172, 176
Trans-1,10-dimethyl-*trans*-9-decalol, determination of, 446–448
Triarylphosphate esters, determination of, 246–247
Tributyltin methyl, determination of, 75–78
Tributyltin oxide, determination of, 74
Trichlorobenzene, determination of, 131
1,1,1-Trichloroethane, g.l.c. of, 292–298
Trichloroethylene, 292–293, 352–361, 371, 375, 384–385
Trichloromethane, g.l.c. of, 29
2,3,5-Trichlorophenol, g.l.c. of, 318–319
2,4,5-Trichlorophenol, g.l.c. of, 318–319
2,4,6-Trichlorophenol, g.l.c. of, 318–319
Trichlorophenols
  g.l.c. of, 162, 318–319
  h.p.l.c. of, 325–327
Triethylene glycol, determination of, 130
Triethyllead compounds
  determination of, 92–93
  t.l.c. of, 92
$\alpha\alpha\alpha$-Trifluoro-4-nitro-m-cresol
  determination of, 445
  g.l.c. of, 445
Triholomethanes, g.l.c. of, 394–397
Trimethyllead chloride, t.l.c. of, 92
Trimethyltin hydride, determination of, 68–71

Triphenyltin compounds, determination of, 81–90
Triphenyltin hydroxide, determination of, 81–90
Tyrosine, g.l.c. of, 198

Ultraviolet spectra of
  fulvic acid, 464–469
  humic acid, 464–469, 479
  lignin, 464–469, 479
  lignosulphonic acid, 464–469
  tannins, 464–469
Ureas
  determination of, 215
  g.l.c. of, 198

Valine, g.l.c. of, 199
Violoxanthin, h.p.l.c. of, 498
Vitamin B_{12}
  determination of, 504–505
  h.p.l.c. of, 504
Vinyl chloride
  determination of, 298–313
  g.l.c. of, 301–313
  g.l.c.-m.s. of, 301–313
Vinylidene chloride, g.l.c. of, 313

Xanthophylls
  determination of, 486
  h.p.l.c. of, 496
2,6-Xylenol, g.l.c. of, 163–167
Xylenols
  determination of, 160–167, 170–171
  g.l.c. of, 160–167